Green Hydrogen Production by Water Electrolysis

The world's largest economies have set clear development plans for hydrogen energy. From an Economy, Energy, and Environment (3E) point of view, hydrogen energy can be considered an ideal technology for enabling the energy transition from fossil fuels, restructuring energy systems, securing national energy sources, accelerating carbon neutralization, and driving the development of technologies and industry.

Green hydrogen production by water electrolysis is the key for hydrogen energy, and this book offers urgently needed guidance on the most important scientific fundamentals and practical applied technologies in this field.

This book:

- Details materials, electrochemistry, and mechanics.
- Covers ALK, PEM, AEM, and SOEC water electrolysis, including fundamentals and applications.
- Addresses trends, opportunities, and challenges.

This comprehensive reference is aimed at engineers and scientists working on renewable and alternative energy to meet global energy demands and climate action goals.

Green Hydrogen Production by Water Electrolysis

Edited by
Junbo Hou and Min Yang

CRC Press
Taylor & Francis Group
Boca Raton London New York

CRC Press is an imprint of the
Taylor & Francis Group, an **informa** business

Designed cover image: © Shutterstock

First edition published 2025
by CRC Press
2385 NW Executive Center Drive, Suite 320, Boca Raton FL 33431

and by CRC Press
4 Park Square, Milton Park, Abingdon, Oxon, OX14 4RN

CRC Press is an imprint of Taylor & Francis Group, LLC

ISBN: 978-1-032-43807-8 (hbk)
ISBN: 978-1-032-43810-8 (pbk)
ISBN: 978-1-003-36893-9 (ebk)

DOI: 10.1201/9781003368939

Typeset in Times
by codeMantra

Contents

About the Editors

Junbo Hou received his BS and MS degrees from Harbin Institute of Technology, Harbin, China in 2003 and 2005, respectively, and his PhD degree from Dalian Institute of Chemical Physics, Chinese Academy of Science, Dalian, China in 2008, all in chemical engineering, particularly electrochemical engineering.

From 2008 to 2010, he was with Montanuniversität Leoben and Erich-Schmid-Institut für Materialwissenschaft (ESI), ÖAW, as a Researcher, working on electroceramic materials and high-resolution transmission electron microscopy (HRTEM). From 2010 to 2012, he was with Institute of Critical Technology and Applied Science (ICTAS), Virginia Tech, as a Research Scientist, while simultaneously engaged in research work at the Department of Mechanical Engineering and the Department of Chemistry, Virginia Tech. In 2013, he started working in a startup company SAFCell, Inc. of California Institute of Technology (Caltech). He developed new techniques to realize the mass production, dramatically reduced the cost by 75% while even improving the cell performance, and helped the company obtain funding from ventures, the US government, and the US Army. In 2016, he was invited to join CEMT as a Chief Scientist responsible for the development and production of proton exchange membrane (PEM) fuel cell engines for vehicles. He established membrane electrode assembly (MEA) production line and metal bipolar plate (BIP) production line. The developed fuel cell engines with outputs of 30–128 kW, all passed the mandatory testing of the National Automobile Testing Center, China. Working with OEMs, a total of eight fuel cell vehicles obtained production certification from the Ministry of Industry and Information Technology, China. Since May 2018, he was appointed as an Associate Professor at Shanghai Jiao Tong University, and he continued to conduct research on fuel cells and batteries. He joined Power System Resources Environmental Technology Co. Ltd. Haiyan, China (HYPSR) in March 2022, a listed company on China SSE STAR Market, as a Chief Engineer responsible for developing hydrogen energy and fuel cells technologies and products. Within 1 year, he finished the research and development of the MEA, BIP, stack and system, and the construction of MEA, BIP, stack and system production lines. The developed fuel cell systems with 85–300 kW passed the mandatory testing of the National Automobile Testing Center, China. He also collaborated with bus OEMs and had 17 fuel cell bus running on the road.

He has published two books and contributed four invited book chapters in the area of electrochemical energy conversion and storage. He published 62 peer-reviewed papers and applied 62 Chinese patents. Currently, he is a member of the National Fuel Cell Standardization Technical Committee, an Outstanding Scientist of Zhejiang, China, a member of the APEC Sustainable Energy Center's domestic expert team, and an expert of International Electrotechnical Commission for the task force of TC 105/ahG 16.

Min Yang obtained her MS degree from Harbin Institute of Technology, Harbin, China in 2004 and her Ph.D. degree from Dalian Institute of Chemical Physics,

Chinese Academy of Science, Dalian, China in 2008, in chemical engineering. She joined Montanuniversität Leoben in 2008, as a postdoc researcher, working on solid oxide fuel cells. From 2010 to 2012, she was with Institute of Critical Technology and Applied Science (ICTAS), Virginia Tech, as a technical staff, focusing on the electrochemical test and analysis of NaBH4 fuel cells and lithium ion batteries. From 2013 to 2016, she was an engineer in PY battery, Inc., focusing on the electrochemical tests and the slurry preparation. In 2016, she was invited to join CEMT, China as a Vice Chief Engineer. In 2019, she became a senior director in Central Academy, Shanghai Electric Group Co., Ltd. Both from academic and industrial, she holds a lot of know-how in applied science and critical technology of fuel cells like SOFC, PEMFC, SAFC, etc. and batteries like Li-ion and Lead acid batteries. She has published one book and two book chapters in the area of electrochemical energy conversion and storage, as well as over 20 peer-reviewed papers, and applied 75 Chinese patents.

She was selected as a member of National Technical Committee 342 on Project Management of Standardization Administration of China, and also a member of NEA/TC34 National Energy Administration/technical committee 34, participating in the formulation of national standards of hydrogen energy and fuel cells. She is one of the writers for the major consulting and research projects of Chinese Academy of Engineering-Research on development strategy of hydrogen energy and fuel cell in China. She was selected as an expert by Shanghai Scientific and Technical Committee.

Foreword by Hongbin Zhang

As a Head of New Energy Division, Shanghai Electric Group Co., Ltd, I am very pleased to write the foreword to this book. Our company, Shanghai Electric Group, is a world-class high-end equipment manufacturer, focusing on smart energy, intelligent manufacturing, and smart infrastructure to provide green and intelligent industrial-grade system solutions. We have a global presence in industries such as new energy, efficient clean energy, industrial automation, medical devices, and environmental protection. With our strong R&D and manufacturing capabilities, brand influence, and broad experience in projects, Shanghai Electric Group upholds the values of open and mutually beneficial collaboration, advocates smart energy and intelligent manufacturing, promotes the development of smart industry and industrialization of service, and supports the growth of the "Energy Internet" and "Industrial Internet." By facilitating industrial transformation through technological advancement, we promote sustainable human and social progress.

First, I would like to warmly congratulate the publishers and the editors, Allison Shatkin, Junbo Hou, and Min Yang, on their initiative work in bringing together the fundamentals of hydrogen energy produced by water electrolysis and the technical challenges of real case studies. The knowledge, know-how, and industrial secrets presented in this book are an increasingly valuable and commercially exploitable commodity in the area of hydrogen energy and fuel cells, particularly for the future hydrogen society. This book will – I am sure – be of great interest to scientists, engineers and their professional advisers, and marketers and administrators, to maximize their commercial value and potential – not only at the national but also at the international level. For me – and also for many others involved in hydrogen energy – a particularly valuable feature of this book is that the reader is able to obtain and compare the different techniques of water electrolysis, the state-of-the-art manufacturing, and future development perspectives that the book covers. This is particularly important when technical staff, whether beginners or experts, marketing staff, sales men, and the executives would like to know more details about hydrogen energy by water electrolysis.

As you may expect from my own professional and business activities in our company, this book will be a very welcome and invaluable resource for my clients and me. And one that I can heartily recommend to all others who – in any way – are involved in the exciting and challenging area of water electrolysis, and developing hydrogen energy and society.

Shanghai, June 2023
Hongbin Zhang

Head of New Energy Division, Shanghai Electric Group Co., Ltd

Foreword by Jin Xiaolong

This book provides a widely useful compilation of ideas, cases, innovative approaches, and practical strategies for enhancing discussion of hydrogen energy by water electrolysis. By taking a new look at different techniques of water electrolysis, fundamentals, and perspectives, Dr. Min Yang and Dr. Junbo Hou provide a substantial resource in the effort to increase water electrolysis learning.

This work would be an important resource since it highlighted the more details regarding each specific technique of water electrolysis. This book goes well beyond just making us aware of the learning setting. It covers and describes all the major factors in building big picture of hydrogen energy by water electrolysis.

This book is an important resource for executives and advisors as it identifies the technology roadmap. First, it provides a new perspective on hydrogen energy by water electrolysis, showing how they can be an important source. Understanding how each technique can more effectively influence the performance, cost, and duration adds a major cache of time to completely grab the core competencies of each technique. Second, this book provides a relevant and constructive set of strategies and ideas to understand the challenges and perspectives. This book should be read by anyone who is interested in hydrogen energy by water electrolysis, because it offers a wide set of practical ideas for material design, device fabrication, and performance evaluation. All the key processes necessary for hydrogen production by water electrolysis can be found in this volume. Throughout this book, Dr. Min Yang and Dr. Junbo Hou set out to make the material accessible to readers interested in reading these details. Each of these ideas can be found in the form of case studies and practical strategies to consider.

Overall, this book offers a variety of academia and industrial leaders a concrete, useful, and in-depth look at ways to design, implement, and evaluate a major resource in the learning of water electrolysis. Clearly written, well organized, and enormously practical, it should be in every hydrogen energy company's professional library.

Shanghai, June 2023
JIN Xiaolong

Vice President of Shanghai Electric Group Co., Ltd.

Preface

Economy developing, energy demanding, and environment protecting (3E) pursued by human society, especially the requirement of carbon peaking and carbon neutralization introduced in Paris Agreement, lead us to seek more green and sustainable energy to achieve such goals. Fortunately, hydrogen energy, particularly green hydrogen, seems an encouraging strategic choice. First, similar to electric energy, hydrogen energy is a common secondary energy which could be provided by the large-scale deployment of renewable energy like wind and solar energy. Therefore, it could be an important part of the whole energy system in the future, realizing flexible conversion of hydrogen-thermal energy-electricity. Second, hydrogen could be widely used in transportation and industry sectors like fuel cell vehicles and metallurgy, which might be a competitive alternative for the high-energy-consuming and high-emission applications. As a result, from end-user point of view, hydrogen energy is capable of effectively reducing emissions and achieving green and low-carbon development. Third, the hydrogen energy industry can be deemed intelligent and technology intensive, which is driven by high innovation, like continuously strengthening the construction of the industrial innovation system and constantly breaking through the bottleneck of core technologies and key materials. Thus, it might become an economic driver and success story as a key direction of emerging industries. As a result, from 3E point of view, hydrogen energy can be considered as an ideal way to rebuild energy transition from fossil fuels to others, restructure the plausibility of the whole energy system, secure the national energy sources, accelerate the carbon neutralization, and drive the development of technologies and industry. The green hydrogen production by water electrolysis is the key for the hydrogen energy.

As early as 1789, the phenomenon of water electrolytic decomposition was discovered, but the fundamentals and principles underlying such phenomenon were established until the development of Faraday's Law in 1833. After the separator, asbestos, was commercialized and applied in alkaline water electrolysis in 1890, and Ni-based electrocatalysts were manufactured and considered as the optimal alternates for alkaline oxygen evolution reaction (OER) and alkaline hydrogen evolution reaction (HER), the modern technology of alkaline water electrolysis was established in the 1920s. Driven by industrial applications like ammonia production and hydrogenation reaction, the technology gradually became mature from the 1920s to the 1970s. Besides the energy crisis happened in the 1970s, the emergence of new materials and technologies, stimulated by the pursue of high performance, high efficiency, long endurance, and low cost, provoked the development and deployment of the new electrode materials, membranes, bipolar plates, and stack configurations. Either from materials design or from practical application, it seems there still exists room for improvement. In Chapters 1 and 2, a general overview of H_2 production by water electrolysis, as well as thermal dynamics and the efficiency of water electrolysis, will be introduced. In Chapters 3–5, more specifically, free-standing electrodes and catalysts for alkaline water electrolysis, the effect of electrolytic gas bubbles on the electrode process of water electrolysis, and alkaline water electrolysis at industrial scale will be comprehensively presented and summarized, along with the suggestions and perspectives for the future research development and deployment (RD&D) of alkaline water electrolysis.

Although alkaline water electrolysis has been existed for over 200 years and deployed in industry at large scale, one might notice that the state-of-the-art alkaline water electrolysis has three intrinsic problems hindering the further boost of the practical efficiency of H_2 production: (1) the low efficiency at low partial load range (<40%), which is due to the O_2 and H_2 crossover to each other through the membrane, especially being severe at low partial load; (2) low operating pressure, which is ascribed to the use of liquid electrolyte and the porous membrane; and (3) low limiting current density, which is attributed to the high internal resistance and low electrode kinetics from gas bubbles covering. To overcome these shortcomings, proton exchange membrane (PEM) or solid polymer electrolyte (SPE) water electrolyzer was first developed by General Electrical (GE) in the 1960s. Due to much lower crossover rate and higher proton conductivity of the PEMs, high-specific-activity catalysts with high electrochemical surface area, 3D porous electrode or catalyst layer, and zero-gap electrode–membrane interface, PEM water electrolysis possesses many advantages including high limiting current density, high efficiency, high gas purity, high operating pressure, compact stack design, fast system response, and superior dynamic operation. However, the acidic condition makes the internal environment in PEM electrolyzer very harsh, especially at the anode side. The high electrode potential, high oxidative atmosphere (O_2), and the existence of liquid water are almost hotbed of chemical and electrochemical corrosions. Therefore, it seems that Ir- and Pt-based catalysts and coatings are only choices at the anode side for maintaining high specific activity, stability, and longevity. However, these noble materials are very expensive and might inhibit the large deployment of PEM electrolysis. Another drawback is the crossover at high operating pressure. Due to the fact that PEM water electrolyzer is usually operated at 3 MPa or higher to chase high efficiency of the stack and low parasitic consumption of the system, the crossover becomes severe at such pressures. In Chapters 6–9, some fundamentals and challenges for electrocatalysis of PEM water electrolysis will be thoroughly investigated. Advances and challenges for acidic OER catalysts and the recent development of acidic hydrogen oxidation reaction (HOR) catalysts will be comprehensively reviewed, together with the degradation phenomena and mitigation strategies for PEM water electrolysis.

Stimulated by fuel cell technology evolution, anion exchange membranes (AEMs) have been developed with the aim to possibly eliminate the use of noble catalysts and the challenge of plates coating. Applying AEM instead of the conventional diaphragm or PEM in water electrolyzer generally results in AEM water electrolysis. Basically, AEM electrolysis holds the advantages from both alkaline water electrolysis and PEM water electrolysis, simultaneously avoiding the disadvantages from those traditional water electrolyzers. For example, AEM water electrolysis can use non-noble electrocatalysts and have high operating pressure. The cost can be largely reduced compared to the PEM ones. Due to the use of solid polymeric membranes, leaking and corrosion can be avoided in AEM water electrolysis, and the crossover might be largely reduced compared to the alkaline ones. Furthermore, the compact stack design could be applied on the AEM ones, and the balance of plant (BOP) might be simply the same as the PEM ones. However, AEM water electrolysis is still at its early stage, and the limiting current density is relatively low. Particularly, the chemical stability and durability of the membrane and ionomer are low. Accordingly, in Chapters 10 and 11, the state-of-the-art AEMs and advanced electrocatalysts for AEM water electrolysis will be deeply studied. Another interesting topic is solid oxide electrolysis (SOEC), which has advantages of efficiency up to 100% and uses non-noble catalysts. Nevertheless, it is still in the laboratory stage. The current status, research trends, and challenges in SOEC water electrolysis will be discussed in Chapter 12.

World's large economies have set clear development plans for hydrogen energy. In 2017, Japanese government issued the Basic Hydrogen Strategy to expand its hydrogen economy and hydrogen production by 20 million tonnes by 2050. In 2020, the US Department of Energy Hydrogen Program Plan was issued to advance the affordable production, transport, storage, and use of carbon-neutral hydrogen across different sectors of the economy. In 2022, the publication of the REPowerEU plan by the European Commission guided the implementation of the European hydrogen strategy and will further push forward renewable hydrogen as an important energy carrier to move away from Russia's fossil fuel imports. In 2022, the Chinese National Development and Reform Commission released the Medium- and Long-Term Plan for the Development of Hydrogen Energy Industry (2021–2035). At this moment, it is very necessary and urgent to present the critical technology and applied science of green hydrogen production by water electrolysis. We expect that this book covers most important scientific fundamentals and practical applied technologies in this field. We hope it would be very helpful for beginners and experts, scientists and engineers, theorists and experimenters. We believe green hydrogen produced by renewable water electrolysis would significantly help meet global energy demand and climate action goal. Finally, we would express our gratitude to the efforts of Editor Allison Shatkin at Taylor & Francis for initiating this project.

List of Contributors

Jiakai Bai
Peric Hydrogen Technologies Co., Ltd.
Handan, China

Fuping Chen
Shanghai Electric Group Co., Ltd.
Central Academe
Shanghai, China

Mengxin Chen
MIIT Key Laboratory of Critical
    Materials Technology for New
    Energy Conversion and Storage
School of Chemistry and Chemical
    Engineering
Harbin Institute of Technology
Harbin, China

Wei Chen
Shanghai Electric Group Co., Ltd.
Central Academe
Shanghai, China

Rui Ding
Peric Hydrogen Technologies Co., Ltd.
Handan, China

Meiling Dou
Beijing University of Chemical
    Technology
Beijing, China

Wen Han
Shanghai Electric Group Co., Ltd.
Central Academe
Shanghai, China

Zongying Han
College of Energy Storage Technology
Shandong University of Science and
    Technology
Qingdao, China

Liang Hu
Research Center of Solid Oxide Fuel
    Cell
China University of Mining and
    Technology
Beijing, China

Pengtao Huang
Shanghai Electric Group Co., Ltd.
Central Academe
Shanghai, China

Yixuan Huang
Beijing University of Chemical
    Technology
Beijing, China

Zhenye Kang
School of Chemistry and Chemical
    Engineering
State Key Laboratory of Marine
    Resource Utilization in South China
    Sea
Hainan University
Hainan, China

Changchun Ke
Institute of Fuel Cells, School of
    Mechanical Engineering
Shanghai Jiao Tong University
Shanghai, China

Bang Li
Institute of Fuel Cells, School of
    Mechanical Engineering
Shanghai Jiao Tong University
Shanghai, China

Guangfu Li
Foshan Xianhu Laboratory
Foshan, China

Liming Li
Peric Hydrogen Technologies Co., Ltd.
Handan, China

Pengxi Li
Peric Hydrogen Technologies Co., Ltd.
Handan, China

Yucong Liao
State Key Laboratory of Advanced
  Technology for Materials Synthesis
  and Processing
Wuhan University of Technology
Wuhan, China

Fuyue Liu
Faculty of Maritime and Transportation
Ningbo University
Ningbo, China

Ying Ma
Peric Hydrogen Technologies Co., Ltd.
Peric, China

Wu Mei
State Power Investment Corporation
Beijing, China

He Miao
Faculty of Maritime and Transportation
Ningbo University
Ningbo, China

Mu Pan
Wuhan University of Technology
Wuhan, China

Dongwei Qiao
Peric Hydrogen Technologies Co., Ltd.
Handan, China

Yufeng Qin
Beijing University of Chemical
  Technology
Beijing, China

Liuli Sun
State Power Investment Corporation
Beijing, China

Daoyuan Tang
Shanghai Electric Group Co., Ltd.
Central Academe
Shanghai, China

Haolin Tang
State Key Laboratory of Advanced
  Technology for Materials Synthesis
  and Processing
Wuhan University of Technology
Wuhan, China

Hongchang Tian
Shanghai Electric Group Co., Ltd.
Central Academe
Shanghai, China

Xinlong Tian
School of Marine Science and
  Engineering
State Key Laboratory of Marine
  Resource Utilization in South China
  Sea
Hainan University
Hainan, China

Qiqi Wan
Institute of Fuel Cells, School of
  Mechanical Engineering
Shanghai Jiao Tong University
Shanghai, China

Jiacheng Wang
Shanghai Institute of Ceramics
Chinese Academy of Sciences
Shanghai, China

Jiahao Wang
Beijing University of Chemical
  Technology
Beijing, China

Letian Wang
State Key Laboratory of Advanced
    Technology for Materials Synthesis
    and Processing
Wuhan University of Technology
Wuhan, China

Longxiang Wang
Beijing University of Chemical
    Technology
Beijing, China

Xiang Wang
Research Center of Solid Oxide Fuel
    Cell
China University of Mining and
    Technology
Beijing, China

Xunlu Wang
Shanghai Institute of Ceramics
Chinese Academy of Sciences
Shanghai, China

Yifan Wang
Beijing University of Chemical
    Technology
Beijing, China

Fei Wei
Shanghai Electric Group Co., Ltd.
Central Academe
Shanghai, China

Ping Xu
MIIT Key Laboratory of Critical
    Materials Technology for New
    Energy Conversion and Storage
School of Chemistry and Chemical
    Engineering
Harbin Institute of Technology
Harbin, China

Min Yang
Shanghai Electric Group Co., Ltd.
Central Academe
Shanghai, China

Zhibin Yang
Research Center of Solid Oxide Fuel
    Cell
China University of Mining and
    Technology
Beijing, China

Qingqing Ye
Beijing University of Chemical
    Technology
Beijing, China

Xianming Yuan
Peric Hydrogen Technologies Co., Ltd.
Handan, China

Yinqiao Zhan
Shanghai Electric Group Co., Ltd.
Central Academe
Shanghai, China

Anran Zhang
Peric Hydrogen Technologies Co., Ltd.
Handan, China

Yufeng Zhao
State Power Investment Corporation
Beijing, China

1 Overview of Hydrogen Energy and General Aspects of Water Electrolysis

Wu Mei, Liuli Sun, and Yufeng Zhao

1.1 INTRODUCTION

1.1.1 GLOBAL ISSUES AND GREEN ENERGY TRANSITION

The world is likely to face new record temperatures in the next 5 years, surpassing 1.5°C above pre-industrial levels, a critical threshold that could have irreversible impacts. It is time to accelerate an energy transition from fossil fuels to clean and sustainable energy resources, such as solar and wind power, which have become more viable and lead the growth in power generation due to their rapid development in the past decade, thanks to their cost competitiveness. It was reported that almost two-thirds of renewable power added in 2021 were cheaper than the cheapest coal-fired options in G20 countries [1].

The global energy transition faces two major challenges. First, renewable power such as solar and wind power generation is inherently intermittent, not constant, and depends on weather conditions. To stabilize the power systems including a high proportion of these variable power sources, large-scale energy storage solutions are indispensable. However, the current energy storage options, such as pumped-storage hydroelectricity and batteries, are not sufficient. Second, it is hard to apply electrification to decarbonize some harder-to-abate processes and activities in industry and transportation that rely on fossil fuels, such as iron and steel, long-haul transportation, heating, petrochemicals, ammonia and iron production, heavy-duty trucks and marine transport.

Hydrogen is the simplest element on earth, with only one proton and one electron. It can be produced by using renewable power to split water, stored in large quantities and used to power fuel cell electric vehicles and households, or converted into chemical fuels such as ammonia and methanol. Hydrogen, as a green energy carrier, can enable large-scale long-term energy storage, spatial and temporal transfer, and clean utilization of renewable power. Therefore, it is considered as the ultimate solution to overcome the challenges posed by the high penetration of renewable power.

DOI: 10.1201/9781003368939-1

1

Many countries have announced ambitious roadmaps and strategies to promote hydrogen as a key contributor to decarbonization goals across sectors in the next 10–30 years [2–7].

1.1.2 Green Hydrogen

The current total market of hydrogen consumption is about US$115 billion, mainly driven by the demand for petroleum refining and ammonia and methanol production. There are three main pathways to produce hydrogen: steam methane reforming, coal gasification, and water electrolysis. The first two methods dominate the global hydrogen production, but they use a lot of fossil fuels and emit CO_2 as a by-product. This type of hydrogen is known as "gray hydrogen". The CO_2 emissions can be reduced by capturing and storing the carbon (CCS or CCUS), but this also increases the production cost and requires strict control of methane emissions. This type of hydrogen is known as "blue hydrogen". The cleanest way to produce hydrogen is to split water using renewable energy sources, such as wind, solar, hydro, geothermal, or nuclear power. This type of hydrogen is known as "green hydrogen" and does not emit any CO_2 [6,8].

The idea of hydrogen economy was first proposed by John Bockris, a chemist and electrochemist, in the 1970s, when the oil crisis renewed the worldwide interest in water electrolysis. Hydrogen was considered as a clean, renewable and versatile energy carrier, which can replace fossil fuels and achieve net-zero emissions of greenhouse gases, thereby enhancing sustainability and energy security. However, in the past 50 years, the initiative of hydrogen economy has not translated into consistent investment and widespread adoption in energy systems, although some developed countries have pursued the development and deployment of hydrogen technologies, such as hydrogen production, storage, delivery and utilization, e.g., the US Department of Energy's Hydrogen and Fuel Cells Program, Japan's Basic Hydrogen Strategy, the European Union's Hydrogen and Fuel Cells Joint Undertaking, and International Energy Agency's Hydrogen Implementing Agreement [1].

In recent years, the interest and momentum have increased again, especially after the 2015 Paris Agreement, which set a goal of reaching net-zero emissions of greenhouse gases by the second half of the 21st century, which has sparked a new wave of interest in the properties and the supply chain scale-up of hydrogen. Green hydrogen is generally regarded as an essential product of green energy transition and will accelerate an industrial restructuring. By the end of 2022, more than 30 governments had released national hydrogen strategies or official roadmaps including: the cost reduction and efficiency improvement of green hydrogen, the development and deployment of hydrogen infrastructure, the integration of hydrogen with renewable energy sources, the development of large-scale hydrogen utilization technologies, the safety and public acceptance of hydrogen, and the international cooperation and coordination on hydrogen policies and standards [12–14].

Industrial chains based on green hydrogen, including the production, transport, storage, and use of hydrogen and its derivatives, such as green methanol and green ammonia, as well as sustainable aviation fuel (SAF), will create a huge market. Hydrogen and its derived products are expected to account for 5%–15% of total final energy used in 2050 and the contribution of clean hydrogen to decarbonization is estimated as ca. 20% of global CO_2 emissions [9–14]. Currently, green hydrogen only accounts for 4%–5% of global hydrogen production, mainly due to its high production cost. The main challenge for clean hydrogen production is cost. The relationships among hydrogen cost, applications and market scales have been estimated [14]; a rapid decrease in hydrogen cost to US\$2/kg and eventually below US\$1/kg is highly expected, which will enable widespread applications of green hydrogen and thus market expansion.

According to the latest report of IEA [7], efforts to stimulate low-emission hydrogen demand are lagging behind what is needed to meet climate ambitions. The cost of hydrogen produced using electrolysis depends on the cost of the electricity used to split water and the capital cost of electrolyzers. These years various techno-economic estimation has been carried out on the production costs of green hydrogen and its derivatives in various renewable power conditions [14]. Considering that the cost of renewable electricity has already dropped significantly in the last decade, it is necessary to develop low-cost electrolyzers through further technology innovation and manufacturing scale-up. In the following sections of this chapter, the basic principles of water electrolysis, water electrolysis electrolyzers, and their technical parameters are introduced. Current status and challenges, as well as the development trends of the main water electrolysis technologies, are summarized.

1.2　BASIC PRINCIPLES OF WATER ELECTROLYSIS

1.2.1　History of Water Electrolysis

Electrolyzers have been known for over two centuries. The phenomenon of electrolytic water decomposition was discovered in 1789 by Paets van Troostwijk and Deiman. In 1800, William Nicholson and Anthony Carlisle used a battery that could produce a steady current to electrolyze water and named the process. With the development of Faraday's law in 1833, the quantitative relationship between the produced hydrogen amount and the used electrical energy was established. The concept of water electrolysis was defined scientifically and acknowledged. In 1869, Zénobe Gramme invented the Gramme machine, a dynamo that could produce cheap and continuous electricity. This made water electrolysis a viable method for hydrogen production. In 1888, Dmitry Lachinov developed a technique for industrial synthesis of hydrogen and oxygen through water electrolysis. By 1902, more than 400 industrial water electrolyzers were already in operation. In 1939, the first large water electrolysis plant came into service with a capacity of 10000 Nm^3 H_2/h. In the first half of the 20th century, a huge demand for hydrogen and the development of water electrolysis technology were driven by the production of ammonia fertilizers and the low cost of hydroelectricity at that time. The economic advantage of water electrolysis faded when hydrocarbon energy started to be applied massively in industry, which enabled large-scale hydrogen production through coal gasification and natural gas reforming at much lower costs [15]. In 1966, the Nafion-based proton exchange membrane (PEM) water electrolysis, also named as solid polymer electrolyte (SPE) water electrolysis, was developed by General Electric for the space and military applications.

1.2.2 FUNDAMENTALS OF WATER ELECTROLYSIS

The electrolyzer is the device where electricity and thermal energies are input and transformed into chemical energy stored in hydrogen. Basically, the overall electrolysis reaction is the electrochemical splitting of water molecules (equation 1.1) by passing an electric current between two electrodes (cathode and anode) which are separated by the separator material with key functions to block gases while conducting ions:

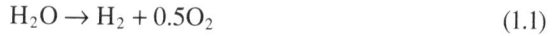

$$H_2O \rightarrow H_2 + 0.5O_2 \tag{1.1}$$

Since hydrogen ions (protons) are monovalent, two moles of electrons are involved in producing one mole of hydrogen. Therefore, an electric amount of 2F is required to produce 1 mole of hydrogen, while F represents Faraday's constant of 96485 C/mol, the electric quantity of 1 mole of electrons. Since 1 m^3 of hydrogen in the standard state is 44.6 moles, the amount of electricity required to produce it is 89.3 Faraday, which is 2393 Ah/Nm^3 in practical units. Since the current efficiency of water electrolysis is high, 2400 Ah is usually considered to be the amount of electricity required to produce 1 Nm^3 of hydrogen. This value is especially applied in many liquid-phase electrolyzers, except for special ones with low current efficiency.

Thermodynamically, considering the overall water-splitting reaction, the total energy demand for electrolysis is related to the enthalpy change ΔH between products and reactants:

$$\Delta H = \Delta G + T\Delta S \tag{1.2}$$

As shown in equation (1.2), at least an electric power equivalent to the Gibbs free energy change ΔG and heat equivalent to the product of temperature and entropy change $T\Delta S$ are required as driving energy to electrolyze water. Based on the minimum required electric energy, the reversible voltage E_{rev} is defined as the thermodynamically required voltage and can be calculated as

$$E_{rev} = \Delta G \, / \, nF \tag{1.3}$$

where n is the number of electrons transferred per reaction (n = 2 for equation (1.3)).

When the applied voltage reaches a specific value that can fulfill the entire energy demand ΔH, the requirement for external heat input or output can be theoretically eliminated. This E_H is called thermoneutral potential or theoretical operating voltage (equation (1.4)):

$$E_H = \Delta H / 2F \tag{1.4}$$

The thermoneutral voltage is different from the theoretical electrolytic voltage, and theoretically, it does not mean that electrolysis cannot be done below this voltage, and it is theoretically possible to electrolyze while absorbing heat from the surroundings below the thermoneutral voltage. In this case, the generated hydrogen has more energy than the supplied power. However, there is no practical electrolyzer that splits liquid-phase water below thermal-neutral voltage.

Under the standard condition of 25°C and 1 bar, ΔG^0 and $T\Delta S^0$ are 237.2 and 48.7 kJ/mol, respectively. The minimum electrolytic voltage (theoretical electrolysis voltage) E_{rev}^0 is 1.23 V. This E_{rev}^0 multiplied by the amount of electricity required to produce the aforementioned 1 m^3 hydrogen is the minimum electrical energy required for electrolysis of hydrogen, 2.94 kWh/Nm^3. E_H^0 is 1.48 V, and the power required for electrolysis at this voltage is 3.54 kWh/Nm^3.

1.2.3 ELECTROLYSIS EFFICIENCY AND OVERPOTENTIALS

Electrolysis is to transform the electric energy to chemical energy. The main parameter is the efficiency of reaction, including current efficiency and voltage efficiency. Current efficiency of 100% means that all the electric power being used for hydrogen production, 2400 Ah/Nm^3 H_2, as calculated according to equation (1.1), instead of being partly consumed in other pathways. Voltage efficiency u is obtained by dividing E^0 by the electrolysis voltage E_{cell} (equation (1.5)):

$$u = E^0 / E_{cell} \tag{1.5}$$

where E^0 can be acquired based on the higher heating value (HHV) or lower heating value (LHV) of E_{rev} or E_H in equations (1.3) and (1.4). The difference between HHV and LHV is the water vapor heat, considering the form of water taking part in the water-splitting reaction and whether a part of the heat is destined to vaporize the water.

Although the total energy demand ΔH and thermoneutral voltage E_H show weak dependences on temperature, except for the sudden change at approximately 100°C owing to water evaporation, the required energy demand ΔG for the electrolysis process changes significantly. In the high-temperature steam electrolysis described later, electrolysis below the thermal-neutral voltage can be realized and show 100% efficiency.

For liquid-phase electrolyzers working under 100°C, E_H is about 1.48 V due to weak dependences on temperature, which is generally regarded as 100% electrolytic efficiency and thus voltage efficiency u is calculated as

$$u = 1.48 / E_{cell} \tag{1.6}$$

The value of 1.48 V also matches the HHV of the hydrogen production. Then, the lower realistic voltage E_{cell} offers a higher energy efficiency for both calculation methods, which can realize a reduction in specific energy demand for hydrogen production and a considerable decrease in cost for large-scale electrolysis systems by alleviating the effect of electricity expenses. Assuming a current efficiency close to 100%, the DC power required for electrolysis at E_{cell} is

$$E_{cell} / 1.48 * 3.54 \text{ kWh} / \text{Nm}^3 = 2.39 * E_{cell} \text{ kWh} / \text{Nm}^3 \qquad (1.7)$$

An electrolyzer working at 1.8 V will require 4.3 kWh electric power to produce 1 Nm^3 hydrogen. In this case, it should be noted that even if the efficiency is 100%, the quality of energy is degraded from the electric power to heat. When electrolyzed at a voltage lower than E_H but higher than E_{rev}, an endothermic reaction occurs, and heat generation occurs at a voltage higher than E_H. The part that exceeds E_{rev} in the actual electrolysis voltage should basically be released as heat and be lost, but as described above, since the reaction is endothermic, if it is below E_H, it will not appear as heat generation and will contribute as energy for hydrogen generation.

The thermal-neutral voltage, E_H, is an important value in the design of an electrolyzer. In the design of an actual electrolyzer, it is ideal that E_{cell} is close to E_H and that the heat generated is almost balanced with the heat loss necessary to maintain the operating temperature.

To increase the electrolysis efficiency, the voltage difference between E_{cell} and E^0 can be carefully divided into various overpotentials resulting from the irreversibility of the electrolysis reaction. The following equation is a typical one:

$$E_{cell} = E^0 + E_{act} + E_{ohm} + E_{mass} \qquad (1.8)$$

Here, the activation overpotential E_{act} reflects the kinetics of transfer between electrodes and chemical species, which is jointly determined by many factors such as the intrinsic catalytic activity and electrode structure.

The ohmic overpotential E_{ohm} can be calculated through Ohm's law:

$$E_{ohm} = I * R_{cell} \qquad (1.9)$$

where I represents the total current and R_{cell} represents the entire effective resistance of the electrolysis cell, involving the resistance of different parts in the cells, such as electrodes and spacer, while the latter two factors are major contributors. The resistance of an electrolysis cell can be measured by using electrochemical impedance spectroscopy (EIS), which has become a popular electrochemical characterization technique because of its superiority in exploring the oxygen evolution reaction (OER)/hydrogen evolution reaction (HER) kinetic process under operating conditions.

The mass transport overpotential E_{mass} usually becomes pronounced at a high current density as starvation or over-accumulation of the participants and products of the water-splitting reaction occur in the vicinity of electrodes. Generally, transportation mechanisms of gas bubbles and water flow through porous structures can be further analyzed.

1.3 WATER ELECTROLYSIS TECHNOLOGY: CURRENT STATUS, CHALLENGES, AND R&D TRENDS

1.3.1 ELECTROLYSIS TECHNOLOGIES AND ELECTROLYZERS

Electrolyzers are the electrochemical devices used to split water into hydrogen and oxygen by the passage of an electrical current. There are typically four types of water electrolysis technologies: alkaline water electrolysis (AWE), proton exchange membrane water electrolysis (PEMWE), anion exchange membrane water electrolysis (AEMWE), and high-temperature solid oxide electrolysis (SOE), and thus four kinds of electrolyzers, which are generally composed of two main parts at a system level (Figure 1.1): the electrolysis stack, where the water-splitting reaction takes place, and the balance of plant (BOP), which includes power supply (usually with transformer and rectifier), water supply, heat exchanger, hydrogen purification, and possibly hydrogen compressor and heater system, which is generally needed for SOE that requires steam supply. An electrolysis stack, as shown in Figure 1.2, usually includes multiple electrolysis cells connected in series, bipolar plates to divide adjacent cells, seals, frames (mechanical support), and end plates (to avoid leaks and collect fluids). Water is supplied into the stack and reaches every cell through manifold. Electrolysis cell is the core of the stack, also the electrolyzer. It is composed of the two electrodes (anode and cathode) divided by a separator, two porous transport layers which facilitate the transport of reactants and removal of products, and the bipolar plates that provide mechanical support and distribute the flow. Water is fed into cell and flows through the bipolar plates and through the porous transport layers. At the electrode, the water is split into oxygen and hydrogen, with ions (typically OH^- or H^+ or O^{2-}) crossing through separator via liquid electrolyte or solid electrolyte. The separators between both electrodes are also responsible for insulating the electron conductivity between two opposite electrodes and keeping the produced gases (hydrogen and oxygen) separated and avoiding their mixture.

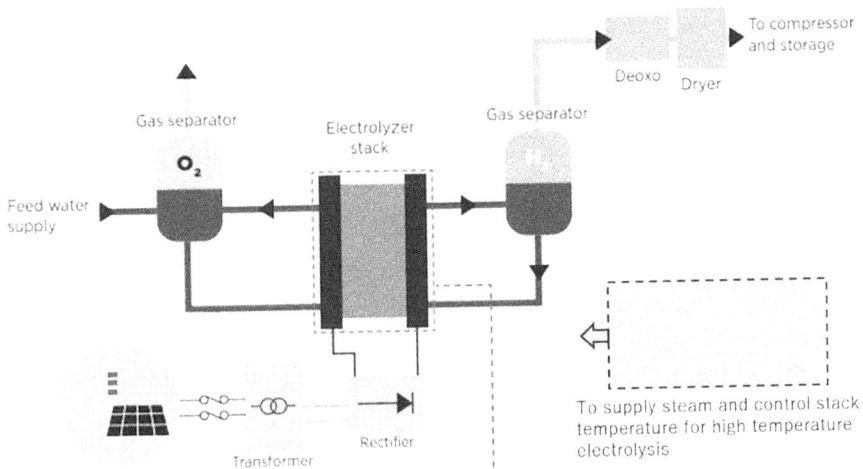

FIGURE 1.1 A schematic of an electrolyzer (system level).

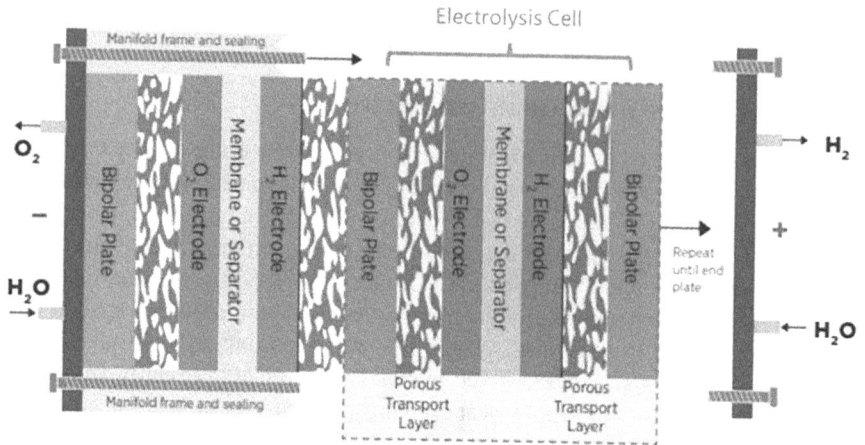

FIGURE 1.2 A schematic of an electrolysis stack.

The principles of these four types of electrolysis cells are shown in Figure 1.3. The main differences among them are: (1) separator, which is generally a porous fabric (called as the diaphragm in AWE) immersed in a liquid electrolyte to conduct OH⁻, or a solid electrolyte membrane (in PEMWE and AEMWE) to conduct or H⁺ or OH⁻, or an oxide electrolyte layer (in SOE) to conduct O^{2-} or H⁺; (2) operation temperature, which guides the selection of materials and components of a stack, ranging from 300°C to 1,000°C for SOE, while being lower than 100°C for the other three types at present; (3) catalysts and electrodes, including anode and cathode, that electrochemically split water and thus directly affect the electrolysis efficiency and durability; and (4) operation current density, which affects the hydrogen production per unit electrode area and thus the size of electrolysis stacks.

The biggest difference between the present AWE and PEMWE is the charge carrier and separator, as the membrane used in PEMWE is non-porous, and only an extremely low amount of gas molecules can diffuse through it. To prevent mixture of hydrogen and oxygen through porous separators, high balanced pressure control of anode and cathode is necessary for AWE.

To improve the performance of various electrolyzers, systematic characterization and evaluation of their key materials and electrolysis stacks are indispensable to obtain insightful information. The main technical parameters are introduced in the following.

1.3.1.1 Parameters for Separators

Main parameters include ionic conductivity, stability, and gas barrier property that is also called as gas crossover or leakage current. Ionic conductivity directly determines the resistance of the electrolysis cell, which affects the voltage efficiency (E_{cell}) and thus the working current density of the cell, which dominates the stack compactness. Gas barrier property affects the hydrogen production efficiency (Faradaic efficiency),

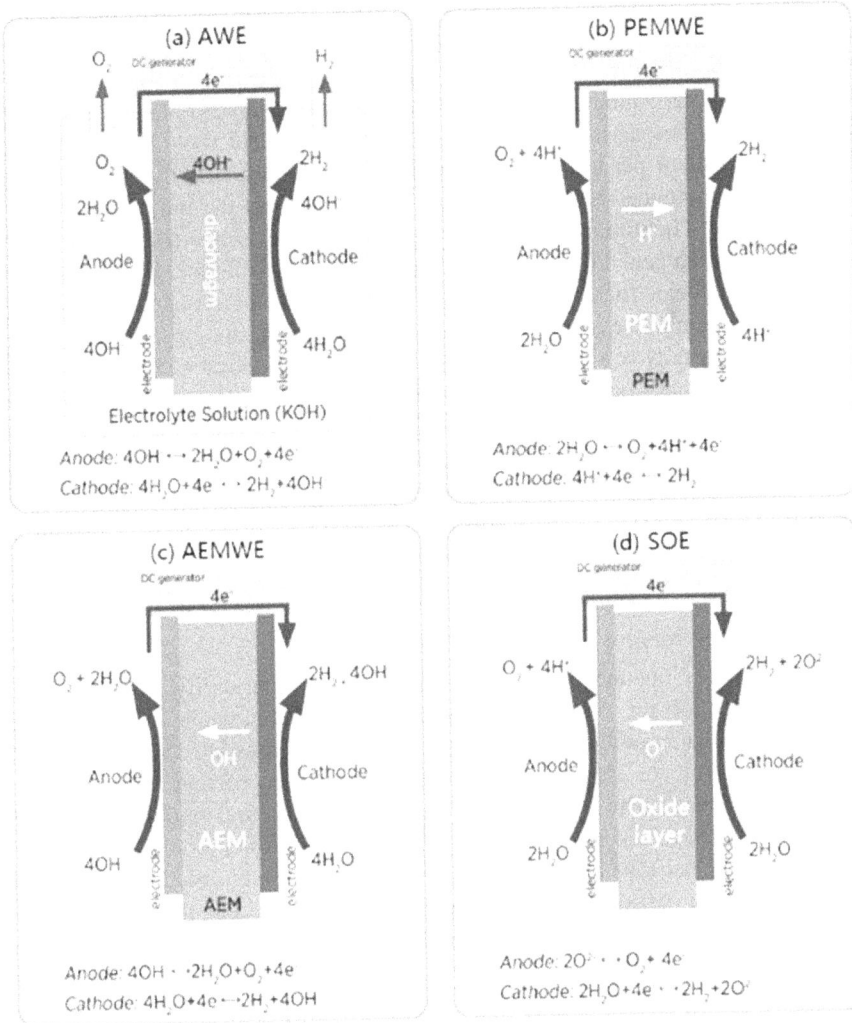

FIGURE 1.3 The principles of the four types of electrolysis cells.

and furthermore, it is essential for the inherent safety of electrolysis stack because the leaked hydrogen reacts with oxygen in the stack. Gas barrier performance of the separators can be monitored by detecting the hydrogen content of oxygen in the stack outlet, which is also a key parameter for electrolyzers.

1.3.1.2 Parameters for Catalysts and Electrodes

Water is split into oxygen and hydrogen on the surface of catalysts, i.e. the "heart" of electrolyzers. The electrolysis efficiency of an electrolyzer is limited by the activity of its catalysts. The performance of a catalyst is generally characterized by its catalytic activity, stability, and Faradaic efficiency. The activity can be evaluated in the real cell using overpotential E_{act} and Tafel slope, which can be used to discern reaction kinetic and the possible reaction mechanism and are easily derived from the measured polarization curve. To assess the detailed analysis of catalytic activity, specific activity and mass activity are used, which are the activity current per unit real surface area or per unit mass of catalyst. The specific activity eliminates the effect from the number of active sites and thus can reflect the intrinsic activity of each active site. The stability of a catalyst is another vital parameter for the practical application, which generally includes mechanical stability, thermal stability, chemical stability, and electrochemical stability. The electrochemical stability is essential for the application in the renewable power conditions. The catalyst measurements in a real cell, especially at a practical current density, are important since the catalyst performance depends on the electronic states of catalyst surface, which is sensitive to electrochemical conditions. Faradaic efficiency of a catalyst is generally used to describe the overall selectivity of an electrochemical process. For the water electrolysis, it is defined as the amount of collected product (H_2 or O_2) relative to the amount that could be produced from the input electrical energy according to Faraday's law of electrolysis, since in some cases other by-products may be produced on the surface of the catalyst.

1.3.1.3   Parameters for Electrolysis Cells/Stacks

Main parameters include efficiency, current density, durability, and flexibility. Generally, electrolyzer efficiency refers to the one of the whole electrolysis systems. The efficiency of electrolysis stack, usually named as DC energy consumption, includes voltage efficiency (u in equation 1.6) and current efficiency (also Faradaic efficiency), which is defined as the ratio of the actual amount of hydrogen produced relative to the theoretical amount of hydrogen that could be produced. In AWE stacks, because all cells share the same electrolyte and thus are in ionic contact, there are parasitic shunt currents flow among the cells through the manifolds and the electrolyte channels. The typical Faradaic efficiency of AWE stacks is reported to be around 90% or higher, which depends on several factors, such as the cell design and the operating conditions [16]. It should be noted that the efficiency strongly depends on the working current density, which affects the compactness of a stack. High efficiency can be achieved simply by setting a low working current density, which can decrease the operational electrical cost but increase the capital cost.

The durability of an electrolysis stack is usually evaluated using its lifespan or lifetime. Life-end of stacks is generally defined as 10% increase of energy consumption needed for hydrogen production, meaning 10% decrease in efficiency. It should be noted that stack durability depends on the quality of the BOP significantly such as the impurities in the water supplied to the stacks and the unsteady electric power. The use of high-purity DI water is generally required to ensure good durability of a PEM stack. A purity as high as <0.1 μS/cm (>10 MΩcm) is usually recommended. It should be noted that the evaluation results are significantly affected by the measurement protocols. Figure 1.4 shows a schematic diagram of the general effects of different measurement methods on the electrolysis properties.

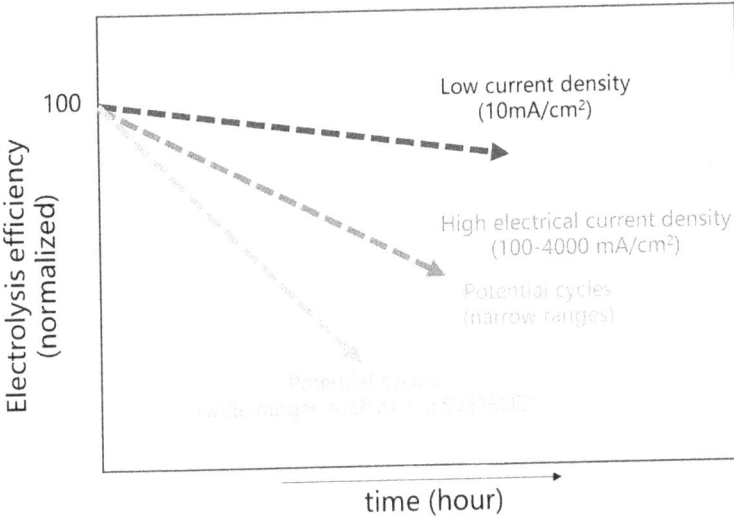

FIGURE 1.4 A schematic of measurement effects on electrolysis efficiency.

For the flexibility of an electrolysis stack, cold start time, hot start time (from standby state), load range, and response rates are generally used to characterize the abilities of a stack to follow fluctuation of renewable power. It should be noted that at present the flexibility of the electrolyzer is probably limited more by BOP rather than by the stack, and the requirements on flexibility vary with working conditions of the electrolyzer.

In the practical evaluation of the key materials of electrolyzers such as catalysts and membranes, the focus of academia and industry is often slightly different. To improve R&D efficiency, it is important to develop systematic evaluation standards that include AST protocols based on degradation mechanisms and realistic renewable power conditions. It is also essential to establish a common database that can facilitate the comparison and sharing of data among different research groups.

Table 1.1 summarizes the main technical features of the four types of water electrolysis technologies. Although AWE and PEMWE electrolyzers have been widely

TABLE 1.1

Main Technical Features of Four Types of Water Electrolysis Technologies

	AWE	PEMWE	AEMWE		SOE
Charge carrier	OH⁻	H⁺	OH⁻	O^{2-}	H⁺
Electrolyte material	Alkaline aqueous	Proton solid polymer electrolyte	Anion solid polymer electrolyte	Oxides	
Separator	PPS fabric, porous composite	PEM (membrane)	AEM (membrane)	Oxide electrolyte	

(Continued)

TABLE 1.1 (*Continued*)
Main Technical Features of Four Types of Water Electrolysis Technologies

	AWE	PEMWE	AEMWE	SOE	
Consumption	Alkaline aqueous	DI water	DI water, Alkaline aqueous	Steam, CO_2	
Catalyst/ electrode material	Metals/ oxides	Precious metals/oxides	Metals/oxides	Mixed ceramic oxides/ composites	
Temperature (°C)	50–90	20–90	20–70	600–1000	300–700
Current density (A/cm²)	0.2–0.6	1–3	0.2–1.5	0.5–1.5	0.5–2
Cell area (m²)	3	0.3	0.03	0.02	0.0025
Stack scale[a] (Nm³ H_2/h)	500–3000 (15 MW)	50–500 (2.5 MW)	0.5 (2.4 kW)	-	-
Specific stack energy consumption (kWh/Nm³)	4.2–5.0	4.2–5.0	4.8	2.5–4	-
Design lifespan (year)	10–20	10–20	>1	-	-
Current density (A/cm²)	0.2–0.6	1–3	0.2–1.5	0.3–2	0.1–4
Current status[a]	Large-scale application	Small-scale application	Prototype stage	Prototype stage	Lab scale
Capital cost range (electrolyzer)[c] (USD/kW)	200–800	700–1400	-	-	-
Advantages	Low capital costs[b]	Quick response, wide load range	Low capital costs[b]	CO_2 electrolysis	Moderate temperature
Disadvantages	Inferior dynamic, corrosive electrolyte	High costs	Low OH^- conductivity, low durability	Low durability, unsuitable for dynamic application	

[a] The current status and stack scale refer to those in the application of green hydrogen production.
[b] When using transition metals as electrodes.
[c] Electrolyzer (>10 MW) cost including BOP.

used at 10 MW-scale nowadays, while AEMWE and SOE are not mature enough to ensure their reliability and long-term durability, it should be noted that each technology faces new challenges in renewable power conditions and is undergoing rapid development, as listed in the disadvantages part in Table 1.1.

In the following four sections, the current status, challenges, and technology trends of the four electrolysis technologies are briefly summarized. More details will be introduced in the following chapters.

1.3.2 ALKALINE WATER ELECTROLYSIS (AWE)

AWE is the oldest and most commercially available electrolysis technology. The AWE systems were originally developed to run under continuous, steady-state operating conditions, to utilize low-cost hydroelectric power to produce hydrogen for ammonia production and thus were large in scale (~100 MW). These years many AWE electrolyzers with 1000 Nm^3 H_2/h stacks have been installed for large-scale green hydrogen production. In China, 2000–3000 Nm^3 H_2/h AWE stacks have been proposed recently [17].

In alkaline electrolyzers, a typical 20–40 wt.% KOH solution is used as electrolyte, and at present, Ni-based transition metal catalysts loaded on current collector are generally used as the working electrodes. The alkaline electrolyzers typically operate with a cell voltage near 1.9–2 V at current densities of about 0.2–0.6 A/cm^2 with a voltage efficiency (HHV) above 70%.

Hydrogen production in the renewable energy scenario is different from the previous scenario, which poses new challenges to AWE, mainly on how to improve its performance and reliability without losing its cost competitiveness. Although some problems can be solved at the system level, breakthrough on key materials is expected. Also, modification of stack structure and operation strategies is necessary to suppress shunt current. Development of diaphragms and electrodes has been focused on recently to decrease cell resistance and gas crossover without sacrificing durability, and thus increase its working current density and dynamic load range. Their current status and R&D trends are briefly introduced below.

1.3.2.1 Diaphragms

Diaphragm is a thin, porous foil that separates the cathode and anode. Low ionic resistance, high gas barrier property, high mechanical strength, and high thermal stability are expected, where there are usually trade-off relationships among these properties. Polyphenylene sulfide (PPS) fabric materials and a composite material of zirconia (ZrO$_2$) and polysulfone (PSU) are generally used. Normally, their main technical parameters are a thickness of 500–1000 μm, an ionic resistance of about 0.1–0.3 Ω cm^2, and a gas crossover of <1%. Major research strategies include controlling material composition and optimizing pore size structure and hydrophilicity to decrease ionic resistance and gas permeability. Various composite materials have been investigated, such as A–X (A: zirconia, alumina, silica, titania; X: polytetrafluoroethylene (PTFE), polyvinylidene fluoride (PVDF), polybenzimidazole (PBI), polypropylene (PP), and polyethylene (PE)) [18,19]. It should be noted that anion exchange membrane (AEM) and hydroxyl-ion exchange membrane (HEM) are also the separator materials which can conduct hydroxide ions (OH$^-$), making them powerful candidates of AWE with excellent ionic conductivity and gas barrier property [18,19].

1.3.2.2 Electrodes

To electrolyze water, HER and OER occur in cathode and anode. In commercial AWE, Ni-based oxides and Ni-based alloys have been widely used as the anodic and cathodic catalysts, respectively, since Ni metal provides excellent balance among activity, stability, and comparably low costs, facing no significant risk of reserve depletion. It should be noted that electrodes containing platinum group metals (PGM) have also been applied in some AWE stacks to achieve high current densities. There are enormous efforts and attempts to clarify the activity mechanisms and explore novel catalysts theoretically and experimentally.

The OER mechanisms are complicated and still debatable in alkaline environments. The conventional mechanism, also called adsorbate evolution mechanism, involves several electron–proton coupled reactions, in which OH^- is oxidized into oxygen molecule and water molecule in alkaline electrolytes. Recently, some new OER reaction pathways are proposed, such as the lattice oxygen mechanism with lower reaction energy barriers [20]. Understanding the underlying OER active site, as well as the reaction routes such as contribution of adsorption energies of intermediates on the surface, is crucial to mechanism investigation. Using descriptors such as the bond energy of metal with OH, the number of electrons in d band, electron occupancy, and the adsorption energy, a volcano-type relation between the OER activity and various specific descriptors has been proposed [21]. Such relation is helpful to predict more active OER catalysts. The volcano-type relation experimentally showed that oxides which are oxidized difficultly or easily are poor catalysts, because their affinity for oxygen is too weak or too strong, respectively.

HER is a multi-step reaction occurring on the electrode surface. The reaction kinetics of HER in alkaline is at least two orders of magnitude lower than that in acidic electrolytes, revealing different mechanisms. The volcano-type relation between HER activity and descriptor is very successful, while the actual HER activity descriptor in alkaline electrolytes remains controversial. Some factors have been proposed, such as the slower transportation rate of OH^- than that of H^+ in solution and the more difficult cleavage of OH bond in water molecules than in hydrated protons, but the exact factor that determines the slow HER rate in alkaline environments is still vague [22].

For the traditional Ni-based catalysts, Raney Ni catalysts (nickel-aluminum or nickel-zinc) are extensively investigated owing to their increasing surface area after leaching of Al or Zn, especially their long-term durability and manufacturing techniques. A schematic of overpotentials in an AWE cell using the Ni-based catalysts is shown in Figure 1.5. The overpotential of anode and that of cathode is above 0.3 V and above 0.1 V, respectively. Considering that E_H^0 is 1.48 V, ca. 20% electricity energy is consumed in the electrodes of this AWE cell.

Great efforts have been made to design and synthesize novel OER/HER catalysts for AWE. Various strategies, such as alloying, phase engineering, and nanostructure engineering, have been employed to improve the electrochemically active surface area (ECSA) and/or intrinsically specific activity of potential candidates [23]. Extensive investigation on transition metals and their alloys, including high-entropy alloys (HEA), oxides, and (oxy)hydroxides, sulfides such as MoS_2 and $NiCoS_2$,

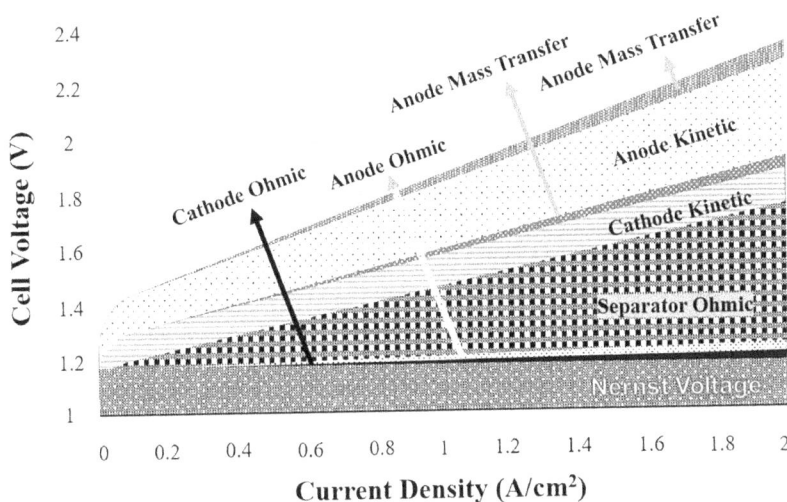

FIGURE 1.5 A schematic of overpotentials in an AWE cell (anode/cathode: Ni/NiMo; separator: porous composite; 80°C).

selenides, nitrides, phosphides, and chalcogenides, has been carried out [24–26]. So far, Pt and Ir oxides are generally regarded as the state-of-the-art catalyst for HER and OER, respectively, in agreement with the volcano-type relation. In alkaline electrolyte, Ru oxides exhibit excellent activity for HER and OER. However, their stability is poor in OER working conditions, due to its transformation into higher oxidation states species at high anodic potential and subsequent dissolution [27]. Among the PGM-free candidates, NiMo shows promising HER activities in alkaline electrolytes, with an overpotential of 15 mV at 10 mA/cm^2 and a low Tafel slope of 30 mV/dec in 1 M KOH, which was comparable to the benchmark Pt catalyst [28]. Some PGM-free catalysts with high OER activity have also been reported, such as NiFe-layered double hydroxides (LDH), in which a current density of 0.2 A/cm^2 at an overpotential of 240 mV has been observed [29].

Considering that the stability test under laboratory conditions (1 mol KOH, 10 mA/cm^2, and room temperature) differs greatly from industrially relevant conditions (6–10 mol KOH, >0.25 A/cm^2, and 60°C–90°C), much attention should be paid to further studies of the catalytic performances under these harsh conditions. Furthermore, applying potential catalyst candidates into practical electrolyzers to confirm their long-term stability, as well as developing their scaled-up synthesis, is required to narrow the gap between academia and industry, thus resulting in OER/HER catalysts for practical industrial application.

To realize the high activity of the novel catalysts in practical electrolyzers, performance (E$_{cell}$) at operation current densities is crucial, where the triple-phase interfaces (i.e., solid catalyst/electrode, liquid electrolyte, and gas product) and the mass transportation in electrodes also play critical roles. Compared with the powder catalyst generally used in PEMEC, catalytically active electrodes are usually applied in

AWE, namely free-standing electrodes which include the electrodes themselves with a surface such as Ni and those with a layer formed on the conductive substrates [30,31].

Understanding the mass transportation in electrodes and optimization of electrode structure is very vital to obtain a low overpotential E_{mass} to reach the maximum activity potential of catalysts. Gas bubble phenomena have been extensively investigated, since in liquid electrolytes the formation of gas bubbles on the electrode surface is hardly avoided and may obstruct the contact between active sites and electrolyte, limit mass transport, and increase liquid resistance, thus leading to low electric energy efficiency. Gas bubbles issue is more obvious at high current density such as 0.5–2 A/cm². Optimization of the electrode to obtain a good balance between adhesion force and buoyant force of the bubbles, as well as construction of super-aerophobic electrode, can reduce bubble adhesive force and accelerate the release of gas, which is conducive to enhancing mass transport and thus boosting the electrode performance. Therefore, it is valuable to benchmark the catalysts at high current density and develop catalysts with consideration on electrode engineering to realize high activity at high current density to satisfy real industrial applications in alkaline water electrolyzers [32,33].

In summary, AWE is the most widespread electrolysis technology to date, largely relying on diaphragms that mainly use PPS and composite ones and Ni-based catalyst/ electrodes. Stimulated by the huge potential market of green hydrogen, numerous R&D activities have been focused on AWE, and progress in current density increase, dynamic response enhancement, and load range expansion are highly expected.

1.3.3 PROTON EXCHANGE MEMBRANE WATER ELECTROLYSIS (PEMWE)

Main features of PEMWE are their use of PEM and precious metal-based catalysts in both electrodes. Traditional market scale of PEM electrolyzers is small, and 1–50 Nm³ H_2/h electrolyzers are generally used. Compared with AWE, PEMWE has the characteristics of lower ohmic resistance, high operating current density, wide load range, and high current efficiency (generally above 98%). These features lead to more compact structure, high safety, fast response speed, easy maintenance, and high hydrogen purity since only pure water is put into electrolysis stacks. Although PEMWE can adapt to the volatility of renewable energy and is regarded to be the most promising technology for green hydrogen production, it is more expensive than AWE mainly due to the precious metal-based catalysts required by the stability in acid environment, as well as its membrane, which is also a cost contributor compared with the fabric-type AWE diaphragm. The main challenge of PEMWE is cost reduction. A simple way is to scale up electrolyzers at the system level, e.g., to use one BOP system to support >10 PEM stacks, considering the scale of a general AWE system. Recently, 10–20 MW electrolyzers consisting of 4–24 PEM stacks have been put into applications [34].

Although scale-up of PEM electrolyzers, also stacks, can achieve rapid cost-down results, extensive research and development have been carried out. Generally, two strategies are usually adopted to reduce the stack costs, increasing operation current density and decreasing the usage of precious metal without sacrificing efficiency and durability. Therefore, it is essential to develop low resistance membrane, high stable catalyst and catalyst layer, low precious metal loading electrodes, and high diffusion bipolar plates. A rapid technology improvement can be expected considering the similarity of PEMWE to PEM fuel cells (PEMFC), which have made significant technical progress in this decade owing to their potential market of fuel cell vehicle (FCV). In the following, the current status, challenges, and main trends of the key materials of PEMWE are summarized.

1.3.3.1 Membranes

At present, the most widely used PEM material for PEM water electrolysis is Nafion, a PFSA (perfluorosulfonic acid) polymer with excellent proton conductivity and chemical stability. Its typical technical parameters are a thickness of 100–300 μm and a resistance of 0.1–0.3 Ω cm^2. Its main drawbacks are high cost, high swelling rate, and high gas crossover when low resistance is required by using its thin types. R&D trends of the PEM include the following aspects: improvement of conductivity; thermal stability of the ionomer; development of novel reinforcement materials and methods; selection of optimal filler type, size, and loading; optimization of thickness, structure, and operating conditions to realize low swelling rate and high mechanical strength; development of process to produce membranes with homogeneous thickness and quality; and membranes with functional additives such as gas recombination catalyst to suppress gas crossover.

Considering the trade-off relationships among various parameters such as proton conductivity, mechanical strength, gas barrier properties, and dimensional stability as well as cost, reinforced membranes have attracted much more attention, in which a reinforcement layer is embedded in the polymer matrix. It is well known that Nafion/ePTFE-reinforced membranes with a thickness of 8–30 μm have been widely used in PEMFCs. For PEMWE, many reinforcement materials have been investigated such as PTFE, PEEK (polyether ether ketone), PVDF, and PBI, which are chemically and thermally stable in acidic media, as well as reinforcement layer structure, such as a woven or non-woven fabric, a microporous film, or a nanofiber mat, which can provide different pore sizes, porosities, and thicknesses for the membrane. Other membrane materials have also been investigated, such as composite membranes, e.g., Nafion/ZrO$_2$, Nafion/TiO$_2$, and Nafion/SiO$_2$, and non-PFSA polymers, e.g., sulfonated poly(ether ketone) (S-PEEK), sulfonated poly(phenylene sulfone) (S-PS), and sulfonated polyimide (S-PI) [35–38].

1.3.3.2 Catalysts

The current state-of-the-art catalysts for PEMWE are IrO$_2$ for the OER and Pt for the HER, as they have excellent electrocatalytic performance and durability in acidic media. Although numerous research works have been carried out, the key challenges are high cost and uncertain degradation rate at high current density and temperature and potential cycling.

The OER and HER activity mechanisms of the catalysts in acidic media are not yet fully understood [39], but it is considered that there are mainly determined by the adsorption and desorption of the reaction intermediates, such as OOH and H, on the catalyst surface. The optimal catalyst should have a moderate binding energy for these intermediates, neither too strong nor too weak, to facilitate the reaction kinetics and avoid poisoning or passivation. Various strategies have been proposed, such as tuning the structure, morphology, composition, and electronic properties of the catalysts via alloying and defects engineering, similar to those applied in PEMFC catalysts. Since the overpotential of the present OER catalyst is far higher than that of HER in acidic media, research has been focused on OER catalysts, mainly about the development of low-Ir technology and PGM-free catalysts. It is revealed that the strong Ir–O bonding is important for the stability of iridium oxide catalysts, although the Ir catalysts in metal state always show high OER activity. To enhance OER activity, many alloys and mix oxides have been explored [40–44]. Although high activity was observed in Ir–Ru oxide, its stability has not yet been well confirmed and dissolution of Ru has been observed during electrolysis. At present, the most powerful candidate of low-Ir technology is the supported Ir oxide catalysts, a type of composite catalyst which has Ir oxide nanoparticles dispersed on a support material, such as TiO_2, doped SnO_2, MnO_2, SiC, and TaC [45–48]. The support and catalyst may exhibit synergistic effects, leading to enhanced catalytic activity [49]. The issue of supported Ir oxide catalysts is stability and durability because generally IrO_2 nanoparticles exhibit coarsening, agglomeration, and drop-down phenomena, as well as the elemental dissolution of metal doped in support materials. Developing novel and improved catalyst supports materials and synthesis methods that can optimize the structure, morphology, and composition of the catalysts, such as solution combustion, hydrothermal, sol–gel, and electrospinning methods.

It should be mentioned that the structure of the catalyst layer is important to reduce the catalyst loading, maximize catalyst utilization, and stabilize catalysts during electrolysis. Catalyst layer engineering usually includes optimization of the catalyst dispersion, distribution, and contact on the electrode and membrane. Apart from the traditional wet chemistry process, other fabrication processes have been explored. Using physical vapor deposition (PVD) process, a novel alternated catalyst layer structure(ACLS) consisting of nano sheet catalysts have been developed [50]. High performance and excellent durability have been reported in an Ir oxide ACLS with an Ir loading as low as 0.15 mg/cm^2 [51], which is considered to be attributable to the sheet-like catalysts with extended surface and the unique catalyst layer structure in the ACLS.

In summary, PEMWE has the highest dynamic potential for green hydrogen production. With the high attention paid on the R&D of membranes and catalysts, operation current densities higher than 2.5 A/cm^2, precious metal usage reduced to <0.3 mg/cm^2, and thus a rapid cost-down of PEM electrolyzers can be highly expected.

1.3.4 ANION EXCHANGE MEMBRANE WATER ELECTROLYSIS

AEMWE technology has attracted attention since it is possible to combine the merits of AWE and that of PEMWE, including (1) electrolysis with pure water feed, avoiding the need for liquid electrolyte circulation and maintenance, and reducing the risk of contamination and corrosion, and (2) using low-cost and abundant materials, such as nickel, cobalt, iron, and stainless steel, for the catalysts, gas diffusion layers, and bipolar plates, as the alkaline environment is less corrosive than the acidic environment in PEMWE electrolysis [52]. However, the breakthrough barrier is high considering the development of AEM-type fuel cells (AEMFC). AEM membrane is essential for AEMWE, since the other materials applied in AEMWE are similar to AWE or PEMWE. There have been various commercial AEMs created and launched in recent years [53]. The main issues are the weaker ionic conductivity of AEMs compared to PEMs and poor durability under alkaline environments. The main degradation mechanism is known as hydroxide (OH⁻) attack on the polymer backbone, which leads to membrane collapse and catalyst layer dissolution within a few days, although some studies have shown that the use of pure water as electrolyte feedstock can lead to a durability beyond 5000 hours, under some operation conditions. The current R&D trends of AEM electrolysis focus on the following aspects: determining the role of supporting electrolyte and the limiting factors behind DI water operation, as well as developing novel and improved polymer/membrane materials that can enhance the OH⁻ conductivity and chemical stability while suppressing gas permeation [54,55]. Some potential candidates have been proposed such as quaternary ammonium-based polymers, poly(aryl piperidinium) polymers, poly(phenylene oxide) polymers, and composite or reinforced membranes with inorganic fillers or fibers [56–59].

Additionally, another advantage of AEMWE is using a neutral-pH electrolyte to avoid the use of undesirable strong acids or bases, which could eliminate environmental and handling problems. However, the development of neutral-pH water electrolysis remains a large challenge due to the lack of highly efficient PGM-free catalysts. In summary, AEMWE is a promising technology that may challenge the PEMWE. More research and development are needed to address the challenges and limitations of the membranes and catalysts in AEMWE, as well as to demonstrate the feasibility and scalability of the technology.

1.3.5 SOLID OXIDE ELECTROLYSIS

SOE has several advantages over other electrolysis technologies and is considered as a promising approach owing to the following technical potentials. SOE can operate at high temperatures (300°C–1000°C), which can enhance the reaction kinetics, increase voltage efficiency close to 100%, and utilize the waste heat from other processes, although heat management for the high-temperature working may induce efficiency decrease of the electrolysis system. SOE can use both steam and carbon dioxide as feedstocks, which can increase the hydrogen yield, reduce the water consumption, and enable the co-production of syngas, a valuable chemical feedstock. SOE can use low-cost and abundant raw materials, such as nickel, for the electrodes as the high-temperature environment can activate these materials for the OER and the HER.

However, SOE faces big challenges and limitations in widespread application. The main drawback is its use of oxide materials, which are intrinsically brittle and thus easily broken if the cell temperature changes [60]. Improvement of reliability and durability is the main research direction, similar to that of solid oxide fuel cell

(SOFC), which includes developing high-property electrolyte/electrode materials, reducing the working temperature, optimizing stack structure and its component materials, scaling up the stack size, and realizing long-term stable operation as well as low-cost fabrication process. Besides the traditional O-SOE, in which oxide ions (O^{2-}) are conducted from the anode to the cathode via a ceramic electrolyte layer, proton-type SOE cells/stacks (p-SOE) have been proposed [61,62]. P-SOE uses a ceramic electrolyte to conduct protons from the cathode to the anode and usually exhibit lower working temperatures such as 600°C. Furthermore, CO_2 electrolysis using SOE has also been focused on recently to expand the range of applications of SOE [63–65]. The CO_2 is put into the cathode, where CO_2 is decomposed into CO and O^{2-} after receiving electrons (CO_2 reduction reaction: CO_2RR), and simultaneously oxygen ions (O^{2-}) are delivered to the anode through the solid electrolyte layer and then converted into O_2 by losing electrons.

The current status, issues, and main R&D trends of SOE are summarized below.

1.3.5.1 Electrolyte Materials

The biggest challenges for electrolyte material of SOE are its reliability and durability considering its degradation resulting from high temperature, steam, carbon dioxide, oxygen radicals, and water-splitting products, as well as contamination from other components in the SOE cell. For an electrolyte material, its crystalline structure stability at the working temperature and material composition, such as dopant ratio, which affects oxygen vacancies and thus ionic conductivity, are important. Generally, the Y_2O_3-stabilized ZrO_2 (YSZ) and (Sr, Mg)-doped $LaGaO_3$-based materials (LSGM) with perovskite structure have been regarded as outstanding electrolyte materials suitable for use at high working temperature and moderate temperature, respectively, owing to their remarkable conductivity, excellent chemical and thermal stability, as well as outstanding mechanical properties, in both oxidizing and reducing atmospheres [66]. Other candidates are Sc_2O_3-stabilized ZrO_2 (ScSZ), Sm-doped CeO_2 (SDC), etc. For p-SOFC, Y-doped $BaZrO_3$ (BZY) and Sc-doped BaZrO3 (BZSc) have been explored to obtain higher proton conductivity, high Faradaic efficiency (FE), long durability, and low electronic leakage at a working temperature at 500°C–600°C [67]. Further understanding of the proton conduction and electronic leakage mechanisms, as well as optimization of synthesis process, is necessary.

1.3.5.2 Electrode Materials

Requirements on electrode materials include high mixed ionic and electronic conductivity, satisfactory physical and electrochemical stabilities, favorable HER/OER activity, and an appropriate thermal expansion coefficient compatible with electrolytes.

For cathode materials, metal-oxide ceramics have been mainly investigated, in which transition metal generally supplies catalytic site for HER. Ni-YSZ cermet has been widely used [68–70]. Its degradation mechanism is regarded to be the coarsening and agglomeration of Ni particles and the oxidation of metal Ni to NiO during the SOE process. A cathode microstructure has been proposed in which fine metal Ni particles uniformly distribute on the YSZ matrix. Addition of other metal elements including precious metal materials such as Pt and Cu has been investigated. Cu has low catalytic activity but has long-term stability due to its resistance to carbon build-up [64].

For the anode materials, since the OER reaction contributes to the main polarization resistance in the electrolysis process in SOE, developing advanced anode materials is crucial to improve the SOE performance. Great strides have been made, including many kinds of anode materials such as single perovskite, double perovskite, Ruddlesden–Popper phase ($A_{n+1}B_nO_{3n+1}$) oxides, spinel oxides, composites with the electrolyte material, heterostructures with noble metals, and various strategies for anode optimization such as combining with ionic conductor, infiltrating OER active species, and structural optimization [71–73]. Some powerful candidates such as $La_{1-x}Sr_xMnO_3$ (LSM), Co-based perovskite oxides such as $La_{1-x}Sr_xCoO_3$ (LSC), LSM-YSZ composites, and Au nanoparticles–loaded LSM-YSZ have been proposed. Further research is necessary to overcome the sluggish kinetics of the OER and the reverse water–gas shift reaction (RWGS) in high-temperature and mixed-gas conditions, to resist sintering, poisoning, and carbon deposition, and to address the issues such as ion segregation and structure failure, which limit the practical application of SOE.

1.3.5.3 Cells/Stacks

A typical SOE cell includes porous anode, dense electrolyte, and porous cathode. In order to maximize the potential of electrolytes and electrode materials, SOE requires highly optimized and integrated designs that can ensure good contact and compatibility among the anode, electrolyte layers, the cathodes, and the interconnects, and minimize the thermal and mechanical stresses, ohmic and mass transport losses, and heat and water management issues in the cells/stacks.

The main challenge of SOE cells/stacks is their durability under high operating temperatures (350°C–1000°C) and dynamic load conditions, which limit its applications or commercialization, although low degradation rates less than 3%/1000 kh and long lifetimes have been reported for some SOE stacks under constant power and well-defined operating conditions [74]. The main degradation mechanism is the thermal cycling, which occurs due to the high operating temperatures and the need to cool down in case of dynamic operation. To ensure homogeneous thermal distribution around SOE cells/stacks, the SOE cells are usually small, and a complex auxiliary system is generally necessary. Deploying SOEC at large scale would require larger cells than currently used, which increases the failure possibility. The degradation phenomena in the cells/stacks are complex since at high working temperature, multi-interactions among the component materials occur easily [75], such as elemental segregation, delamination that occurs at the anode/electrolyte interface during SOE operations, contamination of gaseous chromium species from stainless steel interconnects and sulfur dioxide from the air (sulfur toxicity), silica permeation into the SOE cell from the glass-based sealant materials, and carbon deposition for the CO_2 electrolysis cells/stacks.

To address these issues, further studies include: investigation on various mechanisms, especially those under different practical conditions, selection of appropriate component materials with high chemical, electrochemical and mechanical stability, matching the thermal expansion coefficient of electrolyte layer and both electrodes, and ensuring minimal reactant crossover, interface engineering, as well as developing novel and improved fabrication and methods, such as coating technology of interconnect materials, and optimizing the operating conditions (e.g., temperature and steam concentration).

Modification of stack structures is also important for the reliability and durability of the SOE technology due to their effects on encapsulation, thermal stress, ohmic impedance, the operating temperature, thermal cycling efficiency, and degradation rate. The typical stack shapes are flat plate-type and tubular-type, with a selection of support, e.g., cathode-supported type, electrolyte-supported type, and metal-supported type (MS-SOE), which uses metal electrodes and thus can increase mechanical strength.

In summary, SOE is a promising electrolysis technology with high voltage efficiency and high potential to produce carbonate derivatives besides green hydrogen. To realize widespread industrialization and commercialization, more research and development are needed to further improve the performance of the SOE cells or stacks, especially their reliability and durability, to exhibit a long-term operation under renewable power conditions. Besides, SOE can be converted to reversible SOEC/SOFC cells (RFCs) to be used as hydrogen storage system with a high overall energy transfer efficiency.

1.3.5.4 Summarization

In the worldwide low-carbon energy transition, green hydrogen has been regarded as the most clean and promising energy utilization solution to fundamentally address the issues of large-scale storage and consumption of renewable energy, and thus, a huge market of green hydrogen is highly expected in a near future. The sustainable hydrogen production by water electrolysis is an essential prerequisite of hydrogen economy with zero carbon emission. Although great progress has been made in the development of water electrolysis technology, some challenges still remain since hydrogen production conditions in the renewable energy scenario are different from the traditional scenario. It is important to increase R&D efficiency and accelerate foundational R&D of innovative materials, components, stacks, and systems for advanced water-splitting technologies, as well as their scale-up and low-cost manufacturing process.

There are four types of electrolysis technologies nowadays, although the present AEMWE and SOE may not be sufficient for long-term operation under renewable power conditions, it should be kept in mind that each technology faces new challenges in renewable power conditions, mainly including higher operation current density without sacrificing of electrolysis efficiency, enhancement of reliability and durability, and cost reduction, and it is undergoing rapid development; and finally, various technologies complement each other's strengths, and competition coexists in different scenarios. Since different technologies of electrolyzers have complementary critical material requirements, which can offer protection against disruption in supply of some critical materials and can put strategic value on technology diversification.

ACKNOWLEDGMENTS

The authors thank our colleagues, Dr. Sai Zhang, Dr. Chenxu Li, and Dr. Jialun Tang, for their assistance that greatly improved the manuscript.

REFERENCES

[1] IRENA. Green Hydrogen Cost, Scaling Up Electrolysers to Meet the 1.5°C Climate Goal, 2020/02.

[2] US DOE. U.S. National Clean Hydrogen Strategy and Roadmap (energy.gov), 2022/09.

[3] European Commission. Roadmap and KPI. https://www.clean-hydrogen.europa.eu/knowledge-management/strategy-map-and-key-performance-indicators/clean-hydro-gen-ju-sria-key-performance-indicators-kpis_en, 2022/02.

[4] NEDO, Japan. New Energy and Industrial Technology Development Organization (nedo.go.jp), 2023/02.

[5] China. Medium and Long-Term Plan for the Development of Hydrogen Energy Industry (2021–2035) of China, 2022/03.

[6] IEA. Report prepared by the IEA for the G20, Japan; The Future of Hydrogen - Seizing Today's Opportunities, 2019/06.

[7] IEA. Global Hydrogen Review, Global Hydrogen Review 2023 – Analysis – IEA, 2023/09.

[8] IRENA. Renewable Power Remains Cost-Competitive amid Fossil Fuel Crisis (irena.org), 2022/07.

[9] DNV. 2050 Hydrogen Outlook, 2022.

[10] Hydrogen Council. Path to Hydrogen Competitiveness: A Cost Perspective, 2020/01.

[11] Goldman-Sachs. Carbonomics: The Clean Hydrogen Revolution, 2022.

[12] Deloitte. Green Hydrogen: Energizing the Path to Net Zero (deloitte.com), 2023/07.

[13] BP. Energy –Outlook – 2023 –Edition, 2023/07.

[14] IRENA. Global Hydrogen Trade to Meet the 1.5°C Climate Goal, 2022/07.

[15] D.M.F. Santos, C.A.C. Sequeira, J.L. Figueiredo, Hydrogen production by alkaline water electrolysis, *Química Nova*. 36(8) (2013) 1176–1193.

[16] K. Zeng, D. Zhang, Recent progress in alkaline water electrolysis for hydrogen production and applications, *Progress in Energy and Combustion Science*. 36 (2010) 307–313.

[17] Longqi Hydrogen Company. https://www.longi.com/cn/products/hydrogen/, 2023.9.13.

[18] H. Wendt, H. Hofmann, Ceramic diaphragms for advanced alkaline water electrolysis, *Journal of Applied Electrochemistry*. 19 (1989) 605–610.

[19] N.V. Kuleshov, V.N. Kuleshov, S.A. Dovbysh, et al., Polymeric composite diaphragms for water electrolysis with alkaline electrolyte, *Russian Journal of Applied Chemistry*. 89 (2016) 618–621. https://doi.org/10.1134/S1070427216040157.

[20] X. Tan, M. Zhang, D. Chen, et al., Electrochemical etching switches electrocatalytic oxygen evolution pathway of $IrOx/Y_2O_3$ from adsorbate evolution mechanism to lattice-oxygen-mediated mechanism, *Small*. 19(44) (2023) 2303–2319.

[21] M. Carmo, D.L. Fritz, J. Mergel, D. Stolten, A comprehensive review on PEM water electrolysis, *International Journal of Hydrogen Energy*. 38 (2013) 4901.

[22] B. E. Conway, G. Jerkiewicz, Relation of energies and coverages of underpotential and overpotential deposited H at Pt and other metals to the 'volcano curve' for cathodic H2 evolution kinetics. *Electrochimica Acta*. 45(25–26) (2000) 4075–4083.

[23] Meital Shviro, Bryan Pivovar, Alex Badgett. DOE Hydrogen Program's Annual Merit Review and Peer Evaluation Meeting (AMR) DOE H_2NEW: Hydrogen (H_2) from Next-generation Electrolyzers of Water LTE Task 9: Liquid Alkaline, 2023/04, https://www.hydrogen.energy.gov/docs/hydrogenprogramlibraries/pdfs/review23/p196h_pivovar_2023_p-pdf.pdf.

[24] S. Tanaka, N. Hirose, T. Tanaki, Evaluation of Raney–nickel cathodes prepared with aluminum powder and tin powder. *International Journal of Hydrogen Energy* 25 (5) (2000) 481–485.

[25] A. Grimaud, et al. Double perovskites as a family of highly active catalysts for oxygen evolution in alkaline solution. *Nature Communications* 4(1) (2013) 2439.

[26] Z. Lu, et al. Electrochemical tuning of layered lithium transition metal oxides for improvement of oxygen evolution reaction. *Nature Communications.* 5(1) (2014) 4345.

[27] N. Danilovic, et al. Using surface segregation to design stable Ru-Ir oxides for the oxygen evolution reaction in acidic environments. *Angewandte Chemie International Edition.* 53(51) (2014) 14016–14021.

[28] J. Zhang, T. Wang, P. Liu, et al., Efficient hydrogen production on MoNi(4) electrocatalysts with fast water dissociation kinetics, *Nature Communications.* 8 (2017) 15437.

[29] X. Lu, C. Zhao, Electrodeposition of hierarchically structured three-dimensional nickel-iron electrodes for efficient oxygen evolution at high current densities, *Nature Communications.* 6 (2015) 6616. https://doi.org/10.1038/ncomms7616.

[30] J. Wang, N. Zang, C. Xuan, et al., Self-supporting electrodes for gas-involved key energy reactions, *Advanced Functional Materials.* 31 (2021) 2104620.

[31] N.K. Chaudhari, H. Jin, B. Kim, K. Lee, Nanostructured materials on 3D nickel foam as electrocatalysts for water splitting, *Nanoscale.* 9 (2017) 12231–12247.

[32] L. Wan, Z. Xu, Q. Xu, et al., Overall design of novel 3D-ordered MEA with drastically enhanced mass transport for alkaline electrolyzers, *Energy & Environmental Science.* 15(5) (2022) 1882–1892.

[33] R. Phillips, C.W. Dunnill, Zero gap alkaline electrolysis cell design for renewable energy storage as hydrogen gas, *RSC Advances.* 6(102) (2016) 100643–100651.

[34] Siemens Energy. Siemens Energy to Supply 1.8 GW of Electrolyzers to HIF Global for Texas eFuels Facility - *Green Car Congress*, 2023/3/10.

[35] A. Kusoglu, A.Z. Weber, New insights into perfluorinated sulfonic-acid ionomers, *Chemical Reviews.* 117 (2017) 987–1104.

[36] C. Klose, P. Trinke, T. Böhm, et al., Interlayer with Pt recombination particles for reduction of the anodic hydrogen content in PEM water electrolysis, *Journal of Electrochemistry Society.* 165 (2017) F1271–F1277.

[37] A. Goni-Urtiaga, D. Presvytes, K. Scott, Solid acids as ~ electrolyte materials for proton exchange membrane (PEM) electrolysis, *International Journal of Hydrogen Energy.* 37(4) (2012) 3358–3372.

[38] H. Ito, T. Maeda, A. Nakano, H. Takenaka, Properties of Nafion membranes under PEM water electrolysis conditions, *International Journal of Hydrogen Energy.* 36(17) (2011) 10527–10540.

[39] M. Bernt, A. Siebel, H.A. Gasteiger, Analysis of voltage losses in PEM water electrolyzers with low platinum group metal loadings, *Journal of the Electrochemical Society.* 165(5) (2018) F305–F314.

[40] B. Hinnemann, P.G. Moses, J. Bonde, et al., Biomimetic hydrogen evolution: MoS_2 nanoparticles as catalyst for hydrogen evolution, *Journal of American Chemical Society.* 127(15) (2005) 5308–5309

[41] E. Rasten, G. Hagen, R. Tunold, Electrocatalysis in water electrolysis with solid polymer electrolyte, *Electrochimica Acta.* 48(25–26) (2003) 3945–3952.

[42] M.H. Miles, M.A. Thomason, Periodic variations of overvoltages for water electrolysis in acid solutions from cyclic voltammetric studies, *Journal of the Electrochemical Society.* 123(10) (1976) 1459–1461.

[43] C. Rozain, E. Mayousse, N. Guillet, P. Millet, Influence of iridium oxide loadings on the performance of PEM water electrolysis cells: part II - Advanced oxygen electrodes, *Applied Catalysis B: Environmental.* 182 (2016) 123–131.

[44] S. Zhao, A. Stocks, B. Rasimick, et al., Highly active, durable dispersed iridium nano-catalysts for PEM water electrolyzers, *Journal of the Electrochemical Society.* 165 (2018) F82–F89.

[45] M. Bernt, H.A. Gasteiger, Influence of ionomer content in IrO_2/TiO_2 electrodes on PEM water electrolyzer performance, *Journal of the Electrochemical Society.* 163 (2016) F3179–F3189.

[46] J.I. Cha, C. Baik, S.W. Lee, C. Pak, Improved utilization of IrO_x on Ti_4O_7 supports in membrane electrode assembly for polymer electrolyte membrane water electrolyzer, *Catalysis Today.* 403 (2022) 19–27. https://doi.org/10.1016/j.cattod.

[47] V.K. Puthiyapura, M. Mamlouk, S. Pasupathi, et al., Physical and electrochemical eval-uation of ATO supported IrO_2 catalyst for proton exchange membrane water electroly-ser, *Journal of Power Sources.* 269 (2014) 451–460.

[48] P. Mazur, J. Polonsky, M. Paidar, K. Bouzek, Non-conductive TiO_2 as the anode cata-lyst support for PEM water electrolysis, *International Journal of Hydrogen Energy.* 37 (2012) 12081–12088.

[49] S. Zhao, A. Stocks, B. Rasimick, et al., Highly active, durable dispersed iridium nano-catalysts for PEM water electrolyzers, *Journal of the Electrochemical Society.* 2 (2018) F82.

[50] W. Mei, T. Fukazawa, Y. Nakano, et al., Development of alternated catalyst layer struc-ture for PEM fuel cells, *ECS Transactions.* 50(2) (2013) 1377–1384.

[51] N. Yoshinaga, Y. Kanai, T. Fukazawa, et al., Development of ACLS electrodes for a water electrolysis cell, *ECS Transactions.* 92 (2019) 749–755.

[52] X. Yan, J. Biemolt, K. Zhao, et al., A membrane-free flow electrolyzer operating at high current density using earth-abundant catalysts for water splitting, *Nature Communications.* 12(1) (2021) 4143. https://doi.org/10.1038/s41467-021-24284-5.

[53] K. Chand, O. Paladino, Recent developments of membranes and electrocatalysts for the hydrogen production by anion exchange membrane water electrolysers: a review, *Arabian Journal of Chemistry.* 16(2) (2023) 104451.

[54] X. Zhang, Y. Cao, M. Zhang, et al., Enhancement of the mechanical properties of anion exchange membranes with bulky imidazolium by "thiol-ene" crosslinking, *Journal of Membrane Science.* 596 (2020) 117700.

[55] J. Pan, C. Chen, Y. Li, et al., Constructing ionic highway in alkaline polymer electro-lytes, *Energy & Environmental Science.* 7(1) (2014), 354–360.

[56] C.C. Pavel, ; F. Cecconi, C. Emiliani, et al., Highly efficient platinum group metal free based membrane-electrode assembly for anion exchange membrane water electrolysis, *Angewandte Chemie International Edition in English.* 53(5) (2014) 1378–1381.

[57] G. Merle, M. Wessling, K. Nijmeijer, Anion exchange membranes for alkaline fuel cells: a review, *Journal of Membrane Science.* 377 (1–2) (2011) 1–35. https://doi.org/10.1016/j. memsci.2011.04.043.

[58] L. Yang, Z. Wang, F. Wang, et al., Poly(aryl piperidinium) anion exchange membranes with cationic extender sidechain for fuel cells, *Journal of Membrane Science.* 653 (2022) 120448.

[59] K.H. Lee, , J.Y. Chu, A.R. Kim, et al., Functionalized TiO_2 mediated organic-inorganic composite membranes based on quaternized poly(arylene ether ketone) with enhanced ionic conductivity and alkaline stability for alkaline fuel cells, *Journal of Membrane Science.* 634 (2021) 119435.

[60] M.S. Khan, X. Xu, R. Knibbe, Z. Zhu, Air electrodes and related degradation mech-anisms in solid oxide electrolysis and reversible solid oxide cells, *Renewable and Sustainable Energy Reviews.* 143 (2021) 110918.

[61] H.S.Y. Arachi, O. Yamamoto, Y. Takeda, N. Imanishai, Electrical conductivity of the ZrO_2-Ln_2O_3 (Ln=lanthanides) system, *Solid State Ionics.* 121 (1999) 133–139.

[62] T.G. Fujimoto, V. Seriacopi, L.A.S. Ferreira, et al., Mechanical and electrical characterization of 8YSZ-ScCeSZ ceramics, *Materials Research*. 26 (2023).

[63] Q. Li, Y. Zheng, Y. Sun, et al., Understanding the occurrence of the individual CO_2 electrolysis during H_2O-CO_2 co-electrolysis in classic planar Ni-YSZ/YSZ/LSM-YSZ solid oxide cells, *Electrochimica Acta*. 318 (2019) 440–448.

[64] M. Lo Faro, W. Oliveira da Silva, W. Valenzuela Barrientos, et al., The role of CuSn alloy in the co-electrolysis of CO_2 and H_2O through an intermediate temperature solid oxide electrolyser, *Journal of Energy Storage*. 27 (2020).

[65] Z. Yang, Z. Lei, B. Ge, et al., Development of catalytic combustion and CO_2 capture and conversion technology, *International Journal of Coal Science & Technology*. 8 (2021) 377–382.

[66] S. Ryu, J. Hwang, W. Jeong, et al., A self-crystallized nanofibrous Ni-GDC anode by magnetron sputtering for low-temperature solid oxide fuel cells, *ACS Applied Materials & Interfaces*. 15 (2023) 11845–11852.

[67] G. Tsekouras, J.T.S. Irvine, The role of defect chemistry in strontium titanates utilised for high temperature steam electrolysis, *Journal of Materials Chemistry*. 21 (2011).

[68] R.A. Budiman, T. Ishiyama, K.D. Bagarinao, et al., Dependence of hydrogen oxidation reaction on water vapor in anode-supported solid oxide fuel cells, *Solid State Ionics*. 362 (2021) 115565.

[69] D. Chen, J. Zhang, M. Barreau, et al., Ni-doped CeO_2 nanoparticles to promote and restore the performance of Ni/YSZ cathodes for CO_2 electroreduction, *Applied Surface Science*. 611 (2023).

[70] D. Neagu, G. Tsekouras, D.N. Miller, et al., In situ growth of nanoparticles through control of non-stoichiometry, *Nature Chemistry*. 5 (2013) 916–23.

[71] R. Wang, U.B. Pal, S. Gopalan, S.N. Basu, Chromium poisoning effects on performance of (La, Sr)MnO_3-based cathode in anode-supported solid oxide fuel cells, *Journal of the Electrochemical Society*. 164 (2017) F740–F747.

[72] Z. Liu, W. Wu, Z. Zhao, et al., Electrochemical behaviors of infiltrated (La, Sr)MnO_3 and Y_2O_3-ZrO_2 nanocomposite layer, *International Journal of Hydrogen Energy*. 42 (2017) 5360–5365.

[73] Y. Song, X. Zhang, Y. Zhou, et al., Improving the performance of solid oxide electrolysis cell with gold nanoparticles-modified LSM-YSZ anode, *Journal of Energy Chemistry*. 35 (2019) 181–187.

[74] Y. Tian, J. Li, Y. Liu, et al., Preparation and properties of $PrBa_{0.5}Sr_{0.5}Co_{1.5}Fe_{0.5}O_{5+\delta}$ as novel oxygen electrode for reversible solid oxide electrochemical cell, *International Journal of Hydrogen Energy*. 43 (2018) 12603–12609.

[75] Y. Zheng, H. Jiang, S. Wang, et al., Mn-doped Ruddlesden-Popper oxide $La_{1.5}Sr_{0.5}NiO_{4+\delta}$ as a novel air electrode material for solid oxide electrolysis cells, *Ceramics International*. 47 (2021) 1208–1217.

2 Recent Advances in Non-Precious Metal-Based Electrodes for Alkaline Water Electrolysis

Jiakai Bai, Pengxi Li, Dongwei Qiao, and Xianming Yuan

2.1 INTRODUCTION

Hydrogen production by water electrolysis is a clean energy storage method that can replace fossil fuels [1]. In this field, electrolytic water technology is one of the most promising ways to generate pure hydrogen energy. Cathodic hydrogen evolution reaction (HER) and anodic oxygen evolution reaction (OER) are the two major reactions of electrolysis of water [2]. From this point of view, the creation of electrodes that operate at low overpotential and work steadily for a long time can maximize the energy efficiency of hydrogen production. Therefore, many efforts have been devoted to exploring active catalysts and efficient electrodes with optimal structural characteristics to generate large amounts of H_2.

On the electrode side, the HER performance of a variety of highly efficient electrocatalysts was tested in acidic medium. Despite their excellent electrocatalytic properties, there are still many limitations, such as control of acid mist or electrolyte vapor in batteries and corrosion of electrodes at high temperatures. In contrast, alkaline electrolysis requires low vapor pressure and high temperature conditions. This, compared to the acid catalysis, is more efficient in H_2 production [3]. Similarly, only rare precious metal catalysts can provide high stability and reactivity for OER in acidic media, but considering the scarcity and high price, such catalysts are difficult to be widely used in industry. From this point of view, alkaline water electrolysis is an effective and cost-effective way to achieve industrial hydrogen production by water electrolysis. In order to improve the electrocatalytic performance of HER and OER, current academic research focuses on controlling and optimizing the electronic structure of the electrocatalyst to improve the catalytic performance and thus change the reaction kinetics [4]. Oxides and hydroxides of Group VIII 3d transition metals (Fe, Co, Ni) and their hybrids with other metals have been reported to have

DOI: 10.1201/9781003368939-2

strong OER catalytic activity under alkaline conditions [5]. In addition, these transition metals and their hybrids have been shown to have superior surface and catalytic areas with rapid charge transport properties, especially in the form of nanostructured metals. The following is a brief introduction to HER and OER electrocatalysis, thermodynamic concepts, and an understanding of reaction pathways. Then, the properties, surface characteristics, synthesis strategies, and broader chemical and physical aspects of the electrocatalyst are discussed in detail, including electronic structure tuning, heteroatom doping, oxygen vacancy/defect, and the inclusion of coupled conductive substrates.

2.2 ELECTROCATALYSTS FOR HER

Volmer–Heyrovsky and Volmer–Tafel processes are mechanisms for producing pure hydrogen by HER reaction. These two processes largely depend on the physicochemical properties of the electrocatalyst used. As equation (2.1) shows, the Volmer step of the reaction occurs when a water molecule is adsorbed to the catalytic surface and then expends an electron, splitting into a hydrogen atom and a hydroxide anion. Then, the two hydrogen atoms combine to form a hydrogen molecule, which eventually leaves the catalyst's surface, a process known as the Tafel process (equation 2.2), or under similar circumstances, a hydrogen atom reacts with another water molecule, producing a hydroxide anion and a hydrogen (H_2) molecule by assimilating an electron (Heyrovsky, equation 2.3).

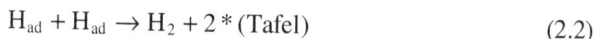

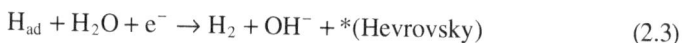

$$H_2O + e^- + * \rightarrow H_{ad} + OH^- \text{ (Volmer)} \tag{2.1}$$

$$H_{ad} + H_{ad} \rightarrow H_2 + 2 * \text{ (Tafel)} \tag{2.2}$$

$$H_{ad} + H_2O + e^- \rightarrow H_2 + OH^- + * \text{(Hevrovsky)} \tag{2.3}$$

where * is the hydrogen adsorption sites on the electrocatalyst.

Hydrogen adsorption and desorption are two key processes in hydroelectrolysis that are both continuous and competitive, and they determine the efficiency of HER process. The hydrogen bond of the electrocatalyst plays a key role here, and if the hydrogen bond strength is weak, sufficient hydrogen adsorption cannot be achieved. If the hydrogen bond strength is too strong, it will make it difficult to release the catalytic product. In either case, the efficiency of HER will be affected. Thus, an optimum balance of forces that are the nature of catalyst plays a critical role in controlling HER kinetics. In this context, Conway et al. investigated the relationship between the maximum exchange current density and metal properties, i.e., M-H, bond energy or hydrogen adsorption free energy. Among them, materials such as Pt occupy the top position indicating the highest efficiency. It can be seen that transition metals near precious metals have higher HER efficiency.

Precious metals such as platinum are still HER electrocatalysts with better performance in alkaline media because of their low overpotential value. However, the production cost of this precious metal catalyst is too high to be widely used. The use of precious metal catalysts in large-scale projects will lead to extremely high production costs, which will greatly reduce the attractiveness of clean energy production. In this context, the new research direction is to produce, study, and configure effective catalysts to achieve suitable HER electrocatalysis and as a suitable alternative to currently used laboratory precious metal catalysts. The potential of transition metals, alloys, oxides, chalcogenides, carbides, nitrides, and phosphates as highly efficient HER electrocatalysts has been reported in many literatures.

2.2.1 TRANSITION METAL AND THEIR ALLOYS

Compared with other transition metals, nickel (Ni) has excellent catalytic properties and a wide range of abundances, making it a potential application prospect of efficient HER electrocatalysts. Nickel has unique electrocatalytic properties, and people have modified it based on this. For example, the surface structure of nickel is modified to obtain a larger surface area, or it is mixed with other transition metals to obtain greater catalytic activity. These approaches have proven to be effective strategies for obtaining high-quality nickel-based electrocatalysts. In general, the formation of a porous surface helps provide more exposed surface area for the catalyst, allowing for greater efficiency. Based on this, the preparation of Raney Ni by co-precipitation of metal nickel with aluminum or zinc has been proven to be an effective method to effectively enhance nickel-based catalysts. Tanaka et al. described the characteristics of Raney Ni-based electrode for HER in 1.0M KOH solution. The authors emphasized that the use of Raney Ni contributed to more specific surface area which is three times higher than conventional Ni-based catalyst. This increment influenced the catalytic capabilities, where hydrogen evolution rate rose significantly concurrently with the improved electrochemical characteristics. In general, the Brewer–Engel valence bond theory suggests alloying transition metal to promote HER. Here, when the transition metal with free or half-empty d orbital is coupled with transition metals with paired d electrons, a synergistic effect may occur, potentially improving the HER [6]. Various Ni alloys with different transition metals have been considered such as Ni–Mo, Ni–Co, Ni–Fe, and Ni–Cr, while the ternary alloys such as Ni–Mo–Fe, Ni–Mo–Cu, and Ni–Mo–Co have also been proposed. The properties of these alloy catalysts match the lattice, and the exposure of more surface sites improves the activity and stability of the HER. A variety of methods have been used to evaluate the performance of nickel-based catalysts. Ming Fang et al. prepared NiMo alloy nanowires supported by Ni foam using hydrothermal method and thermal reduction method. The material has unique hierarchical structure and catalytic activity, and the substrate-supported catalyst shows high HER activity. The geometric current density of the alloy catalyst is −10 mA/cm^2 and the overpotential is as low as −30 mV. Catalytic performance can also be improved by inducing more electrical contact points. In this case, Jiao Deng et al. coupled FeCo alloy with N-doped carbon nanotubes (CNTs), which significantly improved the long-term durability of the catalyst during HER activity [7].

2.2.2 TRANSITION METAL AND METAL OXIDES

Transition metal oxides have good performance in OER. The performance of transition metal oxide catalysts in hydrogen evolution in alkaline solution is not good, but the coupling with pure metals can produce a heterogeneous structure system with synergistic properties. In this regard, Yan et al. synthesized 3D crystalline/amorphous Co/Co_3O_4 core–shell nanostructure to be used as catalyst in HER electrocatalyst in alkaline solution. The coupling enabled high electrical conductivity from the core, while potent catalytic activity could be achieved at the shell [8]. Similarly, Dai's group designed nanoscale nickel oxide/nickel heterostructures over the sidewalls of CNTs [9]. This hetero-interface was capable of showing high HER activity similar to that of Pt-based catalyst. The superiority was attributed to the greater hydrogen atom adsorption sites that promoted the generation of hydrogen molecules during HER in alkaline solution. The high catalytic activity of the discussed catalyst opened up new pathways to produce and explore low-cost, earth-abundant metal catalyst with superior HER activity. Similarly, Weng et al. discussed the capability of Ni/CeO_2-CNT hybrids as HER catalyst. This catalyst possessed an overpotential value of -91 mV with a current density of -10 mA/cm^2 in 1.0 M KOH [10]. Similarly, approaches involve using transition metal alloy/alloy-oxide interfaces with a partially oxidized nanosheet array for HER. These methods have been shown to be effective in improving HER reactivity in alkaline solutions. In short, modifying the surface of the electrode is an effective way to improve HER electrocatalytic potential to a higher efficiency and durability.

2.2.3 TRANSITION METAL CHALCOGENIDES

The reactivity of metal sulfides in HER has also been extensively studied. Some studies have shown that amorphous Cu_2MoS_4 nanocages prepared by hydrolyzing and etching-precipitating method have shown promising results. The hollow structural morphology could attain an overpotential value of 96 mV at 10 mA/cm^2 with a Tafel slope of 61 mV/dec in an alkaline environment [11]. Similarly, the MoS_2 has also shown promising outcomes. In other cases, Feng et al. described the potential application of Ni_3S_2 nanosheet arrays with exposed high-index facets for HER [12].

Transition metal selenides have advantages in structure and relatively low cost. Sun's team reports a simple method to produce in situ grown NiSe nanowire films that are highly reactive and sustainable [13]. The electro-deposited Ni_3Se_2 film on Cu foam has also been promising material based on its efficient HER reactivity and reliable response. Xie's group showed that the lattice control of two phases for $CoSe_2$, i.e., orthorhombic phase $CoSe_2$ (o-$CoSe_2$) and cubic phase $CoSe_2$ (c-$CoSe_2$), could be achieved with an experimentally proven relationship between different Co–Se bond lengths with adsorbed H atoms and water adsorption energy. This study provided an comprehensive understanding of the relationship between the crystal structure and the intrinsic HER electrocatalytic activity, advancing the field of HER process [14].

2.2.4 Transition Metal Phosphides, Nitrides, and Carbides

Compared with precious metals, transition metal phosphide, nitride, and carbide have almost the same d band electron density state. Moreover, materials based on electronic transition state similarity can be used as effective catalysts for HER activity. Among many, the P-$Co_{0.9}Ni_{0.9}Fe_{1.2}$ nanocubes with rough morphology were reported to be an efficient catalyst with an overpotential of −200.7 mV at a current density of 10 mA/cm^2 [15].

Transition metal nitrides have high electrical conductivity. Unlike the layered structure of graphene and its similar transition metal disulfide compounds, the graphene layer is made up of tip Mo atoms sandwiched between nitrogen atoms in the center of each monolayer. The said configuration allows MoN to act as a suitable conductive platform to fabricate highly efficient HER catalyst. The MoN nanosheets prepared using liquid exfoliation could attain atomic thinness, with highly exposed Mo atoms. Such atoms could efficiently act as catalytic active sites for reducing protons into hydrogen.

Metal carbides are another promising catalyst for improving the activity of HER reaction. Molybdenum carbide and tungsten carbide structure is flexible, easy to integrate with other conductive materials (such as graphene or carbon), and obtain relatively high reactivity. Tungsten carbide, for example, has catalytic properties similar to platinum, which has caused extensive research.

2.3 ELECTROCATALYSTS FOR OER

As the key half-reaction of hydrogen production, oxygen reaction, the reaction mechanism of OER, has also attracted great attention. Compared with HER, OER mechanism is more complex. The OER process consists of four electron transfer processes. In an acidic medium, the oxidation reaction of two water molecules produces one oxygen molecule ($2 H_2O \leftrightarrow 4 H^+ + O_2 + 4 e^-$), while in an alkaline medium, the oxidation reaction of four hydroxide ions produces one oxygen molecule ($4 OH^- \leftrightarrow 2 H_2O + O_2 + 4 e^-$). In contrast, the mechanism of OER reaction in alkaline solutions can be explained by the following equation:

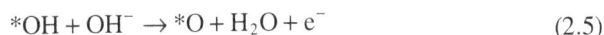

$$OH^- + * \rightarrow *OH + e^- q \tag{2.4}$$

$$*OH + OH^- \rightarrow *O + H_2O + e^- \tag{2.5}$$

$$*O + OH^- \rightarrow *OOH + e^- \tag{2.6}$$

$$*OOH + OH^- \rightarrow * + O_2 + H_2O + e^- \tag{2.7}$$

Here, * represents the catalyst active site, OER involves multiple electron transfer processes, and each O_2 molecule is produced by the transfer of four electrons. At each step (equations 2 4–2.7), the accumulation of energy barriers leads to the lag provided by the dynamics of the OER process. Therefore, in order to overcome the above obstacles and successfully complete the OER process, a large overpotential is needed to provide potential energy. According to this reaction mechanism, the catalyst to be designed needs to be very efficient in obtaining low overpotential. The catalytic community is more interested in equations (2.4–2.6) based on their role in the overall OER process. At present, precious metal catalysts such as Ir and $Ru_{0.5}Ir_{0.5}$ alloy are ideal catalysts. Meanwhile, the search for alternative catalysts that can take this reaction to a larger commercial market continues. In general, the best OER catalyst must have the advantages of low cost, high catalytic activity, sufficient stability, and wide applicability. As shown in equations (2.5 and 2.6), the active site of the catalyst involved must have a variable valence state in order to adsorb the oxygen intermediate and form a bond. Therefore, transition metal–based catalysts with adjustable electron density are expected to be ideal catalyst materials for a new generation of alkaline OER electrolysis.

2.3.1 TRANSITION METAL OXIDES AND HYDROXIDES

Transition metal elements such as Ni, Fe, and Co have variable oxidation state valence states, which can all be found in their oxides and hydroxides. The metal and its hydroxide have excellent metal potential valence states under OER reaction conditions. At the same time, the electron density that accumulates around the metal site can significantly accelerate the response rhythm of OER. The composition and structural adjustment of these metals and their hydroxides achieve outstanding OER activity. It can be emphasized that the Earth's abundant first-row (3d) transition metals such as Fe, Co, Ni, and Mn, as well as their mono, binary, and ternary oxides/hydroxides, are described as highly efficient, and the presence of these elements is described as a key factor in achieving higher efficiency. Among these catalysts, carbon-coated Ni@NiO nanocomposites (Ni@NiO@C) and Co_2AlO_4 nanosheets were reported as highly efficient OER catalyst materials. Ni@NiO@C has an overpotential of 380 mV and a current density of 10 mA/cm^2 [16]. In addition, the voltage value of Co_2AlO_4 nanosheets at a current density of 10 mA/cm^2 is 280 mV, and the slope of Tafel is very small [17]. Compared with single transition metal oxides, composite transition metal oxides exhibit superior catalytic activity due to their inherent bonding ability and open coordination sites. Transition metal/metal oxides based on the mesoporous $Ni_{60}Fe_{30}Mn_{10}$ alloy have been reported the most and can also be found in their oxide and hydroxide forms. Among them, the mesoporous $Ni_{60}Fe_{30}Mn_{10}$ alloy-based transition metal/metal oxide reported by Tolbert et al. showed superior OER catalytic performance in 0.5 M KOH medium with an overpotential value of only 200 mV at 10 mA/cm^2. Interestingly, the catalyst has a η value of 360 mV at 500 mA/cm^2 and a Tafel slope of 62 mV/dec at 1.0 M KOH [18].

In addition to transition metal oxides, hydroxides can also provide low-cost adjustable components. Because hydroxides can obtain higher metal valence states, in their laminates, col-xfeooh and $FeOOH/CeO_2$ have better catalytic properties than common metal oxides. Previous studies have suggested that based on Ni and Fe, binary NiFe-LDH (layered double hydroxides) with ternary NiCoFe-LDH has a synergistic effect to achieve higher catalytic performance for LDH containing Ni and Fe. Recent studies have shown that the above catalytic performance can be improved by doping anions [19], which provides a new solution for customizing the catalytic performance of LDHs. Despite the extremely high performance of LDHs, its low conductivity is still a problem that should not be ignored. In this regard, adjusting the valence states of metal ions in pure LDH has proved to be a promising method. In the work reported by Sunet et al. [20], it was shown that the conductivity of NiCoFe-LDH could be improved by pre-oxidizing Co^{2+} to Co^{3+}, thus obtaining higher catalytic performance of OER. In addition, the integration of metal oxides/hydroxides with conductive metal substrates (primarily Ni foam) to form composite "array" electrodes has also been shown to be effective, based on intrinsically active metal oxides/hydroxides [21]. Conductive substrates can provide fast charge transport channels, while nano-array structures that expose the active sites can effectively prevent the aggregation of catalysts. For this purpose, Schafer et al. prepared NiFeO@stainless steel sheet arrays with η values up to 269.2 mV at 0.1 M KOH and 10 mA/cm² [21].

2.3.2 PEROVSKITE

Perovskite has been recognized as an efficient catalyst for oxygen evolution due to its high stability, low cost, high catalytic activity and electronic tunability. Jan Rossmeisl et al. [22] proved that the free energy difference between the HOO* and HO* is nearly constant on different oxide surfaces, while the variation in the over-potential (η^{OER}) is determined by the adsorption energy of O*, which is denoted as $\Delta G^0_{O*} - \Delta G0HO_*$. When plotting η^{OER} as a function of $\Delta G^0_{O*} - \Delta G0HO_*$ for different classes of materials.

From the diagram drawn, higher catalytic performance can be obtained by optimizing the O-binding ability of the structure and its intermediates. The η values of $LiCo_{0.33}Ni_{0.33}Fe_{0.33}O_2$ [23] and $Ca_{0.9}Yb_{0.1}MnO_{3-x}$ [24] prepared by Yi Cui and Yi Xie were 295 000 mV, respectively, under the condition of 0.1 M KOH and the current density of 10 mA/cm². The $LiCo_{0.8}Fe_{0.2}O_2$ [25], $SrNb_{0.1}Co_{0.7}Fe_{0.2}O_{3-x}$ [26], and $BaCo_{0.7}Fe_{0.2}Sn_{0.1}O_{3-x}$ [27] prepared by Shao Zongping's research group also showed good catalytic performance. Its initial potential ranges from 1.49 to 1.53 V.

2.3.3 TRANSITION METAL CHALCOGENIDES

Transition metal oxides or hydroxides can be used to produce transition metal chalcogenides, such as metal sulfides and metal selenides, by reacting with sulfur and selenium. Transition metal chalcogenides, due to their inherent metallic behavior, which is attributed to the continuous network of metal–metal bonds, confer high electrical conductivity, while at the same time, the high electron density near the metal site can significantly promote the overall OER catalytic activity. Ni_3Se_2, Ni_3Se_2 nanorods, Ni_3Se_2, and Ni_3Se_2 films are highly efficient OER catalyst types in transition metal chalcogenides. In order to obtain higher catalytic activity, in situ oxides/hydroxides driven by selenides have been shown to be effective for OER reactions under oxygen evolution conditions. Fang Song and Xile Hu's group [28] used this method to convert ferric nickel diselenide ($Ni_xFe_{1-x}Se_2$) to ferric nickel oxide/hydroxide under oxygen evolution conditions. This method enables the development of catalysts with η values up to 195 mV in 10 mA cm^2.

2.4 CONVENTIONAL WAYS TO IMPROVE ELECTROCATALYTIC PERFORMANCE

In recent decades, scientists have been working to make non-precious metal-based electrocatalysts obtain higher catalytic activity while maintaining higher stability. Next, this paper will discuss the most widely used methods that are currently considered promising for improving catalytic performance.

2.4.1 INTEGRATING CONDUCTIVE SUBSTRATES

Transition metal–based electrocatalysts can be coupled with different conductive substrate materials to improve their overall electrocatalytic performance.

In the case of OER, carbon-based materials are the most widely considered. Materials include graphene, CNTs, and their derivative materials. Compared with metal substrate, carbon substrate has the advantages of lower cost and higher structural integrity. Therefore, carbon and its derivative materials can be used for in situ doping engineering of oxides, hydroxides, and LDH, thereby contributing to improving overall performance. Fan's team prepared CNTs@FeOOH nanosheets with a η value of 250 mV and a Tafel slope of 36 mV/dec at a current density of 10 mA/cm^2 in 1.0 M KOH electrolyte solution [29].

In the case of HER reaction, NiS_2 nanowires are generated on nickel foam (NF) and NiS_2 is grown on carbon cloth (CC), which can effectively achieve higher catalytic activity. Under the condition of 1.0 M KOH, η value of NiS_2/CC is only −96 mV under the condition of −10 mA/cm^2 current density, while η value of NiS^2/CC under the condition of −10 mA/cm^2 is −243 mV. At 1.0 M PBS, the Tafel slope is 69 mV/dec.

2.4.2 ATOMIC DOPING

Metal oxides are one of the most widely studied hydroelectricity decomposition catalysts, and their P-type properties limit the rapid transfer of electrons and reduce the accessibility of active sites. In this regard, the electrical conductivity of metal oxides can be improved by doping with heterotransition metal atoms such as Co, Fe, and Mn. The $Ni_2Co1@Ni_2Co1Ox$ prepared by Yong Zhao et al. has a core–shell structure, and its OER performance is better than NiO. Under the condition of 0.1 M KOH and 10 mA/cm^2, the initial potential is 1.55 V and the η value is 380 mV [30].

2.4.3 AMORPHOUS MATERIALS

The catalytic performance is related to the catalytic site. Therefore, adjusting the composition of the catalyst is one of the effective ways to achieve higher performance.

It can be seen that a large number of surface defect sites can be introduced by inducing the amorphous structure during catalyst preparation, and these defect sites can be used as active catalytic sites for OER and HER. The scheme is based on a photochemical metal-organic deposition method, which shows that amorphous metal oxide films have a tunable composition. Hybrid metal oxide films exhibit higher catalytic activity than their crystalline counterparts. In addition, a-$Fe_{100-y-z}Co_yNi_zO_x$ can even provide OER activity comparable to that of precious metal oxide catalysts. In addition to OER activity, amorphous catalysts prepared by Fe, Co, and Ni can also improve HER catalytic activity. Sun et al. prepared amorphous CoSe films with high HER activity, with η values up to −121 mV under the condition of 1.0 M KOH electrolyte and −10 mA/cm^2 current density [31].

2.5 SUMMARY AND OUTLOOK

In this paper, the reaction mechanisms of HER and OER in alkaline hydroelectrolysis are introduced. After that, the research progress of non-precious metal-based materials as catalysts in alkaline water electrolysis is reviewed. Based on the previous literature, it is concluded that the performance of transition metal compounds as catalysts for HER and OER is worth studying and has potential properties. Finally, we propose some common effective methods to improve electrocatalytic performance, such as integrating conductive substrates, doping heteroatoms, and amorphous materials.

REFERENCES

1. Chen, D.J., et al., Nonstoichiometric oxides as low-cost and highly-efficient oxygen reduction/evolution catalysts for low-temperature electrochemical devices. *Chemical Reviews*, 2015. **115**(18): p. 9869–9921.
2. Kim, J.S., et al., Recent progress on multimetal oxide catalysts for the oxygen evolution reaction. *Advanced Energy Materials*, 2018. **8**(11): p. 1702774.
3. Lu, F., et al., First-row transition metal based catalysts for the oxygen evolution reaction under alkaline conditions: basic principles and recent advances. *Small*, 2017. **13**(45): p. 1701931.
4. Gong, M., et al., An advanced Ni-Fe layered double hydroxide electrocatalyst for water oxidation. *Journal of the American Chemical Society*, 2013. **135**(23): p. 8452–8455.
5. Gong, M. and H.J. Dai, A mini review of NiFe-based materials as highly active oxygen evolution reaction electrocatalysts. *Nano Research*, 2015. **8**(1): p. 23–39.
6. Zhang, Q., et al., Superaerophobic ultrathin Ni-Mo alloy nanosheet array from in situ topotactic reduction for hydrogen evolution reaction. *Small*, 2017. **13**(41): p. 1701648.
7. Deng, J., et al., Highly active and durable non-precious-metal catalysts encapsulated in carbon nanotubes for hydrogen evolution reaction. *Energy & Environmental Science*, 2014. **7**(6): p. 1919–1923.
8. Yan, X.D., et al., Three-dimensional crystalline/amorphous Co/Co$_3$O$_4$ core/shell nanosheets as efficient electrocatalysts for the hydrogen evolution reaction. *Nano Letters*, 2015. **15**(9): p. 6015–6021.
9. Gong, M., et al., Nanoscale nickel oxide/nickel heterostructures for active hydrogen evolution electrocatalysis. *Nature Communications*, 2014. **5**: p. 4695.
10. Weng, Z., et al., Metal/oxide interface nanostructures generated by surface segregation for electrocatalysis. *Nano Letters*, 2015. **15**(11): p. 7704–7710.

11. Yu, J., et al., Morphological and structural engineering in amorphous Cu_2MoS_4 nano-cages for remarkable electrocatalytic hydrogen evolution. *Science China-Materials*, 2019. **62**(9): p. 1275–1284.

12. Feng, L.L., et al., High-index faceted Ni_3S_2 nanosheet arrays as highly active and ultra-stable electrocatalysts for water splitting. *Journal of the American Chemical Society*, 2015. **137**(44): p. 14023–14026.

13. Tang, C., et al., NiSe nanowire film supported on nickel foam: an efficient and stable 3D bifunctional electrode for full water splitting. *Angewandte Chemie-International Edition*, 2015. **54**(32): p. 9351–9355.

14. Chen, P.Z., et al., Phase-transformation engineering in cobalt diselenide realizing enhanced catalytic activity for hydrogen evolution in an alkaline medium. *Advanced Materials*, 2016. **28**(34): p. 7527–7532.

15. Gao, W.K., et al., In situ construction of surface defects of carbon-doped ternary cobalt-nickel-iron phosphide nanocubes for efficient overall water splitting. *Science China-Materials*, 2019. **62**(9): p. 1285–1296.

16. Xu, D.Y., et al., Fabrication of multifunctional carbon encapsulated Ni@NiO nanocom-posites for oxygen reduction, oxygen evolution and lithium-ion battery anode materials. *Science China-Materials*, 2017. **60**(10): p. 947–954.

17. Wang, J.Y., et al., Synthesis of ultrathin Co2AlO4 nanosheets with oxygen vacancies for enhanced electrocatalytic oxygen evolution. *Science China-Materials*, 2020. **63**(1): p. 91–99.

18. Detsi, E., et al., Mesoporous Ni60Fe30Mn10-alloy based metal/metal oxide com-posite thick films as highly active and robust oxygen evolution catalysts. *Energy & Environmental Science*, 2016. **9**(2): p. 540–549.

19. Hunter, B.M., et al., Effect of interlayer anions on NiFe -LDH nanosheet water oxida-tion activity. *Energy & Environmental Science*, 2016. **9**(5): p. 1734–1743.

20. Qian, L., et al., Trinary layered double hydroxides as high-performance bifunctional materials for oxygen electrocatalysis. *Advanced Energy Materials*, 2015. **5**(13): p. 1500245.

21. Schäfer, H., et al., Stainless steel made to rust: a robust water-splitting catalyst with benchmark characteristics. *Energy & Environmental Science*, 2015. **8**(9): p. 2685–2697.

22. Man, I.C., et al., Universality in oxygen evolution electrocatalysis on oxide surfaces. *Chemcatchem*, 2011. **3**(7): p. 1159–1165.

23. Lu, Z.Y., et al., Electrochemical tuning of layered lithium transition metal oxides for improvement of oxygen evolution reaction. *Nature Communications*, 2014. **5**: p. 4345.

24. Guo, Y.Q., et al., Engineering the electronic state of a perovskite electrocatalyst for synergistically enhanced oxygen evolution reaction. *Advanced Materials*, 2015. **27**(39): p. 5989–5994.

25. Zhu, Y.L., et al., A high-performance electrocatalyst for oxygen evolution reaction: $LiCo_{0.8}Fe_{0.2}O_2$. *Advanced Materials*, 2015. **27**(44): p. 7150.

26. Zhu, Y.L., et al., $SrNb_{0.1}Co_{0.7}Fe_{0.2}O_{3-\delta}$ Perovskite as a next-generation electrocatalyst for oxygen evolution in alkaline solution. *Angewandte Chemie-International Edition*, 2015. **54**(13): p. 3897–3901.

27. Xu, X.M., et al., Co-doping strategy for developing perovskite oxides as highly effi-cient electrocatalysts for oxygen evolution reaction. *Advanced Science*, 2016. **3**(2): p. 1500187.

28. Xu, X., F. Song, and X.L. Hu, A nickel iron diselenide-derived efficient oxygen-evolu-tion catalyst. *Nature Communications*, 2016. **7**(1): p. 12324.

29. Zhang, Y.Q., et al., Ultrathin CNTs@FeOOH nanoflake core/shell networks as efficient electrocatalysts for the oxygen evolution reaction. *Materials Chemistry Frontiers*, 2017. **1**(4): p. 709–715.

30. He, J.L., B.B. Hu, and Y. Zhao, Superaerophobic electrode with metal@metal-oxide powder catalyst for oxygen evolution reaction. *Advanced Functional Materials*, 2016. **26**(33): p. 5998–6004.
31. Liu, T.T., et al., An amorphous CoSe film behaves as an active and stable full water-splitting electrocatalyst under strongly alkaline conditions. *Chemical Communications*, 2015. **51**(93): p. 16683–16686.

3 Free-Standing Electrodes and Catalysts for Alkaline Water Electrolysis

He Miao and Fuyue Liu

3.1 INTRODUCTION

Hydrogen (H_2) is recognized as the most promising alternative to fossil fuels because its combustion product is water [1,2]. Water electrolysis has been regarded as the most promising approach for generating hydrogen with zero-carbon emission. Therefore, electrochemical water splitting, which can convert the generated electricity into storable hydrogen, is an ideal and scalable energy conversion technology [3,4]. However, oxygen evolution reaction (OER) as the anodic half reaction during water splitting involves a four-electron transfer process, resulting in a high energy barrier to drive the reaction [5,6]. At present, the precious Pt-, Ir-, and Ru-based metal catalysts are generally recognized as the best active catalysts for hydrogen evolution reaction (HER) and OER, but the limited earth reserves and high-cost situations seriously restricted their commercial application [7,8]. With the purpose of lowering the cost of catalysts and enhancing the corresponding electrochemical behavior, great efforts have been made to design and synthesize non-precious metal electrocatalysts involving earth-abundant materials as cost-effective alternatives for the OER and HER [9,10]. Transition metal oxides, sulfides, phosphides, metal alloys, selenides, and mixed-metal complexes have been widely studied, demonstrating good performance toward the OER and HER. However, most of these electrocatalysts require overpotentials higher than those required for noble-metal-based catalysts [11,12]. Furthermore, improving the stability of these catalysts remains of utmost importance. Consequently, the design of low-cost and efficient alternative OER and HER electrocatalysts with high activity and long-term stability is urgently needed for efficient water splitting [12,13].

Recently, electrode materials with micro spatial-stereo construction are widely investigated, which show great potential in advanced energy storage and renewable energy conversion applications. Among them, free-standing electrocatalysts (FSECs) show great potential in practical application for electrocatalysis. Thus, what do FSECs refer to? For better understanding and further investigation, we make a direct definition of these kinds of electrocatalysts [14,15]: FSECs are kinds of electrocatalysts which include various active materials directly in situ grown on the solid and conductive substrates without adding any binders or additives.

Compared with the traditional powder catalyst, the growth of catalytically active electrodes on the conductive substrate has the following advantages [3,14]: (1) the

DOI: 10.1201/9781003368939-3

substrate material can disperse the catalyst, which is conducive to the gas adsorption and desorption process; (2) without using the adhesive, the catalytic material can be closely combined with the conductive substrate, which not only simplifies the preparation process, but also ensures the rapid transfer of charge and improves the electrocatalytic activity; and (3) the conductive substrate enables high loading of the active components, providing abundant reactive active sites. Therefore, the FSECs with non-precious metals are a series of electrochemically active materials grown directly on a substrate with remarkable conductivity, sufficient porous structure, and high specific surface areas.

At present, transition metal elements of Fe, Co, Ni, Mn, Zn, Ti, Cu, V, Mo, and W are most widely investigated to construct FSECs due to the low cost and unique physicochemical properties. In addition, the substrates of FSECs mainly include metal foam (such as Ni/Cu/Fe foam) [13,16], carbon cloth (CC) [17,18], carbon fiber paper (CFP) [19,20], graphite plate [21,22] and metal plate (such as Ti/Ni/Fe foil) [23,24], fluorine-doped tin oxide (FTO) [25,26], and others [20,27] (Figure 3.1).

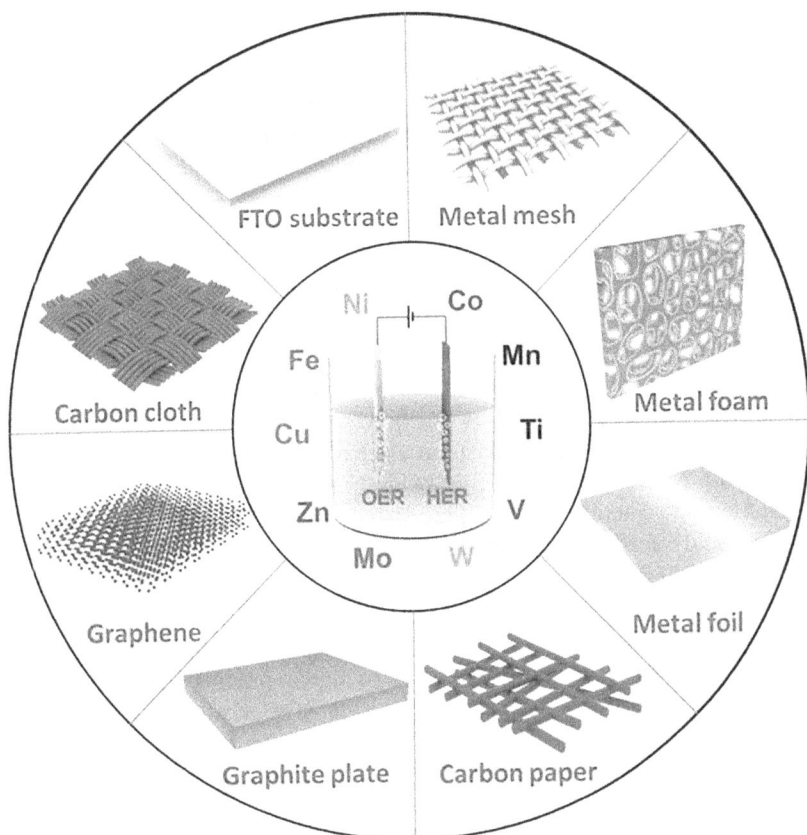

FIGURE 3.1 Substrate materials and transition metal elements that are used to build free-standing electrodes for electrocatalytic water splitting. Adapted with permission from Ref. [10]. Copyright 2019 John Wiley & Sons, Inc.

Conductive materials can not only provide effective electron transport channels, but also increase the surface area of the electrocatalysts. Such materials show great potential in advanced energy storage and conversion [14,28]: (1) the materials may inherit the advantages of two-dimensional (2D) materials, especially unique physical, electrical, and chemical properties; (2) the formation of free-standing structure obviously enhanced the mechanical stability of the materials; and (3) most of the materials featured three-dimensional (3D) framework, which can effectively facilitate ion and electron transfer.

3.2 FSECs FOR OER AND HER

3.2.1 SUBSTRATES OF FSECs

The structural and morphology design plays a key role for the improvement of electrochemical performance by tuning the physical structures. So far, various strategies have been reported to use nickel foam (NF), CC, CFP, metal mesh, etc., as substrates to prepare FSECs with easily modified and defined physical structures. FSECs include metal and non-metal substrates. Metal substrates (such as copper foil, titanium mesh, and Ni foam) have high conductivity, but they have the disadvantages of high price and poor corrosion resistance. Non-metallic collectors mainly include carbon-based substrates, e.g., graphite plate, CFP, and CC. Carbon substrates are widely used in FSECs due to its low price, good flexibility, and simple preparation process. However, carbon-based supports are easily corroded by oxidation in an OER process. To obtain excellent catalytic activity and stability of the electrocatalyst, it is particularly important to select a suitable substrate.

3.2.1.1 Ni Foam

Ni foam (NF) is a kind of commonly applied current substrate in energy storage and conversion systems (Figure 3.2a). Ni foam features 3D network structure with excellent electrical conductivity. It is beneficial for exposing more active sites when used as substrate to prepare catalysts in electrochemical catalysis [29,30]. As the surface of the NF often covers a thin oxide film when exposed to air, it is often immersed in diluted acid before use. Free-standing active materials directly growing on the NF could effectively facilitate the contact between the active materials and the current collector, enhance the mass transportation, and improve the electrocatalytic performance [31]. It should be noted that the flexibility of NF is much poor for its high-purity Ni-metal structure. Even so, many researchers are putting great efforts into NF-based electrocatalysts and have achieved great progress. For instance, Tang et al. [32] prepared the highly active Cu_2S/Ni_3S_2-0.5@NF via two-step hydrothermal method. Chen et al. prepared the high-performance Fe-doped Ni_3S_2 nanosheets on NF by a simple hydrothermal synthesis method [33].

3.2.1.2 Carbon Fiber Paper

CFP has a macro-porous network structure (Figure 3.2b), good chemical inertia, high mechanical strength, and high conductivity, which can be used as 3D substrate for FSECs on a large scale to improve their electrocatalytic performances [34,35].

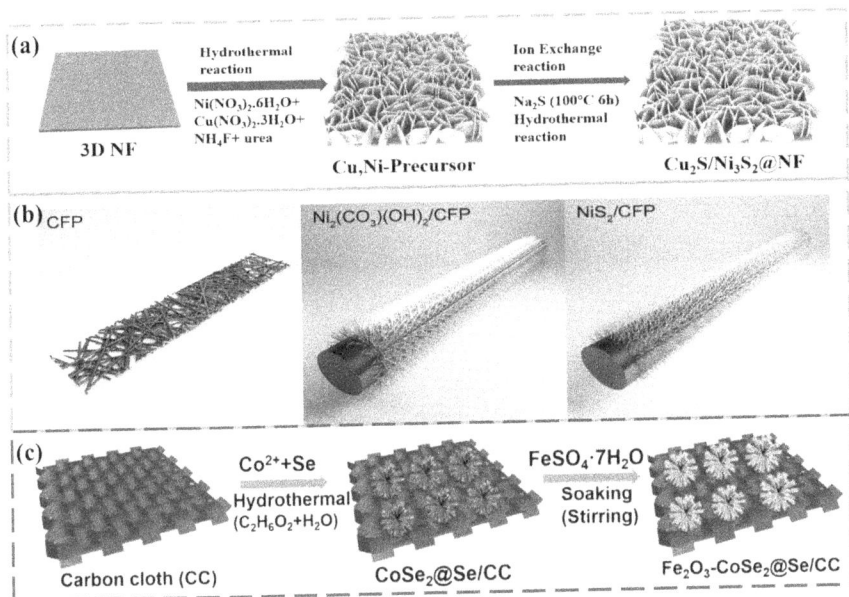

FIGURE 3.2 (a) The synthesis process of Cu_2S/Ni_3S_2@NF hybrids. Adapted with permission from Ref. [32]. Copyright 2021 Hydrogen Energy Publications LLC. (b) Scheme of NiS_2 NWs synthesis process on CFP. Adapted with permission from Ref. [36]. Copyright 2021 Elsevier B.V. (c) Schematic diagram of the synthesis process of Fe_2O_3-$CoSe_2$@Se/CC. Adapted with permission from Ref. [41]. Copyright 2017 Hydrogen Energy Publications LLC.

Guo et al. [36] prepared turf-like NiS nanowires on flexible CFP by two simple methods: hydrothermal method and calcination method. NiS was directly and uniformly grown on the conductive CFP substrate, which not only has large specific surface area, but also has small charge transfer resistance and high conductivity, thus enhancing the electrocatalytic activity. The NiS/CFP catalyst showed excellent electrocatalytic HER and OER performance. For total water splitting, only 1.59 V was required at 10 mA/cm². Li et al. [37] used a two-step hydrothermal method to prepare in situ 3D interconnected Fe-doped NiS nanosheets on CFP (Fe–NiS@CFP) for the OER. The optimal Fe–NiS@CFP showed an η_{100} of 275 mV and maintained the high stability for 50 h in 1.0 M KOH.

3.2.1.3 Carbon Cloth

CC is composed of aligned carbon fibers on micron scale (Figure 3.2c). It features soft texture, light weight, favorable mechanical strength, and high electrical conductivity [18,38]. Moreover, it can be easily cut into various sizes and shapes. It becomes a frequently used substrate for the FSECs. Besides, the macro-porous inter-fiber space also assists the easy flow of aqueous electrolyte. Therefore, CC is well known as a kind of ideal substrate for flexible devices. Qian et al. [39] synthesized Ni_3S_2@Ni/CC via electrodeposition followed by a sulfuration process. The Ni_3S_2@Ni/CC with abundant active sites exhibited a good OER performance with an η_{10} of 290.9 mV as well as a low Tafel slope (101.26 mV/dec) and good stability for 30 h in 1.0 M KOH.

Jiang and co-workers [40] fabricated the Fe_3O_4/NiS nanoplates on CC (Fe_3O_4/NiS@ CC) via two-step carbonization process. Fe_3O_4/NiS@CC exhibited a superior OER catalytic activity with a small η_{10} of 310 mV in 1.0 M KOH due to their large specific surface area (1796 m^2/g) and high conductivity. Selenium-coated cobalt selenide ($CoSe_2$@Se) nanoflake catalyst on CC (Fe_2O_3-$CoSe_2$@Se/CC) was prepared via a hydrothermal synthesis and immersion method (Figure 3.2c) [41]. The unique 3D coral originating from Fe_2O_3-$CoSe_2$@Se/CC provided more abundant electrocatalytic active sites and fluent electrolyte diffusion. The optimized Fe_2O_3-$CoSe_2$@Se/ CC-1.0 h displayed an outstanding OER catalytic performance with an η_{10} of 252 mV.

3.2.1.4 Other Substrates

Metal meshes/plates/foams have excellent conductivity and flexibility [42]. It can be used as a metal source to grow catalysts on the surface [43]. Li et al. [44] constructed NiCoFeS nanosheets with three-metal layered structure on Ti mesh (NiCoFeS/ Ti) via hydrothermal combined vulcanization process. For the OER, NiCoFeS/Ti showed a low η_{10} of 230 mV. Yang et al. [45] prepared 3D hybrid thin-film electrode on copper foil (Co-O@Co-Se/Cu) by electrodeposition. When Se was incorporated, the structure and crystal phase transition occurred. Because copper foil was used as the substrate, the electron transfer resistance was reduced and the gas release was enhanced. The Co-Se species in the film was gradually transformed into Co-O species. Dang et al. [46] used a simple one-step hydrothermal method to directly grow INF-FeCuS nanoparticles on FeNi alloy foams (INF). INF-FeCuS including Cu_7S_4 and $Fe_{0.95}S_{1.05}$ phases showed the nanorod structure covered with tremella-like nanosheets. The optimal INF-FeCuS nanosheets exhibited remarkable OER activity with an η_{100} of 220 mV and a small Tafel slope of 88.1 mV/dec.

The stainless steel (SS), FTO-coated glass substrates, etc. are also widely used [26,47]. SS mesh is a common chemical engineering component with high physical robustness and chemical resistance in both basic and acidic environments, which features high electrical conductivity, flexible characteristic, and mechanical strength. Great efforts have been devoted to growing catalysts on SS mesh to take full advantages of this substrate [48]. Deng et al. [49] synthesized adhesive-free free-standing Co_9S_8@Co_3O_4 core/shell array on SS by simple hydrothermal method and vulcanization process, which exhibited an η_{20} of 260 mV and small Tafel slope of 56 mV/dec.

According to the above discussions, nano-framework materials grown on different substrates are promising candidates for binder-free electrodes, which can effectively increase the electrochemical interfaces, enhance the mass transportation rate, and greatly simplify the constructing process for practical applications. In this regard, it is vital to construct multifunctional electrodes with stable nanostructures directly on conductive substrates to achieve high-efficiency renewable devices. It should be noted that the substrate should be properly chosen in terms of the requirement of the applications.

3.2.2 ACTIVE MATERIALS OF FSECS FOR HER

The electrochemical HER involves multiple steps. In alkaline electrolyte, the first step (Volmer reaction) is reducing protons on catalytic sites (M) to form adsorbed hydrogen (MH_{ads}). At low coverage of H_{ads} on the catalyst surface, H_{ads} will preferably combine with a proton and an electron to generate a H_2 molecule (Heyrovsky reaction). In the case of high H_{ads} coverage, two adjacent H_{ads} atoms bind to form H_2 (Tafel reaction). Mechanistic studies suggest that molecular H_2 forms via Volmer–Heyrovsky or Volmer–Tafel pathway. The Volmer reaction is facile in acid due to abundant available protons, while it is kinetically more sluggish in alkaline media as it involves water dissociation prior to H absorption. The HER process can be described using the following elementary steps [50]:

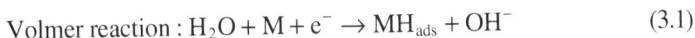

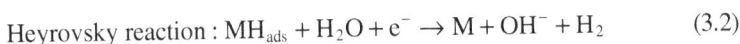

$$\text{Volmer reaction}: H_2O + M + e^- \rightarrow MH_{ads} + OH^- \qquad (3.1)$$

$$\text{Heyrovsky reaction}: MH_{ads} + H_2O + e^- \rightarrow M + OH^- + H_2 \qquad (3.2)$$

$$\text{Tafel reaction}: 2MH_{ads} \rightarrow 2M + H_2 \qquad (3.3)$$

Both Volmer–Heyrovsky and Volmer–Tafel pathways involve the formation of intermediate H_{ads}. The free energy change of H adsorption (ΔG_{H^*}) is thus an important parameter to predict/estimate the activities of HER catalysts. An ideal catalytic site for the HER should have a ΔG_{H^*} near zero. The ΔG_{H^*} is closely associated with the inherent surface chemistry and electronic structure of the materials. In recent years, extensive efforts have been put forward to develop highly active HER catalysts such as transition alloy, transition metal sulfide, selenide, phosphide, carbide, and nitride. Next, we focus our discussion on those active materials of FSECs (Table3.1).

3.2.2.1 Transition Metal or Alloys

Transition metals with optimized ΔG_{H^*} and high intrinsic HER catalytic activity have been extensively investigated. For example, a $MoNi_4$ electrocatalyst embedded in MoO_2 cuboids and supported on NF ($MoNi_4/MoO_2$@Ni) [51] was prepared by reducing a $NiMoO_4$ cuboids precursor at high temperature. The $MoNi_4$ electrocatalyst features a rapid Tafel-step-decided HER catalytic mechanism with a zero onset overpotential, an overpotential of 15 mV at 10 mA/cm², and a low Tafel slope of 30 mV/dec in 1 M KOH, which was comparable to the benchmark Pt/C. Moreover, a high-entropy alloy (HEA) of FeCoNiCuPd thin film with a single face-centered cubic (FCC) structure was deposited on carbon fiber cloth (CFC) by magnetron sputtering (Figure 3.3a) [52]. The newly developed HEA/CFC system exhibited superior HER activity compared with the commercially available catalysts under alkaline conditions, resulting in an outstanding water electrolysis performance with ultralow overpotential as low as 29 mV for HER at a current density of 10 mA/cm² (Figure 3.3b and c). In addition to the Ni–Mo-based alloy, some other transition metal alloys were also demonstrated to be highly active for HER under alkaline condition, such as Cu–Ti bimetallic alloy and Al_7Cu_4Ni@Cu_4Ni core/shell nanocrystals [53,54].

3.2.2.2 Transition Metal Sulfides

As bioinspired HER electrocatalysts, transition metal sulfides have gained extensive interest over a decade. Among them, MoS_2 is the most extensively developed because

TABLE 3.1
Comparison of the Electrocatalytic Activities of Free-Standing Transition Metal Catalysts for HER

Sample	$E_{j=100}$ ($|V|$)	Tafel Slope (mV/dec)	Substrate	Electrolyte	Preparation Method	Reference
H-FeCoNiCuMo	~0.05	34.7	NF	1 M KOH	Pulse current electrodeposition	[7]
MoC–Mo$_2$C-790	~0.17	59	Mo plate	1 M KOH	Electrodeposition	[24]
NiS$_2$ NWs/CFP	0.3	134	CFP	1 M KOH	Hydrothermal and sulfur annealing	[41]
O-NiCu	0.06	34.1	NF	1 M KOH	Electrodeposition	[55]
MoS$_2$-MoO$_{3-x}$/Ni$_3$S$_2$ @NF	~0.15	53.2	NF	1 M KOH	Chemical reaction and electrodeposition	[56]
NiCoNiNi$_x$PiNiCoN	/	139.2	NF	1 M KOH	Phosphorization, nitridation, and solvothermal	[57]
NiMoN@NC-6	~0.06	39.1	NF	1 M KOH	Hydrothermal and pyrolysis	[58]
CFeCoNiP–NF	0.15	31	NF	1 M KOH	Electrodeposition	[59]
CoSe$_2$/CMF	/	67	Carbon foam (CMF)	1 M KOH	Wet impregnation, carbonization, and selenization	[27]
FeNiCo@NC/NF-600	~0.25	82	NF	1 M KOH	In situ growth and annealing	[60]
CoP$_2$/Co$_2$P@CNT-CC	/	61.5	CC	1 M KOH	Phosphorization and calcination	[61]
Co/CeO$_2$@CF	0.17	/	CF	1 M KOH	Hydrothermal and annealing	[8]
(Ni,Co)Se$_2$/CoSe$_2$/NF	0.17	51.4	NF	1 M KOH	Hydrothermal and selenization	[62]
MoNi$_4$/MoO$_2$@Ni	0.03	30	NF	1 M KOH	Hydrothermal and annealing	[51]
Co–B–P/NF	~0.09	42.1	CF	1 M KOH	Electroless deposition	[63]
Ni$_3$Se$_2$@NiFe-LDH/NF	0.17	106.2	NF	1 M KOH	Hydrothermal and electrodeposition	[64]
MoO$_3$/Ni–NiO	0.16	59	NF	1 M KOH	Electrodeposition	[65]
Bi-NP Cu/Al$_7$Cu$_4$Ni@Cu$_4$Ni	/	110	Alloy plate	1 M KOH	Chemical alloying/dealloying	[54]
Mo$_2$C/CC	~0.37	124	CC	1 M KOH	Hydrothermal and annealing	[66]

/, not available.

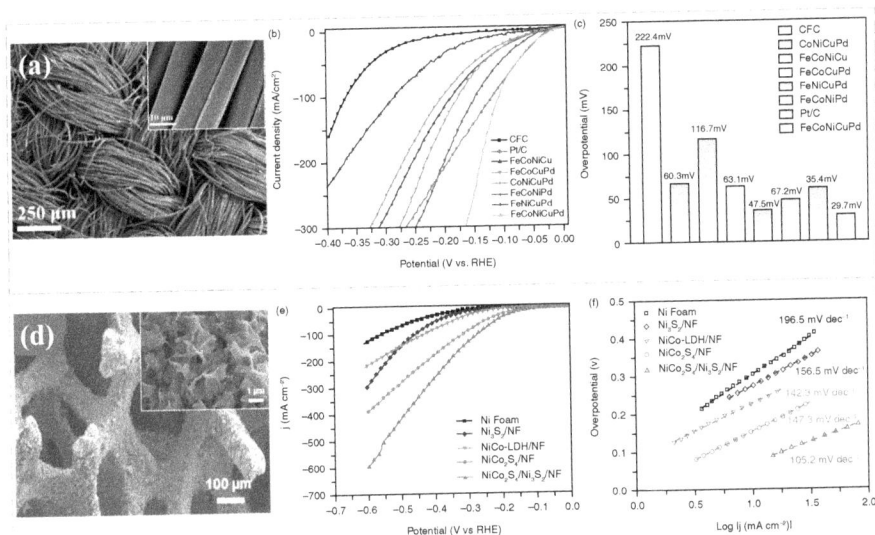

FIGURE 3.3 (a) SEM image of the FeCoNiCuPd/CFC. HER performance evaluation of FeCoNiCuPd in 1 M KOH electrolyte. (b) HER polarization curves. (c) Comparison of over-potential at 10 mA/cm². Adapted with permission from Ref. [52]. Copyright 2022 Elsevier B.V. (d) SEM image of f $NiCo_2S_4/Ni_3S_2/NF$. (e) Polarization curves and (f) corresponding Tafel slopes of $NiCo_2S_4/Ni_3S_2/NF$, NiCo-LDH/NF, $NiCo_2S_4/NF$, Ni_3S_2/NF, and bare Ni foam at a scan rate of 5 mV/s without IR correction. Adapted with permission from Ref. [70]. Copyright 2018 American Chemical Society.

of its unique structural and electronic properties. Both experimental and computational studies have shown that the catalytic activity of bulk MoS_2 is poor and highly dependent on the number of exposed edge sites [67,68]. Thus, it is desirable to design and modulate MoS_2 catalysts with the maximum exposed edge sites. In this regard, Cui et al. [69] prepared a MoS_2 thin film with vertically aligned layers on a flat substrate. The obtained MoS_2 catalyst with rich edges showed superior catalytic performance to the bulk counterpart. Afterward, MoS_2 nanosheet arrays vertically aligned on NF and edge-oriented MoS_2 films loaded on Mo substrate were also developed as free-standing HER catalysts. Similar to MoS_2, WS_2 adopts a layered structure with adjustable electrical properties and exposed edge sites being the active HER centers. Likewise, studies on WS_2 are mainly focused on synthesizing nanostructured WS_2 with more edge sites. The edge-rich WS_2 nanosheets and synergistic effect between the WS_2 nanolayers and heteroatom-doped graphene sheets were proposed to boost the HER activity of the integrated film electrode, which attained a current density of 10 mA/cm² at an overpotential of 125 mV.

The sulfides of Ni, Fe, and Co have also received extensive research. For example, a facile strategy to fabricate a 3D heteromorphic $NiCo_2S_4$ and Ni_3S_2 nanosheets network on Ni foam (denoted as $NiCo_2S_4/Ni_3S_2/NF$) as a free-standing cathode for HER in alkaline solution (Figure 3.3d) [70]. As expected, the optimal $NiCo_2S_4/Ni_3S_2/NF$ electrode exhibited greatly catalytic activity and stability with extremely low onset overpotential of 15 mV and Tafel slope of 105.2 mV/dec in alkaline solution, which outperforms most of those reported non-noble-metal-based HER catalysts (Figure 3.3e and f).

3.2.2.3 Transition Metal Selenides

Generally, transition metal selenides have a higher intrinsic catalytic activity in comparison to the corresponding sulfides [71]. Accordingly, there is an increasing interest to develop the transition metal selenides as HER electrocatalysts. Owing to the homologous structure of MoS_2 and $MoSe_2$, strategies investigated for improving the HER activities of MoS_2 such as increasing conductivity by phase transformation, exposing active sites by fabricating vertical array, and coupling with conductive materials were also beneficial for $MoSe_2$ [72]. As an example, N-doped $MoSe_2$/graphene nanoflake arrays on CC can act as an advanced free-standing HER electrode.

The Co-based selenide catalysts are more frequently reported HER catalysts. Nanoparticulate $CoSe_2$ with cubic-pyrite-type phase on CFP is a representative example in this case [73]. The synthesis followed a two-step process: drop-casting and pyrolysis of the precursor ink to form cobalt oxide nanoparticles on CFP, and post-selenization in Se vapor. The layer of $CoSe_2$ nanoparticles was conformably covered on carbon fibers. The CFP-supported $CoSe_2$ nanoparticle film gave a current density of 100 mA/cm at an overpotential of about 180 mV. Aside from cubic-pyrite-type phase, $CoSe_2$ can also crystallize into an orthorhombic macar-site-type structure. Zhu et al. [62] fabricated a unique heterostructure arrays of (Ni, Co)Se_2 nanowires integrated with the metal–organic frameworks (MOFs)–derived $CoSe_2$ dodecahedra on NF as an effective binder-free electrode for water splitting (Figure 3.4a), and the as-synthesized (Ni, Co)Se_2/$CoSe_2$/NF electrocatalyst exhibits excellent electrochemical performance in alkaline solutions with HER overpotentials as low as 65 and 169 mV at 10 and 100 mA/cm, respectively, as well as high stability (Figure 3.4b and c). The unique heterostructure with abundant active sites and strong synergistic effects between (Ni, Co)Se_2 nanowires and $CoSe_2$ could modulate the electronic structure and enhance the charge transfer, thus contributing to high electrocatalytic activity.

3.2.2.4 Transition Metal Phosphides

Transition metal phosphides with metalloid properties are well known as the efficient electrocatalytic species toward HER. Free-standing structured metal phosphides are considered as one of the most promising ones in hydrogen evolution. For example, Liu and co-workers [74] successfully synthesized 3D porous CoP nanosheet arrays which were vertically distributed on NF through electrodeposition with subsequent phosphatization procedure. Due to the unique 3D structure, the as-prepared CoP nanosheets exhibited high HER activity, and the overpotentials to deliver a current density of 10 mA/cm^2 were 79.5 and 86.6 mV in both acidic and basic media, respectively. In addition, Wen and co-workers [75] successfully prepared vanadium modification of Ni_2P nanosheet arrays on a CC substrate (V-Ni_2P NSAs/CC) via hydrothermal and post-low-temperature phosphorization methods. During the hydrothermal process, VNi-LDH was in situ grown on the fiber structure of CC, which is beneficial to expose enormous surface-active sites. When serving as HER catalyst,

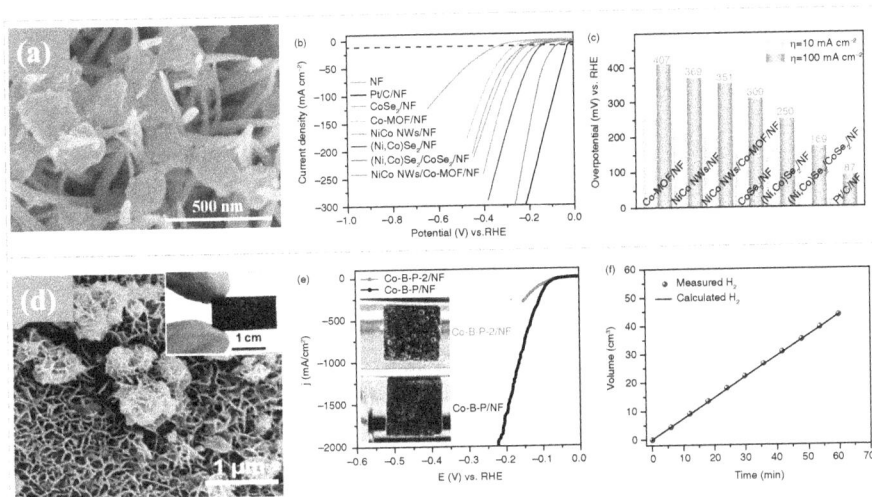

FIGURE 3.4   (a) SEM of (Ni, Co)Se$_2$/CoSe$_2$/NF heterogeneous nanoarrays. (b) Polarization curves of (Ni, Co)Se$_2$/CoSe$_2$/NF and other contrasting electrocatalysts, and (c) comparison of the electrocatalysts at 10 and 100 mA/cm^2. Adapted with permission from Ref. [62]. Copyright 2022 Elsevier B.V. (d) The high-magnification SEM image of synthesized Co–B–P/NF (the inset of (d) shows its digital photo). (e) Polarization curves of Co–B–P/NF and Co–B–P-2/NF in 1 M KOH electrolyte. The inset illustrates the optical images of the electrode during the HER tests (100 mA/cm^2). (f) The measured and theoretical amount of H$_2$ as a function of time on Co–B–P/NF at 100 mA/cm^2. Adapted with permission from Ref. [63]. Copyright 2018 Royal Society of Chemistry.

only a small overpotential value of 85 mV was required to deliver a current density of 10 mA/cm^2.

The introduction of metalloid in the metal phosphides also exerts positive effects. The study on ternary Co–B–P made by Sun et al. [63] indicated that the synergistic effect of P and B favors dissociation of H$_2$O, weakens surface H absorption, and suppresses Co oxidation. A typical Co$_{2.9}$B$_{0.73}$P$_{0.27}$ nanosheet array was synthesized on NF (Co–B–P/NF) via a facile room-temperature one-pot method (Figure 3.4d). The obtained Co–B–P/NF electrode, which combines advantages of B/P synergistic effects and super hydrophilic surface properties, resulted in significantly higher HER catalytic activity than that of binary Co–P and Co–B as well as the hydrophobic Co–B–P. Furthermore, the 3D free-standing architecture contributed to remarkable performance at high current densities up to 2000 mA/cm^2, with a nearly 100% Faradaic efficiency of HER (Figure 3.4e and f).

3.2.2.5   Transition Metal Carbides and Nitrides

Metal carbides and nitrides have been widely investigated as an alternative to Pt due to their Pt-like electronic behavior, high conductivity, and considerable stability in a wide pH range [76]. The synthesis of metal carbides usually requires a high-temperature (above 700°C) carbonization process, which inevitably leads to the sintering and agglomeration of catalysts [77]. FSECs assembly is an effective strategy to address this problem. Zou and co-workers in situ synthesized Mo_2C micro-islands on a flexible CC as a HER electrode [66]. Benefiting from the rich catalytic sites of inland-like Mo_2C, high electrical conductivity, and intimate connection of the catalyst and substrate, the as-prepared binder-free electrode showed respectable HER electrocatalytic performance in terms of a low overpotential of 140 mV at a current of 10 mA/cm^2.

Compared with metal carbides, metal nitrides usually showed an inferior HER catalytic properties. Even so, some metal nitride-based HER catalysts have been exploited by cation doping, which could significantly improve the performance by redistributing the charge to activate the catalytic sites. For example, V-doped Co_4N nanosheets were recently synthesized by growing the V-doped $Co(OH)_2$ nanowires on NF via a hydrothermal reaction, followed by a nitridation treatment in an NH_3 atmosphere [78]. The V doping caused downshift in the d-band center, which favors the H desorption. In alkaline media, the optimized V-doped Co_4N nanosheet arrays showed a lower overpotential of 37 mV at 10 mA/cm^2, lower than the Co_4N counterpart. Besides, the V-doped Co_4N nanosheet displayed excellent stability for more than 27 h.

3.2.3　Active Materials of FSECs for OER

Compared with HER, OER involving complex reaction pathways is more sluggish and is generally considered as the thermodynamically and kinetically rate-determining process in water electrolysis. Essentially, oxygen evolution is the result of the oxidation of a hydroxyl group in alkaline solution. A proposed four sequential proton-coupled electron transfer steps of OER in alkaline solution can be described below [79]:

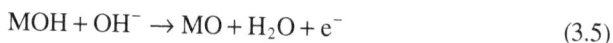

$$M + OH^- \rightarrow MOH + e^- \qquad (3.4)$$

$$MOH + OH^- \rightarrow MO + H_2O + e^- \qquad (3.5)$$

$$MO + OH^- \rightarrow MOOH + e^- \qquad (3.6)$$

$$MOOH + OH^- \rightarrow M + O_2 + H_2O + e^- \qquad (3.7)$$

The four steps of OER are all thermodynamically uphill processes, and the rate-limiting step has the highest energy barrier. In the OER process, the intermediates of MOH, MO, and MOOH are generated in turn as a result of concomitant electron and proton transfer, and the bonding interactions (M–O) within these intermediates are crucial for the catalytic activity. The adsorption energies of intermediates are widely used as descriptors for the electrocatalytic ability. In acidic solution, the oxides of Ru/Ir are the best for OER, while the catalysts derived from transition metal catalyze the OER more favorably in alkaline media. The active materials of FSECs for OER mainly include transition metal alloys, sulfides, hydroxides/oxyhydroxide, oxides, and phosphates (Table 3.2).

TABLE 3.2

Comparison of the Electrocatalytic Activities of Free-Standing Transition Metal Catalysts for OER

Sample	$E_j = 100$ (V)	Tafel Slope (mV/dec)	Electrolyte	Substrate	Preparation Method	Reference
NF15	/	70.8	1 M KOH	NF	Magnetron sputtering	[82]
S-FeOOH/IF	1.54	59	1 M KOH	Fe foil	Solution-phase pathway	[84]
NiCoP@NiMn LDH/NF	1.52	43.7	1 M KOH	NF	Hydrothermal and phosphorization	[85]
CoSn$_2$/NF	1.59	/	1 M KOH	NF	Solvothermal	[86]
H-FeCoNiCuMo	~1.47	34.7	1 M KOH	NF	Pulse current electrodeposition	[7]
NC/CuCo/CuCoOx/	1.52	88	1 M KOH	NF	Hydrothermal and carbonization	[87]
Mo-Ni-Se@NF	1.63	44.9	1 M KOH	NF	Hydrothermal	[88]
CuFe/NF	1.53	54.5	1 M KOH	NF	Hydrothermal	[89]
Fe-Co-S/Cu$_2$O/Cu	1.60	111	1 M KOH	CF	Redox and electrodeposition	[90]
sd-NFF	1.54	49	1 M KOH	NFF	Hydrothermal	[20]
NiNS/NF	1.63	58.8	1 M KOH	NF	Calcination	[91]
H-CoS$_x$@NiFe LDH/NF	1.54	80.0	1 M KOH	NF	Solvothermal and electrodeposition	[92]
CeO$_2$-NiCoP/NCF	~1.59	72	1 M KOH	NCF	Hydrothermal and phosphorization	[93]
NiFeHCH (1:0.2)	~1.56	39	1 M KOH	CC	Co-precipitation	[94]
Mo$_3$S$_4$/Co$_{1-x}$S@CF-8	~1.56	86	1 M KOH	CF	Hydrothermal	[95]
Fe-Ni$_3$S$_2$/FeNi	>1.72	54	1 M KOH	INF	Solvothermal	[96]
1D-CeO$_2$/C	1.62	46	1 M KOH	NF	Hydrothermal and annealing	[97]
NiFe-OOH$_{OV}$	~1.65	38	1 M KOH	NF	Solvothermal and O$_2$ plasma exposure	[98]
(Fe, V, Co, Ni)-doped MnO$_2$	/	104.4	1 M KOH	CFP	Electrodeposition	[99]
Ni$_3$S$_4$@CoS$_x$-NF	1.56	95.2	1 M KOH	NF	Hydrothermal	[100]
Mo$_3$S$_4$/Co$_{1-x}$S@CF-8	~1.56	86	1 M KOH	Co foil	Hydrothermal	[95]
NiFe-OH NS/NF	1.49	52.8	1 M KOH	NF	Ambient redox and hydrolysis co-precipitation	[101]
Fe$_2$O$_3$-CoSe$_2$@Se/CC	1.55	50.2	1 M KOH	CC	In situ hydrothermal and soaking	[36]
NiS$_2$ NWs/CFP	~1.56	94.5	1 M KOH	CFP	Hydrothermal and sulfur annealing	[41]
Ni-650-carbon	/	69	1 M KOH	Nickel-silica	In situ CVD	[102]
SS felt	~1.57 (Cell)	/	1 M KOH	SS	Direct use	[47]

/, not available.

3.2.3.1 Transition Metals or Alloys

Recently, a new class of multi-component catalyst systems, entropy-stabilized materials including alloys and other types, has emerged that demonstrate superior OER catalytic activity for water splitting. In light of the variable and flexible compositions, entropy-stabilized materials provide infinite and enormous potentials for the design of promising electrocatalysts. The HEAs are the alloys whose element numbers are not less than 5 [80]. HEAs can bring abundant and diverse active centers which lead to different selectivity. Both active sites and electronic structures of HEAs can be tuned easily [81]. A large number of unique binding sites on the surface of HEAs cause the proper adsorption energies.

In addition, HEAs often have good strength, corrosion resistance, and oxidation resistance, all of which are important for the stability of electrocatalysts for OER [7]. These characteristics make HEAs have good application prospects in electrocatalytic water splitting. Yang et al. [82] developed a FeCoNiCrMn high-entropy thin films with different sputtering times prepared on NF substrate by magnetron sputtering, and they were adopted as the free-standing electrodes for OER in alkaline water. It is found that the OER electrocatalytic performances were related to the sputtering times. Furthermore, the OER performances of FeCoNiCrMn HEA electrodes could be boosted by surface reconstruction through cyclic voltammetry. The FeCoNiCrMn HEA electrodes reconstructed by cyclic voltammetry could reach an overpotential of 282 mV at 10 mA/cm² and a Tafel slope of 64.3 mV/dec. Moreover, Huang et al. [7] synthesized an equimolar FeCoNiCuMo HEAs on NF, denoted as H-FeCoNiCuMo, with a fast (50 min), simple, low temperature (50°C), and scalable pulse current electrodeposition method (Figure 3.5a). H-FeCoNiCuMo exhibited breakthrough electrocatalytic performances for OER, achieving ultralow, less than 200 mV η_{10} in alkaline electrolytes (194 mV in 1 M KOH), outperforming those of the popular IrO_2 (294 mV in 1 M KOH) and RuO_2 (232 mV in 1 M KOH). However, the metal alloy electrocatalysts can only be referred as pre-catalysts in the OER process, because the hydroxides or oxyhydroxides are confirmed as the real active sites of transition metal electrocatalysts [83]. And converting transition metals to hydroxides or (oxy)hydroxide proactively (surface reconstruction) is an inevitable way to improve their electrocatalytic activity [81]. Therefore, the oxidation state enhancement or ion leaching of transition metal elements may occur during surface reconstructions.

3.2.3.2 Transition Metal Sulfides

Compared with transition metals such as Fe and Co, Ni and their composites have been the most frequently investigated materials for OER and HER electrodes because of their superior stability and excellent resistance toward corrosion in alkaline media. However, employing Ni metal alone leads to relatively low catalytic activity and low resistance to intermittent electrolysis. In this respect, several studies have been performed involving various Ni-based electrodes on NF. In particular, nickel sulfides (Ni_3S_2 and NiS) have been the most studied materials, demonstrating promising electrochemical behavior because of their high electrical conductivity and stability [103]. The metal-vacancy pair composed of Ni atom and sulfur vacancy as the catalytic active site showed the catalytic synergy during OER. Sulfur vacancy improves the OER performance by reducing the energy barrier and optimizing the adsorption free

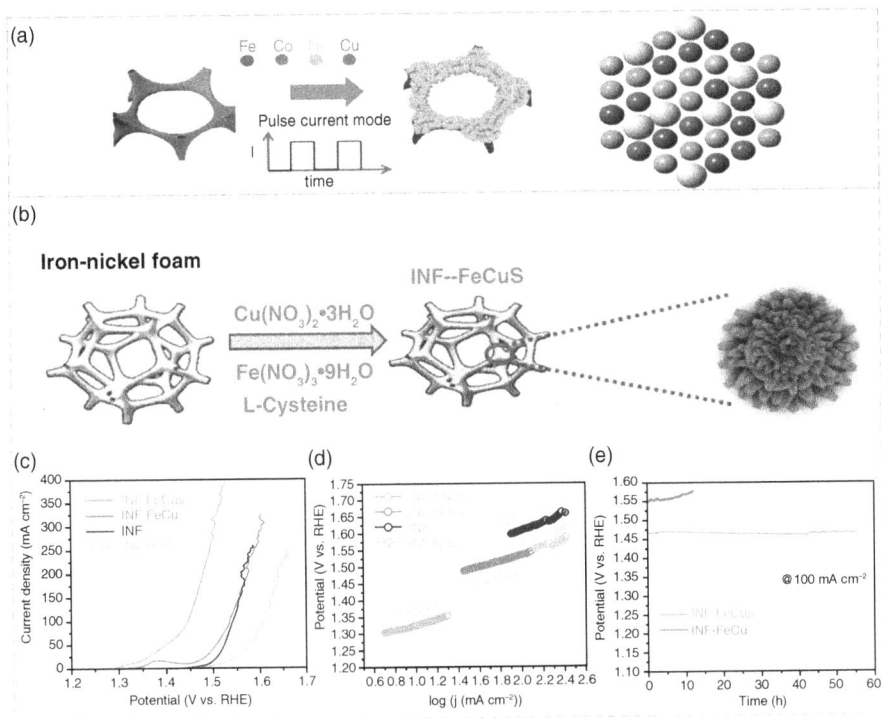

FIGURE 3.5 (a) Schematic of fabrication process for H-FeCoNiCuMo. Adapted with permission from Ref. [7]. Copyright 2022 Elsevier B.V. (b) Preparation schematics of INF-FeCuS. OER properties of INF-FeCuS, INF-FeCu, and INF. (c) OER iR-compensated LSV curves, (d) OER Tafel plots, and (e) OER durability tests of INF-FeCuS at 100 mA/cm² for 200,000 s operation in 1 M KOH. Adapted with permission from Ref. [46]. Copyright 2023 Elsevier B.V.

energy of oxygen-containing intermediates (OH*, O*, and OOH*). Creating oxygen vacancies in electrocatalysts is a common and effective method to promote OER. As revealed by the OER mechanism, all intermediates interact with the transition metal oxide surface through oxygen atoms, and the presence of oxygen vacancies will change the absorption and desorption process of the electrocatalyst with reactants. Zhuang et al. [104] prepared $FeCoO_xVo$-S nanosheet catalyst by heat treatment synthesis strategy. The addition of S atoms modified and stabilized the oxygen vacancy, forming Co–S coordination that effectively regulated the electronic structure of the active site. The FeCoOx-Vo-S electrocatalyst only needed an η_{50} of 240 mV in 1.0 M KOH. Metal cations can adjust the ligand field of the active center and have certain influences on the electronic configuration. For example, Dang et al. [46] constructed a novel metal sulfide heterostructure (INF-FeCuS) composed of Cu_7S_4 and $Fe_{0.95}S_{1.05}$ by a simple hydrothermal method (Figure 3.5b). According to the results of experimental and DFT calculation, the exposed Cu(I) sites in INF-FeCuS are the active sites for OER process, which play the key role in the catalysis of OER process, while the composite systems are also helpful. INF-FeCuS exhibited the superior catalytic property, with the OER potential of 1.45 V, low Tafel slope of 88.1 mV/dec, and good OER stability for 55 h at 100 mA/cm^2 (Figure 3.5c–e), respectively.

3.2.3.3 Transition Metal Hydroxides/Oxyhydroxides

Layered double hydroxides (LDHs) are widely studied electrochemical catalysts, in particular for OER [105]. The composition and structure of LDHs are easily tunable by adjusting either positively charged layers (e.g., Ni^{2+}, Co^{2+}, Cu^{2+}, Mg^{2+}, Zn^{2+}, Ca^{2+}, and Mn^{2+}) or interlayer anions (e.g., CO$_3^{2-}$, NO^{3-}, SO$_4^{2-}$, Cl$^-$, and Br$^-$), leading to unique redox features. However, the poor electrical conductivity of LDHs limits the catalytic performance. Hybridizing LDHs with conductive carbon or exfoliating bulk LDHs to ultrathin nanosheets is widely employed to address this issue [106,107]. To simplify the synthesis and improve the long-term stability, LDHs are often in situ grown on a conductive substrate to form a free-standing electrode. Taking the most widely investigated NiFe-LDH as an example [108], the amorphous mesoporous NiFe hydroxide sheets were supported on an NF substrate by a facile one-step electrodeposition route. The synthesized electrode enabled a current density of 200 mA/cm^2 at an overpotential of 240 mV and a TOF of 0.075/s at an overpotential of 400 mV, which is almost threefold that of Ir/C (0.027/s). To further improve the long-term stability and simplify the preparation process, Wang and co-worker [101] developed a simple strategy to fabricate NiFe LDHs nanosheet arrays on the Ni foam (NiFe-OH NS/NF) through a room-temperature redox and hydrolysis co-precipitation method (Figure 3.6a), in which the Ni foam is directly used as the Ni source. The NiFe-OH NS/NF exhibited excellent OER activity, with an overpotential of 292 mV to reach a current density of 500 mA/cm^2, which was lower than the industrial criterion (300 mV at 500 mA/cm^2). The high stability and large-scale synthesized method made it possible for industrial applications. Similarly, Wang et al. [109] also developed Fe-doped Ni(OH)$_2$ nanosheets on Ni foam using Ni foam as Ni resource, which exhibited high OER activity.

Apart from the LDHs, the transition metal oxyhydroxides have also been widely studied as OER electrocatalysts. For example, Fe-substituted CoOOH nanosheet arrays grown on CC were employed as an efficient OER electrode by in situ anodic oxidation of α-Co(OH)$_2$ nanosheet [110]. X-ray absorption fine spectra demonstrated that the CoO$_6$ octahedral structure in CoOOH was partially replaced by FeO$_6$ octahedrons during the anodic oxidation process. DFT calculation revealed that the active site of FeO$_6$ octahedron had a high catalytic activity for OER. The optimized Fe-substituted CoOOH nanosheet arrays manifested good OER activity, which was superior to most of the reported Co-based OER electrocatalysts.

3.2.3.4 Transition Metal Oxides

Owing to the low cost, high abundance, and considerable anticorrosion properties in an alkaline environment, transition metal oxides were widely developed as OER catalysts [111]. Spinel-type oxide (AB$_2$O$_4$, A and B are 3d transition metals such as Ni, Fe, Co, Cu, Zn, and Mn) is a family of composite oxides attracting extensive interests [112]. As an example of free-standing oxide electrode, rope-like CuCo$_2$O$_4$ nanosheets directly grown on NF were successfully synthesized by Zhang and co-workers for OER [113]. Numerous gaps existing between the sheet-like clusters

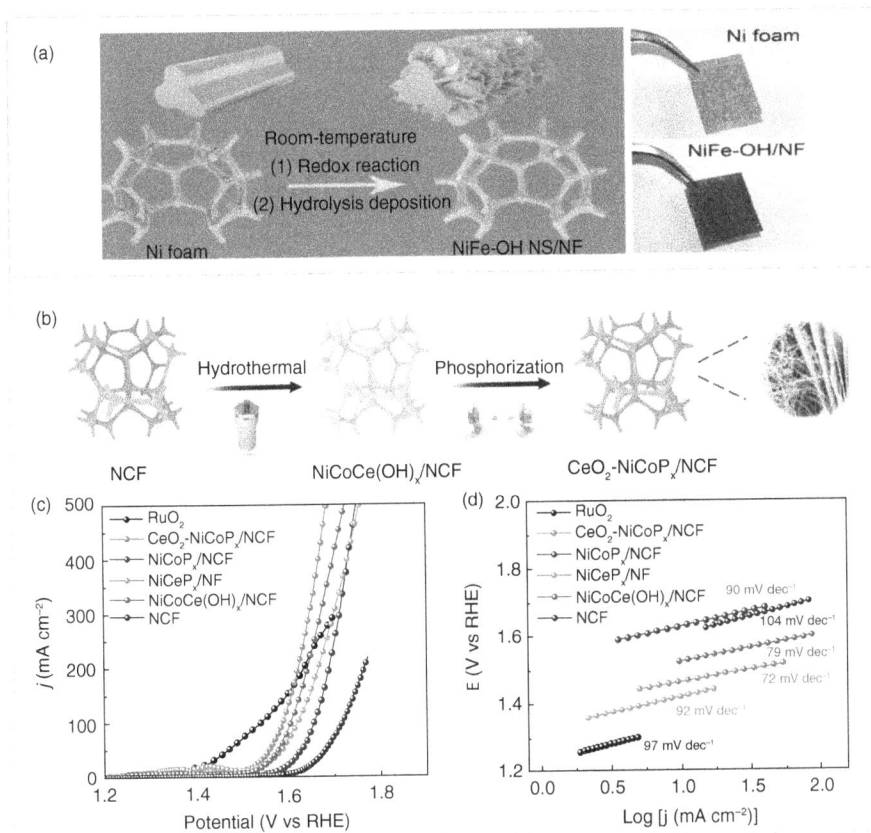

FIGURE 3.6 (a) Schematic diagram of the synthetic process of NiFe-OH NS/NF and comparison of photographs of bare Ni foam and NiFe-OH NS/NF. Adapted with permission from Ref. [101]. Copyright 2022 Elsevier B.V. (b) Fabrication of multiphase CeO_2-$NiCoP_x$ electrocatalyst on NCF substrate. (c) LSV curves and (d) Tafel slopes of the CeO_2-$NiCoP_x$ electrocatalysts. Adapted with permission from Ref. [93]. Copyright 2022 Elsevier B.V.

are beneficial for the contact of $CuCo_2O_4$ OER catalysts with electrolytes. Without binder additives, the nanosheets resulted in low internal resistance, so that rope-like $CuCo_2O_4$ nanosheets showed superior performances, lower OER overpotential, and good long-term stability. Gong and co-workers [114] investigated the OER performance of three spinel structured FSECs, including $MnCo_2O_4$/NF, $ZnCo_2O_4$/NF, and $NiCo_2O_4$/NF. Benefiting from the special homogeneous urchin-like structure and porous property, $NiCo_2O_4$/NF delivered a much smaller overpotential for the OER relative to the other two catalysts.

Apart from spinel oxides, other types of metal oxides were also developed as OER catalysts. For example, nanostructured MoO_2 and MnO_2 have been widely investigated and supported on 3D substrates such as NF and carbon paper [99,115]. The intrinsically high conductivity of MoO_2 and porous nanosheet structures with abundant active sites resulted in a low overpotential of 260 mV at 10 mA/cm². Manganese dioxides are also active for OER but are plagued by low conductivity and high over-potential, which could be well addressed by engineering of Mn/O vacancies and cation doping [116]. Calculation of the density of states (DOS) of an oxygen-deficient MnO_2 nanosheet revealed its half-metallicity property. As a result of enhanced charge transfer, NF-supported ultrathin δ-MnO_2 nanosheet arrays with abundant oxygen vacancies afforded a high OER performance with an overpotential of 320 mV at 10 mA/cm².

3.2.3.5 Transition Metal Phosphates

In 2008, Nocera's group [117] reported a cobalt phosphate (Co–P) catalyst that exhibited unexpected OER catalytic activity in neutral solution. After that, the earth-abundant and low-cost metal phosphates have stimulated particular interests. For preparation of the free-standing OER electrodes, amorphous $FePO_4$ nanosheets were in situ grown on NF by a solvothermal method [118]. Compared with crystalline $FePO_4$, the amorphous $FePO_4$ nanosheets with disordered structure possessed low-energy level of d-band center and smaller Gibbs free energy, which played a positive role in enhancing the OER catalytic activity. In addition to ferrous phosphate, $Fe(PO_3)_2$ catalyst also showed high efficiency in catalyzing water oxidation [119]. $Fe(PO_3)_2$ in situ grown on the surface of a conductive Ni_2P/NF scaffold generated a robust electrode. During the OER electrocatalysis, the $Fe(PO_3)_2$ phase was converted into amorphous FeOOH, which was proposed as the real catalytic sites. Benefiting from the more active FeOOH and the 3D conductive Ni_2P/NF substrate, the electrode yielded a current density of 500 mA/cm² at overpotentials of 265 mV and a TOF value around 0.12 s^{-1} per 3d Fe atom at an overpotential of 300 mV, along with high durability in 1 M KOH. Similar to the case of NiFe-based hydroxide catalysts, the synergistic effect of Fe and Ni in phosphate gives rise to enhanced OER performances [120]. Not surprisingly, iron-doped nickel phosphate in situ grown on NF is a promising OER catalyst in alkaline electrolyte. In addition, Wen et al. [93] developed an innovative hydrothermal synthesis and low-temperature phosphorization method to in situ synthesize CeO_2-$NiCoP_x$ on the NCF matrix containing the incorporated Ce atoms. Featuring the hybrid nanosheet and nanowire morphology, the resulting CeO_2-$NiCoP_x$/NCF catalysts showed high electrocatalytic performance for OER. The nickel and cobalt atoms positioned at the heterostructure interface were the active centers for OER, and the formed CeO_2 promotes the dissociation and adsorption of water, thus causing the fast generation of O_2. The OER overpotentials of nanosized CeO_2-$NiCoP_x$/NCF for transferring j_{10} in alkaline electrolyte are about 260 mV with a low Tafel slope of 72 mV/dec and good stability (Figure 3.5c and d).

In terms of overall discussion, FSECs are widely adopted for the applications of OER and HER. The tight interaction between the substrate and catalysts ensures the integrate structure of the catalysts in a harsh environment, especially the strong alkaline media at high oxidation potential. Therefore, compared with FSECs, the powder catalysts require a binder for a better fix of catalysts on glassy carbon or other conductive substrate, and the binder is suspected to be oxidized at a high working potential. Besides, the carbon-based FSECs with somewhat lower graphitization degree are also facing the same issue at a high working potential. Therefore, the research directions of FSECs should be divided into three aspects: (1) to understand the reaction mechanisms of the electrochemical process; (2) to develop catalysts with high stability, especially stable at harsh conditions; and (3) to improve the OER activity.

3.3 SYNTHESIS METHODS OF FSECs

To date, various preparation techniques have been developed to synthesize electrocatalysts with specific structures and morphology. This chapter describes five types of synthesis strategies: hydrothermal/solvothermal thermal reaction, electrodeposition, vacuum filtration, chemical vapor deposition (CVD) and low-temperature immersion (LTI), depending on the selective substrate and target catalyst components.

3.3.1 HYDROTHERMAL/SOLVOTHERMAL SYNTHESIS

Hydrothermal/solvothermal method is to heat the autoclave with aqueous solution or organic solvent as the solution in a special closed reaction vessel to make the chemical reaction in a high-temperature and high-pressure environment. The hydrothermal/solvothermal method is simple and low cost [95,121], making it an eco-friendly technology. Under the condition of high temperature and high pressure, it is easy to obtain an appropriate grain size, avoiding the possible grain defects and introduction of impurities in the preparation process. This method is mostly used for large area or flexible substrates, which is of great significance for practical application. For example, Yin et al. [100] fabricated a Ni/Co sulfide heterostructure anchored on NF ($Ni_3S_4@CoS_x$-NF) by a facile two-step hydrothermal method (Figure 3.7a). And $Ni_3S_4@CoS_x$-NF showed the highest OER activity with the overpotential of 332 mV at 100 mA/cm^2. Hu et al. successfully prepared catalysts with 3D porous structure on NF ($Ni_3Se_2@NiFeLDH/NF$) by two-step hydrothermal method [64]. NiFe-LDH nanosheets and Ni_3Se_2 nanowires formed in situ on NF were interlaced to form a porous core–shell structure, which provided a large surface area and accelerated electron transport efficiency. The η_{10} values for HER and OER in 1 M KOH were 68 and 222 mV, respectively.

3.3.2 ELECTROCHEMICAL DEPOSITION

Electrochemical deposition synthesis (EDS) is a technology of coating on electrode by electrochemical reaction under the action of external electric field. It has the advantages of simple operation, low synthesis temperature, low cost, and high synthesis efficiency. Electrodeposition is usually used to fabricate free-standing nano films on conductive substrates. Shang et al. [122] synthesized Fe hydroxides film encapsulated in V-doped nickel sulfide nanowire on NF (uFe/NiVS/NF) composites via a controllable electrodeposition (Figure 3.7b). They found that the best OER catalytic performance could be obtained at the electrodeposition time of 15 s. Xu et al. [123] prepared CoPO@C on NF by simple electrodeposition. The effects of different morphologies (cube, octahedron, sphere, and nanoflower) synthesized at different potentials on the OER performance were further studied, exhibiting that the catalyst with sphere morphology showed the best OER catalytic activity among all

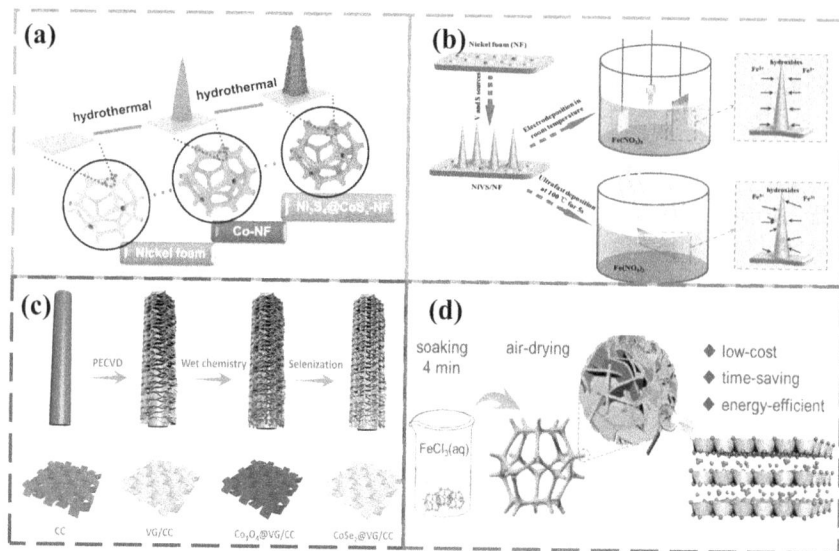

FIGURE 3.7 (a) Schematics of the preparation process of $Ni_3S_4@CoS_x$-NF. Adapted with permission from Ref. [100]. Copyright 2022 Elsevier B.V. (b) Schematic illustration of the synthesis of uFe/NiVS/NF through ultrafast chemical deposition and eFe/NiVS/NF through electrodeposition. Adapted with permission from Ref. [122]. Copyright 2017 Elsevier B.V. (c) Schematic illustration showing the synthetic procedure of $CoSe_2@VG/CC$ array. Adapted with permission from Ref. [125]. Copyright 2019 Elsevier Ltd. (d) Schematic illustration of the synthesis of NF@NiFe-LDH-1.5-4. Adapted with permission from Ref. [126]. Copyright 2021 Wiley-VCH GmbH.

samples. Li et al. [124] prepared NiFe-LDH@Ni NTAs/NF 3D-layered nanoarray on NF via a facile electrochemical dealloying method coupled with the electrodeposition method. Due to the hollow tube-core layer structure, the internal and external electrons were highly dispersed, and a large number of active sites were exposed, which made the catalyst show a low η_{10} of 191 mV for the OER.

3.3.3 VACUUM FILTRATION

The separation of liquid and solid can be realized through a porous substrate by forcing vacuum on the opposite side of the filter using vacuum filtration method. The film thickness can be controlled by changing the concentrations [127]. Although the operation is simple, it consumes a good deal of solvent and time, and therefore, it has not been commonly used. Kong et al. [128] prepared graphene oxide free-standing SnSe thin-film electrode (SnSe-TP@rGO) using a two-step synthesis technology of vacuum filtration and low-temperature annealing. The unique 3D-layered frame structure ensured the good stability of the system and accelerated the electron transfer efficiency. Kader et al. [129] prepared an independently supported PtNLs-MoS_2/rGO graphene oxide paper catalyst which demonstrated high OER catalytic activities by simple vacuum filtration and electrodeposition.

3.3.4   CVD SYNTHESIS

CVD synthesis is usually carried out under atmospheric pressure or low vacuum, and gas–solid growth method is one of the most common chemical vapor synthesis methods. CVD can be used to obtain thin-film coatings with high purity, good compactness, and good crystallization [130]. Generally, argon or hydrogen gas is introduced as the gas phase, and sulfur powder, selenium powder, or other powder raw materials can also be used as the gas phase [131]. CVD is widely used in the preparation of FSECs. Zhou et al. [125] synthesized selenide nanosheet array on CC ($CoSe_2$@vertically oriented graphene (VG/CC)) without any adhesive via an in situ CVD synthesis (Figure 3.7c). The 3D porous VG framework not only provided an electron transmission channel, but also addressed the problems of volume expansion and particle aggregation. Ma et al. [132] synthesized graphene encapsulated in (S, N)-co-doped nanosheets on NF (3DSNG/NF) via an in situ CVD synthesis. The OER properties of 3DSNG/NF with different S and N doping concentrations were further investigated. When the doping contents of N and S were 2.56 and 2.95 at%, respectively, the catalyst showed good catalytic activity. Ali et al. [102] synthesized multi-walled carbon nanotubes (MWCNTs)-graphene hybrid nanomaterials on Ni silica nanocomposites by a simple CVD method. The effects of the combination of Co, Fe, and Ni with silicon matrix on the structure of mixed carbon nanomaterials were studied, showing high OER performances.

3.3.5   LOW-TEMPERATURE IMMERSION

The "low-temperature immersion" (LTI) is a method of preparing electrocatalysts by soaking the sample at a lower temperature [126]. This method is simple and energy-saving, and it generally requires FSECs as a medium [5]. Improving the OER performance of the free-standing non-precious materials with etching-based principle is the main means in LTI engineering. The excellent electrocatalytic performance of catalysts synthesized by LTI engineering has aroused the great interests. Because the rapid growth of primary nanocrystals can be avoided during LTI, the crystallinity of many electrode materials is very low [133,134]. Numerous studies have shown that the amorphous with a large number of active sites and unsaturated sites formed in LTI is the main reason for the excellent OER performances [23,65]. For instance, Li et al. [126] reported a time- and energy-saving approach to directly grow NiFe-layered double hydroxide (NiFe-LDH) nanosheets on NF under ambient temperature and pressure (Figure 3.7d). These NiFe-LDH nanosheets are vertically grown on NF and interdigitated together to form a highly porous array, leading to numerous exposed active sites, reduce resistance of charge/mass transportation, and enhance mechanical stability. As FSECs, the representative sample (NF@NiFe-LDH-1.5-4) shows an excellent catalytic activity for OER in alkaline electrolyte, requiring low overpotentials of 190 and 220 mV to reach the current densities of 100 and 657 mA/cm^2 with a Tafel slope of 38.1 mV/dec. Moreover, Guo et al. [135] prepared a 2D Fe-doped nickel hydroxide electrode (RT–Fe@Ni(OH)$_2$) with high current density for OER under room temperature. The RT–Fe@Ni(OH)$_2$ electrode exhibited an overpotential of 0.312 V at 100 mA/cm^2 and a retention rate of 96.3% after 100 h in 1 M KOH.

3.4 SUMMARY AND OUTLOOK

Exploiting cheap, efficient, and robust HER/OER catalysts is of crucial importance in water-splitting technology for hydrogen energy. Compared with conventional electrode prepared with catalyst powder, FSECs integrating in situ grown catalytically active phase benefit the simplification of electrode preparation, decrease of interface resistance, exposure of abundant active sites and enhancement of stability, making them promising for practical applications. In the past decades, there has been solid developments in free-standing transition-metal-based electrocatalytic materials that range from metals, chalcogenides, phosphides, carbides, and nitrides for HER and OER. Among them, the transition metal phosphides and alloys exhibit higher HER catalytic activity and stability than other compounds. For the OER, the most promising non-noble-metal electrocatalysts are based on LDHs and chalcogenides, which often outperform the benchmark noble-metal catalysts (i.e., Ir and Ru compounds) in alkaline media.

In spite of the substantial progress in materials design/synthesis and properties investigation, the following challenges remain in the further development of FSECs for alkaline water electrolysis:

1. The detail catalytic mechanisms of FSECs and interaction effects of the active materials and support should be clarified concretely.
2. The catalytic properties of FSECs and bonding strength of the active materials and support should be enhanced.
3. The novel structures and morphologies of FSECs should be further designed.
4. The low-cost and facile synthesis methods of FSECs should be developed for their large-scale preparation.

ACKNOWLEDGMENTS

Thanks to the authors and corresponding publishers of all references in this chapter for their support, and we would like to express our special gratitude.

REFERENCES

[1] J. Li, M. Song, Y. Hu, C. Zhang, W. Liu, X. Huang, J. Zhang, Y. Zhu, J. Zhang, D. Wang, A self-supported heterogeneous bimetallic phosphide array electrode enables efficient hydrogen evolution from saline water splitting, *Nano Research*, 16 (2022) 3658–3664.

[2] X. Zhao, K. Tang, X. Wang, W. Qi, H. Yu, C.-F. Du, Q. Ye, A self-supported bifunctional MoNi₄ framework with iron doping for ultra-efficient water splitting, *Journal of Materials Chemistry A*, 11 (2023) 3408–3417.

[3] T. Zhang, J. Sun, J. Guan, Self-supported transition metal chalcogenides for oxygen evolution, *Nano Research*, (2023) 1–28.

[4] D. Yu, Y. Hao, S. Han, S. Zhao, Q. Zhou, C.H. Kuo, F. Hu, L. Li, H.Y. Chen, J. Ren, S. Peng, Ultrafast combustion synthesis of robust and efficient electrocatalysts for high-current-density water oxidation, *ACS Nano*, 17 (2023) 1701–1712.

[5] Y. Xin, Y. Cang, Z. Wang, X. Dou, W. Hao, Y. Miao, Construction of non-precious metal self-supported electrocatalysts for oxygen evolution from a low-temperature immersion perspective, *Chemical Record*, 23 (2023) e202200259.

[6] F. Liu, X. Wu, R. Guo, H. Miao, F. Wang, C. Yang, J. Yuan, Suppressing the surface amorphization of Ba(0.5)Sr(0.5)Co(0.8)Fe(0.2)O(3-delta) perovskite toward oxygen catalytic reactions by introducing the compressive stress, *Inorganic Chemistry*, 62 (2023) 4373–4384.

[7] C.-L. Huang, Y.-G. Lin, C.-L. Chiang, C.-K. Peng, D. Senthil Raja, C.-T. Hsieh, Y.-A. Chen, S.-Q. Chang, Y.-X. Yeh, S.-Y. Lu, Atomic scale synergistic interactions lead to breakthrough catalysts for electrocatalytic water splitting, *Applied Catalysis B: Environmental*, 320 (2023) 122016.

[8] H. Chen, H.B. Huang, H.H. Li, S.Z. Zhao, L.D. Wang, J. Zhang, S.L. Zhong, C.F. Lao, L.M. Cao, C.T. He, Self-Supporting Co/CeO(2) heterostructures for ampere-level current density alkaline water electrolysis, *Inorganic Chemistry*, 62 (2023) 3297–3304.

[9] M. Zubair, M.M. Ul Hassan, M.T. Mehran, M.M. Baig, S. Hussain, F. Shahzad, 2D MXenes and their heterostructures for HER, OER and overall water splitting: A review, *International Journal of Hydrogen Energy*, 47 (2022) 2794–2818.

[10] H. Sun, Z. Yan, F. Liu, W. Xu, F. Cheng, J. Chen, Self-supported transition-metal-based electrocatalysts for hydrogen and oxygen evolution, *Advanced Materials*, 32 (2020) e1806326.

[11] Z.P. Ifkovits, J.M. Evans, M.C. Meier, K.M. Papadantonakis, N.S. Lewis, Decoupled electrochemical water-splitting systems: a review and perspective, *Energy & Environmental Science*, 14 (2021) 4740–4759.

[12] Y. Zhang, J. Wu, B. Guo, H. Huo, S. Niu, S. Li, P. Xu, Recent advances of transition-metal metaphosphates for efficient electrocatalytic water splitting, *Carbon Energy*, 5 (2023) e375.

[13] T. Liu, Y. Xiang, Z. Tan, W. Hong, Z. He, J. Long, B. Xie, R. Li, X. Gou, One-step growth of Ni$_3$Fe-Fe$_3$C heterostructures well encapsulated in NCNTs as superior self-supported bifunctional electrocatalysts for overall water splitting, *Journal of Alloys and Compounds*, 949 (2023) 169825.

[14] J. Wang, N. Zang, C. Xuan, B. Jia, W. Jin, T. Ma, Self-supporting electrodes for gas-involved key energy reactions, *Advanced Functional Materials*, 31 (2021) 2104620.

[15] N.K. Chaudhari, H. Jin, B. Kim, K. Lee, Nanostructured materials on 3D nickel foam as electrocatalysts for water splitting, *Nanoscale*, 9 (2017) 12231–12247.

[16] F. Diao, W. Huang, G. Ctistis, H. Wackerbarth, Y. Yang, P. Si, J. Zhang, X. Xiao, C. Engelbrekt, Bifunctional and self-supported NiFeP-Layer-Coated NiP Rods for electrochemical water splitting in alkaline solution, *ACS Applied Materials & Interfaces*, 13 (2021) 23702–23713.

[17] Y.-Y. Ma, C.-X. Wu, X.-J. Feng, H.-Q. Tan, L.-K. Yan, Y. Liu, Z.-H. Kang, E.-B. Wang, Y.-G. Li, Highly efficient hydrogen evolution from seawater by a low-cost and stable CoMoP@C electrocatalyst superior to Pt/C, *Energy & Environmental Science*, 10 (2017) 788–798.

[18] M. You, X. Du, X. Hou, Z. Wang, Y. Zhou, H. Ji, L. Zhang, Z. Zhang, S. Yi, D. Chen, In-situ growth of ruthenium-based nanostructure on carbon cloth for superior electrocatalytic activity towards HER and OER, *Applied Catalysis B: Environmental*, 317 (2022) 121729.

[19] K. Kordek-Khalil, I. Walendzik, P. Rutkowski, Low-overpotential full water splitting with metal-free self-supported electrodes obtained by amination of oxidised carbon cloth, *Sustainable Energy Technologies and Assessments*, 53 (2022) 102569.

[20] D. Li, Z. Pan, H. Tao, J. Li, W. Gu, B. Li, C. Zhong, Q. Jiang, C. Ye, Q. Zhou, Self-derivation-behaviour of substrates realizing enhanced oxygen evolution reaction, *Chemical Communications*, 56 (2020) 12399–12402.

[21] Y. Yu, C. Yu, Z. Wu, B. Huang, P. Zhou, H. Zhang, W. Liu, Y. Liu, Z. Xiong, B. Lai, Switching the primary mechanism from a radical to a nonradical pathway in electrocatalytic ozonation by onsite alternating anode and cathode, *Chemical Engineering Journal*, 457 (2023) 141340.

[22] L. Wei, S. Huang, Y. Zhang, M. Ye, C.C. Li, Enhancing the coupling effect in a sandwiched FeNiPS$_3$/graphite catalyst derived from graphite intercalation compounds for efficient oxygen evolution reaction, *Journal of Materials Chemistry A*, 10 (2022) 11793–11802.

[23] Y. Liu, X. Liang, L. Gu, Y. Zhang, G.D. Li, X. Zou, J.S. Chen, Corrosion engineering towards efficient oxygen evolution electrodes with stable catalytic activity for over 6000 hours, *Nature Communications*, 9 (2018) 2609.

[24] W. Liu, X. Wang, F. Wang, K. Du, Z. Zhang, Y. Guo, H. Yin, D. Wang, A durable and pH-universal self-standing MoC-Mo$_2$C heterojunction electrode for efficient hydrogen evolution reaction, *Nature Communications*, 12 (2021) 6776.

[25] A.J. Esswein, Y. Surendranath, S.Y. Reece, D.G. Nocera, Highly active cobalt phosphate and borate based oxygen evolving catalysts operating in neutral and natural waters, *Energy & Environmental Science*, 4 (2011) 499–504.

[26] D. Senthil Raja, P.-Y. Cheng, C.-C. Cheng, S.-Q. Chang, C.-L. Huang, S.-Y. Lu, In-situ grown metal-organic framework-derived carbon-coated Fe-doped cobalt oxide nanocomposite on fluorine-doped tin oxide glass for acidic oxygen evolution reaction, *Applied Catalysis B: Environmental*, 303 (2022) 120899.

[27] Y. Wang, J. Yu, Q. Liu, J. Liu, R. Chen, J. Zhu, R. Li, J. Wang, Porous carbon foam loaded CoSe$_2$ nanoparticles based on inkjet-printing technology as self-supporting electrodes for efficient water splitting, *Electrochimica Acta*, 438 (2023) 141594.

[28] K.C.S. Lakshmi, B. Vedhanarayanan, T.-W. Lin, Electrocatalytic hydrogen and oxygen evolution reactions: Role of two-dimensional layered materials and their composites, *Electrochimica Acta*, 447 (2023) 141594.

[29] Y. Wu, F. Li, W. Chen, Q. Xiang, Y. Ma, H. Zhu, P. Tao, C. Song, W. Shang, T. Deng, J. Wu, Coupling interface constructions of MoS$_2$/Fe$_5$Ni$_4$S$_8$ heterostructures for efficient electrochemical water splitting, *Advanced Materials*, 30 (2018) 1803151.

[30] S.-W. Wu, S.-Q. Liu, X.-H. Tan, W.-Y. Zhang, K. Cadien, Z. Li, Ni$_3$S$_2$-embedded NiFe LDH porous nanosheets with abundant heterointerfaces for high-current water electrolysis, *Chemical Engineering Journal*, 442 (2022) 136105.

[31] Y. Wang, Z. Yin, Z. Wang, X. Li, H. Guo, J. Wang, D. Zhang, Facile construction of Co(OH)$_2$@Ni(OH)$_2$ core-shell nanosheets on nickel foam as three dimensional free-standing electrode for supercapacitors, *Electrochimica Acta*, 293 (2019) 40–46.

[32] M. Tang, Y. Liu, H. Cao, Q. Zheng, X. Wei, K.H. Lam, D. Lin, Cu$_2$S/Ni$_3$S$_2$ ultrathin nanosheets on Ni foam as a highly efficient electrocatalyst for oxygen evolution reaction, *International Journal of Hydrogen Energy*, 47 (2022) 3013–3021.

[33] Q. Chen, L. Huang, Q. Kong, X. An, X. Wu, W. Yao, C. Sun, Facile synthesis of self support Fe doped Ni$_3$S$_2$ nanosheet arrays for high performance alkaline oxygen evolution, *Journal of Electroanalytical Chemistry*, 907 (2022) 116047.

[34] N. Cao, S. Chen, Y. Di, C. Li, H. Qi, Q. Shao, W. Zhao, Y. Qin, X. Zang, High efficiency in overall water-splitting via Co-doping heterointerface-rich NiS$_2$/MoS$_2$ nanosheets electrocatalysts, *Electrochimica Acta*, 425 (2022) 140674.

[35] H. Yang, K. Lin, Z. Zhou, C. Peng, S. Peng, M. Sun, L. Yu, Surface phosphorization of Ni-Co-S as an efficient bifunctional electrocatalyst for full water splitting, *Dalton Transactions*, 50 (2021) 16578–16586.

[36] Z. Wan, Q. He, Y. Qu, J. Dong, E. Shoko, P. Yan, T. Taylor Isimjan, X. Yang, Designing coral-like Fe$_2$O$_3$-regulated Se-rich CoSe$_2$ heterostructure as a highly active and stable oxygen evolution electrocatalyst for overall water splitting, *Journal of Electroanalytical Chemistry*, 904 (2022) 115928.

[37] W. Li, H. Zhao, H. Li, R. Wang, Fe doped NiS nanosheet arrays grown on carbon fiber paper for a highly efficient electrocatalytic oxygen evolution reaction, *Nanoscale Advances*, 4 (2022) 1220–1226.

[38] M. Kim, D. Park, J. Kim, A thermoelectric generator comprising selenium-doped bismuth telluride on flexible carbon cloth with n-type thermoelectric properties, *Ceramics International*, 48 (2022) 10852–10861.

[39] H. Qian, B. Wu, Z. Nie, T. Liu, P. Liu, H. He, J. Wu, Z. Chen, S. Chen, A flexible Ni_3S_2/Ni@CC electrode for high-performance battery-like supercapacitor and efficient oxygen evolution reaction, *Chemical Engineering Journal*, 420 (2021) 127646.

[40] S. Jiang, H. Shao, G. Cao, H. Li, W. Xu, J. Li, J. Fang, X. Wang, Waste cotton fabric derived porous carbon containing Fe_3O_4/NiS nanoparticles for electrocatalytic oxygen evolution, *Journal of Materials Science & Technology*, 59 (2020) 92–99.

[41] Y. Guo, D. Guo, F. Ye, K. Wang, Z. Shi, Synthesis of lawn-like NiS_2 nanowires on carbon fiber paper as bifunctional electrode for water splitting, *International Journal of Hydrogen Energy*, 42 (2017) 17038–17048.

[42] S. Ganguli, S. Ghosh, S. Das, V. Mahalingam, Inception of molybdate as a "pore forming additive" to enhance the bifunctional electrocatalytic activity of nickel and cobalt based mixed hydroxides for overall water splitting, *Nanoscale*, 11 (2019) 16896–16906.

[43] Y. Dong, S. Ji, H. Wang, V. Linkov, R. Wang, In-site hydrogen bubble template method to prepare Ni coated metal meshes as effective bi-functional electrodes for water splitting, *Dalton Transactions*, 51 (2022) 9681–9688.

[44] D. Li, Z. Liu, J. Wang, B. Liu, Y. Qin, W. Yang, J. Liu, Hierarchical trimetallic sulfide $FeCo_2S_4$-$NiCo_2S_4$ nanosheet arrays supported on a Ti mesh: An efficient 3D bifunctional electrocatalyst for full water splitting, *Electrochimica Acta*, 340 (2020) 135957.

[45] W.Q. Yang, Y.X. Hua, Q.B. Zhang, H. Lei, C.Y. Xu, Electrochemical fabrication of 3D quasi-amorphous pompon-like Co-O and Co-Se hybrid films from choline chloride/urea deep eutectic solvent for efficient overall water splitting, *Electrochimica Acta*, 273 (2018) 71–79.

[46] J. Dang, M. Yin, D. Pan, Z. Tian, G. Chen, J. Zou, H. Miao, Q. Wang, J. Yuan, Four-functional iron/copper sulfide heterostructure for alkaline hybrid zinc batteries and water splitting, *Chemical Engineering Journal*, 457 (2023) 141357.

[47] B. Chen, A.L.G. Biancolli, C.L. Radford, S. Holdcroft, Stainless steel felt as a combined OER electrocatalyst/porous transport layer for investigating anion-exchange membranes in water electrolysis, *ACS Energy Letters*, 8 (2023) 2661–2667.

[48] M. Xie, Y. Ma, D. Lin, C. Xu, F. Xie, W. Zeng, Bimetal-organic framework MIL-53(Co-Fe): an efficient and robust electrocatalyst for the oxygen evolution reaction, *Nanoscale*, 12 (2020) 67–71.

[49] S. Deng, S. Shen, Y. Zhong, K. Zhang, J. Wu, X. Wang, X. Xia, J. Tu, Assembling Co_9S_8 nanoflakes on Co_3O_4 nanowires as advanced core/shell electrocatalysts for oxygen evolution reaction, *Journal of Energy Chemistry*, 26 (2017) 1203–1209.

[50] Y. Zheng, Y. Jiao, A. Vasileff, S.Z. Qiao, The hydrogen evolution reaction in alkaline solution: from theory, single crystal models, to practical electrocatalysts, *Angewandte Chemie International*, 57 (2018) 7568–7579.

[51] J. Zhang, T. Wang, P. Liu, Z. Liao, S. Liu, X. Zhuang, M. Chen, E. Zschech, X. Feng, Efficient hydrogen production on MoNi(4) electrocatalysts with fast water dissociation kinetics, *Nature Communications*, 8 (2017) 15437.

[52] S. Wang, B. Xu, W. Huo, H. Feng, X. Zhou, F. Fang, Z. Xie, J.K. Shang, J. Jiang, Efficient FeCoNiCuPd thin-film electrocatalyst for alkaline oxygen and hydrogen evolution reactions, *Applied Catalysis B: Environmental*, 313 (2022) 121472.

[53] Q. Lu, G.S. Hutchings, W. Yu, Y. Zhou, R.V. Forest, R. Tao, J. Rosen, B.T. Yonemoto, Z. Cao, H. Zheng, J.Q. Xiao, F. Jiao, J.G. Chen, Highly porous non-precious bimetallic electrocatalysts for efficient hydrogen evolution, *Nature Communications*, 6 (2015) 6567.

[54] J.-S. Sun, Z. Wen, L.-P. Han, Z.-W. Chen, X.-Y. Lang, Q. Jiang, Nonprecious intermetallic Al_7Cu_4Ni nanocrystals seamlessly integrated in freestanding bimodal nanoporous copper for efficient hydrogen evolution catalysis, *Advanced Functional Materials*, 28 (2018) 1706127.

[55] J. Wang, S. Xin, Y. Xiao, Z. Zhang, Z. Li, W. Zhang, C. Li, R. Bao, J. Peng, J. Yi, S. Chou, Manipulating the water dissociation electrocatalytic sites of bimetallic nickel-based alloys for highly efficient alkaline hydrogen evolution, *Angewandte Chemie International*, 61 (2022) e202202518.

[56] M. Luo, S. Liu, W. Zhu, G. Ye, J. Wang, Z. He, An electrodeposited MoS_2-MoO_3−x/Ni_3S_2 heterostructure electrocatalyst for efficient alkaline hydrogen evolution, *Chemical Engineering Journal*, 428 (2022) 131055.

[57] L. Yu, L. Wu, S. Song, B. McElhenny, F. Zhang, S. Chen, Z. Ren, Hydrogen generation from seawater electrolysis over a sandwich-like NiCoN|NixP|NiCoN microsheet array catalyst, *ACS Energy Letters*, 5 (2020) 2681–2689.

[58] R. Zhang, L. Xu, Z. Wu, L. Wang, J. Zhang, Y. Tang, L. Xu, A. Xie, Y. Chen, H. Zhang, P. Wan, Nitrogen doped carbon encapsulated hierarchical NiMoN as highly active and durable HER electrode for repeated ON/OFF water electrolysis, *Chemical Engineering Journal*, 436 (2022) 134931.

[59] F.-T. Tsai, Y.-T. Deng, C.-W. Pao, J.-L. Chen, J.-F. Lee, K.-T. Lai, W.-F. Liaw, The HER/OER mechanistic study of an FeCoNi-based electrocatalyst for alkaline water splitting, *Journal of Materials Chemistry A*, 8 (2020) 9939–9950.

[60] S. Ren, X. Duan, F. Ge, M. Zhang, H. Zheng, Trimetal-based N-doped carbon nanotubes arrays on Ni foams as self-supported electrodes for hydrogen/oxygen evolution reactions and water splitting, *Journal of Power Sources*, 480 (2020) 228866.

[61] R. Guo, J. Shi, L. Hong, K. Ma, W. Zhu, H. Yang, J. Wang, H. Wang, M. Sheng, CoP(2)/Co(2)P encapsulated in carbon nanotube arrays to construct self-supported electrodes for overall electrochemical water splitting, *ACS Applied Materials & Interfaces*, 14 (2022) 56847–56855.

[62] J. Zhu, Y. Lu, X. Zheng, S. Xu, S. Sun, Y. Liu, D. Li, D. Jiang, Heterostructure arrays of (Ni,Co)Se_2 nanowires integrated with MOFs-derived $CoSe_2$ dodecahedra for synergistically high-efficiency and stable overall water splitting, *Applied Surface Science*, 592 (2022) 153352.

[63] H. Sun, X. Xu, Z. Yan, X. Chen, L. Jiao, F. Cheng, J. Chen, Superhydrophilic amorphous Co-B-P nanosheet electrocatalysts with Pt-like activity and durability for the hydrogen evolution reaction, *Journal of Materials Chemistry A*, 6 (2018) 22062–22069.

[64] J. Hu, S. Zhu, Y. Liang, S. Wu, Z. Li, S. Luo, Z. Cui, Self-supported Ni(3)Se(2)@ NiFe layered double hydroxide bifunctional electrocatalyst for overall water splitting, *Journal of Colloid and Interface Science*, 587 (2021) 79–89.

[65] X. Li, Y. Wang, J. Wang, Y. Da, J. Zhang, L. Li, C. Zhong, Y. Deng, X. Han, W. Hu, Sequential electrodeposition of bifunctional catalytically active structures in MoO(3)/Ni-NiO composite electrocatalysts for selective hydrogen and oxygen evolution, *Advanced Materials*, 32 (2020) e2003414.

[66] M. Fan, H. Chen, Y. Wu, L.-L. Feng, Y. Liu, G.-D. Li, X. Zou, Growth of molybdenum carbide micro-islands on carbon cloth toward binder-free cathodes for efficient hydrogen evolution reaction, *Journal of Materials Chemistry A*, 3 (2015) 16320–16326.

[67] W. Chen, J. Gu, Y. Du, F. Song, F. Bu, J. Li, Y. Yuan, R. Luo, Q. Liu, D. Zhang, Achieving rich and active alkaline hydrogen evolution heterostructures via interface engineering on 2D 1T-MoS_2 quantum sheets, *Advanced Functional Materials*, 30 (2020) 2000551.

[68] Y. Li, S. Wang, Y. Hu, X. Zhou, M. Zhang, X. Jia, Y. Yang, B.-L. Lin, G. Chen, Highly dispersed Pt nanoparticles on 2D MoS_2 nanosheets for efficient and stable hydrogen evolution reaction, *Journal of Materials Chemistry A*, 10 (2022) 5273–5279.

[69] D. Kong, H. Wang, J.J. Cha, M. Pasta, K.J. Koski, J. Yao, Y. Cui, Synthesis of MoS_2 and $MoSe_2$ films with vertically aligned layers, *Nano Lett*, 13 (2013) 1341–1347.

[70] H. Liu, X. Ma, Y. Rao, Y. Liu, J. Liu, L. Wang, M. Wu, Heteromorphic $NiCo_2S_4/Ni_3S_2/$ Ni foam as a self-standing electrode for hydrogen evolution reaction in alkaline solution, *ACS Applied Materials & Interfaces*, 10 (2018) 10890–10897.

[71] A. Eftekhari, Molybdenum diselenide ($MoSe_2$) for energy storage, catalysis, and opto-electronics, *Applied Materials Today*, 8 (2017) 1–17.

[72] Y. Yin, Y. Zhang, T. Gao, T. Yao, X. Zhang, J. Han, X. Wang, Z. Zhang, P. Xu, P. Zhang, X. Cao, B. Song, S. Jin, Synergistic phase and disorder engineering in 1T-MoSe(2) nanosheets for enhanced hydrogen-evolution reaction, *Advanced Materials*, 29 (2017) 1700311.

[73] D. Kong, H. Wang, Z. Lu, Y. Cui, $CoSe_2$ nanoparticles grown on carbon fiber paper: an efficient and stable electrocatalyst for hydrogen evolution reaction, *Journal of the American Chemical Society*, 136 (2014) 4897–4900.

[74] P. Guo, Y.-X. Wu, W.-M. Lau, H. Liu, L.-M. Liu, Porous CoP nanosheet arrays grown on nickel foam as an excellent and stable catalyst for hydrogen evolution reaction, *International Journal of Hydrogen Energy*, 42 (2017) 26995–27003.

[75] L. Wen, J. Yu, C. Xing, D. Liu, X. Lyu, W. Cai, X. Li, Flexible vanadium-doped Ni(2) P nanosheet arrays grown on carbon cloth for an efficient hydrogen evolution reaction, *Nanoscale*, 11 (2019) 4198–4203.

[76] X. Zou, Y. Zhang, Noble metal-free hydrogen evolution catalysts for water splitting, *Chemical Society Reviews*, 44 (2015) 5148–5180.

[77] R. Ma, Y. Zhou, Y. Chen, P. Li, Q. Liu, J. Wang, Ultrafine molybdenum carbide nanoparticles composited with carbon as a highly active hydrogen-evolution electro-catalyst, *Angewandte Chemie International*, 54 (2015) 14723–14727.

[78] Z. Chen, Y. Song, J. Cai, X. Zheng, D. Han, Y. Wu, Y. Zang, S. Niu, Y. Liu, J. Zhu, X. Liu, G. Wang, Tailoring the d-band centers enables Co(4)N nanosheets to be highly active for hydrogen evolution catalysis, *Angewandte Chemie International*, 57 (2018) 5076–5080.

[79] J. Yu, Q. He, G. Yang, W. Zhou, Z. Shao, M. Ni, Recent Advances and prospective in ruthenium-based materials for electrochemical water splitting, *ACS Catalysis*, 9 (2019) 9973–10011.

[80] S. Wang, W. Huo, F. Fang, Z. Xie, J.K. Shang, J. Jiang, High entropy alloy/C nanoparticles derived from polymetallic MOF as promising electrocatalysts for alkaline oxygen evolution reaction, *Chemical Engineering Journal*, 429 (2022) 132410.

[81] S.Y. Li, T.X. Nguyen, Y.H. Su, C.C. Lin, Y.J. Huang, Y.H. Shen, C.P. Liu, J.J. Ruan, K.S. Chang, J.M. Ting, Sputter-deposited high entropy alloy thin film electrocatalyst for enhanced oxygen evolution reaction performance, *Small*, 18 (2022) e2106127.

[82] P. Yang, Y. Shi, T. Xia, Z. Jiang, X. Ren, L. Liang, Q. Shao, K. Zhu, Novel self-supporting thin film electrodes of FeCoNiCrMn high entropy alloy for excellent oxygen evolution reaction, *Journal of Alloys and Compounds*, 938 (2023) 168582.

[83] D. Kim, S. Park, B. Yan, H. Hong, Y. Cho, J. Kang, X. Qin, T. Yoo, Y. Piao, Synthesis of remarkably thin Co-Fe phosphide/carbon nanosheet for enhanced oxygen evolution reaction electrocatalysis driven by readily generated active oxyhydroxide, *ACS Applied Energy Materials*, 5 (2022) 2400–2411.

[84] X. Chen, Q. Wang, Y. Cheng, H. Xing, J. Li, X. Zhu, L. Ma, Y. Li, D. Liu, S-doping triggers redox reactivities of both iron and lattice oxygen in FeOOH for low-cost and high-performance water oxidation, *Advanced Functional Materials*, 32 (2022) 2112674.

[85] P. Wang, J. Qi, X. Chen, C. Li, W. Li, T. Wang, C. Liang, Three-dimensional het-erostructured NiCoP@NiMn-layered double hydroxide arrays supported on ni foam as a bifunctional electrocatalyst for overall water splitting, *ACS Applied Materials & Interfaces*, 12 (2020) 4385–4395.

[86] P.W. Menezes, C. Panda, S. Garai, C. Walter, A. Guiet, M. Driess, Structurally ordered intermetallic cobalt stannide nanocrystals for high-performance electrocatalytic overall water-splitting, *Angewandte Chemie International*, 57 (2018) 15237–15242.

[87] J. Hou, Y. Sun, Y. Wu, S. Cao, L. Sun, Promoting active sites in core-shell nanowire array as mott-schottky electrocatalysts for efficient and stable overall water splitting, *Advanced Functional Materials*, 28 (2018) 1704447.

[88] H. Yang, Y. Huang, W.Y. Teoh, L. Jiang, W. Chen, L. Zhang, J. Yan, Molybdenum selenide nanosheets surrounding nickel selenides sub-microislands on nickel foam as high-performance bifunctional electrocatalysts for water splitting, *Electrochimica Acta*, 349 (2020) 136336.

[89] A.I. Inamdar, H.S. Chavan, B. Hou, C.H. Lee, S.U. Lee, S. Cha, H. Kim, H. Im, A robust nonprecious CuFe composite as a highly efficient bifunctional catalyst for overall electrochemical water splitting, *Small*, 16 (2020) e1905884.

[90] J. Sun, P. Song, H. Zhou, L. Lang, X. Shen, Y. Liu, X. Cheng, X. Fu, G. Zhu, A surface configuration strategy to hierarchical Fe-Co-S/Cu2O/Cu electrodes for oxygen evolution in water/seawater splitting, *Applied Surface Science*, 567 (2021) 150757.

[91] Y. Zhao, B. Jin, A. Vasileff, Y. Jiao, S.-Z. Qiao, Interfacial nickel nitride/sulfide as a bifunctional electrode for highly efficient overall water/seawater electrolysis, *Journal of Materials Chemistry A*, 7 (2019) 8117–8121.

[92] Y.J. Lee, S.K. Park, Metal-organic framework-derived hollow CoSx nanoarray coupled with nife layered double hydroxides as efficient bifunctional electrocatalyst for overall water splitting, *Small*, 18 (2022) e2200586.

[93] S. Wen, J. Huang, T. Li, W. Chen, G. Chen, Q. Zhang, X. Zhang, Q. Qian, K. Ostrikov, Multiphase nanosheet-nanowire cerium oxide and nickel-cobalt phosphide for highly-efficient electrocatalytic overall water splitting, *Applied Catalysis B: Environmental*, 316 (2022) 121678.

[94] K. Karthick, S. Anantharaj, S.R. Ede, S. Kundu, Nanosheets of nickel iron hydroxy carbonate hydrate with pronounced OER activity under alkaline and near-neutral conditions, *Inorganic Chemistry*, 58 (2019) 1895–1904.

[95] W. Zhang, M. Liu, H. Liang, L. Cui, W. Yang, J.M. Razal, J. Liu, Flower-like nanosheets directly grown on Co foil as efficient bifunctional catalysts for overall water splitting, *Journal of Colloid and Interface Science*, 587 (2021) 650–660.

[96] C.-Z. Yuan, Z.-T. Sun, Y.-F. Jiang, Z.-K. Yang, N. Jiang, Z.-W. Zhao, U.Y. Qazi, W.-H. Zhang, A.-W. Xu, One-step in situ growth of iron-nickel sulfide nanosheets on FeNi alloy foils: high-performance and self-supported electrodes for water oxidation, *Small*, 13 (2017) 1604161.

[97] N. Nazar, S. Manzoor, Y.u. Rehman, I. Bibi, D. Tyagi, A.H. Chughtai, R.S. Gohar, M. Najam-Ul-Haq, M. Imran, M.N. Ashiq, Metal-organic framework derived CeO2/C nanorod arrays directly grown on nickel foam as a highly efficient electrocatalyst for OER, *Fuel*, 307 (2022) 121823.

[98] Z. Ahmed, Krishankant, R. Rai, R. Kumar, T. Maruyama, C. Bera, V. Bagchi, Unraveling a graphene exfoliation technique analogy in the making of ultrathin nickel-iron oxyhydroxides@nickel foam to promote the OER, *ACS Applied Materials & Interfaces*, 13 (2021) 55281–55291.

[99] Z. Ye, T. Li, G. Ma, Y. Dong, X. Zhou, Metal-ion (Fe, V, Co, and Ni)-doped MnO2 ultrathin nanosheets supported on carbon fiber paper for the oxygen evolution reaction, *Advanced Functional Materials*, 27 (2017) 1704083.

[100] M. Yin, H. Miao, J. Dang, B. Chen, J. Zou, G. Chen, H. Li, High-performance alkaline hybrid zinc batteries with heterostructure nickel/cobalt sulfide, *Journal of Power Sources*, 545 (2022) 231902.

[101] W. Zhu, T. Zhang, Y. Zhang, Z. Yue, Y. Li, R. Wang, Y. Ji, X. Sun, J. Wang, A practical-oriented NiFe-based water-oxidation catalyst enabled by ambient redox and hydrolysis co-precipitation strategy, *Applied Catalysis B: Environmental*, 244 (2019) 844–852.

[102] Z. Ali, M. Mehmood, J. Ahmad, T.S. Malik, B. Ahmad, In-situ growth of novel CNTs-graphene hybrid structure on Ni-silica nanocomposites by CVD method for oxygen evolution reaction, *Ceramics International*, 46 (2020) 19158–19169.

[103] F. Jing, Q. Lv, J. Xiao, Q. Wang, S. Wang, Highly active and dual-function self-supported multiphase $NiS-NiS_2-Ni_3S_2/NF$ electrodes for overall water splitting, *Journal of Materials Chemistry A*, 6 (2018) 14207–14214.

[104] L. Zhuang, Y. Jia, H. Liu, Z. Li, M. Li, L. Zhang, X. Wang, D. Yang, Z. Zhu, X. Yao, Sulfur-modified oxygen vacancies in iron-cobalt oxide nanosheets: enabling extremely high activity of the oxygen evolution reaction to achieve the industrial water splitting benchmark, *Angewandte Chemie International*, 59 (2020) 14664–14670.

[105] Q. Han, Y. Luo, J. Li, X. Du, S. Sun, Y. Wang, G. Liu, Z. Chen, Efficient NiFe-based oxygen evolution electrocatalysts and origin of their distinct activity, *Applied Catalysis B: Environmental*, 304 (2022) 120937.

[106] D. Tang, J. Liu, X. Wu, R. Liu, X. Han, Y. Han, H. Huang, Y. Liu, Z. Kang, Carbon quantum dot/NiFe layered double-hydroxide composite as a highly efficient electrocatalyst for water oxidation, *ACS Applied Materials & Interfaces*, 6 (2014) 7918–7925.

[107] L. Peng, N. Yang, Y. Yang, Q. Wang, X. Xie, D. Sun-Waterhouse, L. Shang, T. Zhang, G.I.N. Waterhouse, Atomic cation-vacancy engineering of NiFe-Layered double hydroxides for improved activity and stability towards the oxygen evolution reaction, *Angewandte Chemie International*, 60 (2021) 24612–24619.

[108] X. Lu, C. Zhao, Electrodeposition of hierarchically structured three-dimensional nickel-iron electrodes for efficient oxygen evolution at high current densities, *Nature Communications*, 6 (2015) 1–7.

[109] P. Wang, Y. Lin, L. Wan, B. Wang, Autologous growth of Fe-doped $Ni(OH)_2$ nanosheets with low overpotential for oxygen evolution reaction, *International Journal of Hydrogen Energy*, 45 (2020) 6416–6424.

[110] S.H. Ye, Z.X. Shi, J.X. Feng, Y.X. Tong, G.R. Li, Activating CoOOH porous nanosheet arrays by partial iron substitution for efficient oxygen evolution reaction, *Angewandte Chemie International*, 57 (2018) 2672–2676.

[111] P.B. Perroni, T.V.B. Ferraz, J. Rousseau, C. Canaff, H. Varela, T.W. Napporn, Stainless steel supported $NiCo_2O_4$ active layer for oxygen evolution reaction, *Electrochimica Acta*, 453 (2023) 142295.

[112] J.O. Olowoyo, R.J. Kriek, Recent progress on bimetallic-based spinels as electrocatalysts for the oxygen evolution reaction, *Small*, 18 (2022) e2203125.

[113] P. Zhang, H. He, Rational rope-like $CuCo_2O_4$ nanosheets directly on Ni foam as multifunctional electrodes for supercapacitor and oxygen evolution reaction, *Journal of Alloys and Compounds*, 826 (2020) 153993.

[114] Y. Gong, H. Pan, Z. Xu, Z. Yang, Y. Lin, M. Zhang, ACo_2O_4 (A=Ni, Zn, Mn) nanostructure arrays grown on nickel foam as efficient electrocatalysts for oxygen evolution reaction, *International Journal of Hydrogen Energy*, 43 (2018) 14360–14368.

[115] Y. Zhao, C. Chang, F. Teng, Y. Zhao, G. Chen, R. Shi, G.I.N. Waterhouse, W. Huang, T. Zhang, Defect-engineered ultrathin δ-MnO_2 nanosheet arrays as bifunctional electrodes for efficient overall water splitting, *Advanced Energy Materials*, 7 (2017) 1700005.

[116] T. Zhang, F. Cheng, J. Du, Y. Hu, J. Chen, Efficiently enhancing oxygen reduction electrocatalytic activity of MnO_2 using facile hydrogenation, *Advanced Energy Materials*, 5 (2015) 1400654.

[117] M.W. Kanan, Y. Surendranath, D.G. Nocera, Cobalt-phosphate oxygen-evolving compound, *Chemical Society Reviews*, 38 (2009) 109–114.

[118] L. Yang, Z. Guo, J. Huang, Y. Xi, R. Gao, G. Su, W. Wang, L. Cao, B. Dong, Vertical growth of 2D amorphous FePO(4) nanosheet on ni foam: Outer and inner structural design for superior water splitting, *Advanced Materials*, 29 (2017) 1704574.

[119] H. Zhou, F. Yu, J. Sun, R. He, S. Chen, C.W. Chu, Z. Ren, Highly active catalyst derived from a 3D foam of Fe(PO(3))(2)/Ni(2)P for extremely efficient water oxidation, *Proceedings of the National Academy of Sciences of the United States of America*, 114 (2017) 5607–5611.

[120] Y. Li, C. Zhao, Iron-doped nickel phosphate as synergistic electrocatalyst for water oxidation, *Chemistry of Materials*, 28 (2016) 5659–5666.

[121] Y. Li, X. Wu, J. Wang, H. Wei, S. Zhang, S. Zhu, Z. Li, S. Wu, H. Jiang, Y. Liang, Sandwich structured Ni_3S_2-MoS_2-Ni_3S_2@Ni foam electrode as a stable bifunctional electrocatalyst for highly sustained overall seawater splitting, *Electrochimica Acta*, 390 (2021) 138833.

[122] X. Shang, K.-L. Yan, S.-S. Lu, B. Dong, W.-K. Gao, J.-Q. Chi, Z.-Z. Liu, Y.-M. Chai, C.-G. Liu, Controlling electrodeposited ultrathin amorphous Fe hydroxides film on V-doped nickel sulfide nanowires as efficient electrocatalyst for water oxidation, *Journal of Power Sources*, 363 (2017) 44–53.

[123] S.-S. Xu, X.-W. Lv, Y.-M. Zhao, T.-Z. Ren, Z.-Y. Yuan, Engineering morphologies of cobalt oxide/phosphate-carbon nanohybrids for high-efficiency electrochemical water oxidation and reduction, *Journal of Energy Chemistry*, 52 (2021) 139–146.

[124] S. Li, J. Sun, J. Guan, Strategies to improve electrocatalytic and photocatalytic performance of two-dimensional materials for hydrogen evolution reaction, *Chinese Journal of Catalysis*, 42 (2021) 511–556.

[125] Z. Xia, H. Sun, X. He, Z. Sun, C. Lu, J. Li, Y. Peng, S. Dou, J. Sun, Z. Liu, In situ construction of $CoSe_2$@vertical-oriented graphene arrays as self-supporting electrodes for sodium-ion capacitors and electrocatalytic oxygen evolution, *Nano Energy*, 60 (2019) 385–393.

[126] X. Li, C. Liu, Z. Fang, L. Xu, C. Lu, W. Hou, Ultrafast room-temperature synthesis of self-supported NiFe-Layered double hydroxide as large-current-density oxygen evolution electrocatalyst, *Small*, 18 (2022) e2104354.

[127] O. Kwon, Y. Choi, E. Choi, M. Kim, Y.C. Woo, D.W. Kim, Fabrication techniques for graphene oxide-based molecular separation membranes: Towards industrial application, *Nanomaterials (Basel)*, 11 (2021) 757.

[128] P. Kong, L. Zhu, F. Li, G. Xu, Self-supporting electrode composed of snse nanosheets, thermally treated protein, and reduced graphene oxide with enhanced pseudocapacitance for advanced sodium-ion batteries, *ChemElectroChem*, 6 (2019) 5642–5650.

[129] E. Topçu, K.D. Kıranşan, Flexible and free-standing $PtNLs$-MoS_2/reduced graphene oxide composite paper: A high-performance rolled paper catalyst for hydrogen evolution reaction, *ChemistrySelect*, 3 (2018) 5941–5949.

[130] K. Qin, L. Wang, S. Wen, L. Diao, P. Liu, J. Li, L. Ma, C. Shi, C. Zhong, W. Hu, E. Liu, N. Zhao, Designed synthesis of NiCo-LDH and derived sulfide on heteroatom-doped edge-enriched 3D rivet graphene films for high-performance asymmetric supercapacitor and efficient OER, *Journal of Materials Chemistry A*, 6 (2018) 8109–8119.

[131] P. Guo, J. Wu, X.-B. Li, J. Luo, W.-M. Lau, H. Liu, X.-L. Sun, L.-M. Liu, A highly stable bifunctional catalyst based on 3D Co(OH)(2)@NCNTs@NF towards overall water-splitting, *Nano Energy*, 47 (2018) 96–104.

[132] J. Zhou, Z. Wang, D. Yang, W. Zhang, Y. Chen, Free-standing S, N co-doped graphene/ Ni foam as highly efficient and stable electrocatalyst for oxygen evolution reaction, *Electrochimica Acta*, 317 (2019) 408–415.

[133] Y. Zhai, X. Ren, J. Yan, S. Liu, High density and unit activity integrated in amorphous catalysts for electrochemical water splitting, *Small Structures*, 2 (2020) 2000096.

[134] Y. Liu, G. Sun, X. Cai, F. Yang, C. Ma, M. Xue, X. Tao, Nanostructured strategies towards boosting organic lithium-ion batteries, *Journal of Energy Chemistry*, 54 (2021) 179–193.

[135] C.X. Guo, C.M. Li, Room temperature-formed iron-doped nickel hydroxide on nickel foam as a 3D electrode for low polarized and high-current-density oxygen evolution, *Chem Communication (Cambridge)*, 54 (2018) 3262–3265.

4 The Effect of Electrolytic Gas Bubbles on the Electrode Process of Water Electrolysis

*Meiling Dou, Yifan Wang, Yufeng Qin,
Longxiang Wang, Jiahao Wang,
Yixuan Huang, and Qingqing Ye*

4.1 INTRODUCTION

To address the issues of increased demand for energy usage worldwide and the need to cut down the carbon emissions, the development of renewable energy with the fastest growing rate, such as solar, tide, and hydro energy, to substitute fossil energy is critical [1,2]. In 2020, the total share of renewable energy power generation globally is reported to be 29% [3,4]. By 2019, Chinese cumulative wind power installed capacity and generated electricity had grown to 209,150 MW and 406,030 GWh, respectively [5]. However, most of these energy sources are unstable with intermittent electricity production, setting obstacles to their integration into the electricity grid system [1]. Owing to its properties of high energy density and carbon-free-emission, hydrogen has been esteemed as a highly efficient energy carrier that can be greenly produced using electrical energy from renewable sources through water electrolysis technologies (*i.e.*, alkaline water electrolysis (AWE), proton exchange membrane water electrolysis (PEMWE), and solid oxide water electrolysis (SOEC)) [6–11].

Among these water electrolysis techniques, AWE is regarded as a potential candidate for the green production of hydrogen due to its relatively mature technology with simple operation, high energy efficiency (60%–80%), large-scale production capacity, and relatively low-cost electrode and membrane materials, different from the PEMWE that is only commercial for small scale [12,13,18] (Table 4.1). In general, the major AWE components consist of the diaphragm, gas diffusion layer, bipolar plate, and end plate. The separator materials, such as polyphenylene sulfide (PPS) fabric and Zirfon-type separator consisting of ZrO_2 nanoparticles, are generally employed as the diaphragms in AWE. The nickel foam is usually employed as the gas diffusion layer, and stainless steel-based materials are generally utilized as the bipolar plate and end plate [14]. AWE generally works at a low temperature of 50–80 °C with approximately 20%–30% KOH (or NaOH) solution as the electrolyte [9,15]. The

DOI: 10.1201/9781003368939-4

TABLE 4.1
Parameter Comparison for AWE and PEMWE

Parameters	AWE	PEMWE
Diaphragm	Zirfon-type separator	Proton electrolyte membrane
Cathode	Nickel-based catalysts	Pt/C
Anode	Nickel-based catalysts	RuO_x or IrO_x
Current collector plate	Nickel	Titanium
Bipolar plate	Stainless steel	Titanium
Electrolyte	20%–30% KOH/NaOH	Pure water
Current density	<0.5 A/cm²	1–2 A/cm²
Gas purity	>99.5%	>99.99%
Lifetime	~100 kh	~50–100 kh

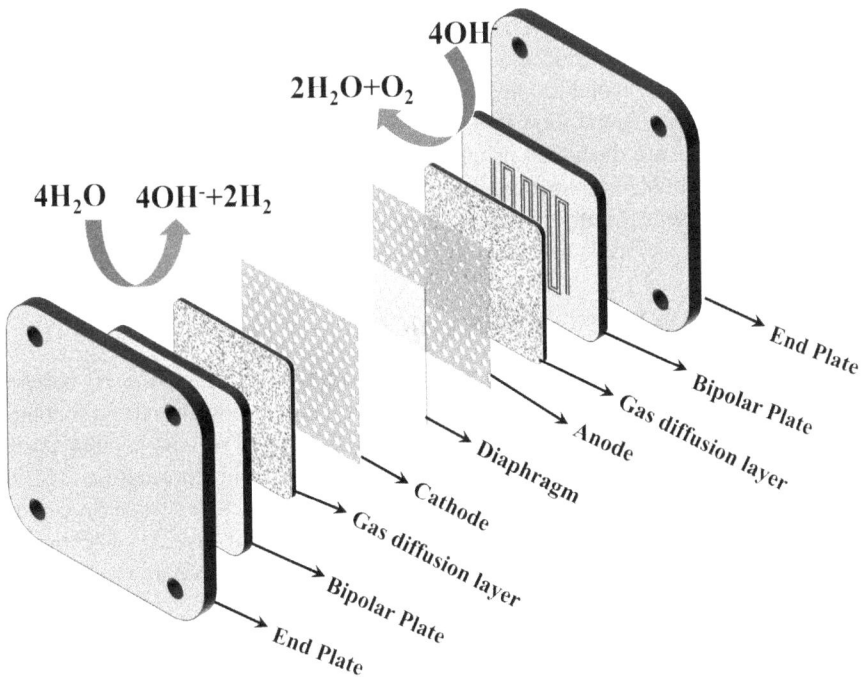

FIGURE 4.1 Structure illustration of AWE.

simple composition and functioning mechanism of the AWE system is illustrated in Figure 4.1 [16].

In AWE, the cathode hydrogen evolution reaction (HER) and the anode oxygen evolution reaction (OER) are the key electrochemical reactions. In the cathode, H_2O molecules are reduced by catalyst to form H_2 and negatively charged hydroxyl ion

(OH⁻) involving four-electron transfer process ($4H_2O + 4e^- \rightarrow 2H_2 + 4OH^-$), while in the anode, OH⁻ are oxidized by the catalyst to generate O_2 and H_2O and release four electrons ($4OH^- \rightarrow O_2 + 2H_2O + 4e^-$) [17,18]. The current research mainly sheds light on the development of electrode materials involving efficient electrocatalysts for HER and OER catalysis offering low overpotential, especially for OER catalysis with sluggish kinetics [19]. As the typical gas evolution reactions, there exists considerable effect induced by the formation of bubbles adhering to the electrode surface during the water electrolysis process [20,21]. However, the impacts of electrolytic gas bubbles on the electrode process and cell performance during practical operations of AWE are usually ignored and are currently understudied. The gas bubbles are reported to promote the local convection inside AWE to some extent but mainly lead to a negative effect on the electrochemical reactions involving the simultaneously occurring evolution and transport processes of bubbles in the practical AWE cell. The presence of bubbles inside an AWE cell probably leads to a series of problems, such as high coverage and accumulation of bubbles on the catalytic sites, as well as slow detachment of bubbles from the catalytic sites. The slow detachment of bubbles can significantly affect the performance of AWE because of the increased activation, ohmic, and concentration overpotentials, resulting in a decreased electrolysis performance and reduced durability.

In this chapter, the evolution mechanism of bubbles in water electrolysis systems and the consequent polarization loss including activation, ohmic, and concentration overpotentials are discussed primally. Moreover, the latest advances regarding the promotion of bubble removal, as well as the challenges and prospects of bubble on AWE, are presented with the aim to provide a valuable guide for the design of high-performance electrode materials for AWE application.

4.2 BUBBLE EVOLUTION DYNAMICS

For the AWE device involving H_2 and O_2 gas evolution reactions, the H_2 bubbles generated in the cathode should be taken into consideration as well as the O_2 bubbles generated in the anode of AWE, which is different from PEMWE in which H_2 gas is able to flow straight out the cell bypassing the process of bubble formation [22,23]. The evolution process of gas bubbles for AWE generally includes three processes: nucleation, growth, and detachment (Figures 4.2 and 4.3a) [24,25]. The bubble dynamic and factors on evolution processes are introduced in this part.

4.2.1 NUCLEATION

The bubble nucleation is related to the increased amount of the dissolved gas adjacent to the electrode surface during water electrolysis, where the gas molecules are continuously generated due to electrochemical reactions. There exists a saturation concentration (C_{sat}) of the dissolved gas in the electrolyte near the electrode surface, which is positively related to the partial pressure P of the gas acting on a liquid surface according to Henry's law (equation 4.1):

$$C_{sat} = PK_H(T) \qquad (4.1)$$

(I) Nucleation   (II) Growth   (III) Detachment

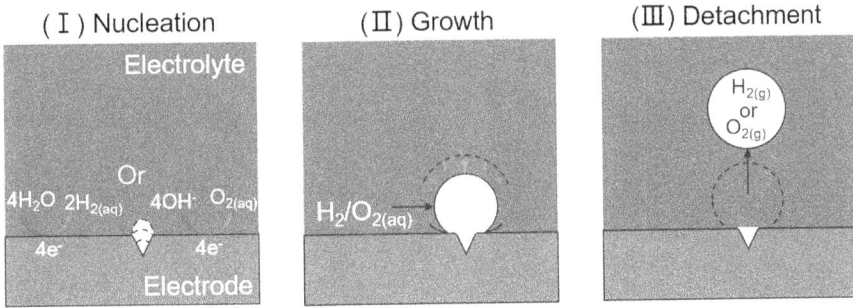

FIGURE 4.2 Various stages of bubble evolution: (I) the nucleation, (II) growth, and (III) detachment of bubbles on electrode surface. Nucleation takes place typically on cracks and crevices in the electrode surface, after which the bubble grows by taking in gas from the dissolved gas boundary layer.

where K_H represents Henry's solubility constant, which is a characteristic of each liquid–gas pair and is also related to temperature (T) with a decreasing function [29–31]. Once the actual dissolved gas concentration (C_0) adhering to the electrode surface exceeds C_{sat}, the gas in the liquid phase is supersaturated. The supersaturation state at pressure P can be expressed by the supersaturation ratio (ζ) to describe the excess amount of dissolved gas (equation 4.2):

$$\zeta = \frac{C_0 - C_{sat}}{C_{sat}} = \frac{C_0}{K_H P} - 1 \tag{4.2}$$

When ζ is larger than 0, the nucleation of bubbles is initiated on the electrode surface [32]. The bubble nucleation in AWE typically occurs at the heterogeneous interface, like the electrode surface with defects (e.g., cracks and splits) and the impure electrolyte, and the bubble formation depends on both the gas–liquid interaction and the solid–gas interaction [26,33,34]. If the value of ζ is lower than 0, bubbles tend to shrink in the unsaturated liquid near the electrode surface.

4.2.2 GROWTH

The bubble growth is defined as the transfer process of dissolved gas from the liquid phase to the gas bubble phase on the interface. When C_0 exceeds the dissolved gas concentration near the interface of a bubble (C_b) [35], with the continuance of electrocatalytic reaction of HER and OER, the bubbles will grow and their growing rate depends greatly on the supersaturation level (i.e., ζ) [23], which can be affected by the geometry feature and wettability of catalyst layer at the same current density [24,36].

For the growth of a bubble, there are three different growth stages that are governed by different forces (Figure 4.3b), including the first stage: nuclei (lasts about 10 ms, controlled by inertia exerted by the liquid around the bubble), the second stage: under critical growth (controlled by the mass transfer of dissolved gases to the bubble), and the third stage: critical growth (controlled by the electrochemical

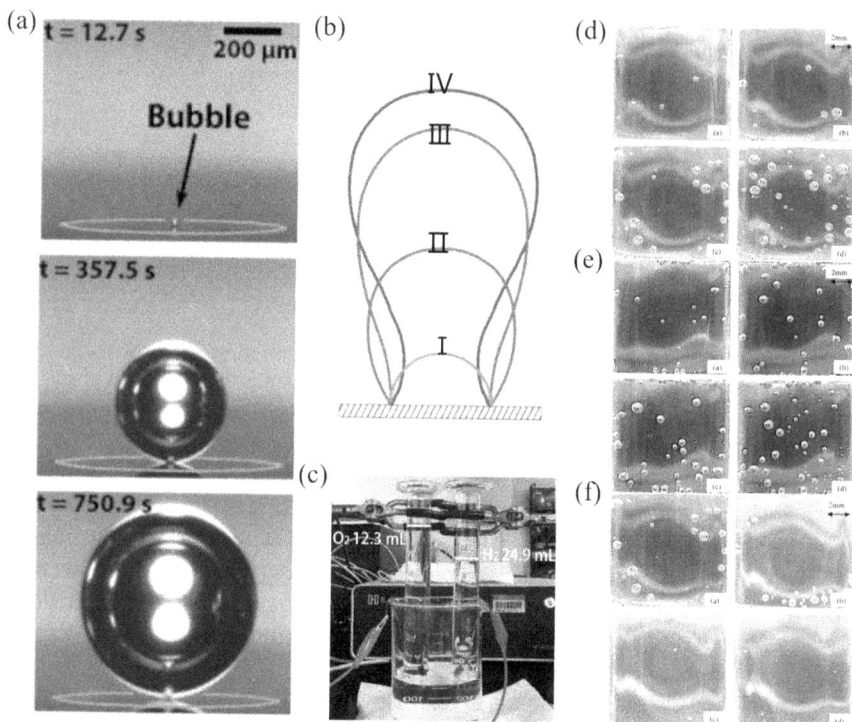

FIGURE 4.3 (a) A H_2 bubble nucleates and grows from the mid of a ring electrode [25]. Copyright 2019, IOP Publishing. (b) A sketch of the stages of bubble growth before detachment from a substrate: stage I: nuclei, stage II: under critical growth, stage III: critical growth, stage IV: necking [26]. Copyright 2018, *Journal of the Electrochemical Society*. (c) Water electrolysis produces twice as much hydrogen as oxygen [27]. Copyright 2020, Royal Society of Chemistry. (d) The critical diameter for H_2 bubble departure increased with increasing current density: (d-a) 0.3; (d-b) 0.45; (d-c) 0.6; and (d-d) 0.75 mA/cm^2 in 0.5 M KOH, at $22 \pm 1°C$ [28]. Copyright 2012, *Industrial & Engineering Chemistry Research*. (e) The critical diameter for O_2 bubble departure increased with increasing current density: (e-a) 0.3; (e-b) 0.45; (e-c) 0.6; and (e-d) 0.75 in 0.5 M KOH, at $22 \pm 1°C$ [28]. Copyright 2012, *Industrial & Engineering Chemistry Research*. (f) The critical diameter for H_2 bubble departure decreases with increasing KOH concentration: (f-a) 0.5 M; (f-b) 1 M; (f-c) 2 M; and (f-d) 4 M at 0.6 mA/cm^2, at $22 \pm 1°C$ [28]. Copyright 2012, *Industrial & Engineering Chemistry Research*.

reaction rate) [22]. The growth of bubbles could affect the performance of AWE, which is due to the coverage effect induced by bubbles with a slower detachment rate, thus covering up active sites for electrochemical reactions [37].

4.2.3 DETACHMENT

Bubble detachment means that bubbles break away from the electrode surface. When the buoyant force (upward force) is larger than that of adhesion force (related with the hydrophilicity and gas–liquid interfacial tension), the bubbles will depart from the electrode surface. The radius of bubble detachment (r_d) can be calculated according to Fitz's formula (equation 4.3) [38,39]:

$$r_d = \left(\frac{3\gamma r_c}{2\rho g} \right)^{\frac{1}{3}} \tag{4.3}$$

where r_c is the radius of contact area between a bubble and solid surface, γ is the surface tension, g is the gravitational constant, and ρ is the liquid density.

The maximum theoretical bubble detachment radius and the detachment rate can be affected by the physicochemical properties of the electrode, such as the micromorphology, hydrophilicity, and the electrostatic interactions between charged bubbles and electrode surface [19]. In AWE, the size of the O_2 bubble is reported to be larger than that of the H_2 bubble, whereas in PEMWE, there is little distinction in the size of H_2 and O_2 bubbles, which can be ascribed to different wettability of electrolyte to the electrode [1]. The wettability of the electrode in an acidic electrolyte is lower compared with that in an alkaline electrolyte. Moreover, the generated H_2 is more than O_2 during water electrolysis, with the H_2/O_2 volume ratio of 2/1 at the same electrolysis current density (Figure 4.3c), which leads to a fast growth rate and detachment rate of H_2 bubbles [40,41].

The critical radius for bubble departure is also reported to be related to the working condition of AWE (e.g., current density and KOH concentration). In Dongke Zhang's work (Figure 4.3d and e), they found that the critical radius for gas bubble departure increases with the increase of current density in 0.5 M KOH, showing the increase from 0.59 to 1.09 mm for H_2 bubbles for current density increase from 0.3 to 0.60 mA/cm^2 and from 0.60 to 1.08 mm for O_2 bubbles when the current density endures an increase from 0.3 to 0.60 mA/cm^2, respectively [28]. The change of bubble size with the rise of current density can be ascribed to the change of interfacial tension force of bubbles. There is an obvious increase in the number of H_2 bubbles at a higher current density over that at a lower current density. Higher current density can result in a larger cell voltage, which can lead to an interfacial tension force growth in the x-coordinate direction. To overcome the interfacial tension force, the bubble buoyancy force needs to increase to a certain extent by enlarging the bubble radius. Furthermore, the critical radius for electrolytic gas bubble detachment has a dependence on the concentration of electrolyte, showing a decrease from 0.59 to 0.27 mm while KOH concentration increases from 0.5 to 4 M, respectively (Figure 4.3f).

4.3 BUBBLES IMPACT ON ELECTROCHEMICAL PROCESSES

According to the free Gibbs energy of water electrolysis, the theoretically reversible voltage is determined to be 1.23 V under typical environment. However, generally, the working voltage for AWE is high because there exists the voltage loss (ΔE) primarily caused by the kinetic energy and mass transfer loss and ohmic drop. ΔE can be determined by the difference between the equilibrium voltage (E_0) of the electrochemical reactions and the operating voltage (E) at a given current density. The voltage loss includes the activation, ohmic, and concentration losses (Figure 4.4), which

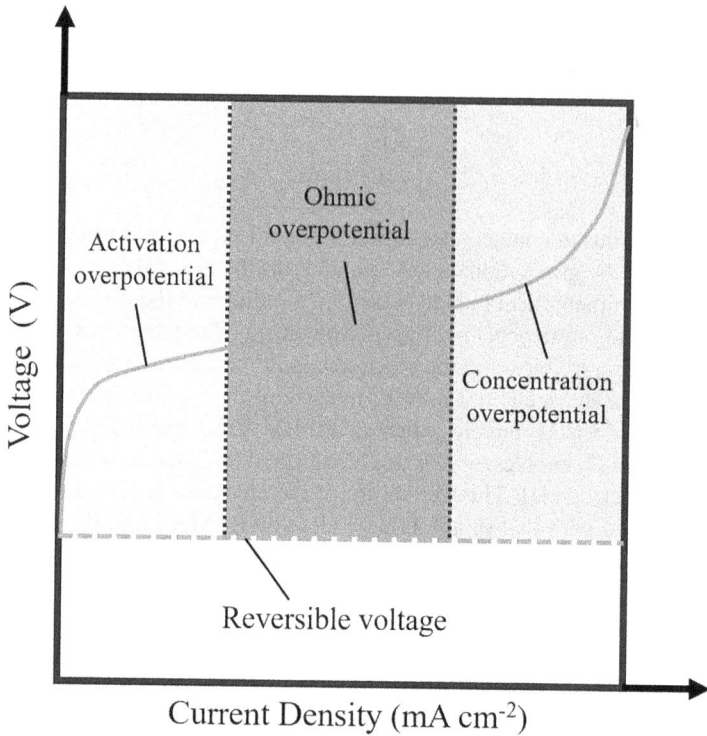

FIGURE 4.4 Three polarization losses observed from the polarization curve of water electrolysis.

are related to the electrochemical reactions and transport process (e.g., diffusion, convection, and migration), expressed as follows (equation 4.4) [42]:

$$E = \eta_{act} + \eta_{ohm} + \eta_{conc} \qquad (4.4)$$

where η_{act} represents the activation overpotential that includes the anodic (η_a) and cathodic (η_c) reaction overpotentials, η_{ohm} represents the ohmic overpotential that is in connection with the ions transport in the electrolyte, and η_{conc} represents the concentration overpotential. Since the water electrolysis involves the gas-evolving electrode, the bubbles will be generated on the electrode surface. The bubble generation and attachment on the electrode will affect the polarization loss during the electrolysis process, which in turn affects the total overpotential losses and thus decreases the efficiency of water electrolysis and even its lifetime, as will be described in detail in the following subsections.

4.3.1 BUBBLE EFFECT ON ACTIVATION OVERPOTENTIAL

The activation overpotential reflects the kinetic loss of electrochemical reactions that take place at the electrode surface during the water electrolysis process that corelate to the activation energy of these reactions. The presence of bubbles adhering to the electrode surface will result in increasing the activation overpotential at a given current density, which is due to the partial coverage by bubble on solid–liquid interface between electrode and electrolyte through blocking part of the catalytic active sites for each of the electrochemical reactions and thus decreasing the electrocatalytic effective active area. Therefore, the coverage of the bubble is generally determined by the proportion of the total area which is covered up. The activation overpotential can be expressed according to the Butler–Volmer equation [42]:

$$\eta_{act} = -\frac{R_g T}{\alpha Z F} \ln(1-\sigma) \tag{4.5}$$

where σ is the percentage of bubble coverage on the electrode surface, R_g is the universal gas constant, T is the temperature, α is the transfer coefficient, Z is the stoichiometric number, and F is the Faraday constant. From this equation, the activation overpotential is mainly determined by the bubble coverage.

In Vogt et al.'s work, they investigated the influence of bubble coverage on current density and found that an increase in bubble coverage is proportional to the current density increase shown with an empirical equation [43,44]. The relation between bubble coverage and current density will provide a guide for the determination of the electrocatalytic active area and thus predict the increased overpotential induced by bubbles. They also found that the percentage of bubble coverage is affected by the velocity of the electrolyte flow in 1 M KOH solution, showing that the bubble coverage decreased with the increase of velocity.

4.3.2 Bubbles Effects on Ohmic Overpotential

The presence of bubbles also leads to the increase in ion conducting resistance, resulting in the rise of ohmic overpotential. The ohmic overpotential generally originates from two aspects: the ion conducting resistance (relates to the ion component flow through the electrolyte) and the electronic resistance (represents the electronic component flow through the external circuit) [45,46]. In AWE electrochemical system, the ohmic overpotential is mainly contributed by the ion conducting resistance, because the electronic resistance accounts for a very low percentage of overall ohmic loss. Since the liquid was used as the electrolyte in AWE, the effect of bubbles in the electrolyte on the ion transport is more significant than that of PEMWE using solid as electrolyte (e.g., perfluorosulfonic acid ionomer) that can eliminate the ion transport blockage effect. In the liquid electrolyte of AWE, both bubbles attached on the electrode and bubbles flow in the electrolyte can hinder the ion transport by reducing the number of available pathways for ions to migrate, thus lowering the effective conductivity of the electrolyte and resulting in an increase in ohmic loss [24].

In addition, bubbles adhering to the electrode surface can also lead to the uneven distribution of current near bubbles, which is indicated in Tobias et al.'s work [47]. The shorter distance between bubbles is more significant of this effect. When bubbles contact the electrode surface at an acute angle, the current density drops to zero, while influences little as bubbles far from the electrode.

Additionally, the ohmic overpotential was found to change with the dynamical processes involving bubbles nucleation, growth, and separation from the electrode surface, all of which have an effect on total resistance of the electrolyte.

4.3.3 Bubble Effects on Concentration Overpotential

The concentration overpotential depends on the concentration gradients of reactants, intermediates, and products. At high current density, the concentration polarization becomes significant, in which the mass transport process during electrochemical reactions becomes the limiting step [48]:

$$\eta_{mass,\,anode} = \frac{RT_{anode}}{nF} \ln \frac{C_{O_2}}{C_{O_2,0}} \tag{4.6}$$

$$\eta_{mass,\,anode} = \frac{RT_{cathode}}{nF} \ln \frac{C_{H_2}}{C_{H_2,0}} \tag{4.7}$$

As mentioned above, the evolution of bubbles is initiated by the gas supersaturation adjacent to the electrode surface. In this premise, the generated bubbles could adsorb the dissolved gas and thus decrease the gas supersaturation levels in the electrolyte (decrease C_{O_2} or C_{H_2} in equations 4.6 or 4.7), thereby facilitating the diffusion of electrolyte and the reduction of concentration overpotential [23].

Apart from the decrease of concentration overpotential caused by bubble growth which has been mentioned above, their growth can also have a declining effect on the concentration overpotential by inducing convective flow effect, which can decrease concentration gradients and thus increase the efficiency of AWE. This particular promotion effect is more apparent when bubble detachment begins, thus inducing turbulence.

The effect of reduced mass transport loss is generally produced by many bubbles, while few bubbles will cause an increase in mass transport loss. For the presence of many bubbles observed on the electrode at high current density, bubble evolution becomes faster to eliminate the excess supersaturation and reduce concentration overpotential. When the dissolved gas in the electrolyte is not released by enough bubble detachment frequency, which means few bubbles are generated, the mass transport loss will increase due to the high supersaturation [49].

4.3.4 Overpotential Fluctuations by Evolving Bubbles

The attachment of bubbles not only increases the activation loss by masking the electrode surface, but also increases the ohmic loss through changing the ionic conduction pathway. Additionally, the detachment of bubbles is reported to be beneficial for the mass transfer of reactants during electrochemical reaction, thus decreasing the concentration loss. The electrochemical parameters, such as the current density, potential, and resistance, are very sensitive to the change in electrode active area and dissolved gas concentration. Under this condition, the formation and detachment of bubbles in AWE will lead to fluctuations in electrochemical parameters, especially potential and current density. Accordingly, by analyzing the electrochemical behavior in the presence of bubbles, the overpotential change under different bubble evolution, growth, and detachment characteristics can be investigated and compared. The total overpotential fluctuation $\Delta\eta_T(t)$ can be simplified and written as follows [50]:

$$\Delta\eta_T(t) = \Delta\eta_{ohm}(t) + \Delta\eta_{act}(t) + \Delta\eta_{conc}(t) \tag{4.8}$$

where $\Delta\eta_{ohm}(t)$, $\Delta\eta_{act}(t)$, and $\Delta\eta_{conc}(t)$ represent the overpotential fluctuation caused by ohmic, activation, and concentration loss during the gas evolution process, respectively. The fluctuating terms of $\Delta\eta_{ohm}(t)$ and $\Delta\eta_a(t)$ are related to the fluctuations of electrode active surface, while $\Delta\eta_c(t)$ can be viewed as induced by the gas evolution fluctuation. Iwata et al. investigated the relation between wettability, bubble kinetics, and transport overpotential, and showed that the bubble kinetics shift dramatically as the slight decrease in hydrophilicity of the electrode, leading to a significant increase in transport overpotential [40]. Under the higher hydrophilicity condition, dense and fine bubbles will be formed on the electrode, leading to lower overpotential and overpotential fluctuations.

Sahar et al. investigated the relation between the bubble diameter and the overpotential [51] and found that the overpotential is proportional to the square of bubble diameter and changes linearly with time. Furthermore, they also found that the bubble detachment from electrode surface has an effect on the steep potential fluctuation during gas evolution through spectroscopic analysis [52]. Likewise, Luo et al. found out that resistance (ΔR), ohmic current (Δi_R), and total current (Δi) endure fluctuation during the formation and detachment of a H_2 bubble (Figure 4.5) [22,45]. The jump in the $\Delta R-t$ trace represented the growth of a bubble which blocks the electrode. When the bubble finally detached, the screened surface would be recovered, resulting in a decreased electrolyte resistance, which eventually gives rise to the recovered current Δi.

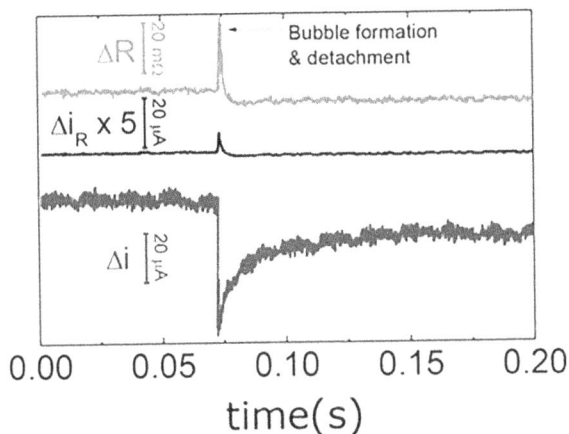

FIGURE 4.5 Fluctuation of total current (Δi) during the formation and detachment of an electrogenerated H_2 bubble [20,45]. Copyright 2019, American Chemical Society.

4.3.5 Mechanical Damage

The presence of bubbles can also affect the catalytic durability of electrode in AWE, because the repeated formation and bubble detachment may cause mechanical damage of catalytic structure and even the shedding of catalyst from the electrode due to the produced stress caused by detachment of accumulated bubbles, especially at high current density [53–55]. Furthermore, the stress of bubbles formed in the pores could tailor the pore structure of the catalytic layer and affect the mass transfer process. Consequently, it is very essential to promote the bubble transport through the catalytic layer with the aim of improving stability [56]. It remains a challenge to have a direct observation of catalyst degradation induced by bubbles in the catalytic layer, and the degradation mechanism caused by bubbles is not fully clarified. Shao-Horn and co-workers investigate the structural changes induced by bubbles of a catalyst in OER by utilizing in situ TEM characterization. The structural oscillations of the catalyst are found, which is due to the generation and rupture of O_2 bubbles inside perovskite oxide particles in the presence of water and e-beam [57,58].

4.3.6 Bubbles-Induced Thermal Losses

The presence of bubbles on the catalytic layer can also result in non-uniform water distribution and localized water scarcity, which leads to the inhomogeneous and unstable distribution of current, voltage, and temperature. Moreover, in the region of local water deficiency, the heat produced by electrochemical reactions cannot be dissipated in time, which leads to localized dryness and "hot spots" on the electrode of AWE.

4.4 STRATEGIES TO MITIGATE BUBBLE PHENOMENA

As mentioned above, the gas bubbles adhering to the electrode surface could mask the electrocatalytic active sites and set an obstacle to the contact between the active sites and the electrolyte, leading to an increase in activation overpotential [59,60]. Furthermore, the bubbles also block the electrolyte diffusion to the active site which results in an increased ohmic overpotential and impeded mass transfer. All these facts consequently lead to a decreased efficiency for AWE and an unsatisfied durability. Under these conditions, developing effective mitigation strategies to reduce the adverse effect of bubble induced is very urgent, especially at high current density. In recent years, a great number of strategies have been explored to inhibit bubble formation or promote bubble removal. The mitigation strategies mainly focus on the material modifications by means of specially designed electrodes to improve bubble management, such as the hydrophilic/hydrophobic and porous structure modification on the electrode and diffusion layer, as well as the surface tension of electrolyte [24,59,61–64]. Although adopting magnetic, electric, and/or sound field treatments embedding in AWE could facilitate bubble transfer by preventing bubble nucleation or induce early separation, the rise in energy consumption suppresses their application in practical condition of AWE and these approaches are generally complex, making them not suitable for practical application of AWE. Therefore, developing highly mass-transferred AWE electrodes to promote bubble detachment from electrode surface is vital and urgent for the practical application of AWE.

4.4.1 INTRODUCTION OF HYDROPHOBIC SITES

Modification of electrode hydrophobicity is reported to be effective in mitigating the adverse effect of bubbles by promoting the nucleation and growth of bubbles. One strategy is to introduce hydrophobic sites on the electrode by generating a hydrophobic location on the electrode surface to control the position for the formation and release of bubbles, in which the nucleation activation energy of bubbles is reduced, thus hindering the coverage of bubbles on the electrocatalytic active sites [61]. Pablo Penas et al. proposed a novel strategy to facilitate gas evolution away from the electrode surface and alleviate the bubble coverage by adopting a ring microelectrode encircling a hydrophobic microcavity where bubbles grow in succession [25]. Results show that the ring microelectrode does not endure the coverage of the bubble under AWE test conditions [65]. The hydrophobic microcavity can induce the nucleation and growth of bubbles during the electrolysis of water because the nucleation energy distribution of bubbles is most favorable there. The bubble formed on the microcavity will reduce the concentration of dissolved gas around it during the growth process, reducing the concentration polarization of the electrode reaction [66]. In addition, the possibility of nucleation of the bubble on the ring working electrode is greatly reduced, so that the reaction site is exposed as much as possible. The presence of bubbles increases the ohmic overpotential, which is alleviated to some extent after the bubbles are removed, as shown in Figure 4.6a, and fluctuates more frequently at high currents [25].

Promoting bubble nucleation by depositing polytetrafluoroethylene (PTFE) islands on the electrode is one of the effective hydrophobic modification strategies. Teschke et al. studied the effect of PTFE partially covered nickel electrodes on the performance of HER catalysis [67]. Figure 4.6b shows that the best modification effect was achieved when PTFE coverage was ~20%, showing the potential lower than that without PTFE modification at high current. The reason is that PTFE draws bubbles from the metallic sites, thus leaving more free active sites for adsorption and reduction of ions in the electrolyte. In 2011, Brussieux et al. adopted PTFE islands with different shapes to modify the Ni and Cu electrodes and used high-resolution photography to show that bubbles do form mainly on the PTFE islands rather than the active sites on the electrode surface [68].

In addition to using PTFE to modify the electrode, other modification methods can also be used, such as hydrophobic modification of the electrode surface that makes the bubble more likely to rupture. Wang et al. used photolithography and wet etching method to construct a series of superhydrophobic "artificial lotus leaves" that mimic the micro/nanolayered structure of lotus leaves and facilitate bubble bursting and separation [69]. Figure 4.6c shows that bubbles burst much faster (13 vs. 220 ms) than bubbles with general microstructure, and the properties of layered structure such as height width and spacing can have an influence on the bubble rupture [70].

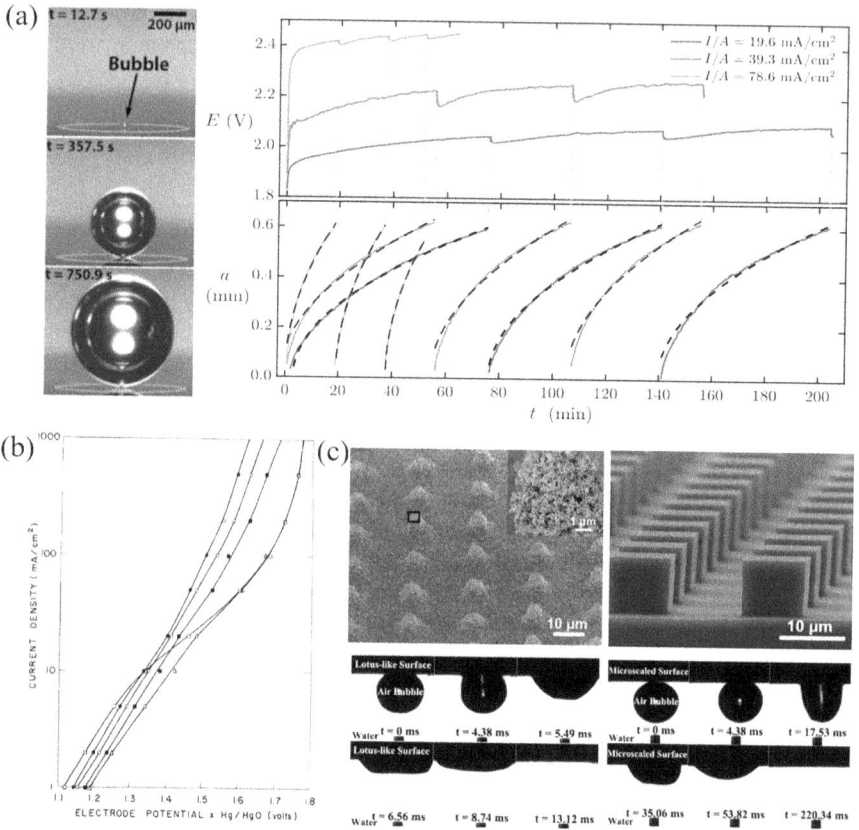

FIGURE 4.6 (a) (Left) Sequence of images of a hydrogen bubble nucleating and growing from the hydrophobic micropit of the SiO_2 substrate. (Right) Cell potential E and bubble radius a of the first three bubbles plotted against elapsed time [25]. Copyright 2019, IOP Publishing. (b) Current vs. voltage measurements using partial covered surface electrodes. □: uncovered; ■: 20% coverage; ○: 40% coverage; ■: 60% coverage [67]. Copyright 1984, IOP Publishing. (c) (left) SEM images of the "artificial lotus leaf" and bubble bursting behavior on surface. (right) SEM images of patterned micropillar array on silicon surfaces and bubble behavior on surfaces [69]. Copyright 2009, American Chemical Society.

The influence of hydrophobic sites on the nucleation and release of bubbles was also proved by Giacomello et al. through molecular dynamics simulation and other calculation methods [71]. Compared with the flat and smooth surface, the superhydrophobic surface increases the nucleation rate of bubbles, and the nucleation rate of bubbles can be further controlled through the careful design of the surface microstructure [72]. In these works, the control over the nucleation sites shows that the prohibition of bubble nucleation near the electrode can be realized, thus minimizing the energy loss induced by bubbles in AWE.

4.4.2 Porous Structure Modification

Designing an AWE electrode with a modified porous structure is one of the effective strategies to mitigate the effect of bubbles on AWE performance by promoting mass transfer capability [73]. By designing gradient porous structures or constructing ordered pore structures to tailor the size and arrangement of pores in the electrode material, the detachment radius of bubbles and the adhesion of bubbles on electrode surface can be effectively reduced, thus promoting the removal of bubbles from the electrode, exposing active sites for electrochemical reactions and the stable operation of electrodes in AWE [74,75]. At present, commercial nickel foam (NF) is one of the suitable diffusion layers owing to its high electrical conductivity and interconnected porous structure that provides diffusion channels for anode OER and cathode HER reactions [76–78]. However, the features of disordered arrangement of NF framework with uneven thickness and density are usually unfavorable for the release of gas bubbles during the continuous electrochemical reaction under AWE working conditions, thus affecting AWE performance. Therefore, constructing electrodes processing pore structure with a designed gradient can promote bubble splitting and minimize bubble residence time inside the electrodes. Yang et al. prepared a well-designed gradient porous NF-based composite substrates with reducing pore size from the middle to two sides of the electrode (SML-LMS) by combining two stack-up gradients porous NF and using a solvothermal treatment to load hierarchically porous coral-shaped MoS_2/Ni_3S_2 heteronanorod electrocatalysts (SML-LMS-HE) [79]. The as-prepared electrode can accelerate the bubble detachment and fully expose catalyst active sites during the process by inducing the splitting of large bubbles. The diameters distribution of hydrogen bubbles shows that bubble detachment diameters on the surface of SML-LMS-HE are generally (80%) smaller than 100 μm, with no bubbles larger than 200 μm, unlike the diameter distribution of LMS-SML-HE possessing gradient porous composite substrate with rising pore size from middle to both sides of the electrode. The average bubble detachment diameter of SML-LMS-HE is 74.5 μm, which is less than LMS-SML-HE (187.7 μm). The mechanism of bubble evolution visualized that the large bubbles have a tendency to get split into bubbles with smaller sizes inside SML-LMS-HE. As a result, the as-prepared electrode with the design of a gradient porous structure offers an ultralow HER overpotential of 83 mV at -10 mA/cm² and can catalyze HER for 18 h.

In addition, by carefully designing the bubble transport path in the electrode, the coalescence of bubbles in the electrode can be reduced and the removal of bubbles can be accelerated, thereby reducing the retention of bubbles on the porous electrode. Recently, 3D printing has been developed as a convenient method to create complicated electrodes with distinct chemical, mechanical, and hollow or gradient pore structures [80–83]. These special structures exhibit unique physical and mechanical properties [84]. The well-designed pore structure in the 3D-printed electrode can optimize the transport path of the bubbles, thereby promoting the further improvement in the performance of the gas evolution electrode. Kou et al. fabricated an ordered periodic porous 3D-printed Ni (3DPNi) through solvent evaporation 3D printing strategy to facilitate bubble transport (Figure 4.7a) [85]. Compared with the randomly scattered irregular pores in the internal space of commercial 3D substrates,

(a)

(b)

(c)

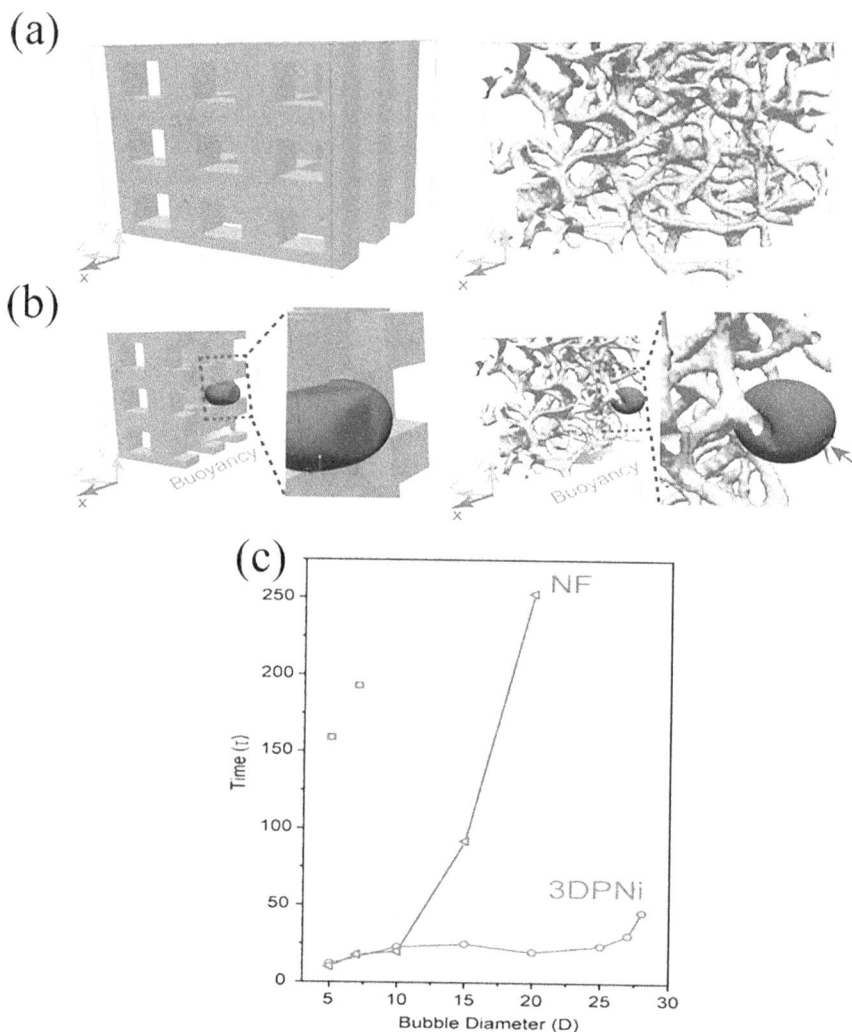

FIGURE 4.7 3DPNi electrode design for solving the bubble trapping. (a) Structure model of 3DPNi and NF. (b) Simulation frames showing bubble shape (d = 20) during transport in 3DPNi and NF. Arrow in the inset highlights an interaction with the NF surface, which is manifested through bubble deformation. (c) Bubble migration time through 3DPNi and NF as a function of bubble diameter [85]. Copyright 2020, *Advanced Energy Materials.*

the ordered micron-scale pores in the 3D-printed periodic structure can effectively reduce the frequency of bubble collision and deformation (Figure 4.7b), thereby achieving rapid bubble release. Besides, the 3DPNi coating the catalyst electrode releases bubbles stably and periodically, with an interruption period of about 124 ms, which is significantly lower than that of commercial disordered NF supporting the catalyst (3131 ms). Figure 4.7c shows the time needed to cross the specific plane of 3DPNi and NF for a bubble with different diameters. Consequently, the release radius for the bubbles in the 3D ordered structure is significantly lower than that of disordered NF, and the catalyst supported on 3DPNi shows a significant advantage over the catalyst supported on ordinary NF in terms of activity. The above changes in bubble behavior once again demonstrate the positive impact of electrode porous structure modification and design on water electrolysis.

4.4.3 SUPERAEROPHOBIC MODIFICATION

The superaerophobic modification is defined as a way to make electrode surface with a high bubble contact angle larger than 150° in water that makes bubbles difficult to attach, resulting in a low adhesion force. In contrast, the superaerophilic surface possesses a low bubble contact angle of ~<10° with a high adhesion force [19]. The superaerophobic modification of the electrode surface is regarded as one of effective strategies to mitigate the adverse effect of bubbles on electrolysis performance by controlling the adhesion behavior of bubbles underwater and promoting bubble dynamics. The superaerophobic properties of the electrode surface can be achieved by adjusting the surface composition and constructing the micro–nano structure of the catalytic layer (e.g., nanoflowers, nanocones, and vertical nanosheets) with a discontinuous three-phase interface (Figure 4.8a), which ensures the rapid separation of bubbles in a small size [74]. In the discontinuous zones of the three-phase interface with a rough surface, the adhesion of bubbles is much lower than that in the continuous zones [59,86].

Bubble adhesion force ($F_{adhesion}$) on the surface can be expressed as follows [87]:

$$-F_{adhesion} = kd\gamma_{lv}\left(\cos\theta_{min} - \cos\theta_{max}\right) \tag{4.9}$$

where γ_{lv} is the surface tension of the liquid, k is the coefficient of solid force, d is the contact width, $\cos\theta_{min} - \cos\theta_{max}$ represents the difference between the cosine of the maximum and minimum static contact angles on the uphill (θ_{max}) and downhill (θ_{min}) upon tilting the substrate at a particular angle, which is called the contact angle hysteresis (CAH) [88].

From this equation, by increasing the aerophobic property of the electrode surface, the contact angle of the bubble can be increased, thus reducing bubble contact width (d) to reduce $F_{adhesion}$ and promote bubble removal from the electrode surface [89]. Wang et al. proposed a method to facilitate bubble escape for water electrolysis by using nonwoven stainless steel fabrics (NWSSFs) as the conductive substrate decorated with flakelike iron nickel-layered double hydroxide (FeNi LDH) nanostructures [70]. Compared with other 3D porous catalytic electrodes, the as-prepared FeNi LDH@NWSSF electrode with flake shape is capable of trapping a continuous water film, resulting in a lower adhesion between the bubble and electrode surface, which is conducive to the fast removal of small bubbles on the electrode. As shown in Figure 4.8b, within 0.025 s, the oxygen bubbles formed can be completely released from the porous structure, and the maximum dragging force released by the bubbles between the NWSSF channels reaches merely 14.29% of that in NF channels. Consequently, it offers overpotentials as low as 210 and 110 mV (@10 mA/cm²) in 1 M KOH for OER and HER, respectively, with a relatively long-term stability

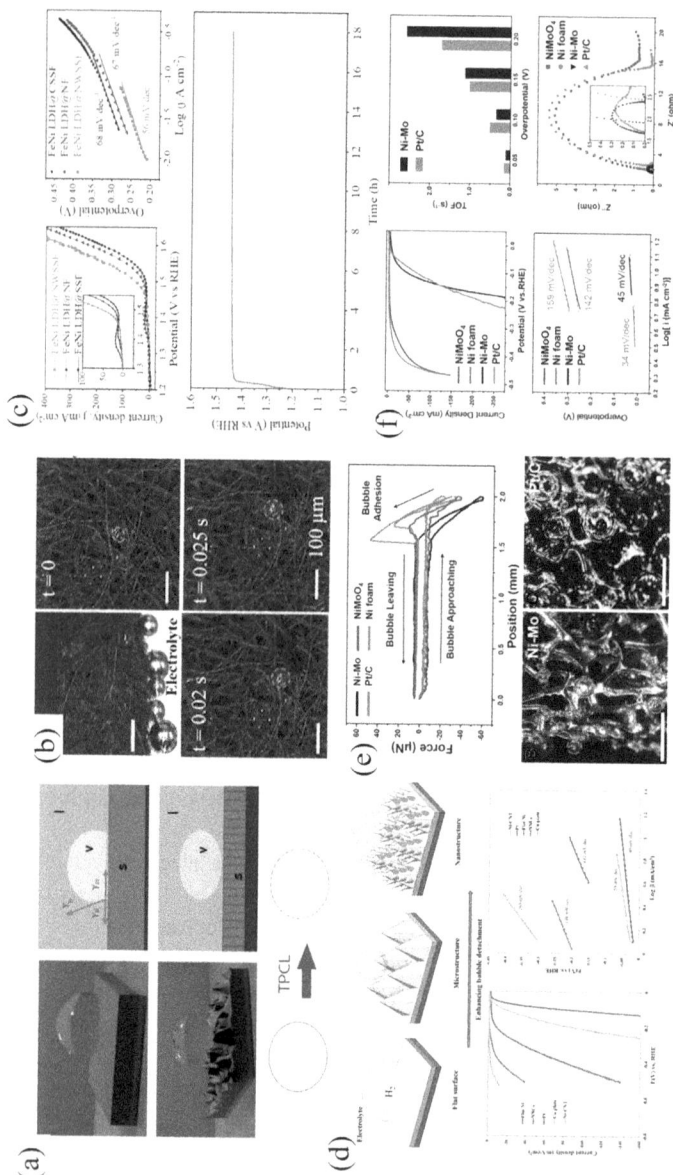

FIGURE 4.8 (a) Schematic representation of the adhesion behavior of air bubbles on smooth and nanostructured films. Smooth film on the left and nano-structured film on the right [59]. Copyright 2022, Journal of Materials Chemistry A. (b) Shapes of gas bubbles at the bottom of FeNi LDH@NWSSF (contact angle is ~169.7, indicating its superaerophobicity) and in situ observations of the oxygen evolution reaction on FeNi LDH@NWSSF. (c) Polarization curves and Tafel plots of FeNi LDH@NWSSF, FeNi LDH@NF, and FeNi LDH@SSF for OER at a scan rate of 5 mV/s. Time dependence of catalytic current density during electrolysis for FeNi LDH@NWSSF at a current density of 10 mA/cm² [70]. Copyright 2017, ACS Applied Materials & Interfaces (d) Schematic of the bubbles detachment behavior of different structures, bubbles images at different intervals during the hydrogen evolution on nanocones structure, and bubbles images at different intervals during the hydrogen evolution on the flat surface and η_{10}, η_{20}, and η_{100} for different electrodes [90]. Copyright 2018, *Journal of Electroanalytical Chemistry*. (e) Adhesive force measurements of gas bubbles on Ni-Mo nanosheets, NiMoO₄ precursor, Pt/C electrode, and Ni foam. Digital images of bubble generation behavior on Ni-Mo nanosheets and Pt/C electrode, scale bar: 500 μm. (f) HER performance of Ni-Mo nanosheets (0.8 mg/cm²), Pt/C (1.6 mg/cm²) powder, NiMoO₄ precursor, and Ni foam in 1 M KOH [91]. Copyright 2017, *Small*.

(Figure 4.8c). Its full water splitting performance is also excellent compared with reported catalysts that offer a voltage of 1.56 V at the current density of 10 mA/cm^2. Barati Darband et al. prepared a new 3D-layered nickel-carbon nanotube (Ni-CNT) nanostructure by electrodeposition by implanting CNT into the Ni nanocones (NNCs) [90]. The Ni-CNT hierarchical nanostructure with a high aerophobic is conducive to bubble separation and effectively reduces the shielding of bubbles to the active site, thus achieving high activity and stability. The diameter of the bubble on the Ni nano-structured surface is ~60 μm, which is smaller than that of the flat surface (300 μm). As a result, the Ni-CNT hierarchical nanostructure accelerates the detachment of bubbles and effectively exposes the electrode area covered by bubbles, affording the HER overpotentials of 82, 116, and 207 mV at the current densities of 10, 20, and 100 mA/cm^2, respectively (Figure 4.8d).

Zhang et al. arranged the assembled two-dimensional Ni-Mo nanosheet structure vertically on a conductive substrate, which helps realize the superaerophobicity of the electrode, thus facilitating bubble release in HER [91]. The facilitated HER process at high current density is related to the boosted mass transfer behavior in comparison with the Pt/C catalyst under the same condition. Comparing NiMoO$_4$ precursor with Pt/C electrode or Ni foam, the results indicate that the bubble adhesion of Ni-Mo nanosheets was the smallest (≈2 μN), while the adhesion of Pt/C electrode prepared by the drip-dry method was almost 15 times that of the Pt/C electrode with a synthesis process of drop drying (≈29 μN). On the Ni-Mo nanosheet electrode, the bubble release rate was faster, with an average bubble diameter less than 95 μm, while on the Pt/C surface, the bubble release size was around 364 μm (Figure 4.8e). The Ni-Mo alloy nanosheets facilitated electron transport and mass transfer, outperforming the state-of-art Pt/C catalyst (Figure 4.8f).

Yang et al. prepared a core–shell structured NiFe nanowire array based OER electrode using Ni$_x$Fe$_{1-x}$ alloy as core and ultrathin amorphous NiFe oxyhydroxide nanowire arrays as shell (denoted as Ni$_{0.8}$Fe$_{0.2}$-AHNA) through a magnetic-field-assisted chemical deposition approach [27]. In the 1 M KOH electrolyte, this electrode shows the overpotentials of only 248 and 258 mV at 500 and 1000 mA/cm^2, respectively, and it can be stable up to 120 h. One of the reasons for excellent performance is that the well-designed structure is able to boost both charge and mass transfer during electrochemical reactions, which also proves the importance of mitigating the effect of bubbles especially under high current density conditions (Figure 4.9). The above conclusion provides us with a favorable approach to strengthen bubble management by designing the electrode, thus avoiding the adverse transfer of the charge and the ion within the electrode and channels, promoting efficient release of the bubbles, and reducing the mechanical damage of the electrode caused by violent bubbles release.

Jong et al. prepared Ni catalysts with controllable surface morphology using the oblique angle deposition (OAD) method [92]. The porosity of Ni catalysts increased with the increase of tilted incidence angle θ, which can significantly boost their aerophobicity, thus tailoring the release behavior of the H$_2$ bubble. When the porosity of the Ni catalyst reaches ~52%, the catalyst has superaerophobicity and exhibits the best HER catalytic activity as well as superior stability. Moreover, the highly porous catalyst that is superaerophobic has the potential to boost the supersaturation of dissolved H$_2$ in the electrolyte, leading to a decreased bubble critical size. Therefore, the

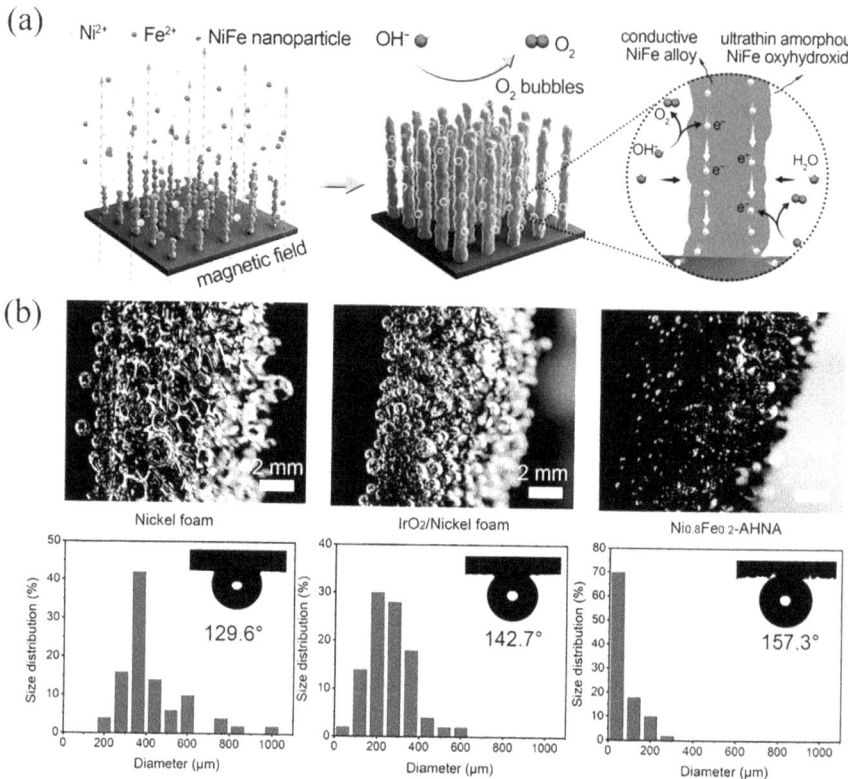

FIGURE 4.9   (a) Schematic diagram of the synthesis of $Ni_{0.8}Fe_{0.2}$-AHNA and its catalytic function for the OER. (b) (Top) Digital photos demonstrating the oxygen bubbles on the surface of nickel foam, IrO_2/nickel foam, and $Ni_{0.8}Fe_{0.2}$-AHNA during the OER process. (Down) The corresponding size distribution statistics of releasing bubbles for fifty bubbles. The insets are the corresponding photos of the bubble/catalyst contact angles under the electrolyte [27]. Copyright 2020, Royal Society of Chemistry.

smaller bubbles that near the surface can expose a more active surface and facilitate the ions transport in the electrolyte, thus boosting the activity of HER. Meanwhile, the mechanical damage of the catalyst induced by larger bubbles can be mitigated, thus enhancing the stability.

Ye et al. synthesized a monolithic 3D hollow foam electrode through a feasible chemical plating-calcination strategy, which can meet the demand of high current density water electrolysis [55]. The prepared electrode is able to endure pressure as high as 2.37 MPa and processes high electrochemical surface area (ECSA) and conductivity as well as the low transfer resistance for gas, all of which favor the catalytic performance boost. Consequently, the electrode offers only 83 and 293 mV at 50 mA/cm^2 for HER and OER, respectively (Figure 4.10). The outstanding aerophobicity of the Ni-Mo-B HF electrode can be observed from the large contact angle of air bubble (158 °). Based on the Cassie–Baxter and Wenzel equations, the wettability of this electrode is in relation to the roughness of micro- and nanosurface structure,

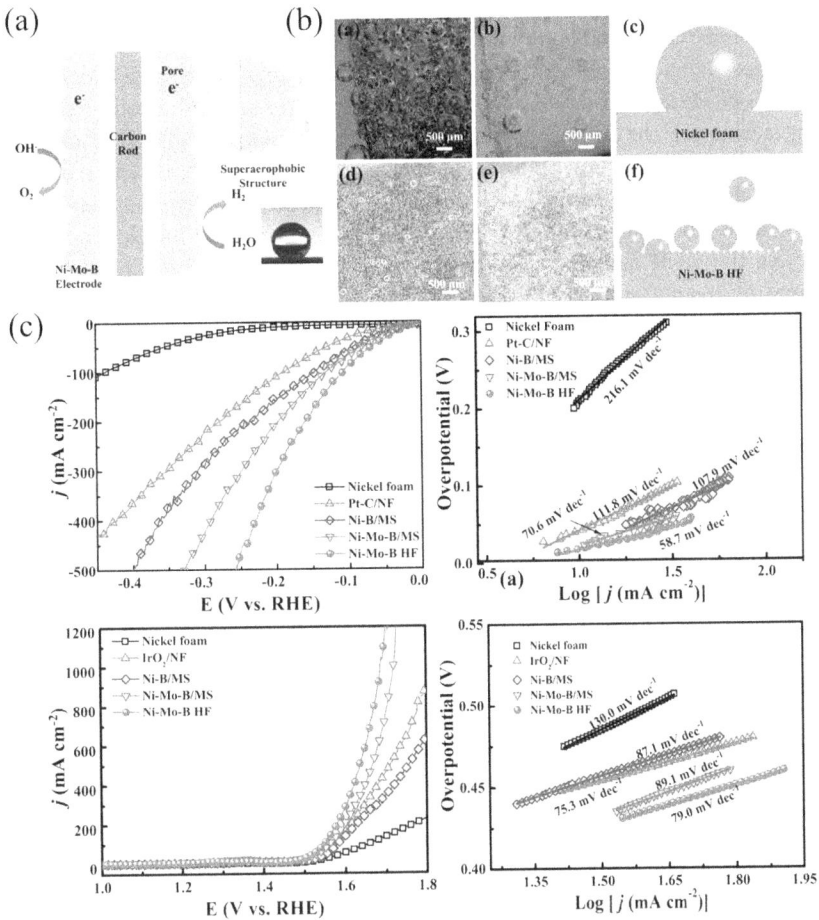

FIGURE 4.10 (a) Schematic diagram of microstructure of Ni-Mo-B HF electrode and its air bubble contact angle in 1 M KOH. (b) Optical photos of H$_2$ bubbles attached to nickel foam and Ni-Mo-B HF at low current density (20 mA/cm²) (b-a, b-d) and large current density (100 mA/cm²) (b-b, b-e). (b-c, b-f) Illustration of bubbles attached to NF and Ni-Mo-B HF electrode. (c) HER and OER performance of Ni-Mo-B HF [55]. Copyright 2021, *Advanced Functional Materials*.

which can bring a reduction of the contact area between the bubbles and electrode [93]. Additionally, the increasing pathway for gas to release can be created by the built-in channel within the foam, thereby facilitating the mass transfer during water electrolysis.

4.4.4 SURFACTANT MODIFICATION

Surfactants, which are amphiphilic molecules that possess both hydrophobic and hydrophilic groups, can be adsorbed on interfaces and self-assemble into different phases in solution [94,95]. Therefore, the introduction of surfactants on the electrode or in electrolyte can also construct a superaerophobic electrode to mitigate bubble effects by changing the surface tension of bubbles, thus influencing their sizes, growth, and detachment behavior. Xie et al. prepared the surfactant modification of a NiFe layered double hydroxide (NiFe-LDH) array electrode for OER by using a cationic (hexadecyl trimethyl ammonium bromide, CTAB) or an anionic (sodium dodecyl sulfate, SDS) surfactant to immerse the NiFe-LDH electrodes followed by infrared baking, showing a surface with superaerophobicity and surface charges to some extent [20]. The surfactants gathered on the electrode surface promote the OER activity by boosting the mass transfer and gas release through decreasing the surface tension of the electrode, showing the lower bubble adhesive force (~1.03 µN for CTAB-modified electrode) and corresponding facilitated small bubbles release during OER. Moreover, the bipolar feature of the CTAB molecule results in bilayer assembly of the surfactants with the polar ends facing the electrode surface and the electrolyte, which leads to charge neutralization on the electrode surface and thus promotes the OH^- transfer during OER catalysis. As a result, the NiFe LDHs-CTAB nanostructured electrode exhibits a high current density increase (9.39 mA/(mV cm²)), which is 2.3 times the number of conventional NiFe-LDH nanoarray electrode. Unfortunately, most surfactants are unstable in alkaline electrolytes due to saponification, which is not beneficial for their practical application.

4.5 SUMMARY AND OUTLOOK

AWE is esteemed as a practical technology for the production of green hydrogen, which has great potential for future large-scale applications driven by renewable energy sources such as solar and wind. In summary, this chapter covers three main aspects of bubbles involved in AWE including the basic evolution dynamics for bubbles, the propounding influences on the performance of AWE, and the practical strategies to resolve the corresponding problems.

First of all, three main processes are introduced: (1) nucleation, (2) growth, and (3) detachment, as well as mathematical formulas and factors that can have an influence on each process are also mentioned, which are affected by the supersaturation of dissolved gases in the electrolyte, transfer process of dissolved gas, and the relationship between bubble adhesion and buoyancy. Second, in terms of the impact of bubbles on the performance of AWE, we summarize six potential aspects of influence on the (1) activation overpotential, (2) ohmic overpotential, and (3) concentration overpotential, as well as bubble evolving induced (4) overpotential fluctuation, (5) mechanical damage, and (6) thermal losses. The increase in activation overpotential is mainly due to the attached bubbles masking the electrodes and decreasing the effective electrocatalytic area. In addition, attached and free bubbles increase the ohmic overpotential due to a blockage of the ion pathways available for current transport. Bubbles may decrease the concentration overpotential by absorbing dissolved gas products and decreasing supersaturation levels in the electrolyte. In addition, in the process of electrode reaction, especially at high current density, the formation and detachment of bubbles in AWE will lead to fluctuations in potential and produced stress caused by the detachment of accumulated bubbles. It should be noted that the local overheating of the electrode at high current density will cause the electrolyte to boil, resulting in a large number of bubbles. For the final mitigation strategy part, we also reviewed different methods to eliminate or reduce the formation and release of bubbles in water electrolysis without introducing additional accessories, including the strategies of (1) introduction of hydrophobic sites, (2) porous structure modification, (3) super-aerophobic modification, and (4) surfactant modification. The summarized strategies here help inspire efficient bubble removal methods in electrochemical systems to enable high-performance AWE. Although a considerable number of literatures have been reviewed, we are still not able to have a comprehensive understanding that would help us make accurate predictions and have control over the bubble impact on electrochemical systems, and there is still much room for improvement of AWE performance with regard to bubble management due to a limited understanding of bubble evolution and transport in AWE. Further development of bubble management in AWE can be focused on, but not limited to, the following areas: (1) in situ characterizations of bubble evolution by a faster camera; (2) simulation of detailed physicochemical processes and its impact on electrochemical performance on computer; and (3) optimization of electrode interface for more bubble nucleation sites, smaller bubble detachment size, and higher bubble detachment frequency.

Looking into the future, the development of advanced characterization technologies is the vital prerequisite that is required for future advancements in bubble management in AWE, owing to the fact that observation and analysis of the bubble behaviors are mainly carried out in three-electrode systems through current technologies. There should be great differences between the experimental results in three-electrode systems and practical results in actual AWE applications, which means the acquired conclusions from experimental data might not be valid in the practical production application. Therefore, specific technologies that can detect bubble behaviors are needed in the future. Moreover, the proposed promising strategies above only involve improvement approaches in terms of developing electrode materials that can mitigate the bubble phenomena, which does not satisfy the current need for the practical application of AWE. However, we believe that with the accumulating theoretical and practical work aimed at understanding the bubble induced phenomena, the development of proper technologies will be facilitated, thus giving rise to more and more feasible mitigation approaches.

REFERENCES

1. Yuan, S., et al., Bubble evolution and transport in PEM water electrolysis: Mechanism, impact, and management. *Progress in Energy and Combustion Science*, 2023. **96**: p. 101075.
2. Chaturvedi, V., Energy security and climate change: Friends with asymmetric benefits. *Nature Energy*, 2016. **1**(6): p. 16075.
3. Grigoriev, S., et al., Current status, research trends, and challenges in water electrolysis science and technology. *International Journal of Hydrogen Energy*, 2020. **45**(49): pp. 26036–26058.
4. IEA, Global Energy Review 2021, 2021: IEA, Paris.
5. Chen, H., et al., Winding down the wind power curtailment in China: What made the difference? *Renewable and Sustainable Energy Reviews*, 2022. 167: p. 112725.

6. Kim, J., et al., Hybrid-solid oxide electrolysis cell: A new strategy for efficient hydrogen production. 2018. **44**: pp. 121–126.

7. Turner, J.A., Sustainable hydrogen production. *Science*, 2004. **305**(5686): pp. 972–974.

8. Aghakhani, A., et al., Direct carbon footprint of hydrogen generation via PEM and alkaline electrolysers using various electrical energy sources and considering cell characteristics. *International Journal of Hydrogen Energy*, 2023. **48**(77): pp. 30170–30190.

9. Brauns, J. and T. Turek, Alkaline water electrolysis powered by renewable energy: A review. *Processes*, 2020. 8(2): p. 248.

10. Hu, K., et al., Comparative study of alkaline water electrolysis, proton exchange membrane water electrolysis and solid oxide electrolysis through multiphysics modeling. *Applied Energy*, 2022. 312.

11. Lee, B., et al., Pathways to a green ammonia future. *ACS Energy Letters*, 2022. **7**(9): pp. 3032–3038.

12. Sebbahi, S., et al., Assessment of the three most developed water electrolysis technologies: Alkaline water electrolysis, proton exchange membrane and solid-oxide electrolysis. *Materials Today: Proceedings*, 2022. 1(66): pp. 140–145.

13. Wang, J., et al., Non-precious-metal catalysts for alkaline water electrolysis: operando characterizations, theoretical calculations, and recent advances. *Chemical Society Reviews*, 2022. 49: pp. 9154–9196.

14. Prabhu Saravanan, M.R.K., C.S. Yee, and N. Dai-Viet, An overview of water electrolysis technologies for the production of hydrogen. *Elsevier*, 2020. **7**: pp. 161–190.

15. Burnat, D., et al., Composite membranes for alkaline electrolysis based on polysulfone and mineral fillers. *Journal of Power Sources*, 2015. **291**: pp. 163–172.

16. Vinodh, R., et al., Recent advancements of polymeric membranes in anion exchange membrane water electrolyzer (AEMWE): A critical review. *Polymers*, 2023. **15**(9): p. 2144.

17. David, M., C. Ocampo-Martínez, and R. Sánchez-Peña, Advances in alkaline water electrolyzers: A review. *Journal of Energy Storage*, 2019. **23**: pp. 392–403.

18. Vincent, I. and D. Bessarabov, Low cost hydrogen production by anion exchange membrane electrolysis: A review. *Renewable and Sustainable Energy Reviews*, 2018. **81**: pp. 1690–1704.

19. Xu, W., et al., Superwetting electrodes for gas-involving electrocatalysis. *Accounts of Chemical Research*, 2018. **51**(7): pp. 1590–1598.

20. Xie, Q., et al., Enhancing oxygen evolution reaction by cationic surfactants. *Nano Research*, 2019. **12**(9): pp. 2302–2306.

21. Kelsall, N.P.B.G.H., Growth kinetics of bubbles electrogenerated at microelectrodes. *Applied Electrochemistry* 1985. **15**: pp. 475–484.

22. Zhao, X., H. Ren, and L. Luo, Gas bubbles in electrochemical gas evolution reactions. *Langmuir*, 2019. **35**(16): pp. 5392–5408.

23. Lubetkin, S.D., The fundamentals of bubble evolution. *Chemical Society Reviews*, 1995. **24**: pp. 243–250.

24. Angulo, A., et al., Influence of bubbles on the energy conversion efficiency of electrochemical reactors. *Joule*, 2020. **4**(3): pp. 555–579.

25. Peñas, P., et al., Decoupling gas evolution from water-splitting electrodes. *Journal of the Electrochemical Society*, 2019. **166**(15): pp. H769–H776.

26. Taqieddin, A., M.R. Allshouse, and A.N. Alshawabkeh, Editors' choice-critical review-mathematical formulations of electrochemically gas-evolving systems. *Journal of the Electrochemical Society*, 2018. **165**(13): pp. E694–E711.

27. Liang, C., et al., Exceptional performance of hierarchical Ni-Fe oxyhydroxide@NiFe alloy nanowire array electrocatalysts for large current density water splitting. *Energy & Environmental Science*, 2020. **13**(1): pp. 86–95.

28. Zhang, D. and K. Zeng, Evaluating the behavior of electrolytic gas bubbles and their effect on the cell voltage in alkaline water electrolysis. *Industrial & Engineering Chemistry Research*, 2012. **51**(42): pp. 13825–13832.

29. Volanschi, A., W. Olthusis, and P. Bergveld, Gas bubbles electrolytically generated at microcavity electrodes (MCE) used for the measrement of the dynamic surface tension in liquids. *Sensors and Actuators*, 1996. **52**: pp. 18–22.

30. Soto, Á.M., et al., The nucleation rate of single O_2 nanobubbles at Pt nanoelectrodes. *Langmuir*, 2018. **34**(25): pp. 7309–7318.

31. Kolasinski, K.W., Bubbles: A review of their relationship to the formation of thin films and porous materials. *Open Material Sciences*, 2014. **1**(1): pp. 49–60.

32. Enríquez, O.R., et al., Growing bubbles in a slightly supersaturated liquid solution. *Review of Scientific Instruments*, 2013. **84**(6): p. 065111.

33. Maris, H.J., Introduction to the physics of nucleation. *Comptes Rendus Physique*, 2006. **7**(9–10): pp. 946–958.

34. Liu, Y. and S.J. Dillon, In situ observation of electrolytic H_2 evolution adjacent to gold cathodes. *Chemical Communications*, 2014. **50**(14): pp. 1761–1763.

35. Shan, J., et al., Regulating electrocatalysts via surface and interface engineering for acidic water electrooxidation. *ACS Energy Letters*, 2019. **4**(11): pp. 2719–2730.

36. Maier, M., et al., Mass transport in PEM water electrolysers: a review. *International Journal of Hydrogen Energy*, 2022. **47**(1): pp. 30–56.

37. Lee, J.K. and A. Bazylak, Bubbles: The good, the bad, and the ugly. *Joule*, 2021. **5**(1): pp. 19–21.

38. Kumar, R. and N.K. Kuloor, The formation of bubbles and drops. *Advances in Chemical Engineering*. 1970. **8**: pp. 255–368.

39. H. Oğuz and A. Prosperetti, Dynamics of bubble growth and detachment from a needle. *Journal of Fluid Mechanics*, 1993. **257**: pp. 111–145.

40. Iwata, R., et al., Bubble growth and departure modes on wettable/non-wettable porous foams in alkaline water splitting. *Joule*, 2021. **5**(4): pp. 887–900.

41. Matsushima, H., et al., Water electrolysis under microgravity. *Electrochimica Acta*, 2003. **48**(28): pp. 4119–4125.

42. He, Y., et al., Insight into the bubble-induced overpotential towards high-rate charging of Zn-air batteries. *Chemical Engineering Journal*, 2022. **448**: p. 137782.

43. Vogt, H. and R.J. Balzer, The bubble coverage of gas-evolving electrodes in stagnant electrolytes. *Electrochimica Acta*, 2005. **50**(10): pp. 2073–2079.

44. Balzer, R.J. and .H. Vogt, Effect of electrolyte flow on the bubble coverage of vertical gas-evolving electrodes. *The Electrochemical Society*, 2003. **150**: p. E11.

45. Gabrielli, C., F. Huet, and R.P. Nogueira, Fluctuations of concentration overpotential generated at gas-evolving electrodes. *Electrochimica Acta*, 2005. **50**(18): pp. 3726–3736.

46. Newman, J., Scaling with Ohm's law; wired vs. wireless photoelectrochemical cells. *Journal of the Electrochemical Society*, 2013. **160**(3): pp. F309–F311.

47. Dukovic, J. and Tobias C.W., The influence of attached bubbles on potential drop and current distribution at gas-evolving electrodes. *The Electrochemical Society*, 1987. **134**: p. 331.

48. Carmo, M., et al., A comprehensive review on PEM water electrolysis. *International Journal of Hydrogen Energy*, 2013. **38**(12): pp. 4901–4934.

49. Taie, Z., et al., Pathway to complete energy sector decarbonization with available iridium resources using ultralow loaded water qlectrolyzers. *ACS Applied Materials & Interfaces*, 2020. **12**(47): pp. 52701–52712.

50. Liu, X., S. Zheng, and K. Wang, Influence of bubble generation on the microchannel electrochemical gas evolution reaction. *Chemical Engineering Journal*, 2023. **463**: p. 142453.

51. Gabrielli, C., F. Huat, M. Keddam, A. Macias, and A. Sahar Potential drops due to an attached bubble on a gas-evolving electrode. *Applied Electrochemistry*, 1989. **19**: pp. 617–629.

52. Gabrielli, C., F. Huat, M. Keddam, and A. Sahar, Investigation of water electrolysis by spectral analysis. I. Influence of the current density. *Applied Electrochemistry*, 1989. **19**: pp. 683–696.

53. Spöri, C., et al., The stability challenges of oxygen evolving catalysts: towards a common fundamental understanding and mitigation of catalyst degradation. *Angewandte Chemie International Edition*, 2017. **56**(22): pp. 5994–6021.

54. Feng-Bin Li, A.R.H., S.D. Lubetkin, and D.J. Roberts, Electrochemical quartz crystal microbalance studies of potentiodynamic electrolysis of aqueous chloride solution: surface processes and evolution of H_2 and C_{12} gas bubbles. *Electroanalytical Chemistry*, 1992. **335**(1–2): pp. 345–362.

55. Liu, H., et al., Monolithic Ni-Mo-B bifunctional electrode for large current water splitting. *Advanced Functional Materials*, 2021. **32**(4): p. 2107308.

56. Lee, H.Y., C. Barber, and A.R. Minerick, Improving electrokinetic microdevice stability by controlling electrolysis bubbles. *Electrophoresis*, 2014. **35**(12–13): pp. 1782–1789.

57. Han, B., et al., Nanoscale structural oscillations in perovskite oxides induced by oxygen evolution. *Nature Materials*, 2016. **16**(1): pp. 121–126.

58. Zeradjanin, A.R., et al., How to minimise destabilising effect of gas bubbles on water splitting electrocatalysts? *Current Opinion in Electrochemistry*, 2021. **30**: p. 100797.

59. Andaveh, R., et al., Superaerophobic/superhydrophilic surfaces as advanced electrocatalysts for the hydrogen evolution reaction: a comprehensive review. *Journal of Materials Chemistry A*, 2022. **10**(10): pp. 5147–5173.

60. Wang, H., et al., Recent progress on layered double hydroxides: comprehensive regulation for enhanced oxygen evolution reaction. *Materials Today Energy*, 2022. **27**: p. 101036.

61. Swiegers, G.F., et al., The prospects of developing a highly energy-efficient water electrolyser by eliminating or mitigating bubble effects. *Sustainable Energy & Fuels*, 2021. **5**(5): pp. 1280–1310.

62. Koj, M., et al., Laser structured nickel-iron electrodes for oxygen evolution in alkaline water electrolysis. *International Journal of Hydrogen Energy*, 2019. **44**(25): pp. 12671–12684.

63. Koj, M., J. Qian, and T. Turek, Novel alkaline water electrolysis with nickel-iron gas diffusion electrode for oxygen evolution. *International Journal of Hydrogen Energy*, 2019. **44**(57): pp. 29862–29875.

64. Chen, Y., et al., A flow-through electrode for hydrogen production from water splitting by mitigating bubble induced overpotential. *Journal of Power Sources*, 2023. **561**: p. 232733.

65. Lake, J.R., Á.M. Soto, and K.K. Varanasi, Impact of bubbles on electrochemically active surface area of microtextured gas-evolving electrodes. *Langmuir*, 2022. **38**(10): pp. 3276–3283.

66. Fernández, D., et al., Bubble formation at a gas-evolving microelectrode. *Langmuir*, 2014. **30**(43): pp. 13065–13074.

67. Teschke, O. and G. Galembeck, Effect of PTFE coverage on the performance of gas evolving electrodes. *The Electrochemical Society*, 1984. 131: p. 1095.

68. Brussieux, C., et al., Controlled electrochemical gas bubble release from electrodes entirely and partially covered with hydrophobic materials. *Electrochimica Acta*, 2011. **56**(20): pp. 7194–7201.

69. Wang, J., et al., Air bubble bursting effect of lotus leaf. *Langmuir*, 2009. **25**(24): pp. 14129–14134.

70. Wang, L., et al., Increasing gas bubble escape rate for water splitting with nonwoven stainless steel fabrics. *ACS Applied Materials & Interfaces*, 2017. **9**(46): pp. 40281–40289.

71. Giacomello, A., M. Amabili, and C.M. Casciola, How to control bubble nucleation from superhydrophobic surfaces. *Journal of Physics: Conference Series*, 2015. **656**(1): p. 012124.

72. Giacomello, A., et al., Geometry as a catalyst: how vapor cavities nucleate from defects. *Langmuir*, 2013. **29**(48): pp. 14873–14884.

73. Li, Y., et al., Selective-etching of MOF toward hierarchical porous Mo-doped CoP/N-doped carbon nanosheet arrays for efficient hydrogen evolution at all pH values. *Chemical Engineering Journal*, 2021. **405**: p. 126981.

74. He, Y., et al., Strategies for bubble removal in electrochemical systems. *Energy Reviews*, 2023. **2**(1): p. 100015.

75. Wang, M., Z. Wang, and Z. Guo, Water electrolysis enhanced by super gravity field for hydrogen production. *International Journal of Hydrogen Energy*, 2010. **35**(8): pp. 3198–3205.

76. Ertürk, A.T., et al., Metal foams as a gas diffusion layer in direct borohydride fuel cells. *International Journal of Hydrogen Energy*, 2022. **47**(55): pp. 23373–23380.

77. Wang, Y., et al., Multi-layer superhydrophobic nickel foam (NF) composite for highly efficient water-in-oil emulsion separation. *Colloids and Surfaces A: Physicochemical and Engineering Aspects*, 2021. **628**: p. 127299.

78. Rocha, F., et al., Effect of pore size and electrolyte flow rate on the bubble removal efficiency of 3D pure Ni foam electrodes during alkaline water electrolysis. *Journal of Environmental Chemical Engineering*, 2022. **10**(3): p. 107648.

79. Yang, Y., et al., Gradient porous electrode-inducing bubble splitting for highly efficient hydrogen evolution. *Applied Energy*, 2022. **307**: p. 118278.

80. Lee, C.-Y., et al., 3D Printing for electrocatalytic applications. *Joule*, 2019. **3**(8): pp. 1835–1849.

81. Aeby, X., et al., Fully 3D printed and disposable paper supercapacitors. *Advanced Materials*, 2021. **33**(26): p. 2101328.

82. Cai, J., et al., 3D printing of a V_8C_7-VO_2 bifunctional scaffold as an effective polysulfide immobilizer and lithium stabilizer for Li-S batteries. *Advanced Materials*, 2020. **32**(50): p. 2005967.

83. Bu, X., et al., Remarkable gas bubble transport driven by capillary pressure in 3D printing-enabled anisotropic structures for efficient hydrogen evolution electrocatalysts. *Applied Catalysis B: Environmental*, 2023. **320**: p. 121995.

84. Xu, X., et al., Highly efficient all-3D-printed electrolyzer toward ultrastable water electrolysis. *Nano Letter*, 2023. **23**(2): pp. 629–636.

85. Kou, T., et al., Periodic porous 3D electrodes mitigate gas bubble traffic during alkaline water electrolysis at high current densities. *Advanced Energy Materials*, 2020. **10**(46): p. 2002955.

86. Su, B., Y. Tian, and L. Jiang, Bioinspired interfaces with superwettability: from materials to chemistry. *Journal of the American Chemical Society*, 2016. **138**(6): pp. 1727–1748.

87. George, J.E., S. Chidangil, and S.D. George, Recent progress in fabricating superaerophobic and superaerophilic surfaces. *Advanced Materials Interfaces*, 2017. **4**(9): p. 1601088.

88. Yu, C., et al., Superwettability of gas bubbles and its application: from bioinspiration to advanced materials. *Advanced Materials*, 2017. **29**(45): p. 1703053.

89. Darband, G.B., M. Aliofkhazraei, and S. Shanmugam, Recent advances in methods and technologies for enhancing bubble detachment during electrochemical water splitting. *Renewable and Sustainable Energy Reviews*, 2019. **114**: p. 109300.

90. Barati Darband, G., M. Aliofkhazraei, and A.S.Rouhaghdam, Three-dimensional porous Ni-CNT composite nanocones as high performance electrocatalysts for hydrogen evolution reaction. *Journal of Electroanalytical Chemistry*, 2018. **829**: pp. 194–207.
91. Zhang, Q., et al., Superaerophobic ultrathin Ni-Mo alloy nanosheet array from in situ topotactic reduction for hydrogen evolution reaction. *Small*, 2017. **13**(41): p. 1701648.
92. Kim, J., et al., Efficient alkaline hydrogen evolution reaction using superaerophobic Ni nanoarrays with accelerated H_2 bubble release. *Advanced Materials*, 2023.
93. Rossky, P.J., Exploring nanoscale hydrophobic hydration. *Faraday Discussions*, 2010. **146**: pp. 13–18.
94. Hosseini, S.R., S. Ghasemi, and S.A. Ghasemi, Effect of surfactants on electrocatalytic performance of copper nanoparticles for hydrogen evolution reaction. *Journal of Molecular Liquids*, 2016. **222**: pp. 1068–1075.
95. Jin, C., et al., Scanning electrochemical cell microscope study of individual H_2 gas bubble nucleation on platinum: effect of surfactants. *Chinese Journal of Analytical Chemistry*, 2021. **49**(4): pp. e21055–e21064.

5 Alkaline Water Electrolysis at Industrial Scale

Anran Zhang, Ying Ma, Rui Ding, and Liming Li

5.1 INTRODUCTION

At the kernel of the global warming dilemma and the ever-increasing depletion of fossil fuels, exploration of renewable energy resources has become the epicenter of intent of researchers comprehensively. Hydrogen, as a clean energy source, is gradually replacing fossil fuels such as oil and coal, becoming an important carrier of global energy. H_2 has long been proposed as an alternative energy vector to fossil fuels to generate power for domestic heating, industrial and transport sectors. In this sense, it has the potential to revolutionize the world's energy economy toward the predicted hydrogen economy/society. Green hydrogen produced by water electrolysis coupled with renewable energy sources has emerged as an advanced and attractive strategy in recent years for storing and providing clean and sustainable energy. The upstream and downstream industrial chain of "green hydrogen" includes renewable energy power supply, hydrogen production systems, auxiliary systems, storage and transportation systems, and downstream applications. Renewable energy hydrogen production is the core of the hydrogen energy industry chain. The excess electrical energy converted from renewable energy enters the electrolytic water hydrogen production device through voltage regulation by the converter, where water electrolysis is carried out to produce hydrogen. The prepared hydrogen is purified and enters the hydrogen storage system. A portion of the gas is regulated on the grid side through a fuel cell power generation system. Another part of the gas enters energy terminals or hydrogen refueling stations through long-distance trailers, liquid hydrogen tank cars, or pipeline transportation to meet downstream hydrogen energy consumption needs in industries such as transportation, power generation, chemical production, and metallurgy.

Water electrolysis hydrogen production equipment, as the core process equipment for the "green electricity-green hydrogen" conversion, has attracted worldwide attention. Many central enterprises and listed companies in China have also actively laid out the manufacturing of water electrolysis hydrogen production equipment and released water electrolysis hydrogen production equipment products. The mainstream technologies for hydrogen production through electrolysis of water include alkaline water (ALK) electrolysis, proton exchange membrane (PEM) electrolysis, solid oxide

DOI: 10.1201/9781003368939-5

electrolysis cell (SOEC), and anion exchange membrane (AEM) electrolysis. Among them, alkaline electrolysis technology is mature, with a single unit scale of up to $1000\ \mathrm{Nm^3/h}\ H_2$; the system has a long lifespan, low cost, and is easy to implement on a large scale. It is currently the mainstream electrolysis technology. Compared to alkaline electrolysis, PEM electrolysis has advantages such as high current density and fast response. However, it is in a relatively early stage and has high costs. Currently, high-power large-scale applications have not been achieved in China, and the green hydrogen demonstration application projects and core products of mainstream enterprises still mainly rely on alkaline electrolysis cells. SOEC and AEM electrolysis are still in the laboratory stage and have not been commercialized. Therefore, in this chapter, we will focus on discussing alkaline electrolytic cell technology.

Alkaline electrolytic cells were commercialized in the mid-20th century; under the action of an electric current, water molecules decompose to produce H_2 and O_2, which are discharged from the anode and cathode, respectively. As a weak electrolyte, pure water has poor conductivity and high resistance, so a 30 wt.% NaOH/KOH solution is usually used as an electrolyte to improve solution conductivity and reduce the internal resistance of the electrolytic cell. From the principle of electrolytic water, it can be seen that the electrolysis process only consumes water, so it is only necessary to supplement the hydrogen system with water through a water pump. The cathode electrode and the anode electrode of electrolytic cell are generally nickel mesh with a catalyst attached to the surface. At present, the catalytic material of industrial equipment is generally Raney nickel. Many universities and enterprises are studying other catalytic functional materials. In addition, as an important component of alkaline electrolyzers, the membrane was initially made of asbestos as the membrane material. However, it has swelling properties in alkaline electrolytes, and asbestos is harmful to human health. It has gradually been replaced by membranes such as polyphenylene sulfide (PPS) with good thermal stability, mechanical strength, and electrochemical performance.

With the development of hydrogen energy becoming a global consensus, various countries' hydrogen production technology routes are based on local hydrogen source potential and future hydrogen industry demand, presenting a cascade development trend from low hydrocarbon, clean hydrogen to renewable hydrogen. This chapter will introduce the alkaline electrolytic water hydrogen production technology in various countries, including the routes and major manufacturers of electrolytic water hydrogen production technology in countries such as the United States, Japan, and Europe. Comparative analysis of the current development status of domestic electrolysis water technology and quantitative comparison of the gap with foreign technology levels were conducted. Based on this, potential technical routes for hydrogen production from electrolysis water in China were analyzed.

5.2 DEVELOPMENT TRENDS OF INTERNATIONAL ALKALI WATER ELECTROLYSIS TECHNOLOGY

From the perspective of development history, alkaline water electrolysis began to achieve industrial application of alkaline water electrolysis hydrogen production technology around the 20th century. After experiencing the development process of unipolar to bipolar, small to large, atmospheric to pressurized, manual control to fully automatic control, alkaline water electrolysis hydrogen production technology has gradually entered a mature industrial application stage.

The research on electrolytic water hydrogen production technology in Europe and America started early. The United States and Europe developed the roadmap for electrolytic water hydrogen production technology in 2011 and 2013, respectively. Among them, the leading companies in electrolytic water hydrogen production are mostly distributed in Europe, including Nel, ITM Power, HydrogenPro, Encapter, Sunfire, Mcphy, and other companies, which have mature applications in alkalinity and PEM.

5.2.1 Thyssenkrupp Nucera

Thyssenkrupp nucera (Germany, https://thyssenkrupp-nucera.com/) is developing a 20 MW alkaline water electrolysis unit which is setting a benchmark in water electrolysis technology worldwide. Table 5.1 provides some technical characteristics. This standardized solution for green hydrogen production offers high current density operation with an optimized footprint. And it matches highest market demands: The prefabricated AWE units can be easily transported, installed, and interconnected to obtain the desired plant capacity, up to several hundred megawatts or even gigawatts as a cost efficient, highly modularized solution for large-scale green hydrogen production.

5.2.2 Nel

Nel ASA (Norway, https://nelhydrogen.com/) has developed the world's most energy-efficient electrolyzers—Atmospheric Alkaline Electrolyzer (150–3880 Nm³/h). The A Series features a cell stack power consumption as low as 3.8 kWh/Nm³ of hydrogen gas produced, up to 2.2 MW per stack. A Series electrolyzers can produce up to 3880 Nm³/h of hydrogen or just over 8 ton/day. The modular concept enhances the

TABLE 5.1
Main Technical Characteristics of the 20 MW Alkaline Water Electrolysis Unit

Technical Characteristics	Performance
Product capacity H_2	4000 Nm³/h
Power consumption at startup	4.5 kWh/Nm³ (DC)
Standard operating range	10%–100%
H_2 product quality at electrolyzer outlet	>99.9% purity (dry basis)
H_2 product pressure at electrolyzer outlet	>300 mbar$_g$

Source: Thyssenkrupp nucera Co.

flexibility of the device by providing customized indoor hydrogen solutions for any application, configuration, and size according to customer requirements. This robust system can be containerized, offering one of the world's smallest footprints for high capacity electrolyzer plants at 200 barg.

5.2.3 McPhy Energy S.A.

McPhy Energy S.A. (France, https://mcphy.com/fr/) launched the revolutionary "Enhanced McLyzer" technology in 2018. Table 5.2 provides some technical characteristics. The "Enhanced McLyzer" electrolytic cell is a true technological breakthrough in the market, combining the reliability and maturity of high-pressure alkaline technology with optimal flexibility while integrating it into the design of ultra-high-capacity (multi-MW) electrolytic platforms, specifically for industrial and heavy transportation sectors. This fully modular solution integrates a 4 MW module design (McLyzer 800-30) and can produce low-carbon hydrogen gas at high pressure (30 bar).

5.2.4 HydrogenPro

HydrogenPro (Norway, https://hydrogen-pro.com/) is committed to developing high-pressure alkaline electrolyzer. The new plating technology acquired recently is able to increase the efficiency of each unit by 14% to reach 93% of the theoretical maximum. The new technology is proven in a small industrial scale unit, and a production facility that can handle full size electrodes is now under construction. Complete assembly lines are being planned in Europe and the United States to satisfy demand for local content. Compared to traditional alkaline systems, HydrogenPro's high-pressure units (up to 30 bar) save compression costs and are superbly suited for variable loads from solar panels and wind turbines.

TABLE 5.2

Main Technical Characteristics of the 20 MW Alkaline Water Electrolysis Unit

Technical Characteristics	Performance
Model	McLyzer 800-30
Pressure (barg)	30
Nominal flow rate H_2 (Nm³/h)	800
Rated power	About 4 MW
Consort. specific direct current at nominal flow rate (kWh/Nm³)	4.5

Source: McPhy Energy S.A. Co.

5.2.5 Sunfire GmbH

Sunfire's (Germany, https://www.sunfire.de/en/) ultra-reliable pressurized alkaline electrolyzer is optimal for applications without or with limited steam availability. With a proven system lifetime of at least 90,000 operating hours, the electrolyzer is their established solution for renewable hydrogen production. The electrolyzer has a scalable system design. The system produces 2230 Nm³/h hydrogen at 30 bar(g) with a power consumption of 4.7 kWh/Nm³.

The demand for green hydrogen construction is strong, and the order size has increased significantly year-on-year. With the acceleration of global green hydrogen construction pace, the demand for electrolytic cells continues to increase rapidly, and the expansion pace of various electrolytic cell giants also keeps up. Among these enterprises, Nel, the leader of electrolytic cells, leads in performance, and its competitors are also closely following. Nel was founded in 1927 and has accumulated over 90 years of alkaline electrolytic cell technology. Through external acquisitions, it has expanded its PEM electrolytic cell and hydrogen refueling station businesses, forming two major business segments: hydrogen electrolytic cell (alkaline electrolytic cell, PEM electrolytic cell) and hydrogen refueling station. Among them, the electrolytic cell business accounts for over 70%, making it the largest electrolytic cell company in Europe. In 2022, Nel's revenue was $94 million, including $30 million for alkaline electrolyzers, a year-on-year increase of +506%, and $40 million for PEM electrolyzers, a year-on-year decrease of −1%. French company Mcphy's revenue in 2022 was $17 million, a year-on-year increase of +22%, with electrolytic cells accounting for 68% and hydrogen refueling station business accounting for 32%; Hydropro's production capacity is currently 0.3 GW. At the end of 2022, HydrogenPro upgraded its manufacturing plant in Tianjin, China, with a goal of reaching 300 MW to deliver purchase orders. The company plans to achieve a global production capacity of 10 GW in the near future. ITM Power currently has a production capacity of 1 GW by the end of 2022, with plans to increase it to 2.5 GW by the end of 2023, and plans to double and increase it to 5 GW by the end of 2024. In addition, Thyssenkrupp, Sunfire, Green Hydrogen Systems, Reliance, and others have all announced expansion plans. It is expected that the overseas electrolytic cell production capacity will reach 8 GW by 2023 (Table 5.3).

In addition, Japan focused on promoting the development of alkaline electrolysis water devices, especially the large-scale electrolysis cell technology of 2000 Nm³/h, through pioneering research and development projects such as hydrogen utilization and hydrogen society construction technology from 2014 to 2018. In 2019, Japan established a 10-year technical breakthrough goal for ALK and PEM water electrolysis technology by benchmarking the development routes of electrolysis water technology in the United States and Europe. It focused on the research of reactor reaction mechanism as well as durability evaluation methods and standardization, and it conducted system level optimization based on various information, such as renewable energy generation prediction, power supply adjustment, and hydrogen demand, to improve current density, efficiency, and durability.

TABLE 5.3

Production Capacity of Major Overseas Electrolytic Cell Companies

Company	Country	Production Capacity in 2022	Notes
Thyssenkrupp nucera	Germany	1 GW	Planned production capacity of 1.5 GW in 2023
Nel	Norway	0.6 GW	Production capacity includes ALK and PEM
HydrogenPro	Norway	0.3 GW	Planned production capacity of 1.3 GW in 2023
Sunfire	Germany	0.3 GW	Planned production capacity of 0.5 GW in 2023
ITM Power	England	1 GW	Planned production capacity of 2.5 GW in 2023
McPhy	France	0.1 GW	
Green Hydrogen Systems	Denmark	0.1 GW	
Reliance Industries	Denmark	/	Planned production capacity of 0.5 GW in 2023

5.3 DEVELOPMENT TREND OF ALKALINE WATER ELECTROLYSIS TECHNOLOGY IN CHINA

China has shown an industrial application status in the field of electrolytic water technology, with ALK hydrogen production as the main technology and PEM hydrogen production as the auxiliary technology. Among them, China's ALK hydrogen production equipment ranks first in the global market share. Due to the high maturity of alkaline water electrolysis technology in China, precious metals are not used as equipment production raw materials, and the unit price is relatively low. Compared to alkaline water electrolysis, although PEM water electrolysis has advantages such as high efficiency, no alkaline solution, and good dynamic response, its cost is still about 5–6 times that of ALK due to the fact that core components such as proton exchange membranes still rely on imports. Therefore, large-scale high-power applications have not yet been achieved in China.

5.3.1 ANALYSIS OF ALKALINE MARKET IN CHINA

At present, demonstration and application projects of renewable energy hydrogen production in China and the core products of mainstream enterprises still mainly rely on alkaline electrolytic cells. According to industry research and release data, the market size of China's electrolytic water hydrogen production equipment exceeded 900 million yuan in 2021, with a shipment volume exceeding 350 MW. In 2022, it is estimated that the annual shipment volume of China's alkaline electrolytic water hydrogen production equipment is about 780 MW, and the total shipment volume of electrolytic cells is about 800 MW, doubling from 2021. The market share of China's top electrolytic water hydrogen production equipment manufacturing enterprises is still relatively high, with a relatively concentrated market. Throughout the year, the delivery amount of top enterprises' equipment exceeded 1 billion yuan, while the contract signing volume exceeded 1.5 billion yuan. The total market share of top three enterprises' electrolytic cells is close to 80%.

According to industry databases, as of the end of 2022, there are over 100 renewable energy electrolysis water hydrogen production projects in China that have been built, under construction, and under planning. More than half of these projects have announced the types and scale of electrolysis water hydrogen production, with a total scale of alkaline electrolysis water hydrogen production exceeding 17 GW. These projects are mainly distributed in the Northwest, North China, and South China regions, with the scale of hydrogen production in the three regions accounting for over 95%. Due to the planning period of the aforementioned project ranging from 2025 to 2035, and taking into account factors such as land, it is preliminarily estimated that the supply of new renewable energy to produce green hydrogen in China will reach approximately 500,000 tons by 2025.

The largest renewable energy hydrogen production demonstration application project that has been built or is currently under construction in China in 2022 is the Sinopec Xinjiang Kuche Green Hydrogen Demonstration Project, and it is also the largest photovoltaic green hydrogen production project under construction in the world. The project has purchased a total of 1000 Nm³/52 sets of alkaline electrolytic cells for hydrogen production, equivalent to a power load of 260 MW; this accounts for nearly one-third of China's water electrolysis hydrogen production delivery this year. After being put into operation, the annual production of green hydrogen can reach 20,000 tons, which is of great significance for promoting the development of the green hydrogen industry chain, promoting the transformation and upgrading of the energy industry, promoting the economic and social development of Xinjiang region, and ensuring national energy security (Table 5.4).

TABLE 5.4
Summary of China's Green Hydrogen Demonstration Projects

Demonstration Projects	Project Status	Scale	Technology Route
Ordos City Scenery Integration Green Hydrogen Demonstration Project	Hydrogen production equipment bidding	390 MW	ALK
Sinopec Nova Oil Company Xinjiang Kuqa Green Hydrogen Demonstration Project	Equipment shipment	52*1000 Nm³/h (260 MW)	ALK
Da'an Wind Solar Production Green Hydrogen Synthesis Hydrogen Integration Demonstration Project	Hydrogen production equipment bidding	39,000 Nm³/h (195 MW)	ALK

(Continued)

TABLE 5.4 (*Continued*)
Summary of China's Green Hydrogen Demonstration Projects

Demonstration Projects	Project Status	Scale	Technology Route
Otok Qianqian 250 MW photovoltaic power station and hydrogen energy comprehensive utilization demonstration project	Hydrogen production equipment bidding	9000 Nm³/h (45 MW)	ALK
The first phase of Guoneng Ningdong Renewable Hydrogen Carbon Emission Reduction Demonstration Zone Project	Hydrogen production equipment bidding	5000 Nm³/h (25 MW)	ALK
State Power Investment Zhejiang Taizhou Dachen Island Hydrogen Energy Comprehensive Utilization Demonstration Project	Put into operation	4,1000 Nm³/h (20 MW)	ALK
Heilongjiang Qitaihe Boli County 200 MW wind power hydrogen production project	Hydrogen production equipment bidding	1500 Nm³/h (7.5 MW)	ALK
300 MW photovoltaic hydrogen production project in Laiyuan County, Hebei	Hydrogen production equipment bidding	2*600 Nm³/h (6 MW)	ALK
Gansu Pingliang 100 MW wind power hydrogen production project	Started construction	5 MW	ALK
Zhongneng Green Power Zhangye Hydrogen Energy Comprehensive Application Demonstration Project	Under construction	1000 Nm³/h (5 MW)	ALK
Baicheng distributed generation hydrogen production and hydrogenation integration demonstration project	Complete startup and put into operation	2*1000 Nm³/h(10 MW-ALK); 1*200 Nm³/h (1 MW-PEM)	ALK/PEM

5.3.2 Inventory of Major Enterprises in China

Driven by dual carbon goals and hydrogen energy industry planning, the green hydrogen industry has emerged as a key focus of new energy development in China. Companies in industries such as wind power, photovoltaic, energy groups, and automobiles have all laid out green hydrogen businesses, involving the upstream, midstream, and downstream of the green hydrogen industry chain. As of the end of 2022, more than a hundred enterprises in China have laid out the production of electrolytic hydrogen production equipment. There are three types of participants in domestic electrolytic cell equipment. The first type is established electrolytic cell enterprises such as PERIC Hydrogen Technologies Co., Ltd., John Cockerill, and Tianjin Mainland Hydrogen Equipment Co., Ltd., which have a deep technological foundation and high market share. The second category is photovoltaic leading enterprises such as LONGI and SUNGROW, with strong financial and technical strength. The photovoltaic business is highly collaborative with the electrolytic water hydrogen production business, each of which has entered the electrolytic water hydrogen production equipment market with technological advantages and order advantages, forming an impact on traditional enterprises.

5.3.2.1 PERIC Hydrogen Technologies Co., Ltd.

PERIC Hydrogen Technologies Co., Ltd. (China, http://www.peric718.com/) is currently a research and production enterprise with a relatively complete domestic hydrogen equipment industry chain. It can produce 350 sets of alkaline hydrogen production equipment and 120 sets of PEM pure water hydrogen production equipment annually, as well as carry out the construction of various types of hydrogen refueling stations. The water electrolysis hydrogen production equipment maintains a leading position in the national market share. The company has been developing pressurized water electrolysis hydrogen production devices using military technology since 1984, and it has now formed four major series with over 20 specifications and a gas production capacity of 0.5–2000 Nm^3/h series of water electrolysis hydrogen production devices has been developed, along with a series of hydrogen purification devices and a series of oxygen purification devices. So far, the company has produced and sold over 1000 sets of water electrolysis hydrogen production devices, including over 400 sets of hydrogen drying, purification devices, pressure swing adsorption devices, and methanol hydrogen production devices, with a cumulative output value of over 3 billion yuan. Users are all over the country and exported to more than 30 countries and regions. On December 16, 2022, hydrogen energy company independently developed a water electrolysis hydrogen production equipment with a single hydrogen production capacity of 2000 Nm^3/h, which was offline in Handan. According to the introduction, the H-type alkaline water electrolysis hydrogen production equipment has fully independent intellectual property rights, achieving key technological breakthroughs such as high current density, wide adjustable range, low operating energy consumption, and high stability.

5.3.2.2 John Cockerill

John Cockerill (China, http://www.cjhydrogen.com/) has undertaken all the personnel and intellectual property rights of Suzhou Jingli Hydrogen Production Equipment Co., Ltd. By increasing research and development capabilities, updating equipment, and expanding production capacity, it focuses on the research and development, production, and sales of alkaline electrolytic water hydrogen production equipment. It is positioned as the headquarters of John Cockerill Group's hydrogen business in China. In 2021, the company produced over 50 units with a hydrogen production capacity of 1000 Nm³/h electrolytic water hydrogen production equipment, participated in 1200 and 1300 Nm³/h R&D, and produced hydrogen production equipment for electrolysis of water. The production capacity will reach 1 GW in 2022, and it is expected to deliver 1500 Nm³/h in the second half of the year hydrogen production equipment for electrolysis of water.

5.3.2.3　Tianjin Mainland Hydrogen Equipment Co., Ltd.

Tianjin Mainland Hydrogen Equipment Co., Ltd. (China, http://www.cnthe.com/) was established in 1994 with a registered capital of 30 million yuan. They mainly produce alkaline water electrolysis hydrogen production equipment and gas purification equipment. The alkaline water electrolysis hydrogen production equipment has formed a series, with a maximum gas production capacity of up to 1000 Nm³/h.

5.3.2.4　LONGI

LONGI (China, https://www.longi.com/cn/) launched the ALK Hi1 series of products in February 2023, which can be as low as 4.3 kwh/Nm^3 under full DC power consumption conditions. Simultaneously launching the ALK Hi1 plus product, the DC power consumption is as low as 4.1 kwh/Nm^3 under full load conditions. At a current density of 2500 A/m^2, it can be as low as 4.0 kwh/Nm^3. In 2022, LONGI ranked third in the country in the shipment of electrolytic water equipment, with a production capacity of 1.5 GW. According to the company's plan, production capacity will be further expanded to 2.5 GW in 23 years and 5–10 GW in 25 years.

5.3.2.5　SUNGROW

SUNGROW's (https://www.sungrowpower.com/) traditional business is photovoltaic inverters, and it is a leading global photovoltaic inverter company. SUNGROW has established a wholly owned subsidiary to produce hydrogen from photovoltaics into local hydrogen energy. At present, SUNGROW has established the first demonstration platform for photovoltaic off-grid hydrogen production and hydrogen storage power generation in China in the fields of platform, technology, and products, as well as the largest 5 MW electrolytic water hydrogen production system testing platform in China and an annual production capacity of GW level hydrogen production equipment factory. SUNGROW can independently produce 1000 standard m^3 of alkaline hydrogen production system, providing a complete system solution including hydrogen production power supply, electrolytic cell, and intelligent hydrogen energy management system. Its electrolytic hydrogen production products have been applied in multiple projects (Table 5.5).

Multiple forces are participating in the competition, and Chinese enterprises are rapidly expanding their production capacity. The production capacity of electrolytic cells such as PERIC Hydrogen Technologies Co., Ltd. and LONGI is globally leading, and domestic enterprises mainly focused on PERIC Hydrogen Technologies Co., Ltd. have sufficient production capacity planning and rapid expansion. With the rapid growth of the market, the large-scale demand in the market has also prompted the

TABLE 5.5

Production Capacity of Mainstream Domestic Electrolytic Cell Enterprises

Province	Electrolytic Water Equipment Enterprise	Production Capacity in 2022	Notes
Hebei	PERIC Hydrogen Technologies Co., Ltd.	1.5 GW	Production capacity includes ALK and PEM
Jiangsu	John Cockerill	1 GW	The company plans to have a production capacity of 1.5 GW by 2023
Tianjin	Tianjin Mainland Hydrogen Equipment Co., Ltd.	1 GW	
Shanxi	LONGI	1.5 GW	The company plans to generate 2.5 GW in 23 years and 5–10 GW in 25 years
Anhui	SUNGROW	1 GW	Production capacity includes ALK and PEM
Guangdong	Kohodo Hydrogen Energy	0.3 GW	The company plans to have a production capacity of 0.5 GW by 2023
Jiangsu	GUOFUHEE	0.5 GW	The company plans to have a production capacity of 1.0 GW by 2023
Beijing	SinoHy Energy	0.5 GW	
Jiangsu	CPU H_2	1 GW	
Beijing	Aerospace Sizhuo Hydrogen Technology Co., Ltd.	0.5 GW	
Guangdong	Kylin-tech	0.5 GW	
Shandong	AUYAN	1 GW	
Guangdong	Sheng Hydrogen Production Equipment Co., Ltd.	/	The company plans to have a production capacity of 0.5 GW by 2023
Neimeng	Yili Hydrogen Field Era Technology Co., Ltd.	0.25 GW	The company plans to have a production capacity of 2.5 GW by 2023
Jiangsu	Shuangliang Eco-Energy	/	The company plans to have 100 sets of 1000 Nm³/h production capacity
Liaoning	Dalian Hydrogen Element Technology Co., Ltd.	/	The company plans to have a production capacity of 1.5 GW by 2023

Source: Trendbank.

continuous development of large-scale water electrolysis hydrogen production equipment. On July 12, 2022, China Huadian Hydrogen Energy Technology Co., Ltd.'s first set of 1200 Nm³/H alkaline electrolytic cell products are offline. On August 18, the new product of Xibeiyou Hydrogen 1400 standard hydrogen production system was launched. On December 16, PERIC Hydrogen Technologies Co., Ltd. held the world's first single unit hydrogen production of 2000 Nm³/h Water electrolysis hydrogen production equipment release ceremony. According to PERIC Hydrogen Technologies Co., Ltd., this device has achieved multiple key technological breakthroughs such as high current density, wide adjustable range, low operating energy consumption, and high stability.

In terms of ALK hydrogen production technology, there is still significant room for improvement in China's hydrogen production efficiency technical indicators. In terms of hydrogen production efficiency and current density, the current electrolysis current density of industrial alkaline electrolytic cells in China is about 0.3 A/cm2@1.84 V. The current density of electrolytic cells in European and American countries is as high as 0.4 A/cm2@1.8 V Above all, there are also problems with low gas production and high electrolysis energy consumption of individual equipment, which leads to high green hydrogen costs and production in China, which is not conducive to the development of the hydrogen energy industry. Therefore, the development of efficient alkaline water electrolysis hydrogen production technology and large-scale alkaline electrolysis hydrogen production equipment is of great significance for achieving low-cost green hydrogen production on a large scale, and it is also in line with the overall development strategy of national energy (Table 5.6).

TABLE 5.6
Manufacturing Scale of Global Alkaline Electrolytic Cell Manufacturers

Company	Hydrogen Production per Stack	Performance Index
PERIC Hydrogen Technologies Co., Ltd.	0.5–2000 Nm³/h	3.2 MPa; 2500–3000 A/m²;
LONGI	0.5–1200 Nm³/h	4.3–4.8 kWh/m³H_2
McPhy	0.4–200 Nm³/h	3.0 MPa; 3000–4000 A/m²; 4.3–4.8 kWh/m³H_2
Cummins Inc.	1.0–15 Nm³/h	1.0 MPa; 4.3–4.8 kWh/m³H_2
Thyssenkrupp	500 Nm³/h	Atmospheric pressure; 4.3–4.8 kWh/m³H_2
Nel	0.4–485 Nm³/h	Atmospheric pressure; 4.4–4.8 kWh/m³H_2

5.4 CONCLUSION

Under the carbon reduction scenario of the "dual carbon" goal, green hydrogen has rich application scenarios. On the one hand, it can cooperate with new energy power plants to play the role of hydrogen energy storage. On the other hand, in the industrial field, hydrogen energy can also be used as a tool for carbon reduction. As the cost of green hydrogen continues to decrease and supply continues to increase, the demand for green hydrogen will significantly expand, with the main increase coming from demonstration projects of carbon reduction by chemical enterprises and large state-owned enterprises in the industrial field. The increase in green hydrogen projects is expected to directly drive the procurement demand for electrolytic cells. Alkaline water electrolysis is currently the main hydrogen production technology suitable for large-scale green hydrogen production. Studying efficient alkaline water electrolysis technology for hydrogen production and promoting the industrial application of technological achievements will provide key equipment for the large-scale production of low-cost green hydrogen, provide rich zero hydrocarbon sources for the development of China's hydrogen energy industry, and generate immeasurable environmental and social benefits.

6 Existing Challenges and Development Directions of PEM Water Electrolysis

Min Yang, Yinqiao Zhan, Fuping Chen, Fei Wei,
Daoyuan Tang, Wei Chen, Wen Han,
Hongchang Tian, Zhibin Yang, and Pengtao Huang

6.1 INTRODUCTION

Due to its high combustion calorific value, sustainability, abundant reserves, and zero pollution, hydrogen energy is known as the cleanest energy in the 21st century, which has great potential to replace fuels and make the energy system greener, cleaner, and more sustainable in the future. In recent years, due to the increasing demand for carbon reduction in various countries, hydrogen energy has received more and more attention as an ideal solution to achieve carbon neutrality. In the hydrogen energy industry, extensive attention has been paid to the development of hydrogen production technology. At present, hydrogen produced by fossil energy and industrial by-products occupy the mainstream market, including hydrogen produced by coal, natural gas, petroleum, and methanol. Due to the dependence on fossil fuels, this method will still emit greenhouse gases such as carbon dioxide, so the hydrogen produced by this method does not belong to clean hydrogen energy. Among the many hydrogen production methods, hydrogen production by electrolytic water is one of the most important green hydrogen production methods, as well as the most promising.

The main principle of hydrogen production by water electrolysis is that water molecules are dissociated under the action of direct current to generate oxygen and hydrogen, wherein hydrogen is generated from the anode of the electrolyzer and oxygen is generated from the cathode. According to different diaphragm materials of the electrolyzer, it can be divided into alkaline water electrolysis (AWE), proton exchange membrane water electrolysis (PEMWE), and high-temperature solid oxide electrolysis cell (SOEC).

The working temperature of AWE is 70°C–90°C and its working pressure is 1–3 MPa. Generally speaking, 30% KOH aqueous solution is used as the electrolyte, and porous materials such as asbestos, polyester cloth, nylon, and ceramics are used as the diaphragm. In terms of electrode materials, commercial electrolyzer products are mainly non-precious metals such as nickel mesh (cloth), supplemented by simple electrode surface roughening or alloying to improve the specific surface area and

DOI: 10.1201/9781003368939-6

activity, while also reducing the electrolytic energy consumption of the electrode to a certain extent. However, in practical applications, there are still many shortcomings in alkaline water electrolysis hydrogen production, including electrolyte pollution, electrode corrosion, low current density, low efficiency, and small load range.

SOEC hydrogen production technology [1–3] uses solid oxide as electrolyte material, with porous cermet Ni/YSZ as cathode material and perovskite oxide and other non-precious metal catalysts as anode material. The commonly used electrolyte is YSZ-based oxygen ion conductor or BZCY-based proton conductor, which needs to operate at high temperatures of 500°C–850°C and high pressure, requiring high stability and durability in the component materials of the electrolyzer, which limits the application of this technology. Most people believe that SOEC is still in the experimental research and development stage, and it is difficult to achieve widespread application in a short time.

Different from AWE and SOEC hydrogen production technology, PEMWE uses perfluorosulfonic acid proton exchange membrane as a solid electrolyte, which has excellent chemical stability, high proton conductivity, and good gas isolation. Compared with alkaline water electrolysis, PEMWE has the characteristics of more compact structure, lower ohmic resistance, higher operating current density and energy efficiency, wide operating temperature (20°C–80°C), high safety, high hydrogen purity, fast response speed, and can adapt to the volatility of renewable energy, etc. Therefore, PEMWE is considered to be the most promising technology for high-purity hydrogen production in future industrial applications.

6.2 BASICS OF PEMWE

6.2.1 WORKING PRINCIPLE OF PEMWE

PEMWE is an advanced hydrogen production technology that can achieve large current density (>1 A/cm^2), high hydrogen purity (>99.99%), and fast response (<5 seconds) toward dynamic electricity input [4,5]. PEMWE refers to the process of converting electrical energy into chemical energy with the help of catalyst and storing it in hydrogen and oxygen. The core of PEMWE is the electrolyzer, which is mainly composed of membrane electrode assembly, current collector, end plate, sealing gasket, etc. The membrane electrode is the core component of the electrolyzer, which is composed of proton exchange membrane (PEM), catalytic layer (CL), and porous transport layer (PTL) from the inside to outside. It is the main place of material transmission and electrochemical reaction of the PEM electrolyzer. Among them, the catalyst layer of MEA provides a three-phase interface for material transport and electrochemical reactions, where the reaction gas, protons, and electrons react with the help of the electrocatalyst. The cathode catalyst layer and the anode catalyst layer are attached to the two sides of the proton exchange membrane, which provides a transmission channel for protons to pass from the anode to the cathode. The porous transport layer is usually in direct contact with the flow channels on the bipolar plate and plays the role of mechanical support, electron conduction, gas diffusion, and drainage. The characteristics and structure of the membrane electrode will directly affect the performance and life of PEM electrolyzer. Figure 6.1 shows

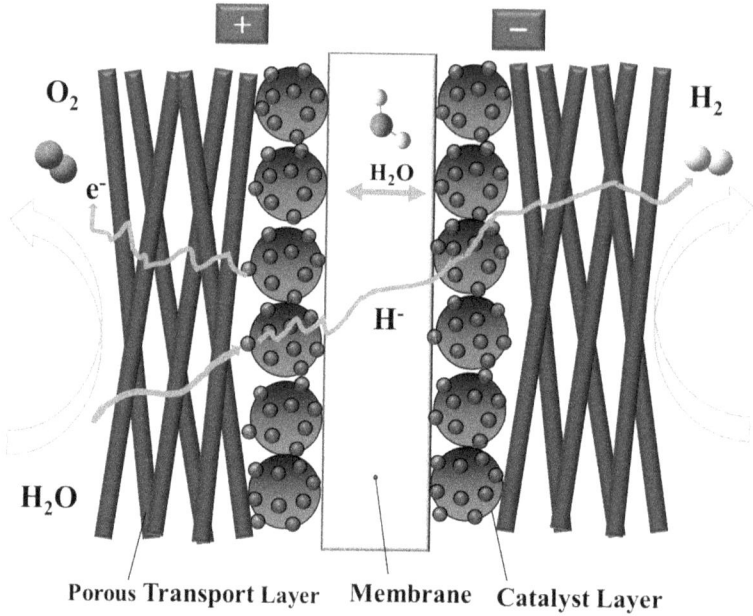

FIGURE 6.1   Working principle of PEMWE hydrogen production.

the working principle diagram of PEM water electrolysis membrane electrode for hydrogen production.

Under the action of an external DC power supply, the water molecules (H_2O) on the anode side will lose electrons to form oxygen molecules (O_2) and hydrogen ions $(H)^+$, and then H^+ will pass through the solid polymer electrolyte membrane (proton exchange membrane or Nafion membrane) to cathode in the form of hydronium ions. Meanwhile, the electrons produced on the anode will travel from the external circuit to the cathode. So, a hydrogen evolution reaction occurs on the cathode side, where electrons react with hydrogen ions to form hydrogen molecules. Unlike alkaline water electrolyzers, the reactant of PEM water electrolyzer is deionized water, and in order to improve the service life time, the resistivity of the deionized water is usually greater than $18.2\,M\Omega*cm$. The reaction formulas of anode and cathode are, respectively:

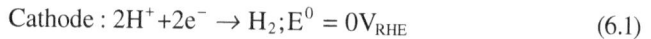

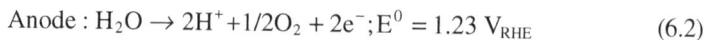

$$Cathode : 2H^+ + 2e^- \rightarrow H_2; E^0 = 0V_{RHE} \tag{6.1}$$

$$Anode : H_2O \rightarrow 2H^+ + 1/2O_2 + 2e^-; E^0 = 1.23\ V_{RHE} \tag{6.2}$$

The actual working voltage (V) [6]:

$$V = E + V_{act} + V_{trans} + V_{ohm} \tag{6.3}$$

where V is the total voltage of the PEM electrolyzer and E is the theoretical decomposition voltage, also known as the reversible cell voltage; V_{act} is the activation overvoltage, which is a voltage loss to overcome the energy barrier formation of electrochemical reactions; V_{ohm} is the Ohmic voltage loss, which represents the energy dissipation related to ohmic drops in the electrolytic cell. These include a number of contributions: electrolyte, electrodes, and electrical connections. V_{trans} is the mass transfer voltage loss, also known as concentration polarization, which is the deviation of electrode potential from equilibrium potential due to the change of reactant concentration during electrochemical reaction.

During hydroelectrolysis, hydrogen evolution reaction (HER) occurs at the cathode and oxygen evolution reaction (OER) occurs at the anode. In theory, a voltage of 1.23 V can drive the PEM electrolytic cell to produce hydrogen. Due to the polarization and energy loss, additional energy needs to be provided to compensate for the energy loss and drive the electrolysis reaction. In fact, only when the electrolytic cell voltage reaches at least 1.481 VRHE can the electrolysis reaction continue to occur, which is considered to be the thermoneutral voltage of electrolysis. In engineering applications, the operating voltage of electrolysis voltage is generally 1.7–2.0 V, which is significantly higher than the thermal neutral voltage. Due to the polarization loss and the restriction of catalytic activity, the hydrogen production efficiency of electrolytic water is low and the energy consumption is large, which limits its wide application. Therefore, many efforts have been made in recent years on catalyst materials, bipolar plate coatings, catalytic layer preparation methods, and design and manufacturing of PEMWE.

6.2.2 STRUCTURE OF PEMWE

The proton exchange membrane (PEM) electrolyzer is a structure composed of numerous electrolytic cells, arranged in series and secured with end plates and bolt sets. Figure 6.2 illustrates the structure of a typical single electrolytic cell, which includes three components: the membrane electrode, anode assembly, and cathode assembly.

The membrane electrode is the site where water is split into oxygen and hydrogen. It typically consists of a proton exchange membrane, which conducts protons, and a catalyst layer on both sides. This catalyst layer is usually created by loading precious metal catalysts such as platinum and iridium onto the surface of the proton exchange membrane through processes like spraying.

Outside the catalyst layer, a porous transport layer is positioned. This layer is typically sintered from titanium fiber or titanium powder and facilitates the transport of electrons, reactants, and products. To minimize the decay rate of the electrolytic cell, it's common to plate its surface. A flow field plate with channels for water and gas transmission is placed outside the porous transport layer. This arrangement ensures efficient operation of the PEM cell.

The output pressure of hydrogen is a crucial parameter in the design of an electrolytic cell. Currently, the output pressure of proton exchange membrane (PEM) hydrolysis hydrogen production equipment available in the market is typically below 5 MPa.

PEM electrolysis
Overview of components

Catalyst coated membrane

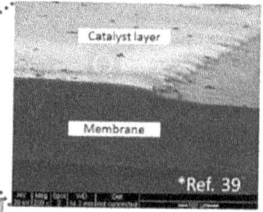

Catalyst layer

Membrane

*Ref. 39

Current collector

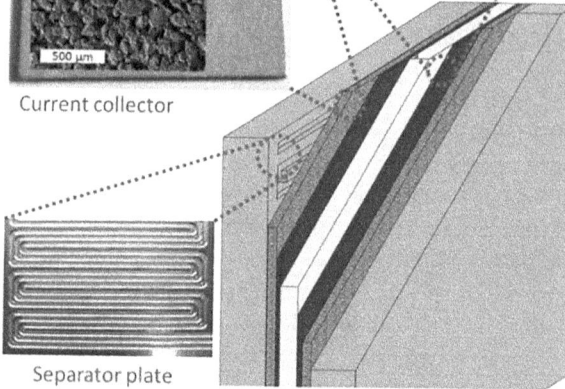

Separator plate

PEM electrolysis R&D

- Reduction or substitution of noble catalysts

- Reduce costs of current collectors and separator plates

- Develop microporous layer

- Improve membrane characteristics

- Enhance long-term durability

- Develop stack concepts (MW)

FIGURE 6.2   Cross section of the electrolytic cell [7].

To minimize the use of noisy and costly hydrogen compressors, or to cater to specific hydrogen applications such as space vehicles and submarines, the PEM hydrolysis cell needs to be capable of outputting high-pressure hydrogen and directly storing it in a cylinder.

In 2015, Honda designed and applied for an invention patent of a 70 MPa high-pressure PEM hydrolysis device, representing the highest level of current PEM high-pressure electrolysis technology. To ensure sealing performance, the hydrogen outlet is positioned at the center of the electrolytic cell, and multiple sets of sealing rings are used for simultaneous sealing. The effectiveness and durability of the electrolytic cell's sealing structure have been verified, and the minimum thickness of the PEM maintaining a high differential pressure of 70 MPa has been confirmed [8].

6.2.3 SYSTEM OF PEMWE

PEM hydrogen production system is generally divided into four parts: water treatment system, gas treatment system, electrolyzer system, and voltage conversion system. The water treatment system is used to produce high-purity deionized water required for electrolysis, which consists of reverse osmosis pure water device, electrodeionization system, etc. The gas treatment system is used to remove oxygen and water impurities from hydrogen to meet the requirements of users with high oxygen content and water content of hydrogen, and it is composed of a deaerator and two dryers to meet the continuous production of stable hydrogen. The electric tank system is the core device of PEM electrolytic water hydrogen production system, which can

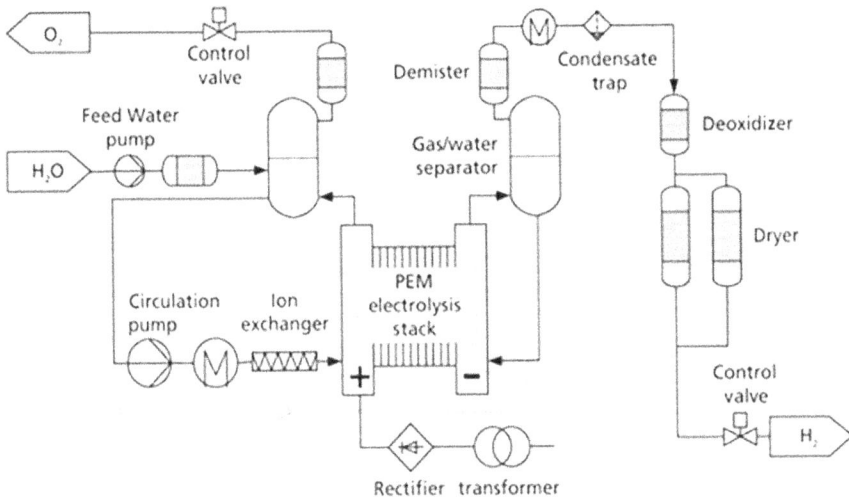

FIGURE 6.3 Schematic diagram of PEMWE system.

be divided into three parts: tank, anode, and cathode. The voltage conversion system provides the power and voltage required for the electrolytic cell to electrolyze water and consists of ACDC and DCDC (Figure 6.3).

At present, the application research of PEM electrolysis water hydrogen production system in wind-solar complementary power generation energy storage system has become a hot spot. According to the characteristics of wind and solar energy, the PEM electrolysis water hydrogen production system is developed with faster dynamic response speed, higher electrolytic efficiency, higher hydrogen production pressure, and smaller equipment footprint.

6.3 KEY COMPONENTS OF PEMWE

6.3.1 ELECTROCATALYST

6.3.1.1 Catalyst of Hydrogen Evolution Reaction

In 1972, Trasatti [9] first plotted a "volcanic" curve for HERs. In an acidic medium, the current density on the surface of the metal catalyst (M) and the bond energy strength (EM-H) of M-H formed during the hydrogen evolution reaction satisfy the "volcanic" curve, as shown in Figure 6.4. The "volcano type" curve is usually used as a basis for understanding the catalytic activity of different catalysts. It is a useful guide [10] for screening or designing suitable catalysts. The best metal catalysts for HER are located near the peak of the volcanic curve, and it is obvious that metal Pt is the preferred catalyst for the hydrogen evolution reaction of the PEMWE. Moreover, compared with other non-precious metal catalysts, it shows good electrochemical performance and long-term stability in acidic media. Therefore, metal Pt has always been the standard electrochemical catalyst on the cathode side of the membrane electrode.

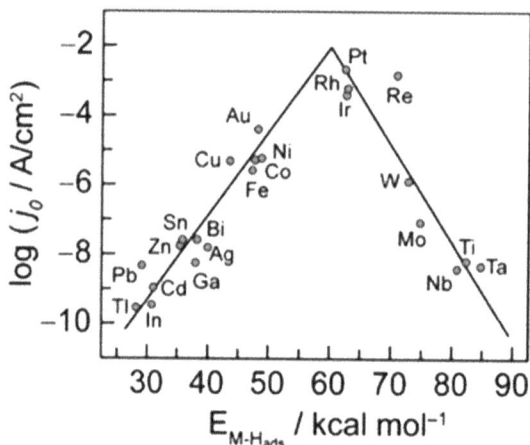

FIGURE 6.4 "Volcanic" relationship between current density on metal surface and M–H bond energy strength [11].

In addition, the catalytic activity of Pt can be improved by changing its surface structure during the preparation process. In order to further improve the life of the catalyst, the graphitized carbon material is usually used as the carrier. A higher degree of graphitization can effectively reduce the oxidation of the carrier under start–stop conditions and high-potential operation. In Lim et al.'s study [12], carbon-coated core–shell Pt/C catalysts containing nitrogen were prepared on carbon nanofibers for hydrogen evolution (cathode reaction of hydroelectrolysis), showing excellent catalytic activity with Pt and providing active sites for HER. Besides, the carbon shell also protects the Pt from dissolution and agglomeration, reducing the loss of catalyst activity during the reaction. Cheng et al. [13] used a carbon defect-driven spontaneous deposition method to construct highly dispersed, ultra-small (<1 nm) and stable Pt atomic clusters (Pt-ACs) supported by defective graphene as the hydrogen evolution electrocatalysts. The strong binding energy between Pt and carbon defects effectively limits the migration of Pt atoms. Meanwhile, the mass-specific activity, utilization efficiency, and stability of Pt catalyst were significantly improved.

Pt is very close to the top of the "volcanic" diagram compared to other transition metals and has excellent catalytic activity. However, it is still not the volcano's peak, which means that there are catalysts with higher catalytic activity than Pt, providing a theoretical basis for designing new catalysts. It is found that the catalytic performance of Pt can be improved effectively after alloying with other transition metals. For example, catalysts of Pt_3M series, because of the change in the surface structure of the d-band's center position, the catalytic ability is also changed. In the Pt_3M series catalysts, the catalytic activity is $Pt < Pt_3Ti < Pt_3V < Pt_3Ni < Pt_3Fe < Pt_3Co$. In addition to binary alloy catalysts, there are also ternary alloy catalysts, such as Pt_6CrNi_{11}, Pt_6CuFe_9, and Pt_6FeNi_{11}, which also show higher catalytic activity than Pt.

Reducing the amount of precious metals in the catalyst and developing non-precious metal hydrogen evolution catalysts adapted to acidic environments are another research focus, in which sulfide, phosphide, carbide, and boride show promising and desirable

HER activity. Xu et al. [14] tested the HER of tungstophosphate heteropolyanion (PWA) hybridized with carbon nanotubes (CNTs), and the results showed that the activity of the new catalyst can reach 20% of that of Pt/CNTs. Hinnemann et al. [15,16] investigated MoS2 and its compounds as alternative catalysts for HER. The results showed that MoS2 exhibited excellent electrocatalytic activity in HER and was a suitable catalyst material for HER. However, the current density is significantly lower than that of conventional Pt catalysts. Therefore, platinum-group precious metals are still the primary raw materials used in the HER side of PEMWE technology in the short term.

6.3.1.2 Catalyst of Oxygen Evolution Reaction

Miles and Thomason [17] studied OER activity in 1 M H_2SO_4 (80°C) by cyclic voltammetry. The order of OER activity was $Ru \approx Ir > Pd > Rh > Pt > Au > Nb$. Another study by Reier et al. found that the OER activity of Ru, Ir, and Pt nanoparticles in 0.1 M $HClO_4$ followed the order $Ru > Ir > Pt$ [18]. It can be seen that the oxides of Ru and Ir precious metals have good catalytic activity, making them indispensable materials in OER catalysts. As shown in Figure 6.5, among metal oxides materials, RuO_2 exhibits lower overpotential [11] under acidic conditions and is near the top of the "volcanic" diagram of OER activity. Theoretically, it is the best catalytic material for oxygen evolution at present.

However, Ru will be seriously corroded during OER process in acidic electrolyte, resulting in lower stability of Ru as OER electrocatalyst. In 1983, Kötz et al. [19] found precipitation and corrosion of metal Ru and Ir electrodes at the anode side in 0.5 M H_2SO_4. At the same time of O_2 precipitation, metal Ru would form a hydrate oxide film with high defects on the metal surface, which would be dissolved from the catalyst layer by serious corrosion. A corrosion model was built for this process in Figure 6.6.

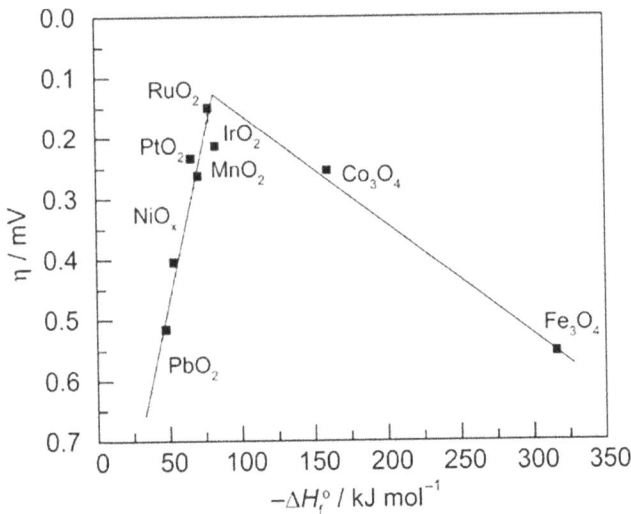

FIGURE 6.5 Volcano-shaped relationship between OER activities on metal oxide surfaces and enthalpy for the transition metal oxides in acidic [11].

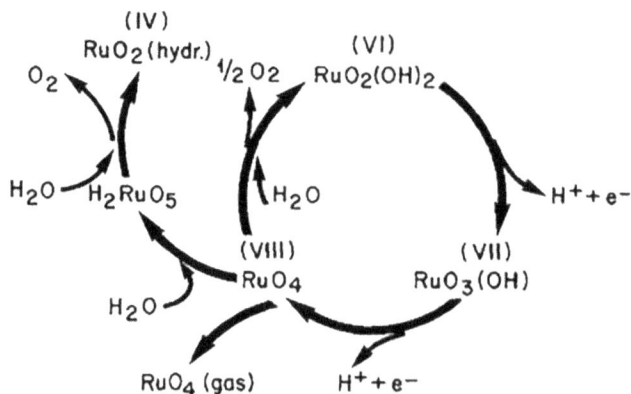

FIGURE 6.6 Model for the oxygen evolution and corrosion on Ru and RuO_2 electrodes [19].

In the research of precious metal oxygen evolution catalysts, the focus is to improve the catalytic performance and increase the durability of catalysts by doping, modifying, and changing the crystal structure and specific surface area [20–22]. Wang et al. [23] doped Rh (Ru mass percentage 21.16%) into graphene-supported RuO_2 through ion exchange adsorption, and Rh stabilized the metal valence state during the OER reaction. The low-valence Ru-O-Rh activity center and oxygen vacancies worked synergistically to optimize the activity and stability of the catalyst, and η10 was 161 mV in 0.5 mol/L H_2SO_4 and stable for 700 hours at a current density of 50 mA/cm². Similar to RuO_2, the introduction of impurity atoms or the formation of solid solutions in IrO_2 is also a common modification method. Huo et al. [24] introduced Re (denoted as Re-IrO_2, Ir atomic percentage is 89%) into IrO_2, and the incorporation of Re does not change the structure of IrO_2 due to the similar ion radii of the two. The surface migration energy of Re-IrO_2 is much greater than that of other elements, such as Ni, Cu, and Zn-doped IrO_2, and the increased surface energy can inhibit the dissolution of Ir. In addition, doped Re increased the recrystallization temperature of IrO_2, thereby increasing OER activity, with η10 being 255 mV in 0.5 mol/L H_2SO_4.

Non-precious metal oxygen evolution catalyst is a kind of potential low-cost oxygen evolution electrocatalyst. By means of heterogeneous element doping or dispersion, anchoring, and protective coating of single atom catalyst, non-noble metal OER catalysts that are stable in acid hydroelectrolysis systems can be prepared. The development of highly active non-precious metal electrolysis catalysts is of great significance for the wide application of hydrogen energy.

Among non-precious metal catalytic materials, only a few amorphous nitrides, Co spinel oxides and Ti alloys, N-doped carbides such as NbNx and ZrNx, and Ni-Mn-SbOx can be applied in acidic OER processes. Gao et al. [25] demonstrate that N atom doping can regulate the electronic structure of single atom catalyst and promote the catalytic activity of OER. The N doping in the Co@GY catalyst induces the redistribution of Co single atoms on the catalyst surface, which greatly enhances the OER activity of the catalyst. Also, the doped N induced the change in charge density on the catalyst surface, which not only increased the number of active sites, but also weakened the adsorption of oxygen-containing species, significantly reducing OER overpotential, as shown in Figure 6.7.

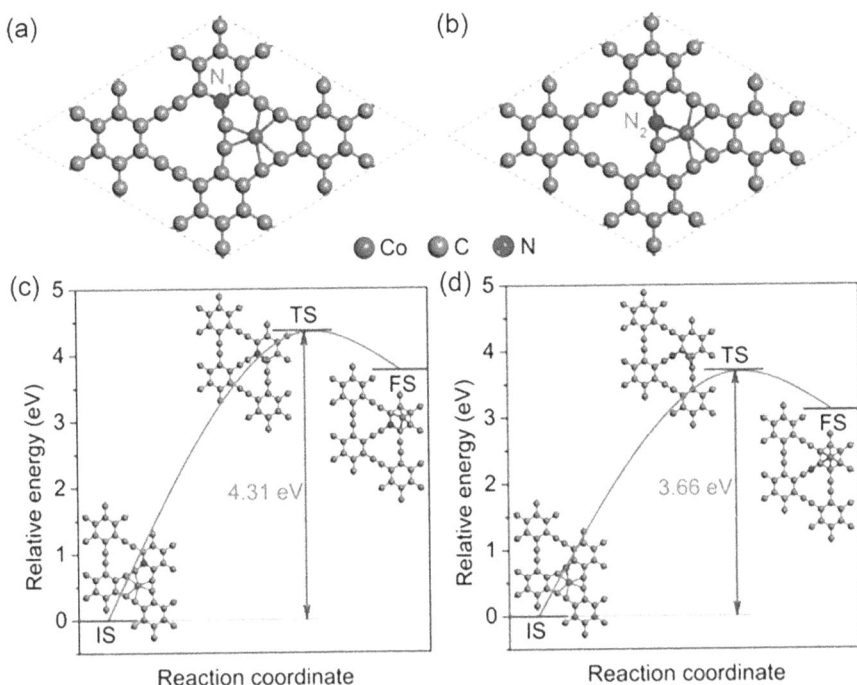

FIGURE 6.7 The optimal structure of Co@N2-GY and Co@N-GY and the minimum energy conversion path of adsorbed species: (a)–(c) Co@N1-GY and (b)–(d) Co@N2-GY [25].

In acidic electrolyte, the initial oxygen evolution overpotential of Mo-Co_9S_8@C is 200 mV, and the slope of Tafel curve is 90.3 mV/dec (Figure 6.8) [26]. For 10 mA/cm^2 current density, the oxidation overpotential is 370 mV. DFT calculation shows that the synergistic interaction between atomically dispersed Mo and Co substrate can effectively change the binding energy of intermediate reactive species on the catalyst surface and reduce the OER overpotential. There is a strong interaction between monatomic metal on Mo-Co_9S_8@C catalyst surface and catalyst support, which significantly improves the stability and electrochemical activity of catalyst. Mo single atom doping has a very important effect on improving the kinetic activity of catalyst OER. The synergistic interaction of Co_9S_8 and Mo atom coupled with charge transfer promoted the conversion of H_2O adsorbed at the active site of Co, and O_2 precipitation reaction occurred. Monatomic modification strategy is beneficial for the preparation of low-cost OER electrocatalysts with stable performance in acidic hydroelectrolysis systems, which will significantly reduce the cost of electrolysis catalysts and has important application value and significance. It is necessary to further understand the catalytic principle of non-noble metal OER catalysts, so as to develop new non-precious metal catalyst materials with high active site density and high stability.

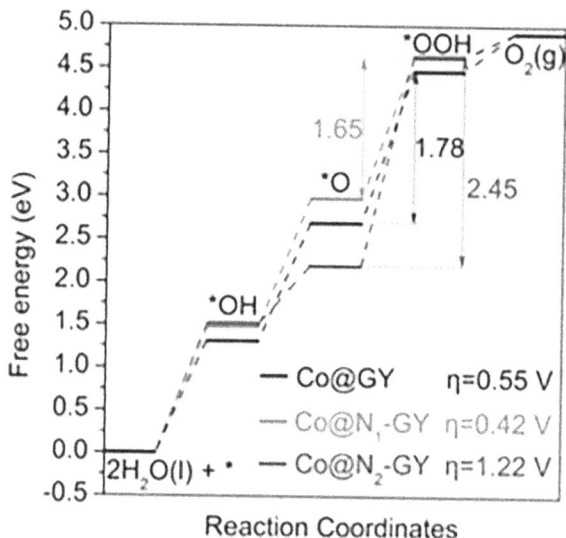

FIGURE 6.8 Calculation of free energy of Co active site for catalysts such as Co@GY, Co@ N1-GY, and Co@N-GY, U = 0 V [25].

6.3.2 Proton Exchange Membrane

The proton exchange membrane used for PEMWE should have high proton conductivity, low permeability, high mechanical strength, good thermal and chemical stability, and high durability. In order to obtain better overall performance in PEMWE, perfluorosulfonic acid (PFSA)-based electrolyte membranes need to be optimized in terms of EW value, thickness, functional additives, and porous support layer.

PFSA membranes are commonly used commercial PEM for hydrogen production. Nafion®, a PFSA solid polymer electrolyte, is the benchmark membrane in PEM electrolyzers due to its chemical stability, mechanical strength, and proton conductivity. The membrane has hydrophobic fluorocarbon backbone and hydrophilic sulfonic acid end group side chain.

Thicker PEMs are generally used in conventional PEMWE because thicker diaphragms possess high mechanical strength and low gas permeation especially hydrogen permeation. It has been found that a new structure of MEA or membrane with an interlayer containing a noble metal catalyst such as Pt between the cathode and anode or in the PEM was proposed. The interlayer allows hydrogen permeating from the cathode side to the anode side to react with oxygen in a recombination reaction, thereby greatly reducing the amount of hydrogen permeating and the hydrogen concentration in the anode. Moreover, the closer the interlayer is to the anode side, the better the effect, because the oxygen concentration on the anode side is high, which is favorable for combining with hydrogen, and increasing the pressure on the anode side is favorable for the functioning of the precious metal interlayer. Mirshekari et al. [27] embedded a Pt sandwich with Pt loading of 0.025 mg/cm² and thickness of 100–200 nm in MEA, which could reduce hydrogen permeation to less than 10% of the low flammability limit, and the composite MEA was operated with a low loading catalyst for 3000 hours without significant degradation.

In order to reduce ohmic losses and their associated energy losses at high current densities, one of the reliable options is to reduce the membrane thickness. However, there are some disadvantages of using thin PEM for PEMWE that need to be improved. First, reducing the thickness of PEM increases the amount of gas permeation and reduces the purity of the hydrogen produced by the cathode, while reaching a certain concentration of hydrogen in the oxygen side poses a safety hazard. Second, the mechanical strength of thin PEM is low. It is easy to form defects and pinholes, which affect the safety of the electrolyzer and reduce the durability. Third, thin PEM is not suitable for high-pressure electrolyzer, which is damaged due to the large pressure difference between two sides. Therefore, the selection of membrane thickness must be considered from the aspects of proton conductivity, permeability, mechanical strength, and so on.

Developing a composite membrane with a supporting layer is one of the methods. Gore used expanded polytetrafluoroethylene (ePTFE) as a reinforcing layer, impregnated with PFSA resin to prepare ultrathin reinforced PEM, which has been applied to fuel cell vehicles from Toyota, Hyundai, Honda, and SHPT (Shanghai Hydrogen Propulsion Technology Co., Ltd.). Fumatech's reinforced short-branched membranes for PWMWE use PEEK as the internal porous layer, and by optimizing the porous support substrate, the membrane thickness is set at 80–150 μm. The decomposition voltage at 80°C is 1.71 V, and it can withstand a pressure difference of 0.5–2 MPa on both sides [20]. Giancola et al. [28] used polysulfone (PSU) nanofibers as the porous support layer to prepare composite membranes impregnated with Aquivion PFSA (PFSA) resin, which could simultaneously ensure the proton conductivity and mechanical strength of the PFSA-based electrolyte membrane. Compared with unsupported membranes, the composite membranes can guarantee the proton conductivity and improve the mechanical strength, effectively reducing the membrane swelling ratio and hydrogen permeation, and making it possible to use thinner and more durable membranes in PEMWE.

In recent years, many studies have been conducted to find a hydrocarbon film as an alternative to PEM hydroelectrolysis. The cost and performance of hydrocarbon polymer electrolytes at high temperature and low humidity are expected to exceed that of commercial fluorinated polymer electrolytes. Researchers have attempted to use hydrocarbon membranes to replace PFSA membranes in PEMWE [29,30] because hydrocarbon membranes have lower gas permeability and higher proton conductivity. At present, the hydrocarbon films that have received attention include sulfonated polyether ketone (sPEEK), sulfonated polyether sulfone (sPAES), and sulfonated polyphenylene sulfide sulfone (sPPS) [31–33].

The key issue in synthesizing hydrocarbon-based polyelectrolytes is the way in which polar groups are introduced. Most of the polar groups are sulfonic acid groups, and the stability of the introduced sulfonic acid group is considered first. In general, desulfurization problem is also important for these sulfonated polyelectrolytes. This problem can be alleviated by alkyl sulfonation and the introduction of electron adsorbent groups, but there is still no fundamental solution, which is one of the reasons for the low chemical stability of hydrocarbon-based electrolytes. There are two common sulfonation reactions for hydrocarbon-based polyelectrolytes. First, direct polymer sulfonation is sulfonation of the base polymer with sulfuric acid or other sulfonating agents. This is a macromolecular reaction that usually takes place in a heterogeneous system, making it difficult to precisely control the sulfonation reaction. Second, sulfonic acid groups are introduced in the monomer stage to protect the monomer, and then prepolymerization and block copolymerization are carried out. This method can be sulfonated in an orderly manner, the monomer can be disulfonated and tetra-sulfonated, and the molecular chain is relatively regular, which is conducive to the comprehensive regulation of the proton conductivity and mechanical properties of the prepared diaphragm.

Block copolymerization is attracting attention as a method to alleviate the problem of high molecular weight to some extent and to significantly improve mechanical and electrochemical properties. As a result, the hydrophilic aggregates form ion channels for efficient proton conduction, while the crystalline aggregates of the hydrophobic moieties maintain the mechanical properties. Similar attempts have been made to synthesize the block copolymers shown in Scheme for hydrocarbon-based polyelectrolytes. Block copolymers are obtained by synthesizing oligomers as block components and further polymerizing them. In practice, it is difficult to precisely synthesize oligomers with high molecular weight, and the reproducibility of synthesis is problematic. In addition, exchange reactions may occur during the copolymerization reaction, making it difficult to precisely control the polymer structure. Since materials with excellent electrochemical or mechanical properties have been obtained, it is necessary to develop a new material with a simple and controllable structure (Figure 6.9).

The development of PEMWE was made possible by technological advances in proton exchange membranes. As a solid electrolyte, the characteristics of the proton exchange membrane not only affect the performance of PEMWE, but also affect the design of Balance of Plant (BOP) and system safety, so the selection and optimization of the proton exchange membrane in the membrane electrodes not only need to be considered to satisfy the requirements of the electrolysis efficiency of the PEMWE, but the life of the performance requirements needs to be taken into account the requirements of the safety aspects.

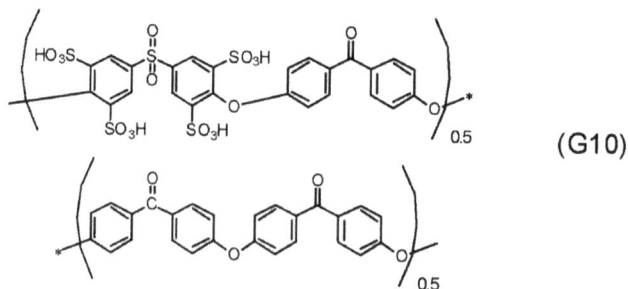

(G10)

FIGURE 6.9 Copolymer electrolytes based on hydrocarbon polymers.

6.3.3   PorOuS TransporT Layer

The PTL of PEM water electrolysis cell mainly plays the role of material distribution and conduction current. The material used as the porous diffusion layer of PEM water electrolysis cell should meet the following requirements: (1) Excellent mechanical properties, which can achieve the effect of supporting the electrolytic cell; (2) high porosity and strong permeability, which can ensure material transmission; (3) good electrical conductivity, low contact resistance with the film electrode, so as to meet the current transmission; and (4) excellent corrosion resistance under high potential. At present, the research work of porous diffusion layer has several directions: (1) The research on the porous diffusion layer material itself, mainly including the development of new materials, coating [34], and modification. (2) Study on the influence of porous diffusion layer on the performance of electrolytic cell, including the durability of porous diffusion layer, corrosion mechanism, bubble formation mechanism, and influence in porous diffusion layer.

6.3.3.1   The Influence of PTL's Structure

The distribution of pore size, porosity, and infiltration in the diffusion layer significantly impacts the electrolyzer's operational efficiency, internal state distribution, and durability.

In general terms, a larger pore size and higher porosity facilitate material transportation [35,36]. However, when the pore size and porosity become too large, the contact between the diffusion layer and the catalyst layer deteriorates [37]. This leads to an increase in resistance [38] and a decrease in the utilization rate of the catalyst layer [39]. Therefore, in practical applications of the diffusion layer, its pore size and porosity have optimal values. According to literature statistics, the optimal pore size is approximately 10 μm, and the optimal porosity is between 30% and 50% [40]. Ito [41,42] utilized titanium felt as the substrate to investigate the impact of pore size and porosity on the diffusion layer. The findings indicate that a decrease in pore size is beneficial to electrolytic performance when the pore size exceeds 10 μm, and changes in porosity have minimal effect on performance when porosity is higher than 0.5. It is discovered through theoretical and experimental studies that the optimal pore size of powder sintered titanium was 12–13 μm, while the optimal particle size of titanium powder was 50–75 μm [35].

In PEM fuel cells, a microporous layer is often added between the diffusion layer and the catalyst layer to enhance contact and promote liquid water discharge in order to prevent waterlogging. This approach has been applied in recent electrolytic cell research. Huget [43] utilized the pore network model to analyze the impact of varying pore size gradients on two-phase transport. The simulation results demonstrated that the diffusion layer structure model, where the pore size increases from the catalyst layer to the flow channel (LTH), has superior overall transport capacity. This structure provides a gas transport path while ensuring smooth water flow. In contrast, in the mode without a pore gradient or with pore size decreasing from the catalyst layer to the flow path (HTL), water becomes segmented into isolated clusters in the diffusion layer region near the flow path. Lee [44] conducted a comparison of the performance of two diffusion layers with opposing porosity gradients. The results showed that when the low porosity region of the diffusion layer was in contact with the catalyst layer, performance improved. Specifically, the voltage was reduced by 29% at 4.5 A/cm², the gas content in the catalyst layer was decreased by 50% as shown by neutron imaging results, and the mass transfer capacity of the electrolytic cell was significantly enhanced. Schuler prepared three types of microporous layers, each with a different pore size. When compared to the traditional single-layer structure, the electrolytic cell utilizing the microporous layer demonstrated superior electrochemical performance. The mass transfer overpotential at 2 A/cm² was reduced by 60 mV. The high specific surface area and low roughness of the microporous layer increased the catalyst utilization rate by more than two times. Additionally, due to the flatter surface and smaller local stress of the microporous layer, the allowable thickness of the proton exchange membrane was also reduced [45]. In addition to improving the gradient in the thickness direction of the diffusion layer, some studies have suggested that in-plane ordering structure designs, such as diffusion layer drilling, can also enhance the mass transfer performance of the electrolytic cell [46–48].

The flow of water and gas in porous media is driven by capillary force [49], which means that the infiltration of the diffusion layer will also influence the mass transfer process of the electrolytic cell. Kang introduced hydrophobic reagents to the diffusion layer to render it hydrophobic. The results show that not only the mass transfer impedance but also the ohmic impedance and activation impedance increase [50]. Lim alternately applied hydrophilic and hydrophobic modification agents to the surface of the diffusion layer to create an amphiphilic diffusion layer. This reduced water and gas interference in the diffusion layer and improved the performance of the PEM fuel cell using this diffusion layer by 4.3 times and the electrolytic cell by 1.9 times [51]. Although hydrophilicity is a crucial property of the diffusion layer, research in this area remains significantly lacking.

6.3.3.2   The Influence of Materials

During operation, the anode potential is higher than the cathode, and the anode is highly corrosive. At present, the porous diffusion layer materials on the anode side of the electrolytic cell are mainly corrosion-resistant materials such as titanium mesh [52], titanium plate, titanium felt [42], and stainless steel [53]. The cathode porous diffusion layer generally uses carbon materials, such as carbon paper, carbon cloth, carbon felt, and some other chemical/physical modified materials. Mo [53] used stainless steel mesh as anode PTL in PEMEC to study the corrosion mechanism of metal migration. Zhang [54] compared the performance of PEM water electrolysis cell using titanium felt and studied titanium microporous layer. It was found that the microporous layer changed the wettability of the titanium diffusion layer, showing super hydrophobicity, and the microporous layer increased the ohmic resistance while improving the catalytic activity. Kang et al. [46] first comprehensively studied the thin titanium-based PTL with straight holes and clear hole morphology. A novel PTL with a uniform spatial distribution of 400 μm pore size and 0.7 porosity was obtained by micromachining technology (lithography and wet etching), as shown in Figure 6.10.

(a)   (b)

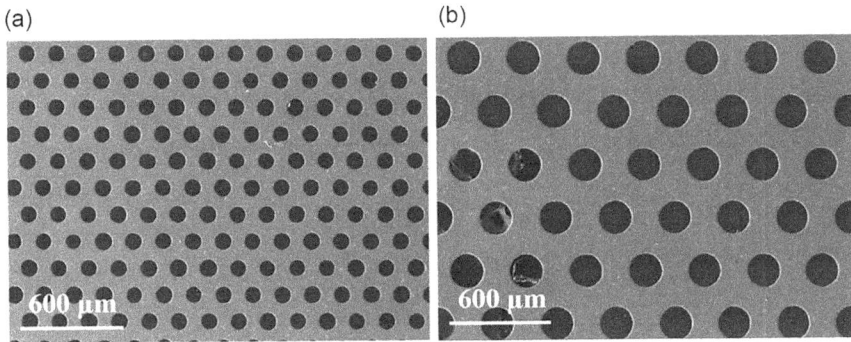

FIGURE 6.10   SEM images of typical thin/well-tunable titanium LGDLs. (a) Pore morphology and structure of sample A1 with approximately 100 µm pore size and 0.30 porosity. (b) Pore morphology and structure of sample A2 with approximately 200 µm pore size and 0.30 porosity [46].

6.3.3.3   The Influence of PTL Coating

However, over time, a "thick" oxide layer tends to grow on the surface of titanium [55]. This will lead to an increase in the interfacial contact resistance (ICR) at the PTL|CCM and BPP|PTL interfaces, which will have a negative impact on the electrolysis efficiency [56]. Researchers have solved this problem by applying precious metal coatings, such as platinum or gold [57]. Rakousky [58] uses Pt-coated PTL, which greatly reduces the degradation rate to only 12 µV/h, indicating that non-corrosive anode PTL is essential for PEM electrolyzers. Liu et al. [59] sputtered a very thin iridium layer onto the titanium PTL to protect the titanium PTL from passivation. Iridium coating is uniformly deposited throughout the internal structure of the PTL. Studies have shown that the iridium layer reduces the overall ohmic resistance of the PTL|CL interface, improves the performance of the battery, and leads to an increase in the durability of the electrolytic cell. Bystron et al. [60] etched the surface of PTL (felt) with 55% HCl at 35°C (the boiling point of the solution is 54°C). This easy-to-implement chemical treatment (compared to the coating process) leads to the formation of titanium hydride on the surface. Compared with the original PTL, the use of etched PTL achieves excellent electrolysis performance (higher current density and lower ohmic resistance). In addition to adding a coating on a titanium-based PTL, the researchers also coated the stainless steel material as a PTL. Stiber et al. [61] used vacuum plasma spraying (VPS) to deposit Nb/Ti on 4-layer reticulated 316L stainless steel PTL. The double layer consists of a thick (about 20–50 µm) titanium film covered with a thin layer of niobium (a few microns thick) to avoid titanium oxidation, as shown in Figure 6.11. Daudt et al. [62] studied the composite material made of porous stainless steel 316L substrate coated with Nb as the PTL of PEMECs. Studies have shown that the use of 316L is expected to reduce the manufacturing cost of PTL, and the addition of niobium layer aims to improve the service life and performance of PEMEC due to its excellent corrosion resistance in an acidic environment.

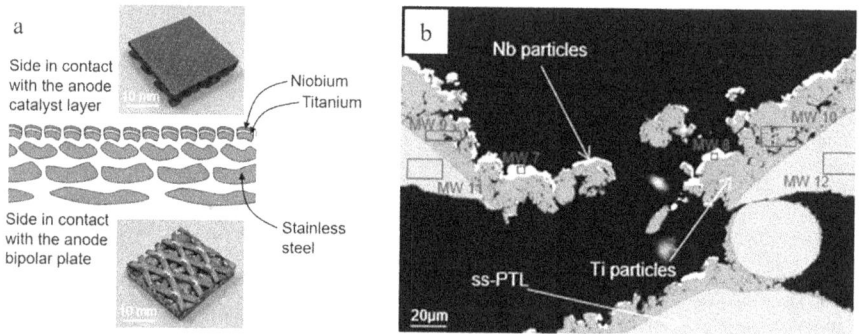

FIGURE 6.11 Thick Nb/Ti bilayer deposited by vacuum plasma spray (VPS) on stainless steel PTL. (a) Schematic description of the coating on the outermost part of the PTL. (b) Cross-section SEM micrograph of the bilayer [61].

6.3.4 BIPOLAR PLATE

Bipolar plates are critical components of PEM electrolysis. Their primary function is to establish electrical connections between adjacent electrolysis cells, transport liquid reactants (H_2O), and gas products (H_2 and O_2), while efficiently conducting heat. They operate in a harsh environment with strong acidity, high voltages, and oxidation (anode) and reduction (cathode) processes. As a result, bipolar plates must possess high electrical conductivity, corrosion resistance, and impermeability. To meet these requirements, appropriate materials are selected for bipolar plates, and surface coatings and treatments are applied. Another challenge lies in cost control since bipolar plates typically account for approximately half of the overall electrolysis cost [63]. Thus, reducing the manufacturing cost of bipolar plates while improving their performance represents a key challenge in bipolar plate engineering applications (Figure 6.12).

6.3.4.1 Substrate Types

Titanium possesses excellent corrosion resistance, low initial electrical resistivity, good mechanical strength, and lightweight characteristics, making it the preferred

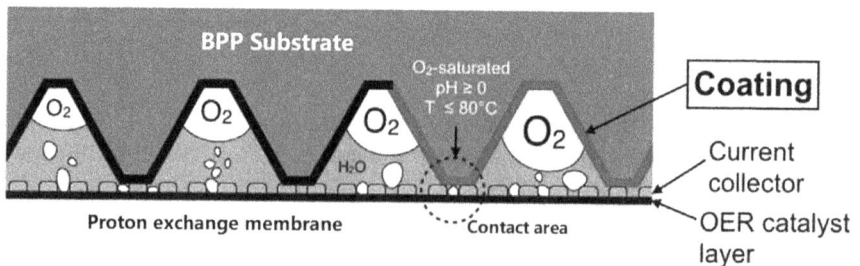

FIGURE 6.12 Cross-sectional scheme of a PEM electrolysis anode. The dashed circle indicates the area of contact between the BPP and the current collector and the oxygen evolution reaction (OER) catalyst layer.

material for bipolar plates in PEM electrolysis [7]. However, titanium plates are prone to passivation corrosion. In high potentials, high humidity, and oxygen-rich environments, the surface of titanium bipolar plates tends to passivate, forming an oxide film. This low-conductivity oxide film significantly increases the contact resistance between the bipolar plates and the current collector [64]. Therefore, titanium plates must undergo coatings and surface treatments to meet durability and performance requirements in high-pressure and oxidative environments.

Lowering the cost of bipolar plates requires finding suitable alternatives to titanium. Stainless steel is one of these alternatives, offering cost advantages and ease of processing. However, stainless steel is prone to rapid corrosion in acidic environments. To mitigate this, a protective layer, typically made of titanium, is applied to prevent pitting corrosion. With the development of composite coating technologies, the differences in material types for bipolar plates may gradually be overlooked, allowing for wider application to relatively active conductive metals such as copper, aluminum, and magnesium alloys [65].

6.3.4.2 Coating Process

Surface coating modification methods have become a research hotspot in the field of bipolar plates. Various preparation processes are employed to apply conductive and corrosion-resistant coatings to the bipolar plate substrate. Coating materials can be divided into precious metal coatings and non-precious metal coatings.

Currently, precious metal coatings remain the most commonly used coating materials in bipolar plate engineering applications. Electroplated gold, platinum, and other precious metal coatings naturally excel in chemical inertness and conductivity. While precious metal coatings are one of the most effective modification methods for enhancing the corrosion resistance of metal bipolar plates, their high cost limits widespread adoption. Current research directions include optimizing coating quality and reducing precious metal usage through techniques like HiPIMS, which allows for controlling platinum sputtering thickness to under 100 nm. Additionally, non-precious metal composite coatings doped with trace amounts of precious metals hold promise in maintaining excellent performance at lower costs.

Non-precious metal coatings include nitride coatings, carbide coatings, oxide coatings, and alloy phase coatings, among others, which have the potential to fully replace precious metal coatings, reducing the cost of bipolar plates. Surface treatment technologies can directly employ oxide films formed on metal sheets, such as TiO_2, as protective coatings. The corrosion resistance and ICR of TiO_2 can be improved through doping with Nb [66]. Sputter deposition of new conductive oxide thin films (e.g., Ti_4O_7) is also under study [67]. Using metal carbides as surface coatings helps achieve cost reduction. The carbides of Ti, Nb, Mo, or Cr can effectively form a dense and uniform protective film on stainless steel sheets. The particulate outer layer is hydrophobic, blocking the penetration of external corrosive solutions, while the robust inner layer enhances adhesion to the metal bipolar plate [68]. Metal nitrides combine the advantages of conductivity and thermal stability. Properly designed, metal nitride coatings can address cost issues. Several mature process technologies, such as Plasma Enhanced Reactive Evaporation (PERE) [69], Plasma Enhanced Atomic Layer Deposition (PEALD) [70], and Multi-arc Ion Plating [71], have been used to deposit protective TiN thin film coatings on metal bipolar plates. MAX phase materials, which combine metal conductivity and ceramic durability, have also been employed in coating preparation. Coatings like TiCN [72] and TiSiN [73], among others, have been successfully developed, with process optimization aiding in the formation of nanocrystalline structures [74], further enhancing performance.

6.3.5   MEMBRANE ELECTRODE ASSEMBLIES (MEAs)

MEAs are the core components of proton exchange membrane (PEM) water electrolysis, which is a promising technology for hydrogen production from renewable energy sources. MEAs consist of a solid polymer electrolyte membrane sandwiched between two CLs and two gas diffusion layers (GDLs). The CLs contain electrocatalysts that facilitate the water splitting reactions at the anode and the cathode, while the GDLs provide electrical conductivity and gas transport. The membrane serves as a proton conductor and a gas separator. The performance and durability of MEAs depend largely on their structure, morphology, and composition, which are influenced by the preparation methods.

The preparation methods of MEAs can be classified into two main categories: direct coating methods and indirect coating methods. Direct coating methods involve the deposition of catalyst inks or slurries onto the membrane or the GDLs, followed by hot pressing to form the MEA. Indirect coating methods involve the fabrication of catalyst-coated membranes (CCMs) or catalyst-coated GDLs (CCGs) separately, followed by lamination or bonding to form the MEA. Each method has its own advantages and disadvantages in terms of cost, scalability, quality control, and performance optimization.

Here, we will review the recent progress and challenges in the preparation methods of MEAs for PEM water electrolysis, with a focus on the slurry dispersion process and the direct and indirect coating methods. The effects of different parameters on the structure and performance of MEAs will be discussed, as well as the future perspectives and opportunities for further improvement.

6.3.5.1   Slurry Dispersion Process

One of the key steps in the preparation of MEAs is the dispersion of catalyst particles in slurries, which affects the homogeneity, porosity, and activity of the CLs. The dispersion state of particles is influenced by various factors, such as the type and concentration of solvents, surfactants, binders, and additives, as well as the mixing and sonication conditions. Different dispersion techniques have been developed to achieve optimal particle size distribution, stability, and viscosity of slurries.

Slurry dispersion process is the process of uniformly dispersing solid particles in a liquid medium, forming a stable suspension. Slurry dispersion process is an important step in many industries, such as manufacturing paints, coatings, pharmaceuticals, cosmetics, and food products. Effective dispersion ensures the uniform distribution of particles and components in the slurry, resulting in improved quality and performance. All ingredients must be carefully matched to optimize mixing and homogenization. The finer the particle size range, the more stable the slurry. Modern dispersion and measurement equipment are used to optimize this process.

Slurry dispersion process mainly involves the following steps:

1. Pretreatment: The raw materials, such as active materials, conductive agents, and binders, are subjected to screening, drying, mixing, and other operations to remove impurities and moisture, and improve uniformity.
2. Wetting: The raw materials are slowly added to the solvent, so that the air or other impurities on the surface of the powder are replaced by the solvent, increasing the affinity between the powder and the solvent. The wetting quality can be expressed by the wetting angle or wetting heat.
3. Dispersing: The wetted powder is subjected to mechanical forces, such as shear, impact, or cavitation, to break up the agglomerates and achieve a fine and uniform particle size distribution. The dispersing quality can be evaluated by the particle size analysis, zeta potential measurement, or rheological measurement.
4. Stabilizing: The dispersed particles are prevented from re-agglomeration or sedimentation by adding stabilizers, such as surfactants, polymers, or electrostatic charges. The stabilizing quality can be assessed by the storage stability test or sedimentation test.

Slurry dispersion process has a significant impact on the structure and performance of various products that use slurries as raw materials or intermediates. For example, in lithium-ion battery manufacturing, slurry dispersion process affects the electrode structure, performance, and lifespan. The slurry dispersion process needs to be optimized for different types of particles, solvents, binders, and applications. Some of the important parameters that affect the slurry dispersion process are:

1. Solvent type and concentration: The solvent affects the solubility of the binder and the surface properties of the particles. The solvent should have a high dielectric constant, low surface tension, low boiling point, and good compatibility with the particle surface. The solvent concentration affects the viscosity and stability of the slurry.
2. Surfactant type and concentration: The surfactant affects the wetting and stabilizing of the particles. The surfactant should have a suitable hydrophilic–lipophilic balance (HLB) value, low critical micelle concentration (CMC) value, and good compatibility with the solvent and the binder. The surfactant concentration affects the zeta potential and rheology of the slurry.
3. Binder type and concentration: The binder affects the adhesion and cohesion of the particles. The binder should have a high molecular weight, good elasticity, low glass transition temperature (T_g), and good compatibility with the solvent and the particles. The binder concentration affects the solids content and the electrical conductivity of the slurry.
4. Mixing and sonication conditions: The mixing and sonication affect the dispersion and homogenization of the particles. The mixing and sonication should provide sufficient shear force, impact force, or cavitation force to break up the agglomerates and achieve a fine and uniform particle size

distribution. The mixing and sonication parameters, such as speed, time, temperature, and power, should be optimized to avoid excessive energy input or damage to the particles.

The future perspectives and opportunities for further improvement of slurry dispersion process are:

1. To develop novel and efficient particles that can improve the quality and performance of various products that use slurries as raw materials or intermediates. For example, nanostructured, alloyed, or composite particles for lithium-ion batteries, paints, and coatings.
2. To optimize the slurry formulation and dispersion process for different types of particles, solvents, binders, and applications, to achieve stable, homogeneous, and well-dispersed slurries with suitable rheological properties for coating or casting.
3. To explore new and green solvents or solvent-free methods for slurry preparation and dispersion, to reduce the solvent consumption and environmental impact, as well as to improve the quality and efficiency of slurry production.
4. To integrate the slurry dispersion process with the product design and operation, as well as to achieve optimal structure and performance of various products that use slurries as raw materials or intermediates.

6.3.5.2 Fabrication Methods

The preparation methods [75,76] of traditional PEM hydrogen production MEA can be divided into two categories based on different CL support materials: one is the CCS (Catalyst Coated Substrate) method, which involves directly coating the catalyst active component onto the GDL to prepare cathode GDL and anode GDL with coated catalyst layers. These two GDLs are then pressed on both sides of the PEM (polymer exchange membrane) using a hot pressing method to obtain the MEA; the other is the CCM method, which involves coating the catalyst active component onto both sides of the PEM and then attaching the cathode and anode GDLs onto the CL on each side, followed by hot pressing to obtain the MEA.

For the CCS fabrication process, the catalysts are first coated onto the porous transport layers (substrate) and form the anode and cathode, which are then fabricated with the membrane on both sides, i.e., the CCS-MEA. This MEA production does not suffer from the swelling of the membrane but is limited by the insufficient cohesion of catalyst layers with membranes. Reasonably, the CCS-MEA has large contact resistance and low proton-transfer conductivity. Moreover, the initial catalyst coating onto the porous transport layers would lead to low catalyst utilization. Compared with CCS method, CCM method can effectively improve the catalyst utilization rate and greatly reduce the proton transfer resistance between membrane and CL, so it has become the mainstream technology for MEA preparation.

Compared with the CCS method, the CCM method has a higher catalyst utilization rate and greatly reduces the proton transfer resistance between the membrane and the catalytic layer. CCM is the most common method to prepare MEAs [77]. In this process, the homogeneous catalyst slurry, which was performed by sonication of a mixture of catalysts, ionomer solution, and solvent, is directly coated on the membrane and hot-pressed at high pressure [77]. The brush coating method, ultrasonic spraying method, screen printing method, sputtering method, electrochemical deposition, and decal method are usually used in CCM methods.

Currently, the main methods for mass production of PEMWE membrane electrodes are coating methods, including ultrasonic spraying, doctor-blading, roll coating, and slot-die coating. Depending on different ways of coating the catalyst layer onto the substrate, there are two main techniques: transfer method and direct coating method. The direct coating technique involves directly applying the catalyst ink onto the proton exchange membrane (PEM). However, due to the solvent swelling of the PEM, this technique presents difficulties in process development and has a narrow process window. Currently, the commonly used method is ultrasonic atomization spraying [78–80], which directly sprays the catalyst layer onto the PEM. The catalyst ink is loaded into an ultrasonic atomizer, which converts the liquid into fine droplets using high-frequency vibrations. The droplets are then sprayed onto the surface of the PEM membrane in a controlled manner. The spray process is typically performed in a controlled environment to ensure uniform deposition of the catalyst ink. The spraying parameters, such as pressure, distance, speed, and angle, can be adjusted to control the thickness and uniformity of the CLs. The advantages of this method are its simplicity, low cost, and flexibility. The disadvantages are its low catalyst utilization, high solvent consumption, and difficulty in scaling up. The advantages of ultrasonic spray coating method for catalyst deposition include high catalyst dispersion, reduced aggregation, reduced nozzle clogging, and uniform catalyst distribution. It enables the effective preparation of thin film coatings with minimal overspray, saving on catalyst usage. It is suitable for laboratory operations and can be easily automated for batch production of MEAs. However, one drawback of the ultrasonic spray coating method is its high energy consumption, which becomes a barrier for large-scale applications.

To solve the issue of membrane swelling, the transfer method is commonly used to prepare membrane electrodes. In the transfer method [81], the PEM does not come into contact with the solvent, effectively avoiding issues such as membrane swelling and wrinkling, making it a reliable method for improving the performance of CCM-type MEAs. However, there are still challenges that need to be overcome in the transfer method: (1) improving catalyst utilization to ensure complete transfer and uniform distribution of the active components from the substrate to the membrane; (2) developing specific transfer substrates and inks that have good affinity during coating and are easy to peel off during the hot pressing process; and (3) avoiding the formation of a thin layer of Nafion (oriented toward the GDL layer) during the preparation process to enhance the mass transfer capability of the MEA.

In addition, there are other methods for preparing CCM, such as screen printing [82] and electrochemical deposition [83,84]. Screen printing involves depositing catalyst ink onto the PEM through a mesh screen using a squeegee. This method allows for high-speed and large-scale production of MEAs. However, it may result in uneven catalyst distribution and relatively thick catalyst layers. Electrochemical deposition, on the other hand, is an efficient, precise, and scalable method for preparing MEAs. It is generally carried out in a three-electrode electroplating bath, where under the action of an external electric field, not only can uniformly distributed catalyst particles be directly deposited into the three-phase reaction zone of the MEA core, but Pt or Pt alloys can also be electrolyzed out of their mixed solution or molten salt and tightly contacted with Nafion.

Each of these methods has its own set of advantages and challenges, and the choice of method depends on factors such as desired performance, scalability, cost, and equipment availability. Researchers and manufacturers continue to explore and develop new techniques to improve the efficiency, durability, and cost-effectiveness of MEA fabrication.

6.3.6 SEALING TECHNOLOGY IN PEMWE

To achieve PEM water electrolysis, all components in the electrolyzer, including the seals, have to be sufficiently durable and reliable to effectively achieve their functions.

The durability and reliability of PEM water electrolysis are critical issues in the process of commercialization. Therefore, great attempts have been made by industry and academia to increase the lifetime of PEMWEs. However, little work on seals has been reported in the literature, while the durability and reliability of electrode and membrane have attracted most attention. In fact, from the durability and reliability viewpoint, any component of the water electrolysis assembly which can cause a failure of the water electrolysis stack is of equal importance. Therefore, seals should be designed effectively in order to ensure a reliable performance of a PEMFC throughout the design lifetime.

It is clear that the main function of the sealing is to prevent the leakage of fluids, the reactants and the by-products, from the PEMWE. To perform this correctly, a suitable seal and a sufficient clamping force should be applied to the PEMWE. Due to the differences between the properties of the MEA and the seal, the thickness of the seal and the amount of the applied clamping force must be optimized to maximize the performance of the PEMWE. Also, due to the viscoelastic properties of most seals for PEMWEs, the lifetime of these seals has to be predicted from the stress relaxation data. In fact, the internal forces within the seal must be greater than the pressure of the fluids inside the PEMWE to prevent leakage.

The seals are affected by a number of conditions inside the PEMWE, i.e. the acidity, temperature, and mechanical stresses. Equally, the seals affect the performance of the PEMWE. Therefore, two trends in the research can be observed:

1. Studies have been conducted to develop sealing materials which can withstand the severe environment of PEMWEs so that the lifetime of the latter is extended. Also, these developed materials should be cost-effective and easy to manufacture. The lifetime of the elastomers used in the manufacture of the sealing materials for PEMWEs has been predicted through studying the compression stress relaxation and mechanical and chemical degradation.
2. Studies have been conducted to predict the effects of the sealing material and the design on the performance of PEMWEs.

There is no doubt that seals have a significant effect on the reliability and durability of the PEMWEs because they depend on each component of the PEMWE.

Poor sealing and surface contact are undesirable in any water electrolysis operation. Proper water electrolysis sealing and surface contact are factors that can have a significant impact on the water electrolysis performance. A leak in the water electrolysis assembly can severely limit the ability of the water electrolysis to deliver the reactants to the reaction surface, drastically reducing cell efficiency, while a lack of sufficient surface contact can cause a significant increase in the electrical resistance and result in voltage losses, thus limiting the power production. However, with proper water electrolysis design and assembly techniques, these issues can be addressed.

Leaks are classified into two modes: external and internal. An external leak is defined as a leak from a sealed interface in the water electrolysis to the surrounding environment. This mode of leakage is most likely caused by an opening in the gasket, created by a tear or improper interface contact. An internal leak is defined as a drastic deviation from the flow path inside the water electrolysis assembly. The internal leak is typically the result of poor surface contact within the water electrolysis, specifically between the PTL and its surrounding solid media. This flow deviation is not to be confused with the flow deviation resulting from the motion into the PTL itself, but rather with the flow into an opening created by a gap between two media. As the flow will follow the path of least resistance, these gaps can cause the flow to completely bypass the reaction surface and significantly limit the cells performance.

However, there is no clear vision about the effect of the sealing design (the cross-section thickness and shape and the physical properties of the sealing material) on the uniformity of the pressure distribution between the multiple solid layers inside the water electrolysis. It is well known that the uniformity of the pressure distribution inside the water electrolysis has real effects on the performance and durability of the PEMWEs. The compressive force used to maintain these multilayers assembled can play significant roles in how effectively the water electrolysis performs its function and the most obvious role is to ensure proper sealing. The compression can have a considerable impact on the water electrolysis performance beyond the sealing aspects, where the compression can manipulate the ability to deliver the reactants and the electrochemical functions of the water electrolysis by altering the properties of the layers.

6.4 SUGGESTED OBJECTIVES

To study the effects of the seals on the performance of PEMWEs:

i. Characterization experiments on seals, such as copolymeric resin (CR), fluorosilicone rubber (FSR), liquid-silicon rubber (LSR), ethylene propylene diene monomer rubber (EPDM), fluoroelastomer copolymer (FKM), reinforced nitrile rubber (NBR), and polytetrafluoroethylene (PTFE), as well as MEAs and BPPs, should be performed. The most common tests for elastomeric sealing materials are uniaxial tensile test, biaxial tensile test, shear test, and compression relaxation test.

ii. Use of a commercial software (ABAQUS, COMSOL, or ANSYS) to simulate the effect of the clamping force on the water electrolysis/and or stack to obtain the optimum case. The presence of the sealing material should be taken into account to see the effect of its physical properties and design on the contact pressure distribution, contact resistance, and the mass transfer through the porous media.

iii. Validate the results from point (ii) by performing experiments on a PEMWE/stack. Using special pressure sensors such as pressure sensitive films is very useful to investigate the uniformity of the pressure distribution between different water electrolysis components. Find the optimum design (material type, cross-section thickness and shape) of the sealing to increase the lifetime of the sealing and the other components.

6.5 SUMMARIZATION

Developing the green hydrogen preparation technology represented by PEM electrolysis, as well as realizing the promotion and application of hydrogen in energy storage, chemical industry, metallurgy, distributed power generation, and other fields, is one of the effective ways to control greenhouse gas emissions and slow down global temperature rise.

PEM electrolytic water hydrogen production technology has the characteristics of high operating current density, low energy consumption, high hydrogen production pressure, adaptation to renewable energy fluctuations, and compact footprint, which has the basic conditions for industrialization and large-scale development. Therefore, the following suggestions are made: (1) Starting from key materials and components such as electrocatalysts, membrane electrodes, and bipolar plates, and reducing costs through scale production and technological growth, so as to support the steady decline of the comprehensive cost of hydrogen production by PEM electrolysis. (2) Improve catalyst activity and catalyst utilization, and effectively reduce the amount of precious metals. (3) Develop efficient mass transfer electrode structure to further improve the running current density of PEM electrolysis. (4) Improve the material properties and surface processes of the bipolar plate, and improve the corrosion resistance while reducing the cost. In addition, developing innovative stack concepts can help address these challenges and improve the scalability of the technology.

REFERENCES

1. Grigoriev, S., et al., Current status, research trends, and challenges in water electrolysis science and technology. *International Journal of Hydrogen Energy*, 2020. **45**(49): pp. 26036–26058.
2. Kim, J., et al., Hybrid-solid oxide electrolysis cell: A new strategy for efficient hydrogen production. *Nano Energy*, 2018. **44**: pp. 121–126.
3. Zheng, Y., et al., A review of high temperature co-electrolysis of H_2O and CO_2 to produce sustainable fuels using solid oxide electrolysis cells (SOECs): Advanced materials and technology. *Chemical Society Reviews*, 2017. **46**(5): pp. 1427–1463.

4. Wang, Y., et al., PEM Fuel cell and electrolysis cell technologies and hydrogen infrastructure development-a review. *Energy & Environmental Science*, 2022. **15**(6): pp. 2288–2328.

5. Chen, Y., et al., Key components and design strategy for a proton exchange membrane water electrolyzer. *Special Issue: Catalysts for Renewable Energy*, 2023. **4**(6): p. 2200130.

6. Carmo, M., D.L. Fritz, J. Mergel, and D. Stolten, A comprehensive review on PEM water electrolysis. *International Journal of Hydrogen Energy*, 2013. **38**(12): pp. 4901–4934.

7. Carmo, M., et al., A comprehensive review on PEM water electrolysis. *International Journal of Hydrogen Energy*, 2013. **38**(12): pp. 4901–4934.

8. Ishikawa, H., et al., Development of 70 MPa differential-pressure water electrolysis stack. *Honda R & D Technical Review*, 2016. **28**(1): pp. 80–87.

9. Trasatti, S., Work function, electronegativity, and electrochemical behaviour of metals: III. Electrolytic hydrogen evolution in acid solutions. *Journal of Electroanalytical Chemistry and Interfacial Electrochemistry*, 1972. **39**(1): pp. 163–184.

10. Mao, Y., et al., Catalyst screening: Refinement of the origin of the volcano curve and its implication in heterogeneous catalysis. *Chinese Journal of Catalysis*, 2015. **36**(9): pp. 1596–1605.

11. Wang, S., A. Lu, and C.-J. Zhong, Hydrogen production from water electrolysis: Role of catalysts. *Nano Convergence*, 2021. **8**(1): p. 4.

12. Yoo, S., et al., Encapsulation of Pt nanocatalyst with N-containing carbon layer for improving catalytic activity and stability in the hydrogen evolution reaction. *International Journal of Hydrogen Energy*, 2021. **46**(41): pp. 21454–21461.

13. Cheng, Q., et al., Carbon-defect-driven electroless deposition of Pt atomic clusters for highly efficient hydrogen evolution. *Journal of the American Chemical Society*, 2020. **142**(12): pp. 5594–5601.

14. Xu, W., et al., A novel hybrid based on carbon nanotubes and heteropolyanions as effective catalyst for hydrogen evolution. *Electrochemistry Communications*, 2007. **9**(1): pp. 180–184.

15. Hinnemann, B., et al., Biomimetic hydrogen evolution: MoS_2 nanoparticles as catalyst for hydrogen evolution. *Journal of the American Chemical Society*, 2005. **127**(15): pp. 5308–5309.

16. Phuruangrat, A., et al., Electrochemical hydrogen evolution over MoO_3 nanowires produced by microwave-assisted hydrothermal reaction. *Electrochemistry Communications*, 2009. **11**(9): pp. 1740–1743.

17. Miles, M.H. and M.A. Thomason, Periodic variations of overvoltages for water electrolysis in acid solutions from cyclic voltammetric studies. *Journal of the Electrochemical Society*, 1976. **123**(10): p. 1459.

18. Reier, T., M. Oezaslan, and P. Strasser, Electrocatalytic oxygen evolution reaction (OER) on Ru, Ir, and Pt catalysts: a comparative study of nanoparticles and bulk materials. *ACS Catalysis*, 2012. **2**(8): pp. 1765–1772.

19. Kötz, R., et al., Oxygen evolution on Ru and Ir electrodes: XPS-studies. *Journal of Electroanalytical Chemistry and Interfacial Electrochemistry*, 1983. **150**(1): pp. 209–216.

20. Poerwoprajitno, A.R., et al., Formation of branched ruthenium nanoparticles for improved electrocatalysis of oxygen evolution reaction. *Small*, 2019. **15**(17): 1804577.

21. Huang, K., et al., Ru/Se-RuO_2 composites via controlled selenization strategy for enhanced acidic oxygen evolution. *Advanced Functional Materials*, 2023. 33(8): 2211102.

22. Shan, J., et al., Transition-metal-doped RuIr bifunctional nanocrystals for overall water splitting in acidic environments. *Advanced Materials*, 2019. **31**(17): 1900510.

23. Wang, Y., et al., Unraveling oxygen vacancy site mechanism of Rh-doped RuO_2 catalyst for long-lasting acidic water oxidation. *Nature Communications*, 2023. **14**(1): p. 1412.

24. Huo, W., et al., Rhenium suppresses iridium (IV) oxide crystallization and enables efficient, stable electrochemical water oxidation. *Small*, 2023. **19**(19): 2207847.

25. Gao, X., et al., Single cobalt atom anchored on N-doped graphyne for boosting the overall water splitting. *Applied Surface Science*, 2020. **502**: p. 144155.

26. Wang, L., et al., Atomically dispersed Mo supported on metallic Co_9S_8 nanoflakes as an advanced noble-metal-free bifunctional water splitting catalyst working in universal pH conditions. *Advanced Energy Materials*, 2020. 10(4): 1903137.

27. Mirshekari, G., et al., High-performance and cost-effective membrane electrode assemblies for advanced proton exchange membrane water electrolyzers: Long-term durability assessment. *International Journal of Hydrogen Energy*, 2021. **46**(2): pp. 1526–1539.

28. Giancola, S., et al., Composite short side chain PFSA membranes for PEM water electrolysis. *Journal of Membrane Science*, 2019. **570–571**: pp. 69–76.

29. Albert, A., et al., Radiation-grafted polymer electrolyte membranes for water electrolysis cells: Evaluation of key membrane properties. *ACS Applied Materials & Interfaces*, 2015. **7**(40): pp. 22203–22212.

30. Shin, D.W., M.D. Guiver, and Y.M. Lee, Hydrocarbon-based polymer electrolyte membranes: importance of morphology on ion transport and membrane stability. *Chemical Reviews*, 2017. **117**(6): pp. 4759–4805.

31. Park, J.E., et al., High-performance proton-exchange membrane water electrolysis using a sulfonated poly(arylene ether sulfone) membrane and ionomer. *Journal of Membrane Science*, 2021. **620**: p. 118871.

32. Wei, G., et al., SPE water electrolysis with SPEEK/PES blend membrane. *International Journal of Hydrogen Energy*, 2010. **35**(15): pp. 7778–7783.

33. Siracusano, S., et al., Electrochemical characterization of a PEM water electrolyzer based on a sulfonated polysulfone membrane. *Journal of Membrane Science*, 2013. **448**: pp. 209–214.

34. Rost, U., et al., Long-term stable electrodes based on platinum electrocatalysts supported on titanium sintered Felt for the use in PEM fuel cells. *IOP Conference Series: Materials Science and Engineering*, 2018. **416**(1): p. 012013.

35. Grigoriev, S.A., et al., Optimization of porous current collectors for PEM water electrolysers. *International Journal of Hydrogen Energy*, 2009. **34**(11): pp. 4968–4973.

36. Lee, C.H., et al., Porous transport layer related mass transport losses in polymer electrolyte membrane electrolysis: a review. In ASME 2016 14th International Conference on Nanochannels, Microchannels, and Minichannels collocated with the ASME 2016 Heat Transfer Summer Conference and the ASME 2016 Fluids Engineering Division Summer Meeting, Washington, DC. 2016.

37. Lee, J.K., C.H. Lee, and A. Bazylak, Pore network modelling to enhance liquid water transport through porous transport layers for polymer electrolyte membrane electrolyzers. *Journal of Power Sources*, 2019. **437**: pp. 226910.

38. Parra-Restrepo, J., et al., Influence of the porous transport layer properties on the mass and charge transfer in a segmented PEM electrolyzer. *International Journal of Hydrogen Energy*, 2020. **45**(15): pp. 8094–8106.

39. Mo, J., et al., Experimental studies on the effects of sheet resistance and wettability of catalyst layer on electro-catalytic activities for oxygen evolution reaction in proton exchange membrane electrolysis cells. *International Journal of Hydrogen Energy*, 2020. **45**(51): pp. 26595–26603.

40. Doan, T.L., et al., A review of the porous transport layer in polymer electrolyte membrane water electrolysis. *International Journal of Energy Research*, 2021. **45**(10): pp. 14207–14220.

41. Ito, H., et al., Experimental study on porous current collectors of PEM electrolyzers. *International Journal of Hydrogen Energy*, 2012. **37**(9): pp. 7418–7428.

42. Ito, H., et al., Influence of pore structural properties of current collectors on the performance of proton exchange membrane electrolyzer. *Electrochimica Acta*, 2013. **100**: pp. 242–248.

43. Vorhauer-Huget, N., et al., Computational optimization of porous structures for electrochemical processes. *Processes*, 2020. 8. DOI: 10.3390/pr8101205.

44. Lee, J.K., et al., Spatially graded porous transport layers for gas evolving electrochemical energy conversion: High performance polymer electrolyte membrane electrolyzers. *Energy Conversion and Management*, 2020. **226**: p. 113545.

45. Schuler, T., et al., Hierarchically structured porous transport layers for polymer electrolyte water electrolysis. *Advanced Energy Materials*, 2020. 10(2): 1903216.

46. Kang, Z., et al., Investigation of thin/well-tunable liquid/gas diffusion layers exhibiting superior multifunctional performance in low-temperature electrolytic water splitting. *Energy & Environmental Science*, 2017. **10**(1): pp. 166–175.

47. De Angelis, S., et al., Unraveling two-phase transport in porous transport layer materials for polymer electrolyte water electrolysis. *Journal of Materials Chemistry A*, 2021. **9**(38): pp. 22102–22113.

48. Kim, P.J., et al., In-plane transport in water electrolyzer porous transport layers with through pores. *Journal of the Electrochemical Society*, 2020. **167**(12): p. 124522.

49. Arbabi, F., et al., Feasibility study of using microfluidic platforms for visualizing bubble flows in electrolyzer gas diffusion layers. *Journal of Power Sources*, 2014. **258**: pp. 142–149.

50. Kang, Z., et al., Effects of various parameters of different porous transport layers in proton exchange membrane water electrolysis. *Electrochimica Acta*, 2020. **354**: p. 136641.

51. Lim, A., et al., Amphiphilic Ti porous transport layer for highly effective PEM unitized regenerative fuel cells. *Science Advances*, 2021. **7**(13): p. eabf7866.

52. Steen, S.M., et al., Investigation of titanium liquid/gas diffusion layers in proton exchange membrane electrolyzer cells. *International Journal of Green Energy*, 2017. **14**(2): pp. 162–170.

53. Mo, J., et al., Electrochemical investigation of stainless steel corrosion in a proton exchange membrane electrolyzer cell. *International Journal of Hydrogen Energy*, 2015. **40**(36): pp. 12506–12511.

54. 张萍俊, et al., 不同材料作为阳极扩散层对质子交换膜水电解池性能的影响. 2019.

55. Bessarabov, D. and P. Millet, Chapter 3 - Performance degradation, in *PEM Water Electrolysis*, D. Bessarabov and P. Millet, Editors. 2018, Academic Press. pp. 61–94.

56. Lædre, S., et al., Materials for Proton Exchange Membrane water electrolyzer bipolar plates. *International Journal of Hydrogen Energy*, 2017. **42**(5): pp. 2713–2723.

57. Rakousky, C., et al., The stability challenge on the pathway to high-current-density polymer electrolyte membrane water electrolyzers. *Electrochimica Acta*, 2018. **278**: pp. 324–331.

58. Rakousky, C., et al., An analysis of degradation phenomena in polymer electrolyte membrane water electrolysis. *Journal of Power Sources*, 2016. **326**: pp. 120–128.

59. Liu, C., et al., Performance enhancement of PEM electrolyzers through iridium-coated titanium porous transport layers. *Electrochemistry Communications*, 2018. **97**: pp. 96–99.

60. Bystron, T., et al., Enhancing PEM water electrolysis efficiency by reducing the extent of Ti gas diffusion layer passivation. *Journal of Applied Electrochemistry*, 2018. **48**(6): pp. 713–723.

61. Stiber, S., et al., A high-performance, durable and low-cost proton exchange membrane electrolyser with stainless steel components. *Energy & Environmental Science*, 2022. **15**(1): pp. 109–122.

62. Daudt, N.F., F.J. Hackemüller, and M. Bram, Powder metallurgical production of 316L stainless steel/niobium composites for proton exchange membrane electrolysis cells. *Powder Metallurgy*, 2019. **62**(3): pp. 176–185.

63. L. Bertuccioli, A. Chan, D. Hart, F. Lehner, B. Madden, and E. Standen, Study on development of water electrolysis in the EU by E4tech Sarl with element energy ltd for the fuel cells and hydrogen joint undertaking. *Clean Hydrogen Partnership*, 2014: pp. 1–160.

64. Jung, H.-Y., et al., Performance of gold-coated titanium bipolar plates in unitized regenerative fuel cell operation. *Journal of Power Sources*, 2009. **194**(2): pp. 972–975.

65. Gago, A.S., et al., Protective coatings on stainless steel bipolar plates for proton exchange membrane (PEM) electrolysers. *Journal of Power Sources*, 2016. **307**: pp. 815–825.

66. Li, W., et al., Enhanced biological photosynthetic efficiency using light-harvesting engineering with dual-emissive carbon dots. *Advanced Functional Materials*, 2018. **28**(44): 1804004.

67. Wakayama, H. and K. Yamazaki, Low-cost bipolar plates of Ti_4O_7-coated Ti for water electrolysis with polymer electrolyte membranes. *ACS Omega*, 2021. **6**(6): pp. 4161–4166.

68. Zhang, H., et al., Electrochemical properties of niobium and niobium compounds modified AISI430 stainless steel as bipolar plates for DFAFC. *Surface Engineering*, 2019. **35**(11): pp. 1003–1011.

69. Wang, Y. and D.O. Northwood, An investigation of the electrochemical properties of PVD TiN-coated SS410 in simulated PEM fuel cell environments. *International Journal of Hydrogen Energy*, 2007. **32**(7): pp. 895–902.

70. Lee, W.-J., et al., Ultrathin effective TiN protective films prepared by plasma-enhanced atomic layer deposition for high performance metallic bipolar plates of polymer electrolyte membrane fuel cells. *Applied Surface Science*, 2020. **519**: p. 146215.

71. Li, T., et al., Surface microstructure and performance of TiN monolayer film on titanium bipolar plate for PEMFC. *International Journal of Hydrogen Energy*, 2021. **46**(61): pp. 31382–31390.

72. Jin, J., et al., Investigation of high potential corrosion protection with titanium carbonitride coating on 316L stainless steel bipolar plates. *Corrosion Science*, 2021. **191**: p. 109757.

73. Peng, S., et al., A reactive-sputter-deposited TiSiN nanocomposite coating for the protection of metallic bipolar plates in proton exchange membrane fuel cells. *Ceramics International*, 2020. **46**(3): pp. 2743–2757.

74. Abbas, N., et al., Direct deposition of extremely low interface-contact-resistant Ti_2AlC MAX-phase coating on stainless-steel by mid-frequency magnetron sputtering method. *Journal of the European Ceramic Society*, 2020. **40**(8): pp. 3338–3342.

75. Wang, T., X. Cao, and L. Jiao, PEM water electrolysis for hydrogen production: fundamentals, advances, and prospects. *Carbon Neutrality*, 2022. **1**(1): p. 21.

76. Shi, L., et al., Proton-exchange membrane water electrolysis: From fundamental study to industrial application. *Chem Catalysis*, 2023. **3**(9): 100734.

77. Immerz, C., et al., Effect of the MEA design on the performance of PEMWE single cells with different sizes. *Journal of Applied Electrochemistry*, 2018. **48**(6): pp. 701–711.

78. Zhang, K., et al., Status and perspectives of key materials for PEM electrolyzer. *Nano Research Energy*, 2022. **1**: p. 9120032.

79. Alia, S.M., et al., The impact of ink and spray variables on catalyst layer properties, electrolyzer performance, and electrolyzer durability. *Journal of The Electrochemical Society*, 2020. **167**(14): p. 144512.

80. Ayers, K.E., et al., Pathways to ultra-low platinum group metal catalyst loading in proton exchange membrane electrolyzers. *Catalysis Today*, 2016. **262**: pp. 121–132.

81. Kang, Z., et al., Studying performance and kinetic differences between various anode electrodes in proton exchange membrane water electrolysis cell. *Materials*, 2022. **15**(20): 7209.
82. Lagarteira, T., et al., Highly active screen-printed IrTi$_4$O$_7$ anodes for proton exchange membrane electrolyzers. *International Journal of Hydrogen Energy*, 2018. **43**(35): pp. 16824–16833.
83. Lee, B.-S., et al., Development of electrodeposited IrO$_2$ electrodes as anodes in polymer electrolyte membrane water electrolysis. *Applied Catalysis B: Environmental*, 2015. **179**: pp. 285–291.
84. Choe, S., et al., Electrodeposited IrO$_2$/Ti electrodes as durable and cost-effective anodes in high-temperature polymer-membrane-electrolyte water electrolyzers. *Applied Catalysis B: Environmental*, 2018. **226**: pp. 289–294.

7 Electrocatalytic Oxygen Evolution Reaction in Acid Media
Mechanism and Interface

Guangfu Li and Mu Pan

7.1 INTRODUCTION

The large-scale production of green hydrogen by proton exchange membrane (PEM) water electrolysis is currently of considerable interest. However, oxygen evolution reaction (OER) occurring in the anode is a primary source of overpotential due to the sluggish kinetics. To realize fast reaction kinetics, it is essential to obtain a facile electron transfer between electrocatalysts and the oxygen intermediates, which is critically related to reaction mechanism and electrode interface. Meanwhile, developing high-performance electrocatalysts for acidic OER is an outstanding challenge since most metal materials are unstable under the strong acidic and oxidative conditions. Even for the promising electrocatalyst candidates, Ir oxides undergo the continuous Ir dissolution due to surface oxidization during OER [1,2]. It is also difficult to establish the structure–performance relationship due to the dynamic changes of catalyst surface (named surface reconstruction). Hence, acquiring an enhanced understanding of the electron transfer pathways and reaction interface is crucial for the design of efficient and robust OER electrocatalysts.

Considering the numerous efforts made on summarizing the electrocatalysts development, this chapter does not attempt an exhaustive review, but rather seeks to emphasize the key concept common of the high-efficiency OER materials. As a complex and multistep process, the OER efficiency and stability of electrocatalysts depends critically on the reaction routes. Detailed introduction of reaction mechanisms is therefore provided in the first section. The reaction interface at the location of OER occurring is reconstructed to form different active centers in the OER process. In the second section, surface and interface electrochemical behaviors are discussed. In particular, the roles of surface reconstruction and the hydrous electronical double layer (EDL) are emphasized due to their dominating influence on electrocatalytic reaction. Finally, with the fundamental knowledge gained in these sections, the remaining challenges toward practical catalytic electrodes for oxygen evolution reaction and related future perspectives are outlined.

DOI: 10.1201/9781003368939-7

7.2 REACTION MECHANISM AND PATHWAYS

A fundamental understanding of the reaction mechanism is crucial to design ideal catalysts for OER. In the acid environment, the OER can be represented as follows:

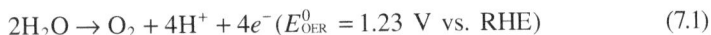

$$2H_2O \rightarrow O_2 + 4H^+ + 4e^- \; (E^0_{OER} = 1.23 \text{ V vs. RHE}) \tag{7.1}$$

where E^0_{OER} is the equilibrium potential under standard conditions (pH = 0, 298.15 K, 1 atm pressure). Note all potentials reported here are relative to the reversible hydrogen electrode (RHE), allowing to directly evaluate overpotential regardless of pH. In principle, the heterogeneous electrocatalytic OER has rather complex elementary steps coupled with the transfer of concerted multiple electrons and protons [3]. There are tremendous notable pioneering efforts associated with studies of OER mechanisms from Bockris, Damjanovic et al., Krasil'shchikov, Conway and Bourgault, and Riddiford [4–9]. Some of the traditional mechanism models for acidic OER are outlined in Table 7.1. To differentiate between the varied reaction paths, Tafel slopes are given at low and high overpotential. In the absence of mass diffusion limitation, the obtained Tafel slope can act as a powerful tool for both a quantitative and mechanistic characterization of an electrocatalytic process [10,11].

In these classic mechanistic schemes, the overall OER process is considered as a sequence of elementary one-electron transfer steps and/or chemical steps with the transfer of total four electrons. It is important to note that most of the proposed mechanisms have the same intermediates including MOH and MO, while the major difference might be around the final steps for O–O bonding formation [12]. Currently, the widely accepted mechanisms mainly involve the conventional adsorbate evolution mechanism (AEM) and the lattice oxygen mechanism (LOM). As illustrated in Figure 7.1, the former links the formation of O–O bond through the direct combination of MO, while the latter involves MOOH intermediate that subsequently

TABLE 7.1
Models for the Classic OER Mechanisms with Tafel Slope Analysis

Oxygen evolution based on LOM	Tafel slope (mV dec⁻¹) Low η	Tafel slope (mV dec⁻¹) High η	Oxygen evolution based on AEM	Tafel slope (mV dec⁻¹) Low η	Tafel slope (mV dec⁻¹) High η
(I) Bockris's oxide Path			**(IV) The metal peroxide path**		
1. $M + H_2O \rightarrow MOH + H^+ + e$	120		1. $M + H_2O \rightarrow MOH + H^+ + e$	120	
2. $MOH \rightarrow MO + H^+ + e$	30	∞	2. $2MOH \rightarrow MO + MH_2O$	30	∞
3. $2MO \rightarrow 2M + O_2$	15	∞	3. $MO + MOH \rightarrow M + MHO_2$	20	∞
			4. $MHO_2 + MOH \rightarrow M + MH_2O + O_2$	15	∞
(II) Bockris's electrochemical Path			**(V) Conway & Bourgault path**		
1. $M + H_2O \rightarrow MOH + H^+ + e$	120		1. $M + H_2O \rightarrow MOH + H^+ + e$	120	
2. $2MOH \rightarrow MO + M + H_2O$	40	120	2. $MOH \rightarrow MO + H^+ + e^-$	40	120
3. $2MO \rightarrow 2M + O_2$	15	∞	3. $MO + MOH \rightarrow MHO_2$	20	60
			4. $MHO_2 + MOH \rightarrow M + MH_2O + O_2$	15	∞
(III) Krasil'shchikov's Path			**(VI) Riddiford path**		
1. $M + H_2O \rightarrow MOH + H^+ + e$	120		1. $M + H_2O \rightarrow MOH + H^+ + e^-$	120	
2. $MOH \rightarrow MO^- + H^+$	60	∞	2. $2MOH \rightarrow MO + MH_2O$	60	∞
3. $MO^- \rightarrow MO + e^-$	40	120	3. $MO + H_2O \rightarrow MHO_2 + H^+ + e^-$	40	120
4. $2MO \rightarrow 2M + O_2$	15	∞	4. $MHO_2 + H_2O \rightarrow MH_2O + O_2 + H^+ + e$	15	∞

**M represents a catalytically active site.

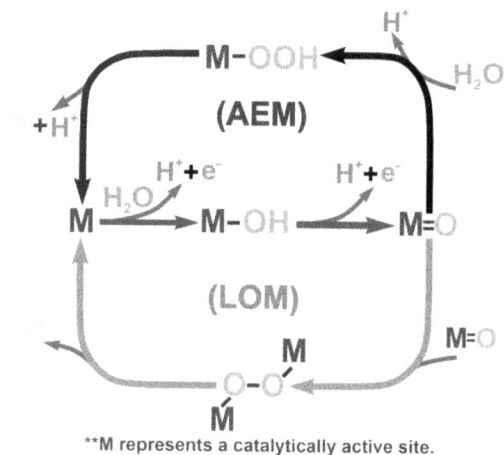

FIGURE 7.1   The OER mechanism under the acid condition.

decomposes to O_2. Herein, we will provide a comprehensive understanding of these two-electron transfer pathways for OER, introducing our perspectives on the further development of the high-performance electrocatalysts and electrodes.

7.2.1 TRADITIONAL ABSORBATE EVOLUTION MECHANISM AND ITS LIMITATION

The AEM pathway for OER is typically assumed as a four concerted proton–electron transfer (CPET) process at a single active site (M), as described in equations (7.2–7.5). The whole reaction steps involve three key O-containing intermediates including MOH, MO, and MOOH. Specifically, the OER sequence in acid media is first initiated by the adsorption of a solvent H_2O molecule to form MOH (equation 7.2) and then deprotonates to produce the key intermediate MO (equation7.3). Afterward, the adsorbed MO is transferred to MOOH through nucleophilic attack by H_2O (equation 7.4). At the final step, the fourth electron transfer and deprotonation lead to O_2 evolution from the reaction interface, releasing the active site for the next cycle (equation 7.5).

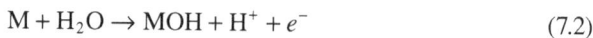

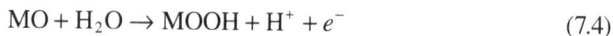

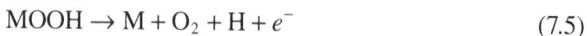

$$M + H_2O \rightarrow MOH + H^+ + e^- \tag{7.2}$$

$$MOH \rightarrow MO + H^+ + e^- \tag{7.3}$$

$$MO + H_2O \rightarrow MOOH + H^+ + e^- \tag{7.4}$$

$$MOOH \rightarrow M + O_2 + H + e^- \tag{7.5}$$

Notably, the entire AEM process takes place between the metal active sites and the specially adsorbed oxygen intermediates, corresponding to a metal redox electrocatalytic reaction [13]. In essence, the OER overpotential can be calculated through the change of reaction Gibbs free energy (ΔG) in equations (7.2–7.5). Each elementary reaction step releases a proton and an electron, implying that ΔG will be potential dependent. The adsorption free energies of the intermediates ΔG_{MOH}, ΔG_{MO}, and ΔG_{MOOH} vary with an external electrode potential (E). Although this AEM pathway takes place in acidic environments, the thermodynamic conclusions referring to RHE are independent of pH as the free energies change in the same way with pH [10,14]. In the theoretical frameworks, ΔG of each CPET step can be derived separately:

$$\Delta G_1 = \Delta G^*_{MOH} - \Delta G_{H_2O} - eE + k_B T \ln a_{H^+} \tag{7.6}$$

$$\Delta G_2 = \Delta G^*_{MO} - \Delta G^*_{MOH} - eE + k_B T \ln a_{H^+} \tag{7.7}$$

$$\Delta G_3 = \Delta G^*_{MOOH} - \Delta G^*_{MO} - eE + k_B T \ln a_{H^+} \tag{7.8}$$

$$\Delta G_4 = \Delta G_{O_2} - \Delta G^*_{MOOH} - eE + k_B T \ln a_{H^+} \tag{7.9}$$

where ΔG^*, k_B, T, and a_{H^+} represent the free energy change after adsorbed O-intermediates, Boltzmann constant, the absolute temperature in Kelvin, and the activity of protons, respectively. Moreover, the total free energy change (ΔG_t) at thermodynamic equilibrium for OER meets the following criterion:

$$\Delta G^0_t = \Delta G^0_1 + \Delta G^0_2 + \Delta G^0_3 + \Delta G^0_4 = 4.92 \text{ eV} \tag{7.10}$$

where ΔG^0 denotes ΔG obtained at the standard conditions and $E=0$. A primary information parameter which can be identified from the ΔG diagram is the thermodynamically least favorable step. This concept is usually referred as the potential determining step (PDS) which has the largest ΔG_i value:

$$\Delta G_{max} = \max\{\Delta G_1, \Delta G_2, \Delta G_3, \Delta G_4\} \tag{7.11}$$

The overall reaction rate relies only on ΔG_{max} without extra barriers from adsorption or dissociation of O_2 or CPET reactions. Meanwhile, the standard onset overpotential (η^{OER}) related to kinetic hindrance of PDS can be defined by the following equation:

$$\eta^{OER} = \Delta G_{max}\big/e - \Delta G^0_t\big/4e = \Delta G_{max}\big/e - 1.23V \tag{7.12}$$

With respect to the traditional AEM, the above analysis implies that the OER activity is strongly correlated with the adsorption energies of O-intermediates. For a given catalyst material, the PDS might be any of the four elementary reaction step equations (7.2–7.5) with the highest kinetic activation barrier. In principle, an ideal OER catalyst requires that the reaction energies for each CPET process have the same magnitude at $E=0$: $\Delta G_1 = \Delta G_2 = \Delta G_3 = \Delta G_4 = 1.23$ eV, giving $\eta^{OER} = 0$ without thermodynamic hindrance. The ideal catalyst trace is schematically plotted in Figure 7.2a. However, in the theoretical frame of AEM, this ideal sample is almost impossible to find since adsorption energies of O-intermediates (including MOH,

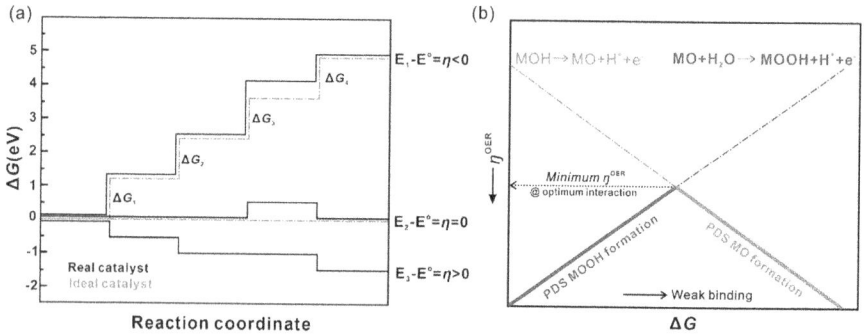

FIGURE 7.2 Schematic representation of (a) the Gibbs free energies for ideal and real catalysts via reaction coordination, and (b) a volcano plot based on AEM as a function of adsorption free energies for O-intermediates, where formation of MOOH and MO is regarded as the PDSs at the left and right volcano legs, respectively.

MO, and MOOH) are linearly related independent of the binding site [15]. In particular, the binding energies of MOOH and MO are strongly linked with a constant difference of approximately 3.2 eV:

$$\Delta G^*_{MOOH} = \Delta G^*_{MO} + 3.2 eV \qquad (7.13)$$

which implies that ΔG^*_{MOOH} and ΔG^*_{MO} can be tuned independently, and the minimum theoretical overpotential is 370 mV. The overpotential wall has been verified by studying a wide range of the benchmarked AEM electrocatalysts, indicating the universal scaling relation between MOOH and MO [16,17].

In the actual OER catalysis, there is generally a CPET step with $\Delta G > 1.23$ eV, leading to $\eta^{OER} > 0$ V. As depicted in Figure 7.2a, the ΔG order on a real catalyst is typically assumed as: $\Delta G_3 > \Delta G_1 = \Delta G_2 > \Delta G_4$, where the formation of MOOH is the PDS arising from the weak binding strength between the active site and O (M-O). On the other hand, the MO formation can become a PDS with increasing M-O bind energy. According to Sabatier's principle, a clear volcano representation has been established, with either too weak or too strong binding strength of M-O resulting in the increase of η^{OER}. As schematically represented in Figure 7.2b, the strong M-O interaction can increase the difficulty of MOOH formation, while the weak interaction can increase the difficulty of reactant activation or MO formation. Consequently, the best catalyst at the top of the volcano plot exists as an optimal M-O interaction governing the intrinsic catalytic activity. Catalyst component, surface structure, adsorbate, and electrolyte solvent can have a strong influence on the M-O bond, thus changing PDS. Indeed, ΔG^*_{MO} has been demonstrated as a good general indicator of the activity trends for a wide variety of OER catalysts, such as rutile, anatase, and transition metal oxides [10,17]. It is a rational approach to obtain high-performance catalysts by modifying the electronic structure to optimize ΔG^*_{MO}. The primary approaches contain substituting with foreign elements [18–21], generating vacancies [22–24], tuning strain [25–27], and engineering the reaction interface [28–31].

In summary, the overall rate in AEM greatly depends on the reaction free energy. The universal scaling relationship predicts that the minimized η^{OER} of 370 mV required even for the state-of-the-act electrocatalysts. Furthermore, many studies confirm that hurdling or decreasing this overpotential wall currently remains a huge challenge without bypassing the conventional AEM approach. Identifying optimal M-O interactions provides valuable insights into predicting catalyst activity, thereby reducing experimental and computational costs.

7.2.2 NOVEL LATTICE OXYGEN MECHANISM AND ITS DEVELOPMENT

As a heterogeneous electrocatalysis reaction, OER is a multistep proton and electron transfer process with the formation of distinct oxygen intermediates. Indeed, this process may take place through varied pathways which determine the catalyst activity and stability. Since AEM cannot surmount the thermodynamic obstacles caused by the universal scaling relationship, increasing efforts have been devoted to studying more efficient routes. An alternative electron transfer pathway with triggering lattice oxygen redox electrochemistry has been widely studied in electrochemistry, e.g., batteries [32,33] and electrolysis. The OER mechanism involving this new electron transfer pathway is known as LOM. The bulk phase oxygen that forms the crystal structure is named lattice oxygen, which frequently involves in thermal oxidative catalysis (i.e., Mars van Krevelen mechanism) and electrochemical redox steps. In the early 1980s, the participation of lattice oxygen atoms in the OER process was first examined using ^{18}O isotope labeling technique [34]. Since then, many notable achievements have been made in the terms of the LOM-based electrocatalysis. Figure 7.3 shows some milestone events in the history of LOM development [19,35–45]. By switching the dominant electron transfer pathway from AEM to LOM, the enhanced activity has been widely demonstrated on Ir/Ru-based oxides, perovskites, transition metal oxyhydroxides, and spinel oxides. Hence, various effective strategies have been proposed to trigger the preferential LOM, including the cation/anion doping, modulating electronic and crystal structure, and surface defect engineering [42,46–52]. According to our recent work, the reaction rate following LOM can be obviously influenced by the interaction strength of specifically adsorbed

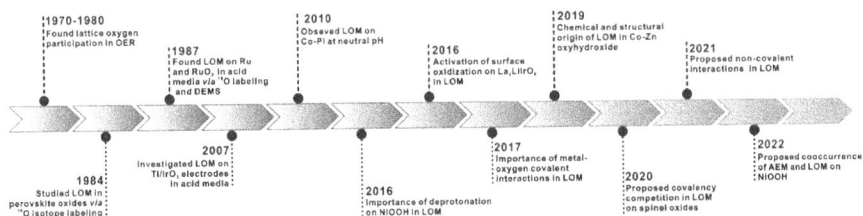

FIGURE 7.3 Timeline-based important accomplishments during the development of LOM electrocatalysis.

intermediates and non-specifically adsorbed alkaline metal cations in the electrical double layer [19].

Recent findings categorize the LOM-based route into different forms according to the varied active centers. Instead of a single metallic active site in AEM, LOM generally takes place in both the metal cation active site and the lattice oxygen. As shown in Figure 7.1, the lattice oxygen (i.e., the bulk phase oxygen) can serve as the origin of O_2 molecules during the OER process. One proposed path of LOM involves nearby bimetallic active sites. Specifically, the initial two elementary steps of LOM follow the same approaches as those of AEM (equations 7.2 and 7.3), whereas in the last steps, the O–O bond formation proceeds via direct coupling of neighboring activated lattice oxygen without an electron transfer to the external circuit, and then the created MOOM species further decomposed to generate O_2. Simultaneously, bimetallic active sites are recovered to electrochemically activated and released from the crystal matrix for the next electrocatalytic cycling. According to the electronic state configuration, two electrons must be removed from the oxygen orbitals to form an O_2 electronic structure, implying that LOM belongs to an oxygen redox reaction. This is obviously different from AEM, where metal bands serve as the redox location. Accordingly, this specific process can be described as the dual-metal-site mechanism, and the final two steps can be simply represented:

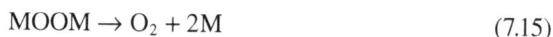

$$MO + MO \rightarrow MOOM \tag{7.14}$$

$$MOOM \rightarrow O_2 + 2M \tag{7.15}$$

In addition to the bimetallic active sites, the LOM pathway may occur at a single active center. Inspired by this, two probable mechanisms have been widely proposed, including oxygen-vacancy-site mechanism [42] and single-metal-site mechanism [43,53]. The former involves the lattice oxygen rather than metal cation as a catalytic site participating in the LOM process. The lattice oxygen atoms can couple with adsorbed oxygen to form O–O bonds followed by a chemical step to release the molecular O_2 and leave two vacant oxygen sites acting as a regenerated "lattice oxygen" for the subsequent cycle. In contrast, the latter takes the single metal site as the catalytic center and follows the deprotonation. In this route, the interfacial reconstruction allows the direct coupling of MO intermediate and the activation of lattice oxygen. Overall, the activation of lattice oxygen ligands is crucial to successfully trigger these distinct LOM pathways. Catalysts with more active lattice oxygen typically lead to lower overpotential and higher OER activity [54]. The interaction between metal cation and oxygen anion is correlated with the reaction mechanism. Triggering the lattice oxygen can improve oxygen ion mobility to form oxygen vacancies and release O_2. The presence of oxygen 2p orbital near the Fermi level is required to activate the lattice oxygen, thus following the LOM. To identify different reaction pathways, it is highly desirable to carry out in situ observations during the entire reaction process, including monitoring the intermediate, product, and catalyst themselves. A variety of advanced characterization techniques thus have been developed to provide straight mechanistic evidences in recent years, such as electrochemical probe, isotope technology, operando X-ray absorption spectroscopy,

Raman spectroscopy, time-resolved optical spectroscopy, and solid-state NMR spectroscopy [13,55]. Additionally, theoretical calculation, particularly based on density functional theory (DFT), is a universal and powerful tool to determine how lattice oxygen ligands can be excited and then participate in oxygen evolution.

In essence, the LOM route involving lattice oxygen has obvious advantages toward facilitating reaction kinetics and lower OER overpotential in comparison to AEM. However, the actual OER overpotential of the LOM-based catalysts is typically in the range of 240–500 mV at a current density of 10 mA/cm^2 [13]. Despite the useful guidance of LOM, there exist three critical issues that hamper the exploitation of efficient electrocatalysts.

First, the oxygen orbital must be around the Fermi energy level, so that the active site oxygen easily loses electrons, leaving holes and rendering lattice oxygen oxidation. Since transition metal oxides are currently the most widely used catalysts in acid media, highly oxidized metal ions are desired to facilitate the direct lattice oxygen coupling. According to Pourbaix diagrams, achieving the deep oxidation state of active metal sites needs a high formation energy due to the thermodynamically unfavorable kinetics, leading to a high overpotential. Furthermore, under an oxidizing potential, the transition metal oxides undergo drastic surface reconstruction to form oxyhydroxide layers which cause sluggish kinetics of overall OER. The OER mechanism would switch to AEM from LOM, arising from the intrinsic lack of oxygen nonbonding states in the formed transition metal oxyhydroxides [13,56–58].

Second, the deprotonation process of MOH (equation 7.3) in LOM is usually the rate-determining step due to the labile bonds with MOH adsorbates. The increased energy required for the deprotonation process makes the OER kinetics more sluggish [33,40,59]. Consequently, these disadvantages make LOM pathways unpredictable and limiting for maximizing the activity improvement. Recently, considerable efforts have been devoted to accelerate the deprotonation process, such as increasing the valence state of metal [60,61], dropping proton acceptors [33,62,63], and tuning electronic structure [64].

Finally, the long-term stability of LOM-based catalysts might be a serious challenge in acid media. As a fact, studies on acidic LOM are relatively lacking due to the poor stability of common OER catalysts compared with those on alkalic LOM. The most widely used catalysts currently concentrate on Ir–Ru-based oxides, since the trade-off between activity and stability represents a limited choice for acidic OER [13,65]. In the LOM pathways, the continuous formation of oxygen vacancies and dissolution of Ir/Ru metal cations can lead to extensive bulk oxygen diffusion, structural collapse, and eventually deactivation [66,67]. Designing highly active and robust Ir/Ru catalysts should suppress the over-oxidation of noble metal cations during the OER. Reliable strategies are generally explored by balancing reaction pathways between the highly stable AEM and the highly active LOM [68]. Alternatively, for LOM-based catalysts, downshifting O 2p centers and increasing the formation energy of oxygen vacancy can facilitate the development of catalysts with both high activity and long-term stability [13].

7.2.3 THE DIFFERENCE BETWEEN AEM AND LOM

To enhance comprehensive understanding of the varied reaction mechanisms, the main difference between AEM and LOM can be further analyzed in terms of the catalytic elementary steps, energy barrier, durability, active sites, and features as illustrated in Figure 7.4a [69]. The CPET process in AEM occurs on metal orbitals, which results in the metal valence increase and the lower-Hubbard bands downshift. This route can thus be considered a kind of metal redox electrocatalytic reaction. Unlike AEM, LOM is a type of the oxygen redox chemistry where the lattice oxygen acts as the electron donor during the entire reaction process. This electron transfer behavior proceeds between the adsorbate and oxygen orbitals. The O_2 is directly generated by the direct coupling of O–O radicals, bypassing the intrinsic limitation (minimal overpotential of ~370 mV) due to the inherent scaling relation between ΔG^*_{MOOH} and ΔG^*_{MO}. As a result, the LOM pathway typically leads to more prominent catalytic activity than the traditional AEM [49–51,70]. In principle, the O 2p band center should be adjacent to the Fermi energy level to facilitate oxygen ion mobility, oxygen vacancy formation, and water adsorption. In this context, effective methods, such as enhancing oxygen vacancy concentration and the orbital coverage of metal cation and oxygen anion, were proposed to switch the electron transfer pathway from AEM to LOM [42,53,71]. Moreover, the DFT calculation can rationalize the remarkable OER via LOM route from the perspectives of both thermodynamics and kinetics [53,72]. In the case of Ruddlesden–Popper-type oxide, it was found that the DFT-calculated theoretical overpotential decreases 490 mV by changing the OER pathway from AEM to LOM [67].

In the actual OER process, these two mechanisms may coexist with some degree of competition (Figure 7.4b) [73]. The LOM usually offers higher OER activity than AEM, because it overcomes the inherent limitation from the universal scaling relation. Unlike AEM with the stable active center, the LOM route involving the increased metal oxidization and the lattice oxygen overflow may lead to metal ion dissolution and structural collapse. When the active LOM steps are triggered for OER, a compensative measure is necessary to utilize the stable AEM route.

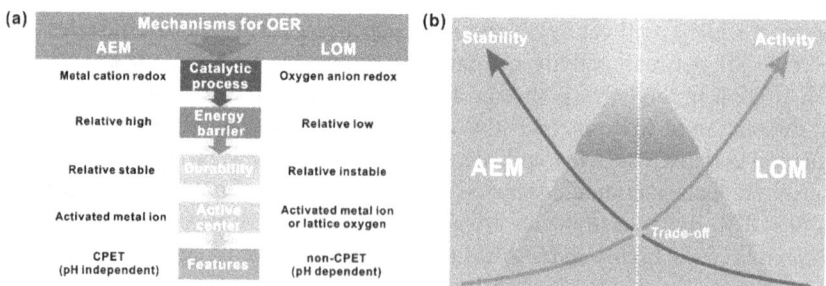

FIGURE 7.4   (a) Comparison of AEM and LOM for oxygen evolution and (b) trade-off between activity and stability.

Constructing catalysts with proper oxygen vacancies or active metal defects has been demonstrated as an effective method to overcome the trade-off between activity and stability, ascribing to synergy between AEM and LOM [74,75].

In addition, LOM involves both the metal cation active center and the lattice oxygen. This route exhibits that the transfer of electrons and protons does not coincide, rendering the OER activity highly rely on electrolyte pH. The OER catalysts favoring LOM have the improved activity with increasing pH in alkaline media [13,19]. In contrast, the pathways for overall AEM elementary steps assume CPET in a single active metal site. The OER overpotential is independent of pH, as observed in Ir oxides [76]. Therefore, different electrocatalytic mechanisms can be identified by establishing the relationship between activity and pH [13]. Furthermore, the applied potential can determine the dominated reaction mechanism to some extent. For Ru–Ni oxides, it is found that the OER process undergoes the alternation from AEM to LOM when the potential is over 1.36 V [77].

7.3   SURFACE AND INTERFACE OER ELECTROCATALYSIS

The electrocatalytic performance of an electrocatalyst is strongly dependent on its electronic configurations and structural characteristics [64,78]. Particularly, the heterogeneous electron transfer reaction occurs only on the catalyst surface. The physicochemical properties of surface will determine the OER behavior by affecting the adsorptions/desorption of reactant, intermediates and products, and the activation process [29,79]. So, rational design and modification of the electrocatalyst surface is a significantly effective strategy to obtain the outstanding OER performance. In an electrochemical system, since there exist overpotentials required for activation, charge transport, and mass diffusion, the applied potential is much more positive than the equilibrium potential (i.e., 1.23 V under the standard conditions). At such high potential, surface metal atoms on an electrocatalyst tend to be further oxidized, leading to the formation of corresponding thin-film oxides or (oxy)hydroxides on the surface. These new oxides or (oxy)hydroxides occupying the original active centers directly participate in electrocatalysis and therefore strongly impact the electrochemical performance. The surficial active sites are dynamic in nature, undergoing continuous reconstruction with serving time. In essence, the reconstruction behavior of an electrocatalyst commonly involves the drastic alternations including composition, morphology, and crystallinity. These newly derived properties under the reaction conditions are considerably different from the initial states. It is difficult to establish a rational correlation between structure and activity without considering construal reconstruction. Therefore, a comprehensive understanding on reconstruction is significantly important to determine the real catalytic active phase and design efficient OER electrocatalysts.

7.3.1   SURFACE INTERFACE RECONSTRUCTION

Surface self-reconstruction is a complex electrochemical behavior with working time dependence. The surface evolution and the degree of surface reconstruction depend on the element composition, geometric and crystal structure. In particular, the widely proposed reconstruction causations involve lattice oxygen evolution and metal leaching in acid media [81]. To trigger LOM in metal oxides, it is necessary that the non-bonding oxygen 2p states should be close to the Fermi level. Meanwhile, the location of nonbonding oxygen 2p states is directly linked with the surface reconstruction and the dissolution rate of active metal [42,82]. The surface metal dissolution occurs under the harsh OER conditions, which eventually leads to the structural collapse. In addition to catalyst intrinsic properties, both reaction and servicing conditions dominate the dynamic interface chemistry process at a given catalyst. The charging of catalyst surface, applied potential, electrolyte pH, property of doping ions and mass transport would obviously influence on the reconstruction behavior and finally decide the trio of OER (i.e., activity, stability, and mechanism).

Designing active and stable catalysts in acid media is more challenging than that in alkaline media. To date, only few materials which have exhibited good potential for the acidic OER typically include the noble metal oxides, i.e., IrO_2 and RuO_2. The applied electrical potential is usually much higher than the oxidation potential of the noble metal catalysts. Therefore, these Ir/Ru-based catalysts would undergo the surface self-reconstruction behavior under the OER conditions, leading to the formation of oxides or (oxy)hydroxides as the true active species, as illustrated in Figure 7.5. Obviously, the reconstruction derived species generally exhibit different electrocatalytic performance in comparison with their initial synthesized counterparts [41,68,83,84]. It is worth noting that monitoring the actual dynamic surface is rather challenging so far, although the composition, structure, and morphology can be real-time characterized in the OER conductions [85]. Therefore, there is a considerable desirability in developing high-resolution in situ and operando characterization techniques to identify the real active location and reaction mechanisms.

The surface self-reconstruction during the LOM-based OER process is highly possible to form high metal-oxygen covalency which was recently demonstrated to be the reaction driving force [86]. For example, Tarascon et al. found the drastic surface reconstruction and Ir migration from the bulk to the surface, because of the gradual formation of hydrous IrO_2 on the surface of La_2LiIrO_6 during the LOM-related OER. Their studies confirmed that the reconstructed surface-active site for oxidized Ir species is a purely oxygen state with the superior activity. However,

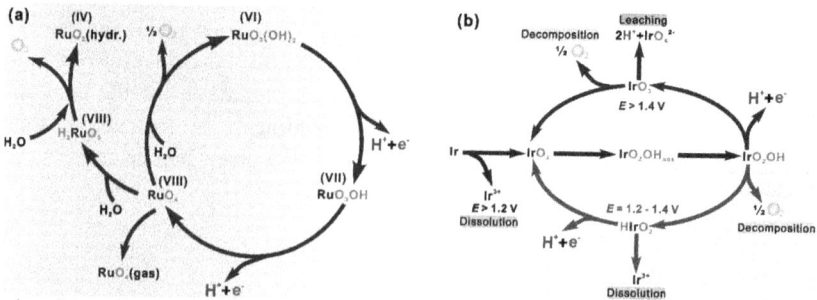

FIGURE 7.5 Schematic representation of universal reconstruction mechanisms for (a) Ru-based [80] and (b) Ir-based catalysts [2].

the Ir-based catalyst suffers from the relatively poor stability in acidic OER due to easy over-oxidation of Ir into dissolution [41]. By coupling a scanning flow cell with inductively coupled plasma and online electrochemical mass spectrometers, Kasian and co-workers found that Ir might dissolve via either the pathway of Ir^{3+}–Ir^{4+} transition or the formation of IrO_3 at high potential, depending on the potential and surface composition [2]. In our early work, repetitive potential cycling provided the direct evidence that the high valence of Ir species was continuously enriched on the catalyst surface within 6000 cycles in the acid electrolyte solution. In addition to the formation of high valence Ir species, the surface reconstruction facilitates the dissolution of surface Sn component and the change of surface roughness [83]. To enhance the stability of Ir-based catalyst, many feasible strategies have been explored to suppress over-oxidation of active metal by doping elements with strong M-OH adsorption ability or by increasing the formation energy of oxygen vacancy [81,87].

Overall, the fundamental origin of surface self-reconstruction is attributed to the potential-driven oxidation process of active metal species during OER. This reconstruction behavior is primarily determined by the electrocatalytic environment as well as the intrinsic chemical and structural characteristics of electrocatalysts. It is expected that characterizing the reconstructed states to reveal the true working catalytic sites can promote new catalyst design.

7.3.2 Electronical Double Layer and Its Function

Rotating disk electrode (RDE) with a thin-film catalyst layer is the most commonly used tool to evaluate OER electrocatalysts [88–90]. The measured results can be utilized to simply predict catalyst performance in a full electrolyzer cell [91]. During the electrochemical measurements, the RDE is immersed in an electrolyte solution, and the species interactions cause the formation of the hydrated electrode–electrolyte interfacial region. This region is known as an EDL resulting from the interfacial reconstruction. The formed EDL is significantly important, since it has a dominant influence on the electron transfer reaction occurring [92,93]. When the electric current flows at a specific working electrode, the double layer can be treated as a pseudocapacitor. To obtain a desired potential at the working electrode, the pseudocapacitor must be first appropriately charged, meaning that a charging capacitive current, not related to the electrochemical reaction, flows in the electrical circuit. Although this pseudocapacitive current would interfere with electrochemical investigations, it carries some information concerning EDL and its structure, and in the case of metal oxide OER electrocatalysts can be used to derive the electrochemical active surface area [94–96].

A classic and simplified EDL structure formed in an electrolyte solution is presented in Figure 7.6. Two compact planes are commonly associated with EDL. Based on the non-covalent interactions with the charging catalyst layer, the specifically adsorbed oxygenate intermediates are in the inner Helmholtz plane (IHP) which plays a decisive role in electrocatalytic processes. There is a layer of adsorbed water molecules on the IHP to separate from the electrode surface. The dipoles of adsorbed

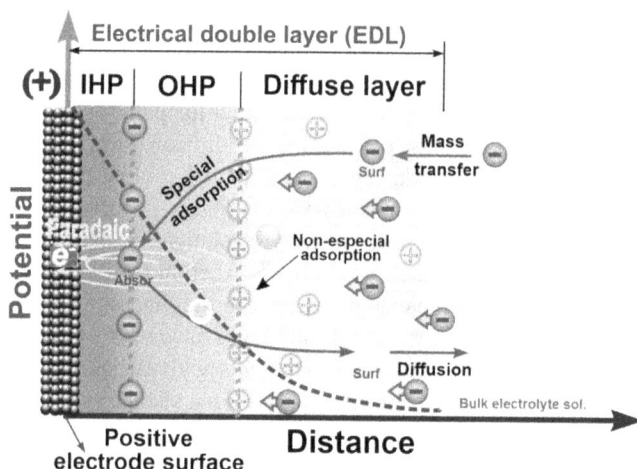

FIGURE 7.6 Schematics of the EDL structure that determine potential distribution of an electrolyzer.

water molecules rely on the charger of electrode surface. To balance the interfacial charge, hydrated ions are non-specifically adsorbed in an outer Helmholtz plane (OHP). The diffuse layer develops outside the OHP. The concentration of ions in the diffuse layer decreases exponentially with the distance from the electrode surface. The species transport resistances in EDL are responsible for the overpotential and increased electrical energy losses. Interfacial ion mobility and adsorption have thus a significant impact on EDL structure, which in turn determines the OER efficiency. In a concentrated electrolyte solution, most of the EDL potential drop occurs in the compact Helmholtz planes. In other words, potential drop across the diffuse layer decreases, which can reduce mass transport resistance and enhance the efficiency of OER. The entire EDL thickness is generally a few hundred angstroms, depending on the applied potential, the interaction between the employed materials and the ionic species, and their repulsive force [92,93,97].

The EDL located at the reaction forefront affects the Faradaic efficiency of OER to great extent. In essence, electrocatalytic OER is always accompanied by the EDL charging/discharging process and continuous reconstruction. However, the reconstructed EDL has an uncertain structure at the atomic level due to the chemical and structural flexibility of oxide materials [92,98]. This uncertainty introduces a challenge into the interpretation of the EDL effects on the electrocatalytic process. For example, depending on the active material, the OER activity detected by the linear scanning voltammetry method exhibits a distinct trend on the potential scan rate. This reason can be critically related to the EDL reconstruction according to our observation [1]. The combination of EDL effects with the optimization of working conditions (e.g. pH, temperature, and potential) will likely enable OER with high reaction efficiency and long-term stability.

7.4 CONCLUSION

The OER is a primary hindrance to the widespread production of green hydrogen via PEM water electrolysis. Addressing the trade-off between activity and stability of electrocatalysts is highly desirable for the large-scale applications. The central challenge for OER research is to understand the mechanistic details and structural features required for efficient electrocatalysis. In general, the LOM route leads to the enhanced activity but the poor durability in comparison with the traditional AEM. Therefore, when triggering the active LOM steps, a compensative strategy is necessary to improve the catalyst stability. Meanwhile, identifying the true structural and chemical properties of the active site is the key to disclosing catalyst design principles. Special attention should be given not only to the initial activity, but also to the structure and performance changes during the long-term serving time. Combination of electrochemical measurements, theoretical calculations, and spectroscopic investigations can provide valuable guidance for designing the better electrocatalysts. However, direct spectroscopic observations of the reaction surface are seriously lacking so far, arising from the drastic interfacial reconstruction of active oxides. In this context, it is encouraged that the ongoing efforts will be made to develop in situ physicochemical characterization tools with high resolution and real-time feedback. With deep understanding as discussed in this chapter, one would see that the future research should center on the advanced interfacial engineering strategies to further drive the development of OER catalysts.

REFERENCES

[1] G. Li, P.-Y.A. Chuang, Identifying the forefront of electrocatalytic oxygen evolution reaction: electronic double layer, *Appl Catal B: Environ*, 239 (2018) 425–432.

[2] O. Kasian, J.-P. Grote, S. Geiger, S. Cherevko, K.J.J. Mayrhofer, The common intermediates of oxygen evolution and dissolution reactions during water electrolysis on iridium, *Angew Chem Int Ed*, 57 (2018) 2488–2491.

[3] L. Negahdar, F. Zeng, S. Palkovits, C. Broicher, R. Palkovits, Mechanistic aspects of the electrocatalytic oxygen evolution reaction over Ni–Co oxides, *ChemElectroChem*, 6 (2019) 5588–5595.

[4] J.O.M. Bockris, Kinetics of activation controlled consecutive electrochemical reactions: anodic evolution of oxygen, *J Chem Phys*, 24 (1956) 817–827.

[5] A.I. Krasil'shchkov, Intermediate stages of anodic oxygen evolution, *Rus J Phy Chem A*, 37 (1963) 273.

[6] A. Damjanovic, A. Dey, J.O.M. Bockris, Kinetics of oxygen evolution and dissolution on platinum electrodes, *Electrochim Acta*, 11 (1966) 791–814.

[7] A. Damjanovic, A. Deya, J.O.M. Bockris, Electrode kinetics of oxygen evolution and dissolution on Rh, Ir, and Pt-Rh alloy electrodes, *J Electrochem Soc*, 113 (1966) 739–746.

[8] B.E. Conway, P.L. Bourgault, Electrochemistry of the nickel oxide electrode: part III. Anodic polarization and self-discharge behavior, *Canadian J Chem*, 40 (1962) 1690–1707.

[9] A.C. Riddiford, Mechanisms for the evolution and ionization of oxygen at platinum electrodes, *Electrochim Acta*, 4 (1961) 170–178.

[10] R.L. Doyle, M.E.G. Lyons, The oxygen evolution reaction: mechanistic concepts and catalyst design, in: S. Giménez, J. Bisquert (Eds.) *Photoelectrochemical Solar Fuel Production: From Basic Principles to Advanced Devices*, Springer International Publishing, Cham, 2016, pp. 41–104.

[11] T. Shinagawa, A.T. Garcia-Esparza, K. Takanabe, Insight on Tafel slopes from a microkinetic analysis of aqueous electrocatalysis for energy conversion, *Sci Rep*, 5 (2015) 13801.

[12] N.-T. Suen, S.-F. Hung, Q. Quan, N. Zhang, Y.-J. Xu, H.M. Chen, Electrocatalysis for the oxygen evolution reaction: recent development and future perspectives, *Chem Soc Rev*, 46 (2017) 337–365.

[13] X. Wang, H. Zhong, S. Xi, W.S.V. Lee, J. Xue, Understanding of oxygen redox in the oxygen evolution reaction, *Adv Mater*, 34 (2022) 2107956.

[14] Á. Valdés, J. Brillet, M. Grätzel, H. Gudmundsdóttir, H.A. Hansen, H. Jónsson, P. Klüpfel, G.-J. Kroes, F. Le Formal, I.C. Man, R.S. Martins, J.K. Nørskov, J. Rossmeisl, K. Sivula, A. Vojvodic, M. Zäch, Solar hydrogen production with semiconductor metal oxides: new directions in experiment and theory, *Phys Chem Phys*, 14 (2012) 49–70.

[15] H. Dau, C. Limberg, T. Reier, M. Risch, S. Roggan, P. Strasser, The mechanism of water oxidation: from electrolysis via homogeneous to biological catalysis, *ChemCatChem*, 2 (2010) 724–761.

[16] J.H. Montoya, L.C. Seitz, P. Chakthranont, A. Vojvodic, T.F. Jaramillo, J.K. Nørskov, Materials for solar fuels and chemicals, *Nat Mater*, 16 (2017) 70–81.

[17] I.C. Man, H.-Y. Su, F. Calle-Vallejo, H.A. Hansen, J.I. Martínez, N.G. Inoglu, J. Kitchin, T.F. Jaramillo, J.K. Nørskov, J. Rossmeisl, Universality in oxygen evolution electrocatalysis on oxide surfaces, *ChemCatChem*, 3 (2011) 1159–1165.

[18] J. Huang, H. Sheng, R.D. Ross, J. Han, X. Wang, B. Song, S. Jin, Modifying redox properties and local bonding of Co_3O_4 by CeO_2 enhances oxygen evolution catalysis in acid, *Nat Commun*, 12 (2021) 3036.

[19] J.A.D. del Rosario, G. Li, M.F.M. Labata, J.D. Ocon, P.-Y.A. Chuang, Unravelling the roles of alkali-metal cations for the enhanced oxygen evolution reaction in alkaline media, *Appl Catal B: Environ*, 288 (2021) 119981.

[20] H. Kim, J. Park, I. Park, K. Jin, S.E. Jerng, S.H. Kim, K.T. Nam, K. Kang, Coordination tuning of cobalt phosphates towards efficient water oxidation catalyst, *Nat Commun*, 6 (2015).

[21] S.R. Ede, Z. Luo, Tuning the intrinsic catalytic activities of oxygen-evolution catalysts by doping: a comprehensive review, *J Mater Chem A*, 9 (2021) 20131–20163.

[22] Y.-Q. Zhou, L. Zhang, H.-L. Suo, W. Hua, S. Indris, Y. Lei, W.-H. Lai, Y.-X. Wang, Z. Hu, H.-K. Liu, S.-L. Chou, S.-X. Dou, Atomic cobalt vacancy-cluster enabling optimized electronic structure for efficient water splitting, *Adv Funct Mater*, 31 (2021) 2101797.

[23] C. Hu, X. Wang, T. Yao, T. Gao, J. Han, X. Zhang, Y. Zhang, P. Xu, B. Song, Enhanced electrocatalytic oxygen evolution activity by tuning both the oxygen vacancy and orbital occupancy of B-site metal cation in $NdNiO_3$, *Adv Funct Mater*, 0 (2019) 1902449.

[24] H. Zhou, Y. Shi, Q. Dong, J. Lin, A. Wang, T. Ma, Surface oxygen vacancy-dependent electrocatalytic activity of $W_{18}O_{49}$ nanowires, *J Phys Chem C*, 118 (2014) 20100–20106.

[25] S. Hao, H. Sheng, M. Liu, J. Huang, G. Zheng, F. Zhang, X. Liu, Z. Su, J. Hu, Y. Qian, L. Zhou, Y. He, B. Song, L. Lei, X. Zhang, S. Jin, Torsion strained iridium oxide for efficient acidic water oxidation in proton exchange membrane electrolyzers, *Nat Nanotechnol*, 16 (2021) 1371–1377.

[26] J.R. Petrie, V.R. Cooper, J.W. Freeland, T.L. Meyer, Z. Zhang, D.A. Lutterman, H.N. Lee, Enhanced bifunctional oxygen catalysis in strained $LaNiO_3$ perovskites, *J Am Chem Soc*, 138 (2016) 2488–2491.

[27] K.A. Stoerzinger, L. Qiao, M.D. Biegalski, Y. Shao-Horn, Orientation-dependent oxygen evolution activities of rutile IrO_2 and RuO_2, *J Phys Chem Lett*, 5 (2014) 1636–1641.

[28] C. Xie, Z. Niu, D. Kim, M. Li, P. Yang, Surface and interface control in nanoparticle catalysis, *Chem Rev*, 120 (2019) 1184–1249.

[29] Y. Yang, M. Luo, W. Zhang, Y. Sun, X. Chen, S. Guo, Metal surface and interface energy electrocatalysis: fundamentals, *Performance Eng Opportunities, Chem*, 4 (2018) 2054–2083.

[30] L. Wu, Z. Guan, D. Guo, L. Yang, X.A. Chen, S. Wang, High-efficiency oxygen evolution reaction: controllable reconstruction of surface interface, *Small*, (2023) 2304007.

[31] W. Yang, Z. Wang, W. Zhang, S. Guo, Electronic-structure tuning of water-splitting nanocatalysts, *Trends Chem*, 1 (2019) 259–271.

[32] G. Assat, J.-M. Tarascon, Fundamental understanding and practical challenges of anionic redox activity in Li-ion batteries, *Nat Energy*, 3 (2018) 373–386.

[33] Y. Wang, X. Ge, Q. Lu, C. Bai, C. Ye, Z. Shao, Y. Bu, Accelerated deprotonation with a hydroxy-silicon alkali solid for rechargeable zinc-air batteries, *Nat Commun*, 14 (2023).

[34] D.B. Hibbert, C.R. Churchill, Kinetics of the electrochemical evolution of isotopically enriched gases. Part 2.-18O16O evolution on $NiCo_2O_4$ and $LiCo_3$-O_4 in alkaline solution, *J Chem Soc Faraday Trans1* 80 (1984) 1965–1975.

[35] E. Yeager, An overview of the electrochemical interface and optical spectroscopic studies, *J Phys Colloques*, 38 (1977) C5-1–C5-17.

[36] J.O.M. Bockris, T. Otagawa, Mechanism of oxygen evolution on perovskites, *J Phys Chem*, 87 (1983) 2960–2971.

[37] M. Wohlfahrt-Mehrens, J. Heitbaum, Oxygen evolution on Ru and RuO_2 electrodes studied using isotope labelling and on-line mass spectrometry, *J Electroanal Chem Interfac Electrochem*, 237 (1987) 251–260.

[38] S. Fierro, T. Nagel, H. Baltruschat, C. Comninellis, Investigation of the oxygen evolution reaction on Ti/IrO_2 electrodes using isotope labelling and on-line mass spectrometry, *Electrochem Commun*, 9 (2007) 1969–1974.

[39] Y. Surendranath, M.W. Kanan, D.G. Nocera, Mechanistic studies of the oxygen evolution reaction by a cobalt-phosphate catalyst at neutral pH, *J Am Chem Soc*, 132 (2010) 16501–16509.

[40] O. Diaz-Morales, D. Ferrus-Suspedra, M.T.M. Koper, The importance of nickel oxyhydroxide deprotonation on its activity towards electrochemical water oxidation, *Chem Sci*, 7 (2016) 2639–2645.

[41] A. Grimaud, A. Demortière, M. Saubanère, W. Dachraoui, M. Duchamp, M.-L. Doublet, J.-M. Tarascon, Activation of surface oxygen sites on an iridium-based model catalyst for the oxygen evolution reaction, *Nat Energy*, 2 (2016) 16189.

[42] A. Grimaud, O. Diaz-Morales, B. Han, W.T. Hong, Y.-L. Lee, L. Giordano, K.A. Stoerzinger, M.T.M. Koper, Y. Shao-Horn, Activating lattice oxygen redox reactions in metal oxides to catalyse oxygen evolution, *Nature Chem*, 9 (2017) 457–465.

[43] Z.-F. Huang, J. Song, S. Du, S. Xi, S. Dou, J.M.V. Nsanzimana, C. Wang, Z.J. Xu, X. Wang, Chemical and structural origin of lattice oxygen oxidation in Co-Zn oxyhydroxide oxygen evolution electrocatalysts, *Nat Energy*, 4 (2019) 329–338.

[44] Y. Sun, H. Liao, J. Wang, B. Chen, S. Sun, S.J.H. Ong, S. Xi, C. Diao, Y. Du, J.-O. Wang, M.B.H. Breese, S. Li, H. Zhang, Z.J. Xu, Covalency competition dominates the water oxidation structure-activity relationship on spinel oxides, *Nat Cataly*, 3 (2020) 554–563.

[45] X. Wang, S. Xi, P. Huang, Y. Du, H. Zhong, Q. Wang, A. Borgna, Y.-W. Zhang, Z. Wang, H. Wang, Z.G. Yu, W.S.V. Lee, J. Xue, Pivotal role of reversible NiO_6 geometric conversion in oxygen evolution, *Nature*, 611 (2022) 702–708.

[46] F. Wang, P. Zou, Y. Zhang, W. Pan, Y. Li, L. Liang, C. Chen, H. Liu, S. Zheng, Activating lattice oxygen in high-entropy LDH for robust and durable water oxidation, *Nat Commun*, 14 (2023) 6019.

[47] Y. Zhu, H.A. Tahini, Z. Hu, Y. Yin, Q. Lin, H. Sun, Y. Zhong, Y. Chen, F. Zhang, H.-J. Lin, C.-T. Chen, W. Zhou, X. Zhang, S.C. Smith, Z. Shao, H. Wang, Boosting oxygen evolution reaction by activation of lattice-oxygen sites in layered Ruddlesden-Popper oxide, *EcoMat* 12021.

[48] C. Roy, B. Sebok, S.B. Scott, E.M. Fiordaliso, J.E. Sørensen, A. Bodin, D.B. Trimarco, C.D. Damsgaard, P.C.K. Vesborg, O. Hansen, I.E.L. Stephens, J. Kibsgaard, I. Chorkendorff, Impact of nanoparticle size and lattice oxygen on water oxidation on NiFeOxHy, *Nat Catal*, 1 (2018) 820–829.

[49] D.A. Kuznetsov, M.A. Naeem, P.V. Kumar, P.M. Abdala, A. Fedorov, C.R. Müller, Tailoring lattice oxygen binding in ruthenium pyrochlores to enhance oxygen evolution activity, *J Am Chem Soc*, 142 (2020) 7883–7888.

[50] Y. Pan, X. Xu, Y. Zhong, L. Ge, Y. Chen, J.-P.M. Veder, D. Guan, R. O'Hayre, M. Li, G. Wang, H. Wang, W. Zhou, Z. Shao, Direct evidence of boosted oxygen evolution over perovskite by enhanced lattice oxygen participation, *Nat Commun*, 11 (2020) 2002.

[51] N. Zhang, X. Feng, D. Rao, X. Deng, L. Cai, B. Qiu, R. Long, Y. Xiong, Y. Lu, Y. Chai, Lattice oxygen activation enabled by high-valence metal sites for enhanced water oxidation, *Nat Commun*, 11 (2020) 4066.

[52] A. Krishnan, R. Ajay, J. Anakha, U.S.K. Namboothiri, Understanding defect chemistry in TMOS involved electrocatalytic OER; an analysis for advancement, *Surfaces Interfaces*, 30 (2022) 101942.

[53] J.T. Mefford, X. Rong, A.M. Abakumov, W.G. Hardin, S. Dai, A.M. Kolpak, K.P. Johnston, K.J. Stevenson, Water electrolysis on $La_{1-x}Sr_xCoO_{3-\delta}$ perovskite electrocatalysts, *Nat Commun*, 7 (2016) 11053.

[54] Y. Zhu, Y. Sun, X. Niu, F. Yuan, H. Fu, Preparation of La-Mn-O perovskite catalyst by microwave irradiation method and its application to methane combustion, *Catal Lett*, 135 (2010) 152–158.

[55] N. Zhang, Y. Chai, Lattice oxygen redox chemistry in solid-state electrocatalysts for water oxidation, *Energy Environ Sci*, 14 (2021) 4647–4671.

[56] A. Landman, S. Hadash, G.E. Shter, A. Ben-Azaria, H. Dotan, A. Rothschild, G.S. Grader, High performance core/shell Ni/Ni(OH)2 electrospun nanofiber anodes for decoupled water splitting, *Adv Funct Mater*, 31 (2021) 2008118.

[57] X. Liu, J. Meng, J. Zhu, M. Huang, B. Wen, R. Guo, L. Mai, Comprehensive understandings into complete reconstruction of precatalysts: synthesis, applications, and characterizations, *Adv Mater*, 33 (2021) 2007344.

[58] X. Zheng, B. Zhang, P. De Luna, Y. Liang, R. Comin, O. Voznyy, L. Han, F.P. García de Arquer, M. Liu, C.T. Dinh, T. Regier, J.J. Dynes, S. He, H.L. Xin, H. Peng, D. Prendergast, X. Du, E.H. Sargent, Theory-driven design of high-valence metal sites for water oxidation confirmed using in situ soft X-ray absorption, *Nat Chem*, 10 (2018) 149–154.

[59] R. Zhang, B. Guo, L. Pan, Z.-F. Huang, C. Shi, X. Zhang, J.-J. Zou, Metal-oxoacid-mediated oxyhydroxide with proton acceptor to break adsorption energy scaling relation for efficient oxygen evolution, *J Energy Chem*, 80 (2023) 594–602.

[60] Z. Shi, Y. Wang, J. Li, X. Wang, Y. Wang, Y. Li, W. Xu, Z. Jiang, C. Liu, W. Xing, J. Ge, Confined Ir single sites with triggered lattice oxygen redox: Toward boosted and sustained water oxidation catalysis, *Joule*, 5 (2021) 2164–2176.

[61] H. Wang, T. Zhai, Y. Wu, T. Zhou, B. Zhou, C. Shang, Z. Guo, High-valence oxides for high performance oxygen evolution electrocatalysis, *Adv Sci*, 10 (2023) 2301706.

[62] F. Zhu, W. Zhang, J. Xun, B.-J. Geng, Q.-M. Liang, Y. Yang, Proton donor/acceptor effects on electrochemical proton-coupled electron transfer reactions at solid-liquid interfaces, *Curr Opinion Electrochem*, 42 (2023) 101377.

[63] T. Takashima, K. Ishikawa, H. Irie, Induction of concerted proton-coupled electron transfer during oxygen evolution on hematite using lanthanum oxide as a solid proton acceptor, *ACS Catal*, 9 (2019) 9212–9215.

[64] H. Wang, K.H.L. Zhang, J.P. Hofmann, V.A. de la Peña O'Shea, F.E. Oropeza, The electronic structure of transition metal oxides for oxygen evolution reaction, *J Mater Chem A*, 9 (2021) 19465–19488.

[65] A. Lončar, D. Escalera-López, S. Cherevko, N. Hodnik, Inter-relationships between oxygen evolution and iridium dissolution mechanisms, *Angew Chem Int Ed*, 61 (2022) 2114437.

[66] F.-Y. Chen, Z.-Y. Wu, Z. Adler, H. Wang, Stability challenges of electrocatalytic oxygen evolution reaction: From mechanistic understanding to reactor design, *Joule*, 5 (2021) 1704–1731.

[67] Y. Zhu, H.A. Tahini, Z. Hu, Y. Yin, Q. Lin, H. Sun, Y. Zhong, Y. Chen, F. Zhang, H.J. Lin, C.T. Chen, W. Zhou, X. Zhang, S.C. Smith, Z. Shao, H. Wang, Boosting oxygen evolution reaction by activation of lattice-oxygen sites in layered Ruddlesden-Popper oxide, *EcoMat*, 2 (2020) 12021.

[68] Z. Lin, T. Wang, Q. Li, Designing active and stable Ir-based catalysts for the acidic oxygen evolution reaction, *Indus Chem Mater*, 1 (2023) 299–311.

[69] Z. Shi, X. Wang, J. Ge, C. Liu, W. Xing, Fundamental understanding of the acidic oxygen evolution reaction: mechanism study and state-of-the-art catalysts, *Nanoscale*, 12 (2020) 13249–13275.

[70] X. Liu, Z. He, M. Ajmal, C. Shi, R. Gao, L. Pan, Z.-F. Huang, X. Zhang, J.-J. Zou, Recent advances in the comprehension and regulation of lattice oxygen oxidation mechanism in oxygen evolution reaction, *Trans Tianjin Univ*, 29 (2023) 247–253.

[71] N. Zhang, Y. Xiong, Lattice oxygen activation for enhanced electrochemical oxygen evolution, *J Phys Chem C*, 127 (2023) 2147–2159.

[72] K.S. Exner, On the lattice oxygen evolution mechanism: Avoiding pitfalls, *ChemCatChem*, 13 (2021) 4066–4074.

[73] Q. Ma, S. Mu, Acidic oxygen evolution reaction: Mechanism, catalyst classification, and enhancement strategies, *Interdisciplinary Mater*, 2 (2023) 53–90.

[74] Y.-H. Wang, L. Li, J. Shi, M.-Y. Xie, J. Nie, G.-F. Huang, B. Li, W. Hu, A. Pan, W.-Q. Huang, Oxygen defect engineering promotes synergy between adsorbate evolution and single lattice oxygen mechanisms of OER in transition metal-based (oxy) hydroxide, *Adv Sci*, 10 (2023) 2303321.

[75] S. Hao, M. Liu, J. Pan, X. Liu, X. Tan, N. Xu, Y. He, L. Lei, X. Zhang, Dopants fixation of Ruthenium for boosting acidic oxygen evolution stability and activity, *Nat Commun*, 11 (2020) 5368.

[76] T. Nakagawa, C.A. Beasley, R.W. Murray, Efficient electro-oxidation of water near its reversible potential by a mesoporous IrOx nanoparticle film, *J Phys Chem C*, 113 (2009) 12958–12961.

[77] K. Macounova, M. Makarova, P. Krtil, Oxygen evolution on nanocrystalline RuO_2 and $Ru_{0.9}Ni_{0.1}O_{2-\delta}$ electrodes - DEMS approach to reaction mechanism determination, *Electrochem Commun*, 11 (2009) 1865–1868.

[78] P. Babar, J. Mahmood, R.V. Maligal-Ganesh, S.-J. Kim, Z. Xue, C.T. Yavuz, Electronic structure engineering for electrochemical water oxidation, *J Mater Chem A*, 10 (2022) 20218–20241.

[79] X. Cui, L. Gao, X. Xu, R. Ma, C. Tang, Y. Yang, Z. Lin, Surface self-reconstruction of catalysts in electrocatalytic oxygen evolution reaction, in: K. Wandelt, G. Bussetti (Eds.) *Encyclopedia of Solid-Liquid Interfaces* (First Edition), Elsevier, Oxford, 2024, pp. 316–327.

[80] R. Kötz, H.J. Lewerenz, S. Stucki, XPS studies of oxygen evolution on Ru and RuO_2 anodes, *J Electrochem Soc*, 130 (1983) 825.

[81] H. Zhong, Q. Zhang, J. Yu, X. Zhang, C. Wu, Y. Ma, H. An, H. Wang, J. Zhang, X. Wang, J. Xue, Fundamental understanding of structural reconstruction behaviors in oxygen evolution reaction electrocatalysts, *Adv Energy Mater*, 13 (2023) 2301391.

[82] S. Geiger, O. Kasian, M. Ledendecker, E. Pizzutilo, A.M. Mingers, W.T. Fu, O. Diaz-Morales, Z. Li, T. Oellers, L. Fruchter, A. Ludwig, K.J.J. Mayrhofer, M.T.M. Koper, S. Cherevko, The stability number as a metric for electrocatalyst stability benchmarking, *Nat Catal*, 1 (2018) 508–515.

[83] G. Li, H. Yu, X. Wang, S. Sun, Y. Li, Z. Shao, B. Yi, Highly effective $IrxSn_{1-x}O_2$ electrocatalysts for oxygen evolution reaction in the solid polymer electrolyte water electrolyser, *Phys Chem Phys*, 15 (2013) 2858–2866.

[84] R. Zhang, N. Dubouis, M. Ben Osman, W. Yin, M.T. Sougrati, D.A.D. Corte, D. Giaume, A. Grimaud, A dissolution/precipitation equilibrium on the surface of iridium-based perovskites controls their activity as oxygen evolution reaction catalysts in acidic media, *Angew Chem Int Ed*, 58 (2019) 4571–4575.

[85] Z. Kou, X. Li, L. Zhang, W. Zang, X. Gao, J. Wang, Dynamic surface chemistry of catalysts in oxygen evolution reaction, *Small Sci*, 1 (2021) 2100011.

[86] H.N. Nong, L.J. Falling, A. Bergmann, M. Klingenhof, H.P. Tran, C. Spöri, R. Mom, J. Timoshenko, G. Zichittella, A. Knop-Gericke, S. Piccinin, J. Pérez-Ramírez, B.R. Cuenya, R. Schlögl, P. Strasser, D. Teschner, T.E. Jones, Key role of chemistry versus bias in electrocatalytic oxygen evolution, *Nature*, 587 (2020) 408–413.

[87] L. Chong, J. Wen, E. Song, Z. Yang, I.D. Bloom, W. Ding, Synergistic Co−Ir/Ru composite electrocatalysts impart efficient and durable oxygen evolution catalysis in acid, *Adv Energy Mater*, 13 (2023) 2302306.

[88] P. Stonehart, P.N. Ross, The use of porous electrodes to obtain kinetic rate constants for rapid reactions and adsorption isotherms of poisons, *Electrochim Acta*, 21 (1976) 441–445.

[89] K. Shinozaki, J.W. Zack, S. Pylypenko, B.S. Pivovar, S.S. Kocha, Oxygen reduction reaction measurements on platinum electrocatalysts utilizing rotating disk electrode technique: II. Influence of ink formulation, catalyst layer uniformity and thickness, *J Electrochem Soc*, 162 (2015) F1384–F1396.

[90] M. Shao, Q. Chang, J.-P. Dodelet, R. Chenitz, Recent advances in electrocatalysts for oxygen reduction reaction, *Chem Rev*, 116 (2016) 3594–3657.

[91] S.S. Kocha, Y. Garsany, D. Myers, *Testing Oxygen Reduction Reaction Activity with the Rotating Disc Electrode Technique*, NREL, DOE, 2013.

[92] Z. Stojek, The electrical double layer and its structure, in: F. Scholz, A.M. Bond, R.G. Compton, D.A. Fiedler, G. Inzelt, H. Kahlert, Š. Komorsky-Lovrić, H. Lohse, M. Lovrić, F. Marken, A. Neudeck, U. Retter, F. Scholz, Z. Stojek (Eds.) *Electroanalytical Methods: Guide to Experiments and Applications*, Springer Berlin Heidelberg, Berlin, Heidelberg, 2010, pp. 3–9.

[93] A.J. Bard, L.R. Faulkner, *Electrochemical Methods. Fundamentals and Applications* (2nd Edition), John Wiley & Sons, Inc. 2001.

[94] S. Ardizzone, G. Fregonara, S. Trasatti, Inner and outer active surface of RuO_2 electrodes, *Electrochim Acta*, 35 (1990) 263–267.

[95] A. Eftekhari, From pseudocapacitive redox to intermediary adsorption in oxygen evolution reaction, *Mater Today Chem*, 4 (2017) 117–132.

[96] S. Trasatti, O.A. Petrii, Real surface area measurements in electrochemistry, *J Electroanal Chem*, 327 (1992) 353–376.

[97] J.H. Bae, J.-H. Han, T.D. Chung, Electrochemistry at nanoporous interfaces: new opportunity for electrocatalysis, *Phys Chem Phys*, 14 (2012) 448–463.

[98] J. Chakhalian, A.J. Millis, J. Rondinelli, Whither the oxide interface, *Nat Mater*, 11 (2012) 92–94.

8 Advances in Surface Reconstruction of Electrocatalysts for Oxygen Evolution Reaction

Mengxin Chen and Ping Xu

8.1 INTRODUCTION

Along with the intensified global energy crisis and climate change, it is particularly crucial to accelerate the transformation of the energy structure and gradually increase the percentage of new energy [1–3]. Currently, electrochemical water splitting is widely regarded as one of the most promising hydrogen production technologies, with considerable implications for tackling the challenge of global warming and achieving the goal of "carbon neutrality" [4–6]. The oxygen evolution reaction (OER) is a crucial anode reaction for water splitting, metal-air batteries, and renewable fuel cells [7–10]. However, the OER process contains four proton–electron transfer steps, resulting in a slow kinetics, which has long been the bottleneck [11–13]. In general, noble metals and their oxides (RuO_2 and IrO_2) are considered to be promising catalysts for OER, but the high price and scarce resources restrict their wide-scale application [14–16]. Hence, it is crucial to develop low-cost, high-activity, and stable catalysts for improving the efficiency of water splitting [17–19]. Recently, significant attention has been dedicated to non-noble transition metal materials as promising alternatives for water splitting [20,21]. Gaining a comprehensive understanding of the intrinsic catalytic mechanism and identifying the active sites of catalysts will greatly benefit the rational design and effective application of high-efficiency catalysts [22–25].

With the continuous advancement of in situ characterization techniques, an increasing number of studies have revealed that the surface sites of the so-called "pre-catalysts" undergo dynamic reconstruction during the reaction process and transform into the actual reactive species [26,27]. The development of in situ characterization techniques has revealed that the process of in situ reconstruction can effectively modulate electrocatalytic behaviors such as adsorption, activation, and desorption, leading to enhanced catalytic performance [28–33]. Building on this knowledge, many researchers have successfully obtained a wide range of active

DOI: 10.1201/9781003368939-8

species for catalytic reactions through pre-reconstruction of electrocatalysts [34,35]. It has been observed that the inherent properties of pre-catalysts, including composition, atomic arrangement, porosity, crystallinity, and others, significantly affect the rate, extent, and catalytic activity of the reconstruction process [36–40]. Therefore, optimizing the reconstruction process to generate a large number of active sites with high intrinsic activity represents a promising strategy for improving the catalytic performance of electrocatalysts.

Therefore, this chapter presents a comprehensive review of the latest research advancements in the surface reconstruction of OER catalysts. It covers three main aspects: the mechanism of the OER reaction, surface reconstruction phenomena occurring during the OER process and the regulation strategies, and characterization method of in situ monitoring reconstruction. This chapter highlights the relationship between catalyst structure and activity by summarizing the phenomenon, origin, and process of OER catalyst surface remodeling. It also suggests future research directions for catalysts and offers valuable guidance for the design and optimization of new electrocatalysts. These insights are also applicable to other important electrochemical reactions, such as the oxygen reduction reaction (ORR), hydrogen evolution reaction (HER), and carbon dioxide electrochemical reduction reaction (carbon dioxide RR).

8.2 OER MECHANISMS

The OER is a crucial anodic reaction in electricity-driven water splitting. However, its complex four-electron transfer kinetics makes the reaction sluggish and requires high overpotential to drive it [41]. Therefore, it is essential to develop low-cost and robust OER catalysts to overcome this kinetics challenge [42]. The design of efficient catalysts relies on a fundamental understanding of the OER mechanism, which is strongly linked to the catalyst surface structure [43].

OER exhibits different mechanisms in different media, as shown in Figure 8.1. Currently, two types of mechanisms are recognized: adsorbate evolution mechanism (AEM) and lattice oxygen-mediated mechanism (LOM). Most transition metals show thermodynamic stability in an alkaline environment. Therefore, in this section,

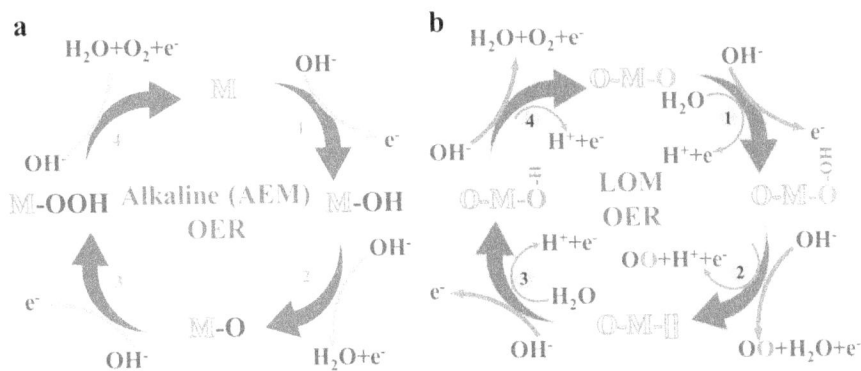

FIGURE 8.1 Schematic illustration of OER mechanisms [56]. (a) Adsorbate evolution mechanism (AEM) and (b) lattice oxygen-mediated mechanism (LOM).

we will focus on the OER mechanism and surface remodeling phenomenon under alkaline conditions. Understanding the mechanism and surface structure of the catalyst is crucial for the rational design of efficient OER catalysts. By optimizing the active sites and their accessibility, as well as controlling the surface chemistry, it is possible to enhance the catalytic performance and reduce the overpotential required for OER.

8.2.1 Adsorbate Evolution Mechanism

The AEM involves four coordinated proton–electron transfer reactions centered on metal ions [44]. As depicted in Figure 8.1a, each step of the AEM involves the transfer of an electron from the reactive site and the simultaneous release of a proton [43]. In an alkaline environment, an OH^- ion is adsorbed on the catalyst's active site* on the surface, forming the intermediate *OH through a reaction. The *OH intermediate then undergoes single-electron oxidation while simultaneously desorbing a proton into the electrolyte, resulting in the formation of *O. The *O intermediate can take two paths to generate O_2. One path involves the direct combination of two *O atoms to produce O_2. The other path involves *O absorbing another OH^- ion and desorbing H^+ to form the intermediate *OOH. Finally, the OOH intermediate releases oxygen through single-electron oxidation, returning to the original active site.

Whether in alkaline or acidic environments, the catalyst transfers the same number of electrons and protons, thus experiencing three key intermediate states: *OH, *O, and *OOH [43,45]. The OER reaction involved in the AEM is a process of adsorption and desorption, and the reaction's activity is related to the adsorption energy of oxygen on the catalyst surface. According to Sabatier's principle, the binding strength between the reaction intermediate and the reaction site cannot be too strong or too weak, ensuring the dynamic balance of the reaction process [44,46,47].

Studies have shown that there is a scalar relationship between the adsorption energies of the key intermediates in the OER reaction (*OH, *O, *OOH), with a linear correlation among them [48]. DFT calculations have found that *OH and *OOH are bound to the catalyst surface by an oxygen single bond, resulting in a constant energy difference of 3.2 eV, which can be used as an OER activity descriptor [49]. The transitions from *OH to *O and from *O to *OOH are both adsorption processes, with a constant energy difference in the adsorption energy of the three intermediates, making *O a key intermediate in the reaction. To better understand the OER reaction, various descriptors have been applied, including the occupancy of e_g orbitals, the $2p$ band center of O, and charge transfer energy [50–52]. However, the OER reaction is a heterogeneous reaction, making the structural and physical property changes of the pre-catalyst during the reaction highly complex. It is challenging to predict the OER reaction using the above descriptors alone. Tracking the structural evolution of catalysts requires proposing multiple possible mechanisms to describe and understand this intricate surface reconstruction phenomenon.

8.2.2 Lattice Oxygen-Mediated Mechanism

To improve the OER performance, researchers have explored the oxygen-containing intermediate adsorption energy scaling relationship in the AEM. It has been found that this relationship limits the OER activity, resulting in a theoretical overpotential of no more than 0.37 eV [53]. By adjusting the electronic structure of the catalyst surface, the proportionality between the active energies can be improved, leading to better OER performance.

In recent years, the understanding of the OER reaction mechanism has expanded beyond the traditional AEM. In 1976, a mechanism involving lattice oxygen was proposed and was officially named the LOM in 2015 [54]. Unlike the AEM, the LOM involves the OER reaction at two adjacent metal sites, and the catalytic active center is not limited to a single metal atomic site.

The LOM starts with the adsorption of OH⁻ ions on the catalyst surface, forming *OH intermediates (Figure 8.1b). The reaction of two *OH molecules deprotonates to form *O. *O then directly couples with lattice oxygen in the catalyst, forming an O–O bond and releasing oxygen. During this process, lattice oxygen is consumed, creating oxygen vacancies, and OH⁻ in the solution migrates to fill these vacancies. Unlike the AEM, the LOM does not involve the formation of *OOH intermediates, thus breaking the limitation of OER activity caused by the proportional relationship of oxygen-containing intermediates in the conventional AEM. The driving force of the LOM is the oxidation of lattice oxygen.

The LOM competes with the AEM [55]. DFT calculations have demonstrated that the LOM exhibits higher OER activity than the AEM at the active sites of benchmark catalysts such as RuO_2 and IrO_2. For perovskite-type OER catalysts, if the metal d band is located above the oxygen p band, the metal center of the oxide acts as the adsorption site and redox reaction center, following the AEM. However, if the d band energy is lower than the p band energy, the metal-oxygen covalence increases, and the mechanism can be switched from AEM to LOM [56]. Additionally, the presence of dopants and metal ion vacancies can also enable more involvement of the LOM. Both AEM and LOM can occur simultaneously in OER reactions, and a rational strategy to precisely adjust the local electronic structure of the catalyst can be employed to achieve the desired OER performance [57].

8.3   RECONSTRUCTION OF ELECTROCATALYSTS

Reconstruction refers to the evolution in structures of electrocatalysts during the reaction process. With the development of in situ characterization techniques, the reconstruction process can now be observed under electrochemical conditions [58–60]. During electrochemical reconstruction, pre-catalysts are transformed into amorphous or low-crystallinity active species [61].

8.3.1   RECONSTRUCTION PHENOMENON

Dynamic reconstruction can tune the catalytic performance of catalysts, significantly reducing the energy required for water oxidation [62–64]. In addition, the pre-catalysts could also undergo reconstruction under HER condition, causing local atomic rearrangement and benefiting the HER process. The reconstruction reaction can be evaluated in terms of triggering condition, reconstruction rate, and conversion degree [65]. Triggering condition refers to the potential or electrolyte concentration at which reconstruction occurs. Reducing the starting potential can initiate HER/ OER at low overpotentials. Accelerating the reconstruction rate can create rich active sites quickly, thus improving water electrolysis efficiency. Enhancing the degree of reconstruction can convert more pre-catalyst components into active species, leading to deeper reconstructed layer, a large number of active sites, and a high utilization rate of pre-catalyst. Tuning the intrinsic properties of the pre-catalyst and applying reasonable reconstruction strategies, such as surface activation, defect engineering, partial dissolution, ionic doping, heterostructure construction, and deep reconstruction, could boost the catalytic activity of the reconstructed layers [65].

Strategies to promote reconstruction include designing catalysts with high surface area, high structural feasibility, or high electronic/ionic transport properties. These strategies could deliver different reconstruction results as they focus on the modification of different properties of the pre-catalysts, and they would possess distinct impacts on the reconstruction kinetics, pathway, and degree [66].

8.3.2 Reconstruction Strategies for OER Electrocatalyst

The reconstruction of pre-catalysts is key to forming active catalytic sites that ultimately determine the performance of the OER. By modifying the reconstruction process, it's possible to precisely adjust the intrinsic structural properties of the reconstructed layers and consequently influence their electrocatalytic activity. This section will systematically explore various modification strategies, such as surface activation, defect engineering, partial dissolution, ionic doping, heterostructure construction, and deep reconstruction [23,67]. The goal is to uncover how these strategies can be effectively applied to fine-tune both the reconstruction process and the structure of reconstructed species in order to achieve high catalytic activity.

8.3.2.1 Surface Activation

The surface reconstruction of pre-catalysts can be accomplished through electrochemical activation [68,69]. Electrochemical activation is a widely used method to induce structural modifications in pre-catalysts, including surface oxidation, ion leaching, and phase transformation [70]. The specific evolution of the surface is influenced by the electrochemical operating conditions employed, such as continuous cycling, galvanostatic, and potentiostatic states [71–73]. Xu et al. used electrochemical cycling to achieve the transformation of the pre-catalyst $LiNiO_2$ into the highly active species NiOOH [74]. Figure 8.2a shows that under the electrochemical corrosion process, the Ni–O bonds contract on the surface, forming a layer of NiOOH. Electron energy loss spectroscopy (EELS) and X-ray absorption spectroscopy (XAS) were used to characterize the surface evolution of $LiNiO_2$ with increasing CV cycles. The EELS results in Figure 8.2b indicate that the intensity of Ni in $LiNiO_2$ treated with 500 cycles (LNO-500) is higher than that in the original $LiNiO_2$, suggesting that Ni accumulates on the surface during the activation cycle. Figure 8.2c shows that with increasing activation cycles, the increasing valence state of Ni and the contraction of Ni–O bonds indicate the formation of γ-NiOOH on the activated $LiNiO_2$ surface. In the OER process, the oxidized $LiNiO_2$ produces abundant Li vacancies, which promote the deprotonation of NiOOH and the formation of NiOO* as an

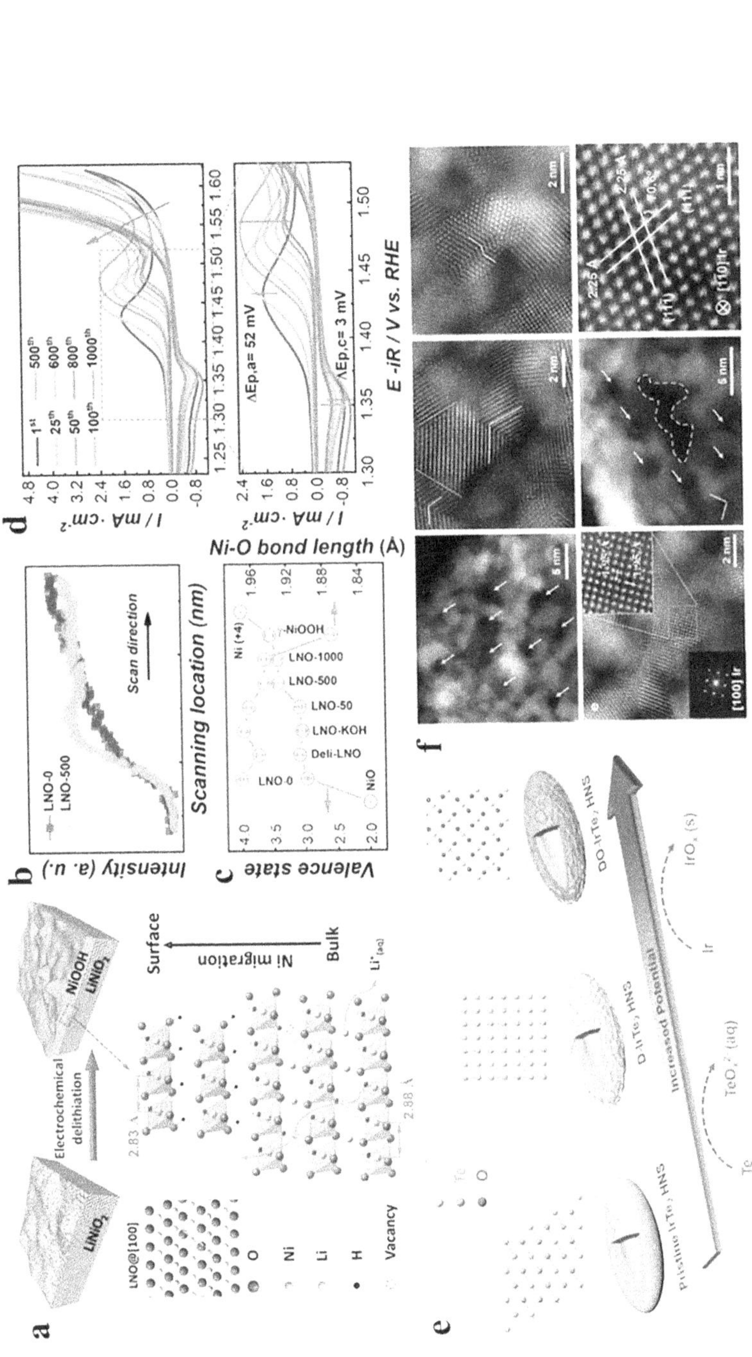

FIGURE 8.2 (a) Schematic illustration of the surface reconstruction of $LiNiO_2$ in the OER process. (b) Ni EELS intensity on the near-surface $LiNiO_2$ before and after activation [74]. (c) Valence state and Ni–O bond length of various Ni-based samples. (d) CV curves of $LiNiO_2$ collected at different CV activation cycles. Reproduced with permission. (e) Schematic illustration of the evolution of $IrTe_2$ HNS as applied potential increases. (f) HRTEM images of D-$IrTe_2$ HNS (top) and DO-$IrTe_2$ HNS (bottom) [76].

electrophilic center to enhance OER. It can be seen that the OER current density of $LiNiO_2$ increases from the first to the 500th cycle and then decreases from the 500th to the 1000th cycle (Figure 8.2d), indicating that the activation cycle has a double-edged effect on the OER performance. Therefore, during the CV activation process, attention should be paid to the influence of the number of cycles on the reconstructed species catalyst's structure and catalytic activity. This work reveals that the release of Li ions during CV activation is favorable for promoting the surface reconstruction of Ni-based catalysts.

Xia et al. conducted cycling at a specific potential range and observed the formation of $FeH_9(PO_4)_4$ (FePi) on the surface of $Ni_{1.4}Fe_{0.6}P$ decorated on reduced graphene oxide ($Ni_{1.4}Fe_{0.6}P@rGO$) after 10 cycles [75]. The formation of a crystalline phosphate layer between $Ni_{1.4}Fe_{0.6}P$ and rGO shell was confirmed by HRTEM imaging, which resulted in enhanced OER activity due to the synergy between FePi and $Ni_{1.4}Fe_{0.6}P$. At a higher potential range, the phosphate layer disappeared and NiFe-OH formed instead.

Similarly, Huang et al. adjusted the applied potential during continuous cycling to control the dealloying of $IrTe_2$ hollow nanoshuttles (HNSs) [76]. At low potentials, partial leaching of Te occurred, leading to the formation of $IrTe_2$ HNSs with a metallic Ir shell and abundant defects ($D-IrTe_2$). At higher potentials, the surface Ir was further oxidized into IrO_x. HRTEM images revealed various surface defects on the reconstructed catalysts, such as vacancies, grain boundaries, stacking faults, and rearrangement of residual atoms, which optimized the local coordination environment and electronic structure of Ir and enhanced OER catalytic performance (Figure 8.2e and f).

The formation of an amorphous shell of active species on the surface of pre-catalysts through electrochemical oxidation allows for the formation of heterostructures, enabling rapid electron transfer at the interface for catalytic reactions. Wang et al. reported the successful transformation of the surface of NiFe and NiCo alloys into corresponding oxides under galvanostatic electrochemical oxidation, resulting in the formation of alloy/hydroxide core–shell structures [69]. The resulting NiFe/NiFe-OH and NiCo/NiCo-OH heterostructures exhibited higher OER and HER activity than the parent alloys due to the accelerated electron transfer at the interface, which reduced the charge transfer resistance of the heterostructure. Stable water splitting electrolysis at a large current density of 1 A/cm^2 for 300 hours was achieved. Similarly, Zheng et al. developed nanostructured NiCo alloys with an oxide layer formed through the activation of $NiCo-SiO_2$ composite along with the dissolution of SiO_2 under continuous CV scans [77]. The presence of the metal core improved the conductive oxide layer, while the coated oxide layer contributed to the increased stability of the catalyst. In summary, the electrochemical activation of metal alloys and composites can lead to the formation of heterostructures with improved catalytic activity and stability.

8.3.2.2 Defect Engineering

Surface defect engineering of pre-catalysts, including step edges, vacancies, and amorphousness, plays a crucial role in regulating the electronic structure and local binding environment of metal sites. This engineering enables efficient electron transfer and enhances the adsorption of oxygen-containing intermediates, thus creating favorable conditions for rapid reconstruction [78–80]. Specifically, the presence of inherent oxygen vacancies can facilitate reconstruction during catalytic reactions by increasing the electron density surrounding metal atoms and reducing the oxidation states of metal cations.

Zhou et al. have uncovered the role of cation vacancy defects in NiFe-LDH, specifically in the form of V_M (M = Ni/Fe) [81]. These vacancies induce surface crystalline $Ni(OH)_x$ to undergo a reconstruction process at low potential, transitioning from an ordered state to a disordered state and subsequently transforming into local NiOOH structures at relatively higher voltages. As depicted in Figure 8.3a, this remodeling of the active component can be attributed to the evolution of cation vacancies from V_M to V_{MOH-H} as the voltage increases. Furthermore, the presence of cation vacancies reduces the formation energy of the reconstructed state.

Furthermore, Song et al. have found that Ni vacancies play a pivotal role in optimizing the electronic properties of $Ni(OH)_2$, thereby facilitating the formation of active γ-NiOOH species [82]. This observation is consistent with the trend observed in the LSV curve, where the oxidation peak of α-$Ni(OH)_2$ increases with an elevated concentration of Ni vacancies (Figure 8.3b). Density functional theory calculations indicate that a higher content of Ni vacancies can induce a partial distribution of charge density near the Fermi level, leading to a reduction in the theoretical formation energy of the reconstructed γ-NiOOH structure (Figure 8.3c and d).

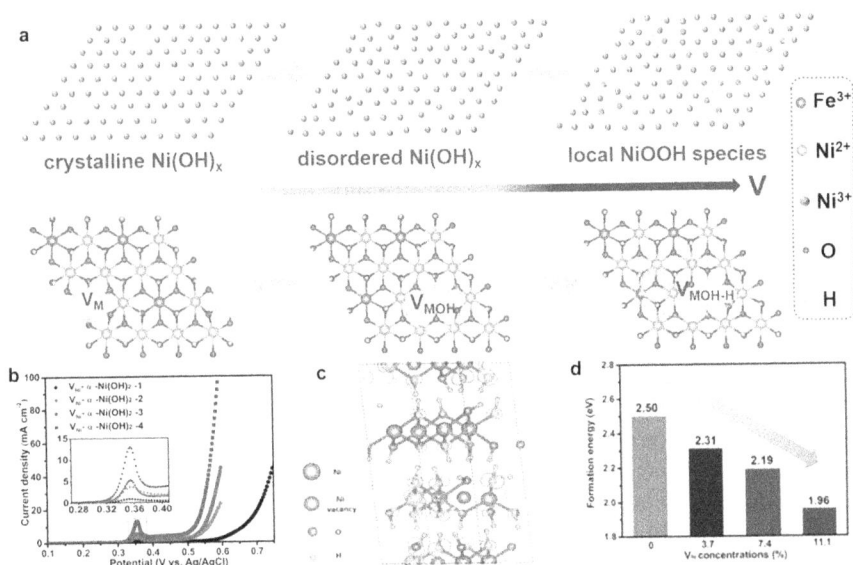

FIGURE 8.3 (a) Evolution of $Ni(OH)_x$ crystal morphology and cation defects in NiFe-LDH. (b) LSV curves of α-$Ni(OH)_2$ at different Ni vacancy concentrations [81]. (c) Simulated distribution of partial charge density of Fermi level induced by V_{Ni} incorporation. (d) Calculation of generating energy of γ-NiOOH at different Ni vacancy concentrations [82].

The transformation of crystalline materials into an amorphous phase has been proven to be an effective method for enhancing the degree of defects in pre-catalysts. Amorphous electrocatalysts possess several advantages, such as abundant active sites, unsaturated electronic configuration, and structural flexibility [83–85]. These properties accelerate the adsorption of reaction intermediates and promote the electron transfer between metal sites and intermediates, thus enabling the rapid reconstruction of pre-catalysts into active species. In a study conducted by Yu et al., they observed that this rapid reconstruction occurs on amorphous NiFeMo oxides (a-NiFeMo) (Figure 8.4a) [86]. Based on the in situ Raman spectra (Figure 8.4b and c) of a-NiFeMo and the crystalline counterpart (c-NiFeMo), the pair of peaks at 474 and 551 cm^{-1} are attributed to surface-generated NiOOH at a low potential of 1.5 V. Meanwhile, the characteristic peaks assigned to MoO_3 vanished, suggesting that the surface evolution of a-NiFeMo is faster than that of c-NiFeMo, which retains the Mo-O structure at a high potential of 1.8 V. The amorphous structure tends to introduce more vacancies in pre-catalysts during the reconstruction process, clarifying the origin of the promoted reconstruction and the enhanced OER activity.

8.3.2.3 Partial Dissolution

Electrocatalysts that contain electrochemically unstable species, including perovskites, metal phosphates, and fluorides, are prone to experiencing partial dissolution during the process of OER catalysis. This dissolution ultimately leads to the reconstruction of the catalyst. Perovskite structures, such as inorganic ABO_3 or $AB(OH)_6$ perovskites, typically consist of alkaline-earth metals and lanthanides occupying the A-site, while various transition metals like Ni, Co, and Ir occupy the B-site [87–89]. During the reconstruction process, cationic leaching of A/B-site elements occurs, which triggers the formation of unique active species on the catalyst's surface. These active species can include active hydroxyl groups and reactive oxygen ligands, both of which contribute to enhancing the OER activity of the catalyst. The presence of these active sites promotes the adsorption of reaction intermediates and facilitates efficient electron transfer processes, leading to improved catalytic performance in the OER reaction [88,89].

In their study, Tileli et al. successfully synthesized $Ba_{0.5}Sr_{0.5}Co_{0.8}Fe_{0.2}O_{3-\delta}$ (BSCF) electrocatalyst with a Co/Fe-rich surface (Figure 8.5a) [90]. They discovered that the CoFe spinel-like surface underwent a conversion into a highly active Co(Fe)OOH phase. This transformation resulted in significantly enhanced electrocatalytic properties for the OER. The cationic composition of the perovskite material plays a vital role in determining the catalytic activity of the reconstructed species (Figure 8.5b–d). Markovic et al. conducted research on $La_{1-x}Sr_xCoO_3$ perovskite compounds and demonstrated that the presence of Sr^{2+} ions in the A-site had a direct impact on the OER activity of the resulting Co hydr(oxy) oxide (CoO_xH_y) after surface reconstruction (Figure 8.5e) [91]. They found that an increased Sr-doping level led to the generation of more oxygen vacancies in the active layer on the surface, thereby enhancing the overall OER performance. This highlights the importance of promoting the reconstruction of perovskite catalysts to facilitate the formation of more active species and improve their OER catalytic performance (Figure 8.5f–k).

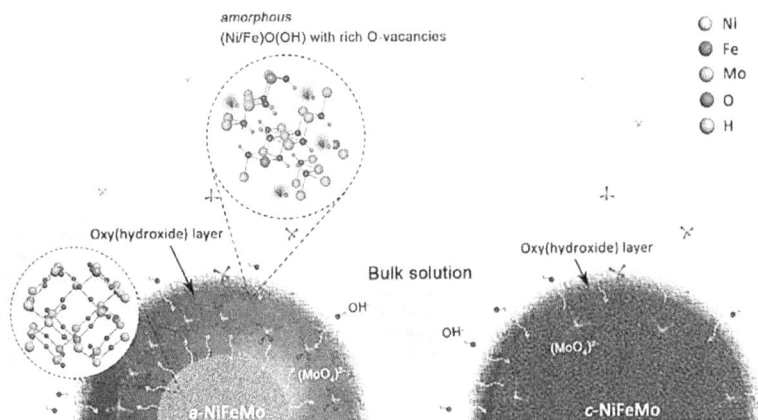

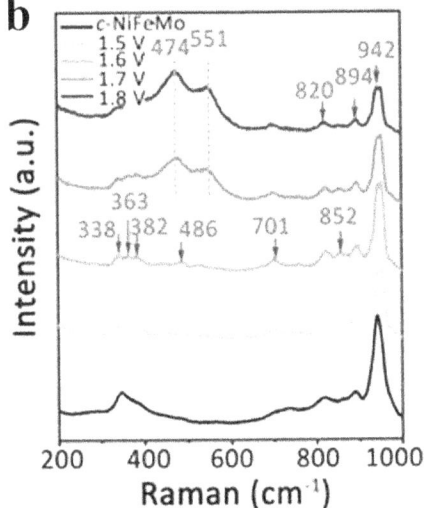

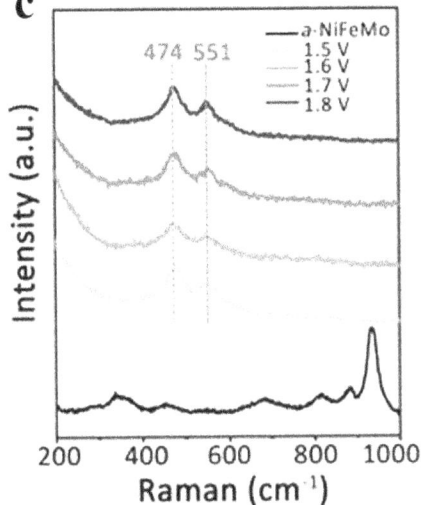

FIGURE 8.4 (a) Schematic illustration of the surface reconstruction of NiFeMo. (b, c) Operando Raman spectroscopy measurements of c-NiFeMo and a-NiFeMo catalysts, respectively [86].

As mentioned earlier, perovskite catalysts undergo a reconstruction process that involves the partial dissolution of pre-catalysts. This process introduces oxygen vacancies to the newly formed species, thereby enhancing their catalytic activity. It is worth noting that similar phenomena can also be observed in non-perovskite catalysts [92,93]. Chen et al. introduced a novel method to promote the surface reconstruction of amorphous FeB catalysts through W-P co-doping, with the aim of

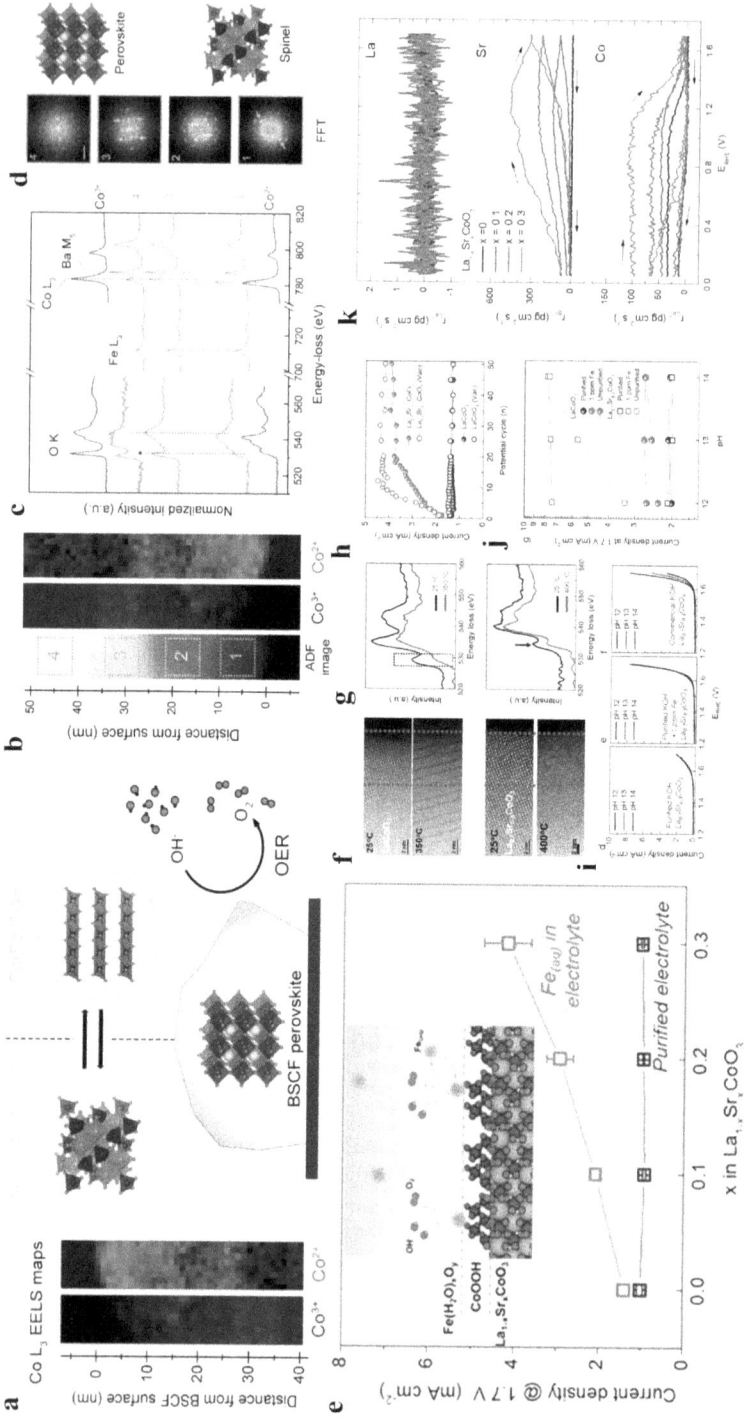

FIGURE 8.5 (a) Surface reconstruction diagram of $Ba_{0.5}Sr_{0.5}Co_{0.8}Fe_{0.2}O_{3-\delta}$. (b–d) EELS and electron diffraction analysis of BSCF surface [90]. (e) Surface reconstruction diagram of $La_{1-x}Sr_xCoO_3$. (f–j) Effect of O-vacancy and pH changes in the OER mechanism on LSCO-x samples measured in KOH media. (k) In situ measured La, Sr, and Co dissolution rates as a function of the electrode potential for all four LSCO samples at 0%, 10%, 20%, and 30% Sr-doping levels [91].

optimizing the alkaline OER activity (Figure 8.6a) [94]. During the OER process, the presence of W-doped iron oxyhydroxides (W-FeOOH) on the catalyst's surface is accelerated by the etching of B and P components. This etching process leads to the exposure of abundant coordinatively active sites, which play a crucial role in facilitating the OER. Furthermore, the dissolution of B and P elements results in the generation of additional oxygen vacancies, thereby tuning the surface electron properties of the reconstructed W-FeOOH species. The introduction of W-P co-doping and the subsequent surface reconstruction process not only enhances the abundance of active sites but also modifies the electronic structure of the catalyst, leading to improved catalytic performance in the alkaline OER (Figure 8.6b–f). This study highlights the significance of surface reconstruction and the creation of oxygen vacancies in optimizing catalytic activity, not only in perovskite-based catalysts but also in other types of catalysts such as amorphous FeB.

Overall, these studies emphasize the significance of surface reconstruction in electrocatalysts, particularly perovskite-based materials, for achieving enhanced OER performance. The composition of the catalyst, such as the presence of specific cations, and the introduction of proton-assisted reconstruction techniques are crucial factors in promoting the formation of more active species and improving the overall catalytic activity.

8.3.2.4   Ionic Doping

The incorporation of ions into pre-catalysts plays a crucial role in optimizing their electronic structure and regulating dynamic restructuring processes [95]. One effective approach to induce the reconstruction of pre-catalysts is through the doping of metal or non-metal heteroatoms, which can trigger significant reorganization of the atomic arrangement [96,97]. Metal ion doping involves introducing metallic elements into the pre-catalyst's lattice, either by substituting existing atoms or occupying interstitial sites. This process can lead to changes in the electronic properties and chemical reactivity of the catalyst. The introduction of metal ions can alter the valence state, modify the band structure, and create new active sites, thereby enhancing catalytic activity. Non-metal heteroatom doping, on the other hand, involves incorporating non-metallic elements into the pre-catalyst's structure [98,99]. These non-metal dopants can include elements such as nitrogen, phosphorus, sulfur, boron, and carbon. By introducing these dopants, the local chemical environment and electronic properties of the catalyst can be modified. This can result in the formation of new bonding configurations, changes in surface charge, and enhanced catalytic performance [100].

These metal and non-metal heteroatom dopants can induce structural rearrangements in the pre-catalyst, leading to the formation of new phases or the alteration of existing crystal structures. The doping process can introduce strain, lattice distortion, or changes in the coordination environment, which can have a profound impact on the catalytic behavior of the material [101]. The induced reconstruction of pre-catalysts through metal or non-metal heteroatom doping offers opportunities to tailor the catalyst's properties according to specific catalytic requirements. It allows for the optimization of electronic structure, enhancement of surface activity, and regulation of surface reactivity [102]. This strategy has been widely explored and applied in

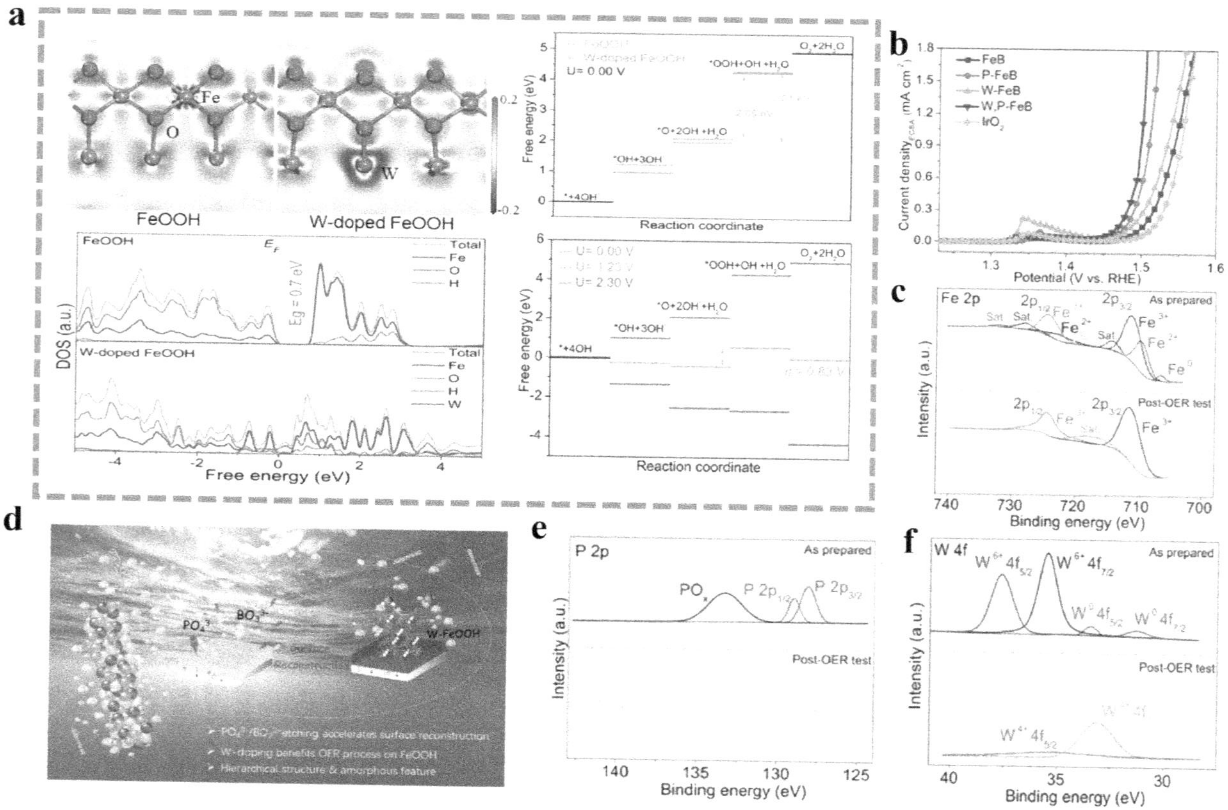

FIGURE 8.6 (a) The DFT calculation of W-doped FeOOH. (b) LSV curves normalized with respect to ECSA for W, P-FeB, W-FeB, P-FeB, FeB, and the IrO$_2$ catalyst. (c, e, f) High-resolution XPS scans of W, P-FeB in the Fe 2p, P 2p, and W 4f, respectively. (d) Illustration of the proposed mechanism for OER activity enhancement [94].

various catalytic systems, ranging from heterogeneous catalysis to electrocatalysis and photocatalysis, with the aim of improving catalytic efficiency and selectivity.

In their study, Lim et al. investigated the effects of Cl doping on $LiCo_2O_4$ pre-catalyst ($LiCoO_{1.8}Cl_{0.2}$) [103]. As Li and Co undergo electrochemical oxidation and leaching, the pre-catalyst undergoes a rapid phase transition into an active amorphous hydroxide. Figure 8.7a demonstrates that as the Cl content increases, the OER performance of the pre-catalyst improves gradually. Additionally, as shown in Figure 8.7b, the normalized Co K-edge XANES spectrum of $LiCoO_{2-x}Cl_x$ shifts toward lower energy regions with increased Cl content, indicating a decrease in the valence state of Co due to Cl^- substituting O^{2-}. Co^{2+} in $LiCoO_{1.8}Cl_{0.2}$ is oxidized to Co^{3+} at a lower positive potential, thereby initiating the reconstruction of $LiCoO_{1.8}Cl_{0.2}$ below 1.4 V. In contrast, the reconfiguration of $LiCoO_2$ starts after 1.4 V due to the Co^{3+}/Co^{4+} transition during OER. Furthermore, Cl doping enables the pre-catalyst to trigger its reconfiguration at a lower potential. Figure 8.7c illustrates that $LiCoO_{2-x}Cl_x$ transforms into an OER-active substance, containing Cl^--modified hydroxyl cobalt oxide, at a lower potential. This stands in stark contrast to $LiCoO_2$ without Cl, which converts to the less active $Li_{1\pm x}Co_2O_4$. Overall, this work presents a novel approach of adjusting surface reconstruction through in situ leaching for the rational design of electrocatalysts.

In a recent study, Ma et al. proposed a novel fluorination strategy for NiFe Prussian blue analog (NiFe-PBAs) [104]. They replaced the conventional CN^- ligand with F^- anions to obtain the fluoride product (NiFe-PBAs-F), which undergoes electrochemical reconstruction to form F-doped NiFeOOH. The migration of F during electrochemical reconstruction was found to be responsible for this transformation. This approach resulted in enhanced OER performance due to the accumulation of F dopants on the surface. This, in turn, facilitated faster adsorption of oxygen-containing intermediates, thus accelerating the OER process. Similarly, Ju et al. also reported on the promotion of F^- doping in CoOOH nanosheets through F^- migration during the electrochemical oxidation of CoF_2 nanowires [105]. The resulting accumulation of F dopants on the surface was shown to enhance the hydrophilicity of the material. This, in turn, boosted the adsorption of oxygen-containing intermediates and improved the OER performance. These findings shed light on the significance of F^- doping as an effective strategy to optimize the electrocatalytic activity of materials for OER applications.

8.3.2.5 Heterostructure Construction

The design of heterostructures by combining two or more dissimilar materials has emerged as an effective strategy to enhance the reconstruction process and improve the structural properties of catalysts [106,107]. This approach has been found to offer mutual advantages and can significantly boost catalytic performance. By introducing a new component into the heterostructure, the reconstruction of pre-catalysts is facilitated, leading to the formation of abundant active species. These active species play a crucial role in enhancing the catalytic performance of the material [108]. The combination of dissimilar materials in the heterostructure creates synergistic effects, allowing for improved electron transfer, enhanced stability, and optimized surface properties. The heterostructure design provides unique opportunities to tailor the

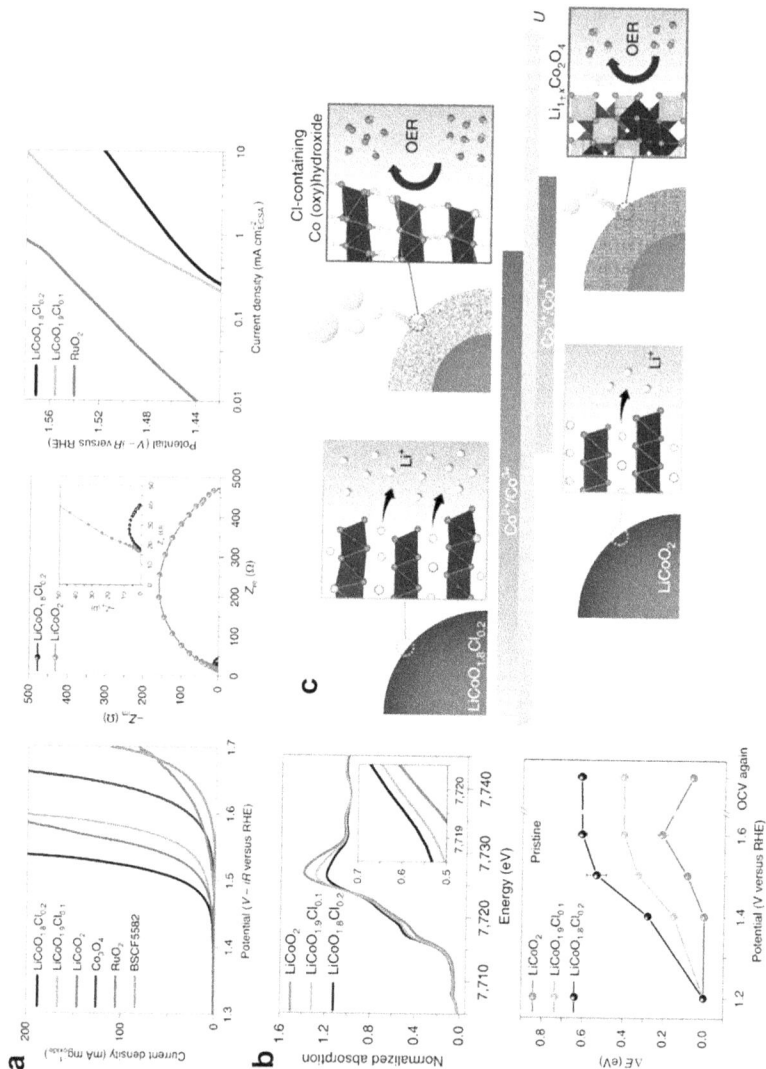

FIGURE 8.7 (a) OER activity of the pre-catalyst of $LiCoO_{2-x}Cl_x$. (b) The XANES spectra of Co-K edges recorded at OCV (top) and the displacement of $LiCoO_{2-x}Cl_x$ (x = 0, 0.1, or 0.2) recorded at different potentials (bottom). (c) Schematic diagram of in situ surface reconstruction processes of $LiCoO_2$ and $LiCoO_{1.8}Cl_{0.2}$ during OER [103].

properties and functionalities of the catalyst, allowing for the exploration of new catalytic mechanisms and the development of highly efficient electrocatalysts [109,110]. This strategy holds great promise for advancing the field of catalysis and accelerating the development of next-generation energy conversion and storage technologies [108].

Kou et al. have developed a novel approach to synthesize two-dimensional Co-based heterostructures composed of Co and Mo_2C nanoparticles as efficient OER pre-catalysts [111]. An anion exchange method was used to fabricate the heterostructures, which showed improved kinetics during the initial stage of OER in an alkaline solution. The presence of Mo_2C enables Co to rapidly transform into γ-CoOOH, leading to the formation of a defect CoOOH rich in Mo. This enhances the kinetics of the OER reaction (Figure 8.8a). In situ Raman spectroscopy and ex situ scanning transmission electron microscopy (STEM) studies revealed that the phase transition to gamma-CoOOH and the reconstruction of the Mo-rich surface were potential-dependent and accelerated at 1.4 V (Figure 8.8b and c). Potential-related X-ray photoelectron spectroscopy (XPS) and methanol oxidation experiments further confirmed that the enrichment of Mo on the defective CoOOH surface promoted electron flow from Mo to the Co site through bridging oxygen. This greatly facilitated the electrostatic adsorption of OH^- ions and improved the performance of the heterostructure as an OER catalyst. These findings demonstrate that the rational design of heterogeneous structures can accelerate electron transfer and regulate the electronic structure of active substances, thereby promoting the adsorption/desorption of intermediates on the surface. This, in turn, enhances the kinetics of the OER reaction. The study highlights the potential of using heterogeneous structures composed of dissimilar materials to develop highly efficient electrocatalysts for energy conversion and storage applications.

The design of catalysts with controllable local bonding environments around the metal center is key to achieving high electrocatalytic activity. The interface effect between dissimilar materials can facilitate the conversion of active species during surface reconstruction, improving the efficiency of catalytic reactions. Huang et al. recently reported the development of a Co_3O_4/CeO_2 heterostructure for acidic OER, in which the redox properties of Co_3O_4 are altered in the presence of CeO_2 [112]. Their study revealed that the Co_3O_4/CeO_2 heterostructure promotes the fast formation of Co(IV) active species without the formation of dimeric Co(IV)Co(IV) (Figure 8.8d). This was evident from the absence of a redox peak attributed to the transition of dimeric Co(III)Co(IV) to Co(IV)Co(IV) in the CV curve of Co_3O_4/CeO_2 at high potential. Furthermore, Co K-edge XANES spectra showed a negative shift in the Co_3O_4/CeO_2 heterostructure, indicating an increase in Co oxidation valence owing to electron transfer from Co_3O_4 to CeO_2. Importantly, Huang et al. used Co K-edge EXAFS to demonstrate that the introduction of CeO_2 modified the local bonding environment of Co_3O_4/CeO_2. This resulted in shorter Co–O and Co–Co_{oct} (octahedral Co cation) bonds and longer Co–Co_{tet} (tetrahedral Co cation) bonds in the heterostructure, suggesting electronic redistribution induced by CeO_2. Such modifications in local bonding facilitated the translation of Co sites to Co(IV) at a lower potential and prevented charge accumulation under the OER process at large potential. Consequently, the Co_3O_4/CeO_2 heterostructure exhibited rich active

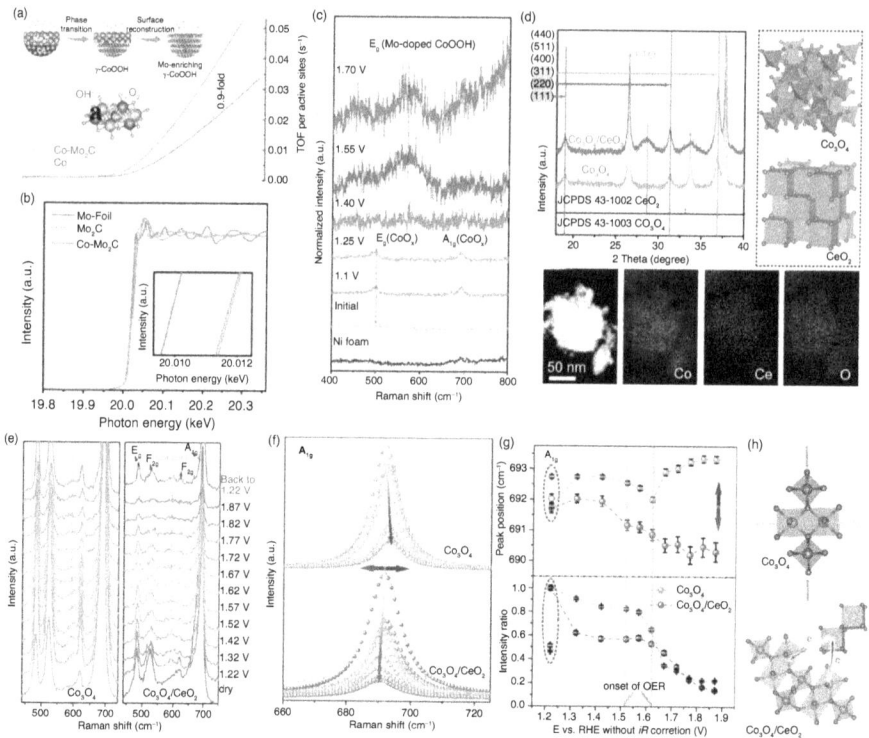

FIGURE 8.8 (a) Schematic diagram of the reconstruction process from Co-Mo$_2$C to rich Mo γ-CoOOH before and after reconstruction and TOF curve. (b) Mo K-edge XANES spectra of Mo-Foil, Mo$_2$C, and Co-Mo$_2$C. (c) In situ Raman spectroscopy of Co-Mo$_2$C [111]. (d) The basic structural characterization of Co$_3$O$_4$/CeO$_2$. (e) The in situ Raman spectra of Co$_3$O$_4$ (left panel) and Co$_3$O$_4$/CeO$_2$ (right panel) at various constant potential. (f) The Raman A$_{1g}$ peaks of Co$_3$O$_4$ (top) and Co$_3$O$_4$/CeO$_2$ (bottom) were fitted with Lorentzian function to extract the peak positions, intensity, and FWHM (dashed lines: raw spectra; dots: fitting results). (g) The Raman A$_{1g}$ peak positions (upper panel) and intensity ratio with respect to the initial intensity at 1.22 V (lower panel) plotted against the applied potential. (h) Schematic illustrations of the local bonding environment changes in Co$_3$O$_4$ before and after OER testing and the hypothesized electronic modifications in Co$_3$O$_4$/CeO$_2$ [112].

Co(IV) with improved catalytic activity. Overall, this study highlights the importance of controlling local bonding environments in designing efficient electrocatalysts for energy conversion and storage applications (Figure 8.8e–h). The use of heterogeneous structures composed of dissimilar materials can help to regulate local bonding environments and enhance catalytic performance.

8.3.2.6 Deep Reconstruction

During the surface reconstruction process, the pre-catalyst can develop an amorphous shell, which is crucial for enhancing the performance of the OER. However, this structural transformation usually occurs only on the catalyst's surface, leading to limited utilization of the internal components, restricted electrochemically active areas, and hindered mass transfer processes. Therefore, it is crucial to promote the reconstruction process of the pre-catalyst components and maximize the production of active species in order to significantly improve the utilization rate and specific activity of these components. The extent of reconstruction is closely related to material properties, structural modifications, and reaction conditions [7,113,114].

To facilitate the reconstruction process, rapid co-leaching of multiple components in the pre-catalyst can be employed, which loosens the catalyst's surface, allowing electrolyte penetration and further etching of the internal material. Recently, Mai et al. discovered that introducing 2-methylimidazole (2-mim) into NiFe polyoxomolybdate (Fe_xNi-POMo) induces complete remodeling under OER conditions [115]. The high reconfigurability of Fe_xNi-POMo is attributed to its low crystallinity and porosity, which facilitates the co-leaching of MoO_4^{2-} and 2-mim ligands, promoting the formation of active NiOOH species for OER (Figure 8.9a).

In addition, Mai et al. reported that the co-solubilization of MoO_4^{2-} and crystal water induces the complete reconfiguration of $NiMoO_4 \cdot xH_2O$ [116]. During the electrochemical oxidation process, amorphous NiOOH layers gradually form on the surface of $NiMoO_4 \cdot xH_2O$, creating a porous surface structure that facilitates alkaline electrolyte penetration into the internal structure. This, in turn, promotes the continuous co-leaching of MoO_4^{2-} with crystal water. As a result, the reconstruction of $NiMoO_4 \cdot xH_2O$ is completed within 20 cyclic voltammetry (CV) cycles. High-resolution transmission electron microscopy (HRTEM) images reveal that the reconstructed species consist of interconnected ultrafine nanoparticles, leading to the formation of abundant interfacial pores.

The reconstruction process can be monitored through in situ Raman analysis, as depicted in Figure 8.9b. In the initial stage of electrochemical oxidation, the vibration peak of MoO_4^{2-} and the tensile peak of Mo-O-Ni decrease significantly with increasing applied potential, while peaks corresponding to NiOOH appear at high potential. In contrast, the Raman peak of $NiMoO_4$ remains observable throughout the reaction. These findings indicate that the co-immersion of MoO_4^{2-} and crystal water plays a crucial role in the rapid and profound reconfiguration of $NiMoO_4 \cdot xH_2O$. The slow reconstruction of $NiMoO_4$ may be attributed to the dense active layer formed on its surface, which hinders further electrolyte penetration. Moreover, the complete reconstruction of $NiMoO_4$ can be achieved through alkaline electrolysis at an industrial operating temperature of 51.9°C, highlighting the temperature dependence of MoO_4^{2-} dissolution. By understanding and leveraging these reconstruction processes, scientists can advance the design of efficient catalysts for the OER reaction, bringing us closer to the realization of sustainable energy conversion and storage systems.

8.4   IN SITU/OPERANDO CHARACTERIZATION OF THE SURFACE RECONSTRUCTION PROCESS

The mechanism of the OER is intricate and not fully understood. However, it involves a four-electron transfer process and the formation of M–O, M–OH, and M–OOH intermediates. At high potentials, catalyst surfaces may undergo self-reconstruction behavior. To characterize the self-reconstruction process of catalyst surface structures, a general strategy involves setting up multiple test control groups and conducting

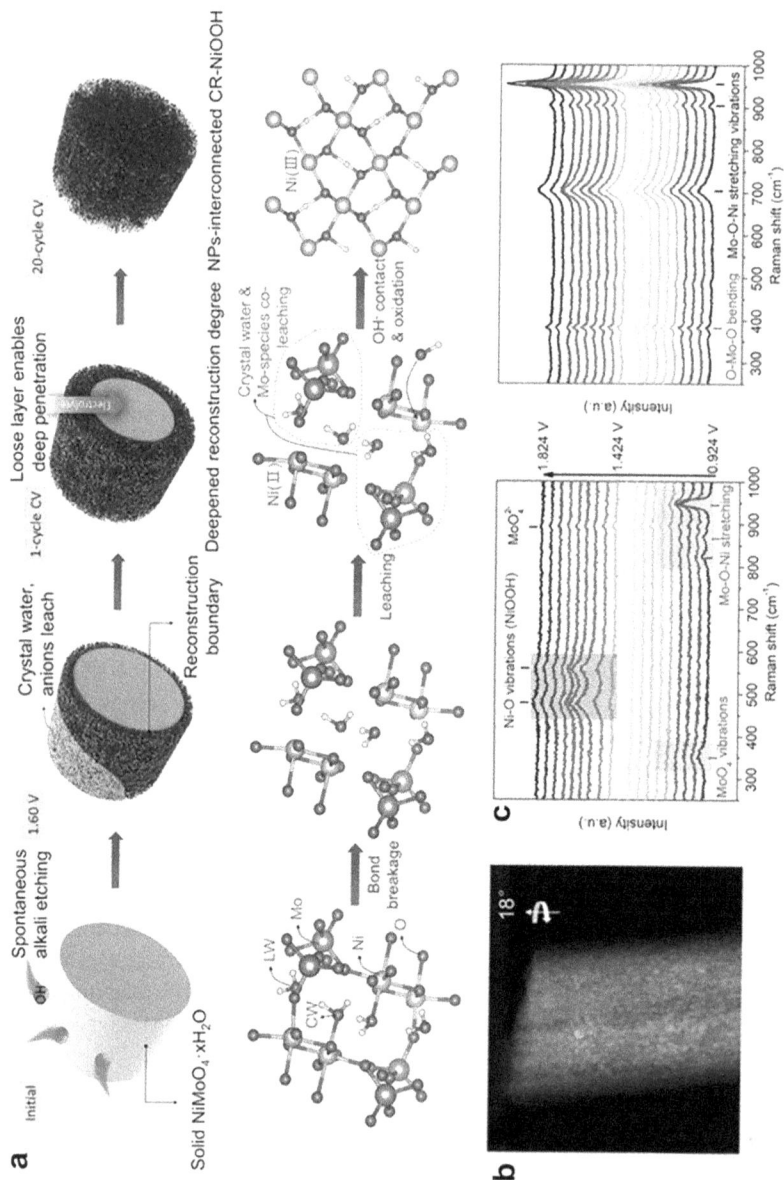

FIGURE 8.9 (a) Schematic diagram of the reconstruction process of $NiMoO_4 \cdot xH_2O$ and the corresponding crystal evolution process [115]. (b) HRTEM images after complete reconstruction of $NiMoO_4 \cdot xH_2O$. (c) In situ Raman spectra of $NiMoO_4 \cdot xH_2O$ (left) and $NiMoO_4$ (right) [116].

electrocatalytic reactions for different durations under identical conditions [117,118]. Subsequently, the electrocatalysts can be characterized as a function of reaction time using various ex situ techniques such as scanning electron microscopy (SEM), transmission electron microscopy (TEM), X-ray diffraction (XRD), Raman spectroscopy, and XPS [119,120]. Finally, the mechanisms underlying the self-reconstruction process of catalyst surfaces can be elucidated by analyzing the diverse characterization results obtained from multiple control catalysts. Nevertheless, the process of catalyst surface reconstruction is intricate and dynamic, with a very short timeframe for complete reconstruction. Traditional characterization techniques performed outside of the reaction environment are unable to precisely capture the actual active sites involved in the OER or provide valuable insights into the mechanism behind it [121,122]. In response to this challenge, in situ/operando characterization techniques have been developed. These techniques enable non-destructive monitoring of the OER process by detecting reaction intermediates, thereby uncovering the catalyst's reconstruction behavior and catalytic mechanism.

8.4.1 In situ/Operando X-Ray Absorption Spectroscopy

In situ/operando XAS provides insightful information about the electronic and geometric structures of electrocatalysts, revealing the catalytic active sites of OER electrocatalysts [123]. The applicable range and experimental setup of in situ/operando XAS for electrocatalysts are shown in Figure 8.10a and b [124,125]. In situ/operando XAS consists of two regions: X-ray Absorption Near Edge Structure (XANES) and Extended X-ray Absorption Fine Structure (EXAFS) [126,127]. XANES is the main absorption spectrum of in situ/operando XAS, corresponding to the 1s–4p states of elements. By multiple scattering of photoelectrons between atoms, information about electronic transitions and local structures can be obtained. EXAFS corresponds to the absorption spectrum outside the XANES region [128]. By single scattering of photoelectrons, atomic-level structural information, including bond length, coordination number, and structural disorder, can be obtained about the absorbing atom's local environment.

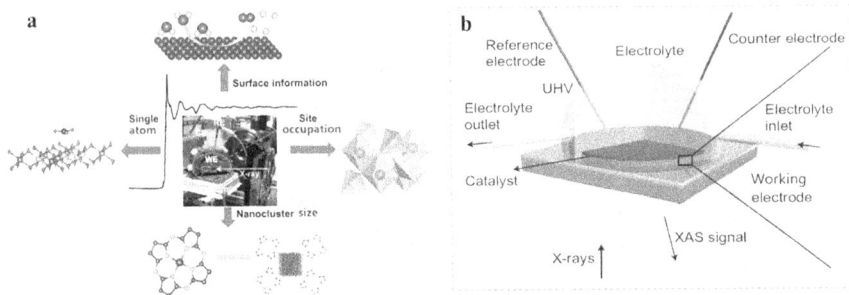

FIGURE 8.10 (a) The applicable range of in situ XAS [125]. (b) The setup scheme for in situ XAS of electrocatalysts [124].

In summary, XANES determines the oxidation state of the absorbing atom based on the position of the K-edge, while EXAFS detects the local geometric structure of the absorbing atom by analyzing the scattering of photoelectrons from neighboring atoms. Although in situ/operando XAS can monitor changes in oxidation states during the OER process, due to the bulk sensitivity of XAS, it can only provide information about bulk and surface phases and cannot directly determine the specific oxidation states of metals. The surface reconstruction process of catalysts often accompanies the redox process, where most precursor catalysts are first reconstructed into metal oxide/hydroxide intermediates before being further reconstructed into metal hydroxides. The entire surface reconstruction process is accompanied by an increase in the oxidation state of the metal ions. The in situ/operando XAS setup shown in Figure 8.8b has been utilized to monitor the changes in metal oxidation states and structures of various precursor catalysts, and to analyze the formation of transitional intermediates from real-time XAS spectra, which can be used to determine whether surface reconstruction occurs.

8.4.2   IN SITU RAMAN SPECTROSCOPY

In situ Raman spectroscopy is an operational molecular characterization technique suitable for multiphase catalytic systems [129]. When combined with in situ XAS, it can reveal the structural changes of catalysts, active sites of catalysts, and intermediates formed during the OER process [130]. The schematic diagrams of the on-site Raman monitoring system and its combination with electrochemical measurements are shown in Figure 8.11a and b [131,132]. When a beam of light illuminates the sample, the incident photons undergo inelastic scattering; the energy difference between the incident photons and the scattered photons is measured to obtain information about the "fingerprint" of the system, including its vibrational and rotational properties. Since OER involves electron transfer, in situ Raman spectroscopy can obtain real-time spectral information of the oxygen evolution process with the assistance of electrochemical testing techniques, making it easier to understand the OER mechanism. However, due to the real-time nature of in situ Raman spectroscopy

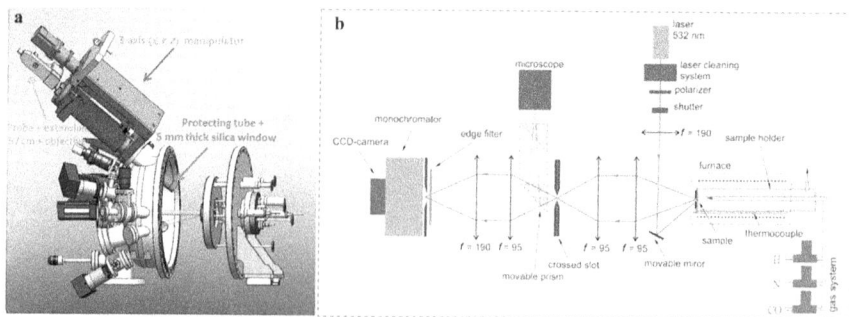

FIGURE 8.11   (a) Schematic diagram of the in situ Raman monitoring system [131]. (b) Schematic diagram of the in situ Raman setup combined with electrochemical measurements [132].

data, XANES and Fourier transform EXAFS are usually required to collect the data. Additionally, due to its excellent molecular specificity, Raman spectroscopy has high selectivity for low-frequency vibrations such as M–OH, M–OH$_2$, and M=O. Coincidentally, the main product observed after pre-catalyst surface reconstruction is hydroxide. During the surface reconstruction process, the pre-catalyst first combines with H$_2$O to form M–OH$_2$ in the OER process. Then, an O–H bond breaks, forming M–OH. Subsequently, another O–H bond breaks, forming M=O, which finally combines with O–H on the electrode to form M–OOH. By monitoring the low-frequency vibration changes in the real-time Raman spectra generated by in situ Raman spectroscopy, the evolution of the entire surface reconstruction process can be determined. Therefore, in situ Raman spectroscopy is suitable for the determination of intermediate products in the OER process and the revelation of surface reconstruction of OER catalysts.

8.4.3 IN SITU FOURIER TRANSFORM INFRARED SPECTROSCOPY

In situ FTIR spectroscopy is used to characterize various molecules present in the catalyst during the OER process, in order to determine the changes in the active species and identify the reaction pathways [133]. In situ FTIR offers advantages such as wide applicability, minimal sample consumption, simple operation, fast analysis speed, and high sensitivity [120]. The schematic diagram of the in situ FTIR setup is shown in Figure 8.12 [134]. By irradiating the sample with infrared light of different frequencies, the vibrations and rotations of the sample molecules cause a change in the dipole moment, leading to the transition of vibrational and rotational energy levels from the ground state to the excited state [135]. The intensity of transmitted light in the absorption region is reduced, resulting in the molecular absorption spectrum. From the molecular absorption spectrum, characteristic absorption frequencies and peak intensities can be determined, enabling qualitative analysis of the material. Due to the distinct infrared spectra of different molecules, infrared spectroscopy can directly identify specific molecules. During the catalyst surface reconstruction process, the precursor catalyst evolves into a reconstructed intermediate,

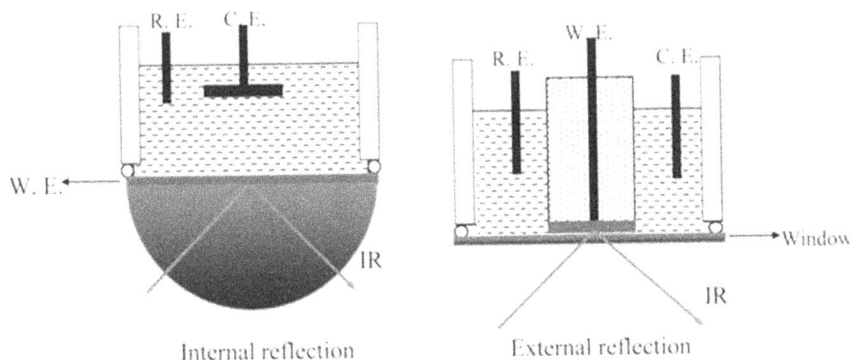

FIGURE 8.12 Simple structure schematic diagram of the in situ FTIR setup [134].

which then transforms into the final reconstructed product. Changes in the types of molecules are accompanied by vibrations and rotations, causing variations in the infrared absorption frequencies and yielding different molecular absorption spectra. The strong characteristics and sensitivity of in situ FTIR enable easy detection of changes in the precursor catalyst molecules, which can help determine whether catalyst surface reconstruction has occurred and what intermediate OER products and surface reconstruction products have formed.

8.4.4 In Situ X-Ray Photoelectron Spectroscopy

In situ XPS can track the surface chemical composition and oxidation state of a catalyst in real time, revealing potential mechanisms of the OER process [136]. The most commonly used in situ ambient pressure XPS (AP-XPS) is typically employed to detect solid interfaces, as illustrated in Figure 8.13 [137]. Traditional non-in situ XPS techniques require ultra-high vacuum conditions, which are not practical for monitoring OER in situ since it occurs under atmospheric conditions. Unlike traditional XPS, AP-XPS is not limited by ultra-high-pressure conditions and can monitor the surface chemistry of electrocatalysts without requiring ultra-high-pressure conditions [138,139]. First, the sample is placed in a high-pressure chamber and irradiated with X-rays in the reaction atmosphere. The electron lens focuses on the sample to

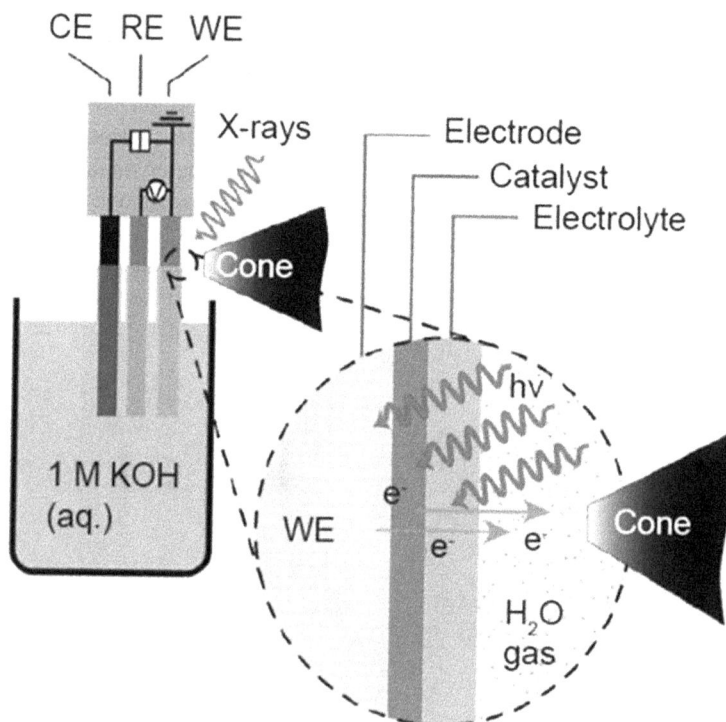

FIGURE 8.13 Schematic diagram of the in situ AP-XPS setup (1 M = 1 mol/L) [137].

produce photoelectrons, and the surface chemistry of the sample is obtained by analyzing the signal of photoelectrons collected by the analyzer, followed by further processing. However, under ultra-high-pressure conditions, scattered photoelectrons on the surface of the sample weaken the signal, making it necessary for AP-XPS to be used in conjunction with synchrotron accelerators to obtain more accurate information. The surface reconstruction process of OER electrocatalysts is dynamic, wherein the precursor catalyst dynamically reconstructs itself into hydroxides and amorphous substances. Correspondingly, this dynamic self-reconstruction process is accompanied by changes in the chemical and oxidation states of the catalyst, such as the increase in metal ion oxidation state and changes in chemical phases. Understanding the evolution of catalyst surface reconstruction is crucial in studying the process, including changes in catalyst species. Therefore, using in situ XPS technology to monitor the dynamic changes in the chemical and oxidation states of the catalyst during the OER process can help understand the evolution of precursor catalyst surface reconstruction.

8.5   SUMMARY AND OUTLOOK

OER is an important semi-reaction that participates in many electrochemical reactions, making it necessary to develop OER catalysts with high catalytic activity, stability, and low overpotential. To achieve this goal, understanding the mechanism of the OER reaction is a necessary prerequisite for the rational design of OER catalysts. This chapter provides a detailed introduction to the response mechanism of OER, including the AEM and the LOM. The limiting conditions for the proportion in the AEM and other activity descriptors within the OER reaction are proposed, which aids in the rapid screening of efficient OER catalysts. In addition, the surface reconstruction of OER is discussed, and methods for regulating OER reconstruction are introduced, such as ion doping, defect engineering, and heterostructure construction. Regulating reconstruction can effectively enhance OER activity by adjusting the adsorption ratio between oxygen intermediates or stabilizing critical reaction intermediates. The phase separation in heterostructure engineering and the addition of soluble compounds in pre-catalysts protect the catalyst matrix. Ion leaching is another method for regulating surface-controlled reconstruction. With appropriate methods, targeted leaching of anions or cations can achieve unexpected catalytic effects. Finally, four commonly used in situ characterization and testing methods in the OER reaction are briefly introduced to guide research on surface reconstruction processes in OER.

In summary, a series of effective strategies that link catalytic activity with reconstruction degree have been summarized, aiming to achieve better catalytic performance. Although progress has been made in the controllable surface reconstruction of catalysts, integrating mechanisms with surface-controlled reconstruction to achieve more precise regulation remains a major challenge. Efforts need to be made in the following aspects: (1) The scale relationship of the AEM limits the performance of OER. Hence, more precise strategies should be applied to optimize the free energy of each step in the OER reaction, approaching the ideal 1.23 V equilibrium potential. (2) The LOM can break the proportional relationship of critical reaction intermediates. Hence, more strategies should be adopted in the OER process to achieve a synergistic effect between the AEM and the LOM. (3) While the degree of catalyst surface reconstruction seems to be positively correlated with catalytic activity in the OER process, excessive destruction of the substrate can affect performance. Therefore, it is necessary to design a self-terminating reconstruction strategy based on understanding the self-termination reconstruction mechanism to achieve appropriate surface reconstruction. In conclusion, controllable surface reconstruction is crucial for the development of a new generation of efficient and cost-effective OER electrocatalysts, providing more possibilities for the precise regulation of catalyst surface electrons. This chapter reports a series of scientific information that is expected to provide theoretical guidance for the design of efficient electrocatalysts.

REFERENCES

1. Huang, X., et al., Vertical CoP nanoarray wrapped by N,P-doped carbon for hydrogen evolution reaction in both acidic and alkaline conditions. *Advanced Energy Materials*, 2019. **9**(22): p. 1803970.
2. Hunter, B.M., H.B. Gray, and A.M. Müller, Earth-abundant heterogeneous water oxidation catalysts. *Chemical Reviews*, 2016. **116**(22): pp. 14120–14136.
3. Jiao, F., et al., Sliver nanoparticles decorated Co-Mo nitride for efficient water splitting. *Applied Surface Science*, 2021. **553**: p. 149440.
4. Li, X., et al., Water splitting: from electrode to green energy system. *Nano-Micro Letters*, 2020. **12**(1): p. 131.
5. Lin, Y., et al., Co-Induced electronic optimization of hierarchical NiFe LDH for oxygen evolution. *Small*, 2020. **16**(38): p. 2002426.
6. Liu, P.F., et al., Activation strategies of water-splitting electrocatalysts. *Journal of Materials Chemistry A*, 2020. **8**(20): pp. 10096–10129.
7. Liu, X., et al., Comprehensive understandings into complete reconstruction of precatalysts: synthesis, applications, and characterizations. *Advanced Materials*, 2021. **33**(32): p. 2007344.
8. Wang, J., et al., Non-precious-metal catalysts for alkaline water electrolysis: operando characterizations, theoretical calculations, and recent advances. *Chemical Society Reviews*, 2020. **49**(24): pp. 9154–9196.
9. Yao, R.-Q., et al., Nanoporous surface high-entropy alloys as highly efficient multi-site electrocatalysts for nonacidic hydrogen evolution reaction. *Small Structures*, 2021. **31**(10): p. 2009613.
10. Zhang, X., et al., $Co_3O_4/Fe_{0.33}Co_{0.66}P$ interface nanowire for enhancing water oxidation catalysis at high current density. *Advanced Materials*, 2018. **30**(45): p. 1803551.
11. Li, G., et al., The synergistic effect of Hf-O-Ru bonds and oxygen vacancies in Ru/HfO_2 for enhanced hydrogen evolution. *Nature Communications*, 2022. **13**(1): p. 1270.
12. Luo, M., et al., PdMo bimetallene for oxygen reduction catalysis. *Nature*, 2019. **574**(7776): pp. 81–85.
13. Fang, Y., et al., Semiconducting polymers for oxygen evolution reaction under light illumination. *Chemical Reviews*, 2022. **122**(3): pp. 4204–4256.
14. Hua, Z., et al., Intrinsic ion migration dynamics in a one-dimensional organic metal halide hybrid. *ACS Energy Letters*, 2022. **7**(11): pp. 3753–3760.
15. Zhang, L., et al., Sodium-decorated amorphous/crystalline RuO_2 with rich oxygen vacancies: A robust pH-universal oxygen evolution electrocatalyst. *Angewandte Chemie International Edition*, 2021. **60**(34): pp. 18821–18829.

16. Wei, B., et al., Rational design of highly stable and active MXene-based bifunctional ORR/OER double-atom catalysts. *Advanced Materials*, 2021. **33**(40): p. 2102595.
17. Wu, Z.-P., et al., Non-noble-metal-based electrocatalysts toward the oxygen evolution reaction. *Advanced Functional Materials*, 2020. **30**(15): p. 1910274.
18. Guo, D., et al., A CoN-based OER electrocatalyst capable in neutral medium: atomic layer deposition as rational strategy for fabrication. *Advanced Functional Materials*, 2021. **31**(24): p. 2101324.
19. Guo, D., et al., TiN@$Co_{5.47}$N composite material constructed by atomic layer deposition as reliable electrocatalyst for oxygen evolution reaction. *Advanced Functional Materials*, 2021. **31**(10): p. 2008511.
20. Wang, J., et al., Surface reconstruction of phosphorus-doped cobalt molybdate microarrays in electrochemical water splitting. *Chemical Engineering Journal*, 2022. **446**: p. 137094.
21. Wang, C., et al., Advances in engineering RuO_2 electrocatalysts towards oxygen evolution reaction. *Chinese Chemical Letters*, 2021. **32**(7): pp. 2108–2116.
22. Binninger, T. and M.-L. Doublet, The Ir-OOOO-Ir transition state and the mechanism of the oxygen evolution reaction on IrO_2(110). *Energy & Environmental Science*, 2022. **15**(6): pp. 2519–2528.
23. Hou, J., et al., Surface reconstruction of Ni doped Co-Fe Prussian blue analogues for enhanced oxygen evolution. *Catalysis Science & Technology*, 2021. **11**(3): pp. 1110–1115.
24. Lei, H., et al., Promoting surface reconstruction of NiFe layered double hydroxide for enhanced oxygen evolution. *Advanced Energy Materials*, 2022. **12**(48): p. 2202522.
25. Zhang, K. and R. Zou, Advanced transition metal-based OER electrocatalysts: Current status, opportunities, and challenges. *Small*, 2021. **17**(37): p. 2100129.
26. Lyu, Y., et al., Identifying the intrinsic relationship between the restructured oxide layer and oxygen evolution reaction performance on the cobalt pnictide catalyst. *Small* 2020. **16**(14): p. 1906867.
27. Chen, R.R., et al., SmCo5 with a reconstructed oxyhydroxide surface for spin-selective water oxidation at elevated temperature. *Angewandte Chemie International Edition*, 2021. **60**(49): pp. 25884–25890.
28. Liu, P., et al., Tip-enhanced electric field: A new mechanism promoting mass transfer in oxygen evolution reactions. *Advanced Materials*, 2021. **33**(9): p. 2007377.
29. Huang, W., et al., Ligand modulation of active sites to promote electrocatalytic oxygen evolution. *Advanced Materials*, 2022. **34**(18): p. 2200270.
30. Wu, B., et al., A unique NiOOH@FeOOH heteroarchitecture for enhanced oxygen evolution in saline water. *Advanced Materials*, 2022. **34**(43): p. 2108619.
31. Guo, T., L. Li, and Z. Wang, Recent development and future perspectives of amorphous transition metal-based electrocatalysts for oxygen evolution reaction. *Advanced Energy Materials*, 2022. **12**(24): p. 2200827.
32. Xu, H., et al., Carbon-based bifunctional electrocatalysts for oxygen reduction and oxygen evolution reactions: Optimization strategies and mechanistic analysis. *Journal of Energy Chemistry*, 2022. **71**: pp. 234–265.
33. Du, X., et al., Toward enhanced oxygen evolution on NaBH4 treated $Ba_{0.5}Sr_{0.5}Co_{0.8}Fe_{0.2}O_{3-\delta}$ nanofilm: Insights into the facilitated surface reconstruction. *Materials Today Energy*, 2022. **27**: p. 101046.
34. Yin, J., et al., Self-supported nanoporous $NiCo_2O_4$ nanowires with cobalt-nickel layered oxide nanosheets for overall water splitting. *Nanoscale*, 2016. **8**(3): pp. 1390–1400.
35. Hu, C., L. Zhang, and J. Gong, Recent progress made in the mechanism comprehension and design of electrocatalysts for alkaline water splitting. *Energy & Environmental Science*, 2019. **12**(9): pp. 2620–2645.
36. Yuan, X., et al., Controlled phase evolution from Co nanochains to CoO nanocubes and their application as OER catalysts. *ACS Energy Letters*, 2017. **2**(5): pp. 1208–1213.

37. Yang, C., et al., Revealing pH-dependent activities and surface instabilities for Ni-based electrocatalysts during the oxygen evolution reaction. *ACS Energy Letters*, 2018. **3**(12): pp. 2884–2890.

38. Yin, J., et al., NiO/CoN porous nanowires as efficient bifunctional catalysts for Zn-air batteries. *ACS Nano*, 2017. **11**(2): pp. 2275–2283.

39. Yin, J., et al., Iridium single atoms coupling with oxygen vacancies boosts oxygen evolution reaction in acid media. *Journal of the American Chemical Society*, 2020. **142**(43): pp. 18378–18386.

40. Yin, J., et al., Ni-C-N nanosheets as catalyst for hydrogen evolution reaction. *Journal of the American Chemical Society*, 2016. **138**(44): pp. 14546–14549.

41. Jiao, Y., et al., Design of electrocatalysts for oxygen- and hydrogen-involving energy conversion reactions. *Chemical Society Reviews*, 2015. **44**(8): pp. 2060–2086.

42. Song, J., et al., A review on fundamentals for designing oxygen evolution electrocatalysts. *Chemical Society Reviews*, 2020. **49**(7): pp. 2196–2214.

43. Gao, J., H. Tao, and B. Liu, Progress of nonprecious-metal-based electrocatalysts for oxygen evolution in acidic media. *Advanced Materials*, 2021. **33**(31): p. 2003786.

44. Shi, Z., et al., Fundamental understanding of the acidic oxygen evolution reaction: Mechanism study and state-of-the-art catalysts. *Nanoscale*, 2020. **12**(25): pp. 13249–13275.

45. Xu, H., et al., Current and future trends for spinel-type electrocatalysts in electrocatalytic oxygen evolution reaction. *Coordination Chemistry Reviews*, 2023. **475**: p. 214869.

46. Moysiadou, A., et al., Mechanism of oxygen evolution catalyzed by cobalt oxyhydroxide: cobalt superoxide species as a key intermediate and dioxygen release as a rate-determining step. *Journal of the American Chemical Society*, 2020. **142**(27): pp. 11901–11914.

47. Chen, J., et al., Recent advances in the understanding of the surface reconstruction of oxygen evolution electrocatalysts and materials development. *Electrochemical Energy Reviews*, 2021. **4**(3): pp. 566–600.

48. Li, J., Oxygen evolution reaction in energy conversion and storage: design strategies under and beyond the energy scaling relationship. *Nano-Micro Letters*, 2022. **14**(1): p. 112.

49. Man, I.C., et al., Universality in oxygen evolution electrocatalysis on oxide surfaces. *ChemCatChem*, 2011. **3**(7): pp. 1159–1165.

50. Grimaud, A., et al., Double perovskites as a family of highly active catalysts for oxygen evolution in alkaline solution. *Nature Communications*, 2013. **4**(1): pp. 2439.

51. Hong, W.T., et al., Toward the rational design of non-precious transition metal oxides for oxygen electrocatalysis. *Energy & Environmental Science*, 2015. **8**(5): pp. 1404–1427.

52. Kuznetsov, D.A., et al., Tuning redox transitions via inductive effect in metal oxides and complexes, and implications in oxygen electrocatalysis. *Joule*, 2018. **2**(2): pp. 225–244.

53. Montoya, J.H., et al., Materials for solar fuels and chemicals. *Nature Materials*, 2017. **16**(1): pp. 70–81.

54. Binninger, T., et al., Thermodynamic explanation of the universal correlation between oxygen evolution activity and corrosion of oxide catalysts. *Scientific Reports*, 2015. **5**(1): p. 12167.

55. Zhang, R., et al., A dissolution/precipitation equilibrium on the surface of iridium-based perovskites controls their activity as oxygen evolution reaction catalysts in acidic media. *Angewandte Chemie*, 2019. **131**(14): pp. 4619–4623.

56. Grimaud, A., et al., Anionic redox processes for electrochemical devices. *Nature Materials*, 2016. **15**(2): pp. 121–126.

57. Pan, Y., et al., Direct evidence of boosted oxygen evolution over perovskite by enhanced lattice oxygen participation. *Nature Communications*, 2020. **11**(1): p. 2002.

58. Ma, Q., et al., Identifying the electrocatalytic sites of nickel disulfide in alkaline hydrogen evolution reaction. *Nano Energy*, 2017. **41**: pp. 148–153.
59. Bergmann, A., et al., Unified structural motifs of the catalytically active state of Co (oxyhydr) oxides during the electrochemical oxygen evolution reaction. *Nature Catalysis*, 2018. **1**(9): pp. 711–719.
60. Huang, J., et al., Identification of key reversible intermediates in self-reconstructed nickel-based hybrid electrocatalysts for oxygen evolution. *Angewandte Chemie*, 2019. **131**(48): pp. 17619–17625.
61. Jiang, J., et al., Nanostructured metallic $FeNi_2S_4$ with reconstruction to generate FeNi-based oxide as a highly-efficient oxygen evolution electrocatalyst. *Nano Energy*, 2021. **81**: p. 105619.
62. Kim, J.-C., C.W. Lee, and D.-W. Kim, Dynamic evolution of a hydroxylated layer in ruthenium phosphide electrocatalysts for an alkaline hydrogen evolution reaction. *Journal of Materials Chemistry A*, 2020. **8**(11): pp. 5655–5662.
63. Du, W., et al., Unveiling the in situ dissolution and polymerization of Mo in Ni_4Mo alloy for promoting the hydrogen evolution reaction. *Angewandte Chemie International Edition*, 2021. **60**(13): pp. 7051–7055.
64. Fan, K., et al., Surface and bulk reconstruction of CoW sulfides during pH-universal electrocatalytic hydrogen evolution. *Journal of Materials Chemistry A*, 2021. **9**(18): pp. 11359–11369.
65. Zeng, Y., et al., Surface reconstruction of water splitting electrocatalysts (Adv. Energy Mater. 33/2022). *Advanced Energy Materials*, 2022. **12**(33): p. 2270141.
66. Wu, L., et al., High-efficiency oxygen evolution reaction: controllable reconstruction of surface interface. *Small*, 2023. **19**(49): p. 2304007.
67. Zhang, B., et al., High-valence metals improve oxygen evolution reaction performance by modulating 3d metal oxidation cycle energetics. *Nature Catalysis*, 2020. **3**(12): pp. 985–992.
68. Mathankumar, M., et al., Potentiostatic phase formation of β-CoOOH on pulsed laser deposited biphasic cobalt oxide thin film for enhanced oxygen evolution. *Journal of Materials Chemistry A*, 2017. **5**(44): pp. 23053–23066.
69. Zhu, W., et al., NiCo/NiCo-OH and NiFe/NiFe-OH core shell nanostructures for water splitting electrocatalysis at large currents. *Applied Catalysis B: Environmental*, 2020. **278**: p. 119326.
70. Sivanantham, A., et al., Surface activation and reconstruction of non-oxide-based catalysts through in situ electrochemical tuning for oxygen evolution reactions in alkaline media. *ACS Catalysis*, 2019. **10**(1): pp. 463–493.
71. Lee, W.H., et al., Electroactivation-induced IrNi nanoparticles under different pH conditions for neutral water oxidation. *Nanoscale*, 2020. **12**(27): pp. 14903–14910.
72. Chu, H., et al., In-situ release of phosphorus combined with rapid surface reconstruction for Co-Ni bimetallic phosphides boosting efficient overall water splitting. *Chemical Engineering Journal*, 2022. **433**: p. 133523.
73. Gao, X., et al., Synergizing aliovalent doping and interface in heterostructured NiV nitride@ oxyhydroxide core-shell nanosheet arrays enables efficient oxygen evolution. *Nano Energy*, 2021. **85**: p. 105961.
74. Ren, X., et al., Constructing an adaptive heterojunction as a highly active catalyst for the oxygen evolution reaction. *Advanced Materials*, 2020. **32**(30): p. 2001292.
75. Miao, M., et al., Surface evolution and reconstruction of oxygen-abundant FePi/NiFeP synergy in NiFe phosphides for efficient water oxidation. *Journal of Materials Chemistry A*, 2019. **7**(32): pp. 18925–18931.
76. Pi, Y., et al., Selective surface reconstruction of a defective iridium-based catalyst for high-efficiency water splitting. *Advanced Functional Materials*, 2020. **30**(43): p. 2004375.

77. Wu, L.-K., et al., A nanostructured nickel-cobalt alloy with an oxide layer for an efficient oxygen evolution reaction. *Journal of Materials Chemistry A*, 2017. **5**(21): pp. 10669–10677.

78. Xiao, Z., et al., Operando identification of the dynamic behavior of oxygen vacancy-rich Co_3O_4 for oxygen evolution reaction. *Journal of the American Chemical Society*, 2020. **142**(28): pp. 12087–12095.

79. He, L., et al., Molybdenum carbide-oxide heterostructures: in situ surface reconfiguration toward efficient electrocatalytic hydrogen evolution. *Angewandte Chemie*, 2020. **132**(9): pp. 3572–3576.

80. Yu, P., et al., Rational synthesis of highly porous carbon from waste bagasse for advanced supercapacitor application. *ACS Sustainable Chemistry Engineering*, 2018. **6**(11): pp. 15325–15332.

81. Wu, Y.J., et al., Evolution of cationic vacancy defects: A motif for surface restructuration of OER precatalyst. *Angewandte Chemie International Edition*, 2021. **60**(51): pp. 26829–26836.

82. He, Q., et al., Nickel vacancies boost reconstruction in nickel hydroxide electrocatalyst. *ACS Energy Letters*, 2018. **3**(6): pp. 1373–1380.

83. Liang, H., et al., Amorphous NiFe-OH/NiFeP electrocatalyst fabricated at low temperature for water oxidation applications. *ACS Energy Letters*, 2017. **2**(5): pp. 1035–1042.

84. Kim, J.S., et al., Amorphous cobalt phyllosilicate with layered crystalline motifs as water oxidation catalyst. *Advanced Materials*, 2017. **29**(21): p. 1606893.

85. Zhou, Y., et al., Single-layer CoFe hydroxides for efficient electrocatalytic oxygen evolution. *Chemical Communications*, 2021. **57**(62): pp. 7653–7656.

86. Duan, Y., et al., Scaled-up synthesis of amorphous NiFeMo oxides and their rapid surface reconstruction for superior oxygen evolution catalysis. *Angewandte Chemie International Edition*, 2019. **58**(44): pp. 15772–15777.

87. Sun, H., Y. Zhu, and W. Jung, Tuning reconstruction level of precatalysts to design advanced oxygen evolution electrocatalysts. *Molecules*, 2021. **26**(18): p. 5476.

88. Fang, Y., et al., In situ surface reconstruction of a Ni-based perovskite hydroxide catalyst for an efficient oxygen evolution reaction. *Journal of Materials Chemistry A*, 2022. **10**(3): pp. 1369–1379.

89. Zhao, J.-W., et al., Regulation of perovskite surface stability on the electrocatalysis of oxygen evolution reaction. *ACS Materials Letters*, 2021. **3**(6): pp. 721–737.

90. Shen, T.-H., et al., Oxygen evolution reaction in $Ba_{0.5}Sr_{0.5}Co_{0.8}Fe_{0.2}O_{3-\delta}$ aided by intrinsic Co/Fe spinel-like surface. *Journal of the American Chemical Society*, 2020. **142**(37): pp. 15876–15883.

91. Lopes, P.P., et al., Dynamically stable active sites from surface evolution of perovskite materials during the oxygen evolution reaction. *Journal of the American Chemical Society*, 2021. **143**(7): pp. 2741–2750.

92. Cao, X., et al., Proton-assisted reconstruction of perovskite oxides: Toward improved electrocatalytic activity. *ACS Applied Materials Interfaces*, 2021. **13**(18): pp. 22009–22016.

93. Chen, Y., et al., Exceptionally active iridium evolved from a pseudo-cubic perovskite for oxygen evolution in acid. *Nature Communications*, 2019. **10**(1): p. 572.

94. Chen, Z., et al., Tuning electronic property and surface reconstruction of amorphous iron borides via WP co-doping for highly efficient oxygen evolution. *Applied Catalysis B: Environmental*, 2021. **288**: p. 120037.

95. Tang, T., et al., Electronic and morphological dual modulation of cobalt carbonate hydroxides by Mn doping toward highly efficient and stable bifunctional electrocatalysts for overall water splitting. *Journal of the American Chemical Society*, 2017. **139**(24): pp. 8320–8328.

96. Zhao, C., et al., Surface reconstruction of $La_{0.8}Sr_{0.2}Co_{0.8}Fe_{0.2}O_{3-\delta}$ for superimposed OER performance. *ACS Applied Materials Interfaces*, 2019. **11**(51): pp. 47858–47867.

97. Lin, Y., et al., Sulfur atomically doped bismuth nanobelt driven by electrochemical self-reconstruction for boosted electrocatalysis. *The Journal of Physical Chemistry Letters*, 2020. **11**(5): pp. 1746–1752.

98. Wang, L., et al., Surface reconstruction engineering of cobalt phosphides by Ru induce-ment to form hollow Ru-RuPx-CoxP pre-electrocatalysts with accelerated oxygen evolution reaction. *Nano Energy*, 2018. **53**: pp. 270–276.

99. Li, M., H. Liu, and L. Feng, Fluoridation-induced high-performance catalysts for the oxygen evolution reaction: a mini review. *Electrochemistry Communications*, 2021. **122**: p. 106901.

100. Sun, Y., et al., A-site management prompts the dynamic reconstructed active phase of perovskite oxide OER catalysts. *Advanced Energy Materials*, 2021. **11**(12): p. 2003755.

101. Zhao, X., et al., Phosphorus-modulated cobalt selenides enable engineered reconstruc-tion of active layers for efficient oxygen evolution. *Journal of Catalysis*, 2018. **368**: pp. 155–162.

102. Trotochaud, L., et al., Nickel-iron oxyhydroxide oxygen-evolution electrocatalysts: The role of intentional and incidental iron incorporation. *Journal of the American Chemical Society*, 2014. **136**(18): pp. 6744–6753.

103. Wang, J., et al., Redirecting dynamic surface restructuring of a layered transition metal oxide catalyst for superior water oxidation. *Nature Catalysis*, 2021. **4**(3): pp. 212–222.

104. Ma, F., et al., Surface fluorination engineering of NiFe prussian blue analogue deriva-tives for highly efficient oxygen evolution reaction. *ACS Applied Materials Interfaces*, 2021. **13**(4): pp. 5142–5152.

105. Chen, P., et al., Dynamic migration of surface fluorine anions on cobalt-based materi-als to achieve enhanced oxygen evolution catalysis. *Angewandte Chemie International Edition*, 2018. **57**(47): pp. 15471–15475.

106. Liu, Y., et al., Helical van der Waals crystals with discretized Eshelby twist. *Nature*, 2019. **570**(7761): pp. 358–362.

107. Li, N., et al., Identification of the active-layer structures for acidic oxygen evolution from $9R-BaIrO_3$ electrocatalyst with enhanced iridium mass activity. *Journal of the American Chemical Society*, 2021. **143**(43): pp. 18001–18009.

108. Wang, C., et al., Engineering lattice oxygen activation of iridium clusters stabilized on amorphous bimetal borides array for oxygen evolution reaction. *Angewandte Chemie International Edition*, 2021. **60**(52): pp. 27126–27134.

109. Zeng, Y., et al., Construction of hydroxide pn junction for water splitting electrocataly-sis. *Applied Catalysis B: Environmental*, 2021. **292**: p. 120160.

110. Gao, X., et al., Synergizing in-grown Ni_3N/Ni heterostructured core and ultrathin Ni3N surface shell enables self-adaptive surface reconfiguration and efficient oxygen evolu-tion reaction. *Nano Energy*, 2020. **78**: p. 105355.

111. Kou, Z., et al., Potential-dependent phase transition and Mo-enriched surface recon-struction of γ-CoOOH in a heterostructured $Co-Mo_2C$ precatalyst enable water oxida-tion. *ACS Catalysis*, 2020. **10**(7): pp. 4411–4419.

112. Huang, J., et al., Modifying redox properties and local bonding of Co_3O_4 by CeO_2 enhances oxygen evolution catalysis in acid. *Nature Communications*, 2021. **12**(1): p. 3036.

113. Rui, K., et al., Hybrid 2D dual metal-organic frameworks for enhanced water oxidation catalysis. *Advanced Functional Materials*, 2018. **28**(26): p. 1801554.

114. Huang, L., et al., Self-dissociation-assembly of ultrathin metal-organic framework nanosheet arrays for efficient oxygen evolution. *Nano Energy*, 2020. **68**: p. 104296.

115. Liu, X., et al., Ligand and anion co-leaching induced complete reconstruction of poly-oxomolybdate-organic complex oxygen-evolving pre-catalysts. *Advanced Functional Materials*, 2021. **31**(31): p. 2101792.

116. Liu, X., et al., Complete reconstruction of hydrate pre-catalysts for ultrastable water electrolysis in industrial-concentration alkali media. *Cell Reports Physical Science*, 2020. **1**(11): p. 100241.

117. Chen, G., et al., A universal strategy to design superior water-splitting electrocatalysts based on fast in situ reconstruction of amorphous nanofilm precursors. *Advanced Materials*, 2018. **30**(43): p. 1804333.

118. Liu, D., et al., Review of recent development of in situ/operando characterization techniques for lithium battery research. *Advanced Materials*, 2019. **31**(28): p. 1806620.

119. Zhu, K., X. Zhu, and W. Yang, Application of in situ techniques for the characterization of NiFe-based oxygen evolution reaction (OER) electrocatalysts. *Angewandte Chemie International Edition*, 2019. **58**(5): pp. 1252–1265.

120. Li, X., et al., In situ/operando techniques for characterization of single-atom catalysts. *ACS Catalysis*, 2019. **9**(3): pp. 2521–2531.

121. Yuan, Y., et al., The absence and importance of operando techniques for metal-free catalysts. *Advanced Materials*, 2019. **31**(13): p. 1805609.

122. Zhao, T., et al., Design and operando/in situ characterization of precious-metal-free electrocatalysts for alkaline water splitting. *Carbon Energy*, 2020. **2**(4): pp. 582–613.

123. Dong, C.L. and L. Vayssieres, In situ/operando X-ray spectroscopies for advanced investigation of energy materials. *Chemistry-A European Journal*, 2018. **24**(69): pp. 18356–18373.

124. Zheng, X., et al., Theory-driven design of high-valence metal sites for water oxidation confirmed using in situ soft X-ray absorption. *Nature Chemistry*, 2018. **10**(2): pp. 149–154.

125. Wang, M., et al., In situ X-ray absorption spectroscopy studies of nanoscale electrocatalysts. *Nano-Micro Letters*, 2019. **11**: pp. 1–18.

126. van Oversteeg, C.H., et al., In situ X-ray absorption spectroscopy of transition metal based water oxidation catalysts. *Chemical Society Reviews*, 2017. **46**(1): pp. 102–125.

127. Li, J., et al., Frontiers of water oxidation: the quest for true catalysts. *Chemical Society Reviews*, 2017. **46**(20): pp. 6124–6147.

128. Fabbri, E., et al., Operando X-ray absorption spectroscopy: A powerful tool toward water splitting catalyst development. *Current Opinion in Electrochemistry*, 2017. **5**(1): pp. 20–26.

129. Deng, Y. and B.S.J.A.C. Yeo, Characterization of electrocatalytic water splitting and CO_2 reduction reactions using in situ/operando Raman spectroscopy. *ACS Catalysis*, 2017. **7**(11): pp. 7873–7889.

130. Wang, Y.H., et al., Probing interfacial electronic and catalytic properties on well-defined surfaces by using in situ Raman spectroscopy. *Angewandte Chemie*, 2018. **130**(35): pp. 11427–11431.

131. Miro, S., et al., Monitoring of the microstructure of ion-irradiated nuclear ceramics by in situ Raman spectroscopy. *Journal of Raman Spectroscopy*, 2016. **47**(4): pp. 476–485.

132. Eliseeva, G., et al., In-situ studies of processes at fuel electrode of solid oxide fuel cells (SOFC) by Raman spectroscopy. *Chemical Problems*, 2019. **3**: p. 18.

133. Kouotou, P.M. and Z.-y. Tian, In situ Fourier transform infrared spectroscopy diagnostic for characterization and performance test of catalysts. *Chinese Journal of Chemical Physics*, 2017. **30**(5): pp. 513–520.

134. Ye, J.-Y., et al., In-situ FTIR spectroscopic studies of electrocatalytic reactions and processes. *Nano Energy*, 2016. **29**: pp. 414–427.

135. Christensen, P. and Z. Mashhadani, In situ FTIR studies on the oxidation of isopropyl alcohol over SnO_2 as a function of temperature up to 600°C and a comparison to the analogous plasma-driven process. *Physical Chemistry Chemical Physics*, 2018. **20**(14): pp. 9053–9062.

136. Crumlin, E.J., et al., X-ray spectroscopy of energy materials under in situ/operando conditions. *Journal of Electron Spectroscopy Related Phenomena*, 2015. **200**: pp. 264–273.

137. Lu, J., et al., In situ UV-vis studies of the effect of particle size on the epoxidation of ethylene and propylene on supported silver catalysts with molecular oxygen. *Journal of Catalysis*, 2005. **232**(1): pp. 85–95.

138. Wang, L., et al., Catalysis and in situ studies of $Rh1/Co_3O_4$ nanorods in reduction of NO with H_2. *ACS Catalysis*, 2013. **3**(5): pp. 1011–1019.

139. Li, X., et al., In situ/operando characterization techniques to probe the electrochemical reactions for energy conversion. *Small Methods*, 2018. **2**(6): p. 1700395.

9 Degradations in PEMWE

Zhenye Kang and Xinlong Tian

9.1 DEGRADATION OVERVIEW IN PEMWE

PEMWE for hydrogen production has been widely accepted as one of the most prominent and feasible methods for renewable energy storage at various scales, which could facilitate the connection of fluctuating renewable energies to the electrical grid [1–3]. Performance, durability, and cost are the three critical aspects of PEMWEs [4]. Among these, performance has been significantly improved in the past decades and superior high current density has been achieved (up to 20 A/cm² by 3M) [5]. Much effort has been dedicated to reducing costs by decreasing noble metal usage, reducing machining/fabrication expenses, and developing low-cost membranes [6–8]. However, developing long-lifetime PEMWE systems that meet the basic requirements of the United States of America Department of Energy (DOE) (>80000 hours and <2.0 µV/h) is still the main challenge [9]. Furthermore, increasing the PEMWE lifetime also reduces the cost to some extent. Therefore, the design and development of materials and components for PEMWEs should be constantly improved to increase their performance and lifetime [10].

The PEMWE is a complex system and its lifetime is influenced by many factors. A PEMWE stack, which involves stacking single cells together, is usually used in commercialized hydrogen production systems. Therefore, understanding the degradation of a single PEMWE device could explain most of the existing challenges and help in the development and progress of PEMWE systems.

Figure 9.1 shows the typical PEMWE components that contribute to the degradation of a single PEMWE device. A proton exchange membrane or polymer electrolyte membrane (PEM; also called a solid electrolyte membrane) is located at the center of the PEMWE device and is sandwiched between the anode and cathode. The anode and cathode include catalyst layers (CLs), porous transport layers (PTLs; also called liquid/gas diffusion layers), flow field plates (FFPs; also called bipolar plates (BPs) in PEMWE stacks), insulating layers, and endplates. PEMWE devices degrade during their operation for multiple reasons, including harsh environments, overpotential, and electrochemical reactions. However, different materials and components experience different environments, which result in complicated degradation mechanisms, including both physical and chemical degradation.

The anode and cathode in PEMWE devices vary significantly. Abundant liquid water and oxygen gas, along with a high overpotential, exist at the anode, leading to a harsh environment that is more conducive to degradation. Hydrogen is produced at the cathode, where a mild environment is created. Therefore, the degradations in the anode and cathode also vary significantly, leading to different requirements for materials and components. The components in the anode must have high corrosion

DOI: 10.1201/9781003368939-9

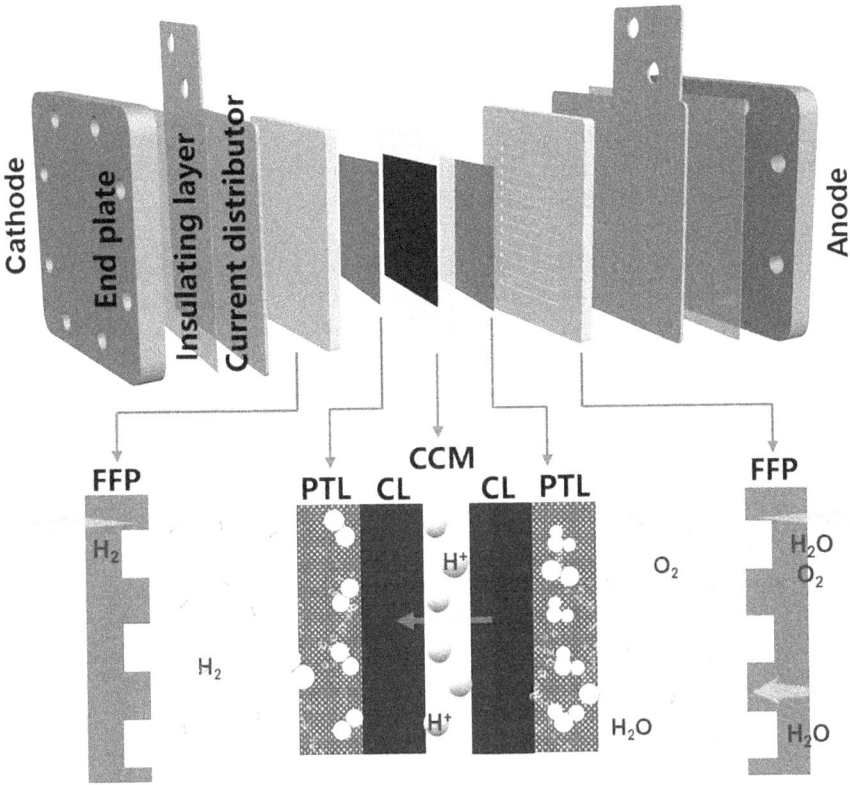

Degradations in PEMWE: corrosion, passivation, membrane thinning, ion contamination, catalyst dissolution/agglomeration, hydrogen embrittlement, mechanical forces

FIGURE 9.1 Schematic illustration of a PEMWE and the origins of degradation. CCM represents catalyst-coated membrane.

resistance to sustain the long-term operation of PEMWEs, and titanium-based PTLs/BPs with noble metal coatings have been widely adopted. However, carbon-based materials can also be used as cathodes.

Therefore, the degradations in PEMWEs are discussed and analyzed on each component, and the diagnostics of degradations are also introduced in the following sections.

9.2   DEGRADATION PHENOMENON AND MECHANISM

This section introduces the degradation of each component of the PEMWE and analyzes their mechanisms. The most critical factors that lead to degradation are also discussed, which will benefit strategies for mitigating PEMWE degradation.

9.2.1 Membranes

Membranes used in PEMWEs contribute to the largest ohmic resistance of the entire cell, which significantly impacts performance and stability. In PEMWE device, PEM always has a thickness of 90–175 μm, which is significantly thicker than the membranes used in PEM fuel cells. Using a thinner membrane can result in a smaller ohmic resistance and better performance; however, its degradation and gas crossover rates will increase.

During the operation of PEMWE devices, the PEM degrades for both physical and chemical reasons. The physical degradation of PEMs is usually due to structural failure, which originates from cracking, tearing, puncturing, mechanical stresses, and non-uniform pressure [11]. Defects such as membrane pinholes, tears, abrasions, and perforations can potentially affect the durability of PEMs [12]. PEM chemical degradation occurs during long-term operation because of the loss of fluoride and metal ions released from the PEMWE metal parts. The PEM degradation rate is closely related to the operating cell voltage.

Pinholes and cracks, which may be caused by the PEM treatment, CL fabrication, PTL spike, compression of the PEMWE, and high hydraulic pressure, are the most common physical failures of PEMs. Once the pinholes are formed, as shown in Figure 9.2, oxygen and hydrogen mix, leading to a dangerous gas mixture that decreases the Faradaic efficiency [13,14]. In some circumstances, liquid water may seal the pinholes and eliminate gas crossover. This sealing effect depends on the pinhole size and pressure gradient on both sides. As previously reported, tiny pinholes can be occupied by water; however, the size of the pinhole has a threshold below which it may not be detectable [12]. In most cases, pinholes or cracks result in significant degradation of the PEM. The regions close to the edges of the flow channel and uneven PTL protrusions are more prone to microcrack fractures [15]. These issues may result in non-uniform current, potential, mass, and temperature distributions. Therefore, PTLs should have good mechanical properties to sustain flat PEMs and restrict PEM deformation.

The PEM in PEMWE devices is sandwiched between two PTLs under compression and is mechanically supported on these PTLs. The roughness of the PTL surface or nonideal PTL properties significantly impact the PEM stability, especially when the PEMWE is operated under high differential pressures. Puncture of the PEM by PTL fibers and spikes may lead to electrical shorts within the PEM, resulting in its failure. By contrast, PTLs with large pores may cause severe degradation because the PEM may swell during operation and extrude into the fluid spaces, resulting in internal stress and mechanical degradation.

Furthermore, PEMs can be dry- or wet-assembled in PEMWEs. Dry-assembled PEMs will deform when water is supplied to the PEMWE, which may cause mechanical stress on the PEM and result in potential failure. This situation is not encountered in wet-assembled PEMs. PEMs should be maintained under wet conditions at all the times as they will shrink upon drying, which will damage the PEM structure and result in mechanical failure [11]. The state of the PEM is a factor that is usually overlooked in terms of its degradation, but it does affect the PEMWE stability and should be considered [16].

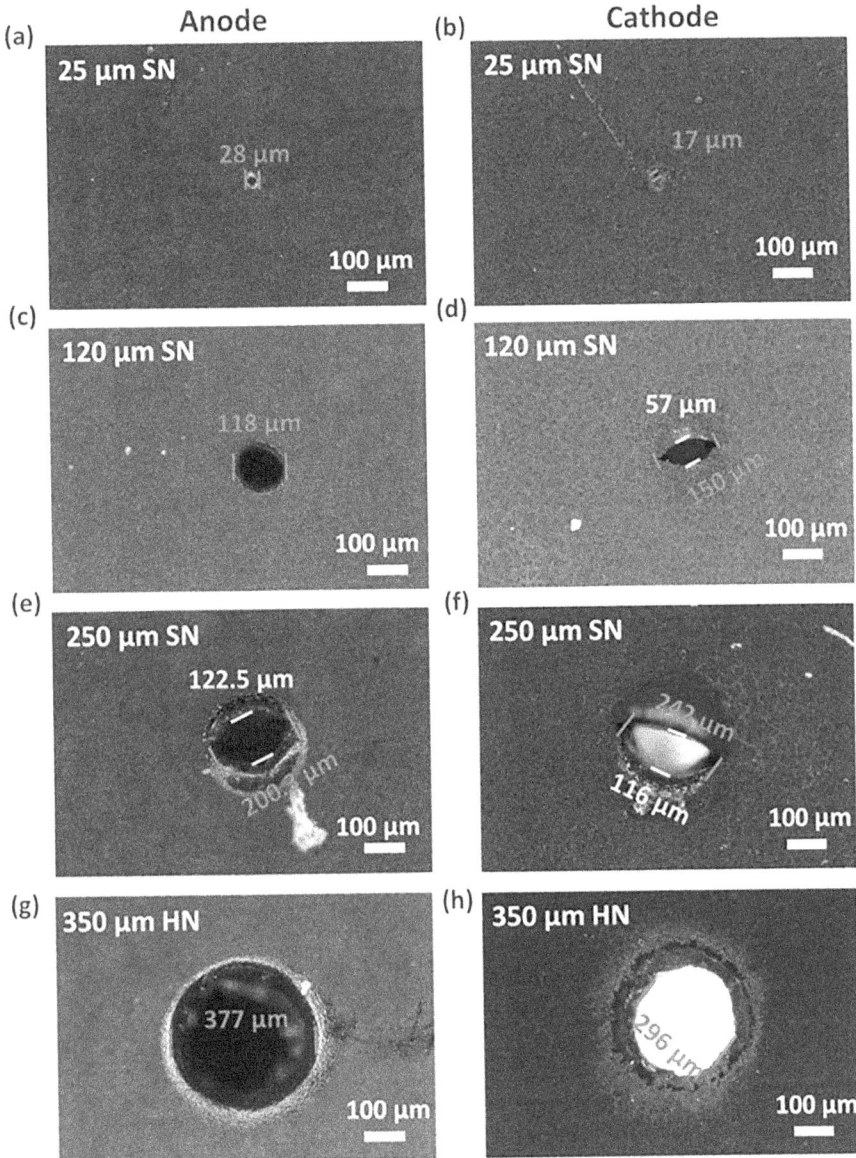

FIGURE 9.2 Images of pinholes in PEMs to study their effects on degradation. SN stands for solid needle and HN stands for hollow needle. Optical microscopy of anode side of CCMs (a) 25 µm SN; (c) 120 µm SN; (e) 250 µm SN; (g) 350 µm HN; cathode side of CCM (b) 25 µm SN; (d) 120 µm SN; (f) 250 µm SN; (g) 350 µm HN. Reproduced from Ref. [12].

In the chemical degradation of PEM, oxidizing species, such as hydroperoxyl, hydroxyl, and hydrogen peroxide, will cause chain breakage, releasing fluoride from the polymer main group and side chain and sulfur ions from the end group [17]. Consequently, the perfluorosulfonic acid backbone degrades and the PEM thickness decreases, which is referred to as membrane-thinning [18]. The membrane thickness is reportedly reduced by as much as 75% after a 5500 hours durability test [19]. Oxygen crossover from the anode to the cathode competes with the hydrogen evolution reaction (HER), and hydrogen peroxide is formed under the platinum catalyst in the cathode, which has been demonstrated to be the main reason for PEM degradation at the cathode [20]. Although more hydrogen can cross through the PEM to the anode, the reaction between hydrogen and oxygen is inhibited due to the high overpotential. Therefore, membrane-thinning degradation usually occurs at the cathode side of the PEM and is more pronounced at higher temperatures.

Purified water with high resistivity and limited amount of ions must be applied during PEMWE operation; however, water may contain impurities under real conditions [21]. The metal ions that originate from corrosion and the water gradually penetrate the PEM under diffusion and electrical-field effects and occupy the ion exchange sites, resulting in lower proton conductivity due to the low mobility of these metal ions, as shown in Figure 9.3a [22]. The catalyst or other ions (such as titanium ions from the PTLs and BPs) migrate from the anode through the PEM to the cathode, which can also lead to degradation owing to electrical shorting. Subsequently, the ohmic losses through the PEM, along with higher Joule heating and accelerated degradation, increase. However, it is interesting to note that this degradation can mostly be recovered by treating the contaminated PEM with an acidic solution, as shown in Figure 9.3b [23]. Additionally, non-metallic ions, such as NH_4^+, can also contaminate PEMs, leading to reduced conductivity.

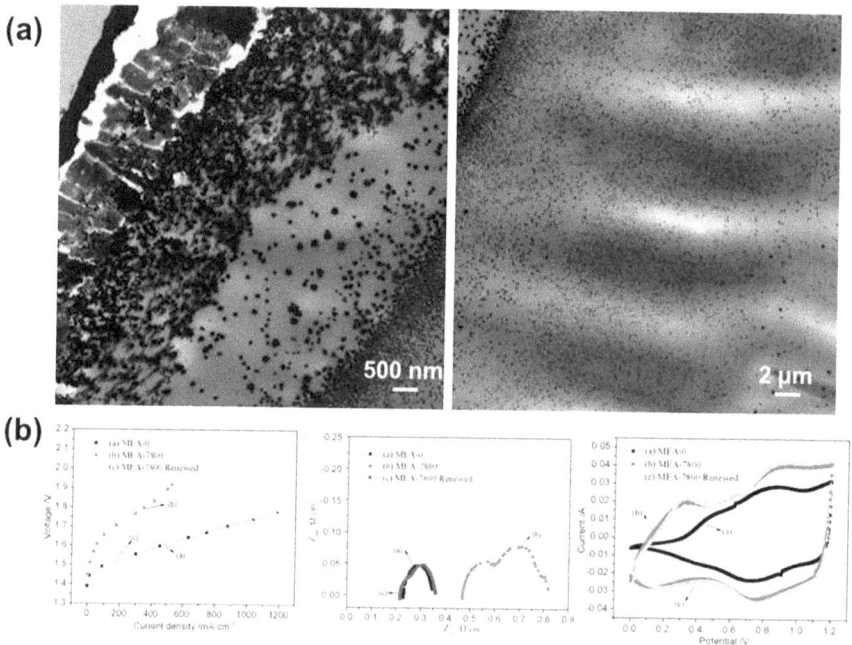

FIGURE 9.3 (a) Degraded PEM due to ion migration and (b) degraded PEM recovered using an acidic treatment. Reproduced from Refs. [19,23].

9.2.2 Catalyst Layers

9.2.2.1 Catalyst Materials

The catalyst is one of the most critical factors that directly determines the electrochemical performance of water electrolysis in PEMWE devices, where the oxygen evolution reaction (OER) occurs at the anode and the HER occurs at the cathode. Currently, platinum black and platinum/carbon are the most widely accepted HER catalyst materials because of their good performance and durability under low loadings, and because the fast kinetics of the HER occur more easily than the OER in the anode. However, catalyst materials must meet special requirements for their adoption as anode OER catalysts because of the harsh environment. Iridium and its oxides are the most widely used electrocatalysts in PEMWE anodes, which are mainly due to their catalyst stability. Ruthenium and its oxides exhibit better kinetic performances than iridium and its oxides; however, their rapid degradation hinders their use. Therefore, numerous studies have focused on improving the stability of OER electrocatalysts. This section introduces the degradation mechanisms of anode/cathode catalyst materials. The main electrocatalyst degradation mechanisms include dissolution, leaching, agglomeration, amorphization, migration, and poisoning, as illustrated in Figure 9.4.

Catalyst dissolution is one of the most critical causes of catalyst degradation, especially at high overpotentials. Iridium dissolution occurs slowly, resulting in the loss of catalyst materials, which eventually reduces their performance. During the OER, IrO_2 transforms into soluble iridium at a voltage >1.8 V [24,25]. As shown in Figure 9.5, iridium in PEMWEs dissolves and then diffuses from the anode to the cathode. Part of the dissolved iridium forms an oxide band at the anode CL/PEM interface, and some of the iridium is redeposited at the cathode in the metallic state, accounting for approximately 42% of its losses [26,27]. Platinum degradation occurs even at a cathodic potential of <0 V vs. RHE, and platinum dissolves when its oxide is reduced during startup at the cathode. Platinum degradation is more severe during

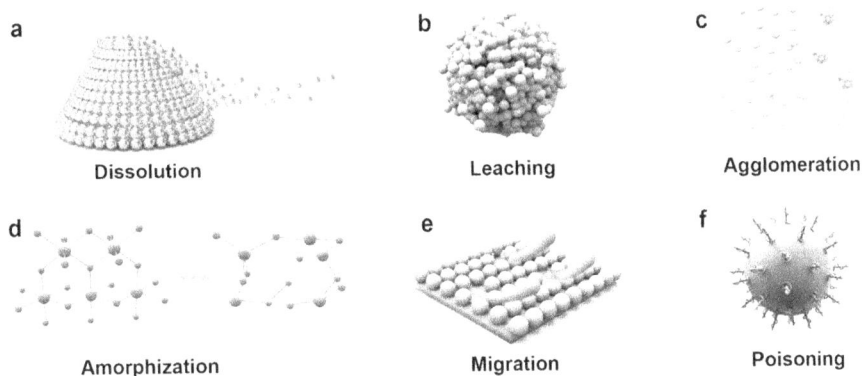

a	b	c
Dissolution	Leaching	Agglomeration

d	e	f
Amorphization	Migration	Poisoning

FIGURE 9.4 Schematic illustration of the catalyst degradation phenomenon.

End-of-test CCM

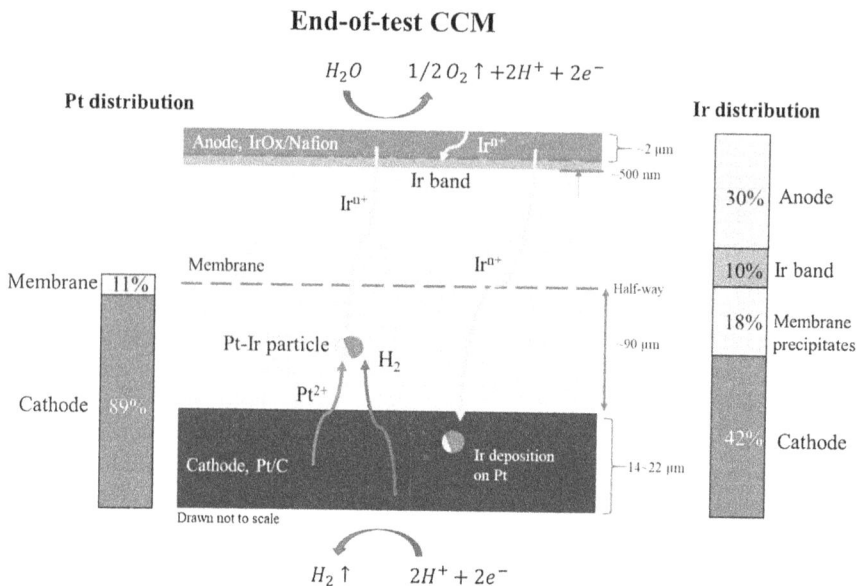

FIGURE 9.5   Schematic illustration of catalyst migration and its distribution after a 4500-h test at 1.8 A/cm². Reproduced from Ref. [27].

the shutdown process because the PEMWE may work in a fuel cell mode owing to the presence of remaining hydrogen and oxygen, and platinum particles will migrate from the cathode CL into the PEM [28]. The dissolved iridium and platinum form iridium–platinum precipitates in the PEM, which may also block the proton transport path and active sites of both the anode and cathode catalysts that are close to the PEM/CL interfaces [27]. Therefore, the dissolution rate is a critical parameter for assessing degradation under this effect. The oxidizability of electrocatalyst materials has been reported to be closely related to the dissolution rate, and dissolution occurs faster under high current densities or overpotentials. The dissolution of electrocatalytic materials can be inhibited by lowering the oxidation state of the catalyst upon interaction with the support material.

With respect to the catalyst nanomaterial design, the electrocatalyst particles should have smaller sizes, which provide a larger active surface area than large bulk materials. However, these nanoparticles are prone to growth and agglomerate into large particles owing to the Gibbs free energy of the clusters. The catalyst nanoparticles agglomerate via three main mechanisms, namely, Ostwald ripening, reprecipitation, and coalescence [29]. For example, platinum particles tend to agglomerate at the cathode under more negative potentials when the hydrogen coverage rate on the platinum surface increases. Consequently, the van der Waals forces between platinum and the carbon support are weakened, which causes platinum particle agglomeration owing to coalescence and redeposition. The platinum crystal size can reportedly double during this process (from 3.5 to 7.8 nm) [30].

Anions and cations in the reactant water may poison the catalysts and reduce their active surface areas. These ions include, but are not limited to, Cl^-, Na^+, Ca^{2+}, Cu^{2+}, Ni^{2+}, Pb^{2+}, Fe^{3+}/Fe^{4+}, and Ti^{4+} [29–32]. The cation impurities may block the catalyst surface, which is under the potential deposition effect, increasing the charge-transfer resistance and decreasing the intrinsic activity of the catalyst. Notably, the poisoning by some cations can be almost reversed by cleaning and removing the cations in acid [23].

Although catalyst materials may undergo various changes, the loss of kinetics may not be directly related to them. Iridium-based materials have been proposed to undergo a change in the oxidation state during the OER in PEMWEs, leading to a low ability to adsorb intermediate oxygen species and dissociate adsorbed water molecules [30].

Support materials are another factor that can cause degradation. A catalyst support is adopted to increase the overall active surface area of the catalyst by dispersing catalyst nanoparticles on the support, which also inhibits agglomeration and provides improved electrical conductivity. However, support materials have a greater impact on the catalyst stability. Oxides and ceramics materials, such as TiO_2, TaB_2, TiC, SnO_2, SbO_x, and TaC, are the most widely used support materials [33–35]. Nanosize electrocatalysts are usually anchored on the support materials. The interactions between the catalysts and supports range from weak electrostatic attractions to strong chemical bonds or the formation of an overlayer on the support. Therefore, an appropriate support could contribute to better performance and stability, whereas a nonideal support may initially enhance the performance but accelerate degradation.

9.2.2.2   Catalyst Layers

CLs are categorized into two types, namely, porous and thin-film CLs. A porous CL is always composed of a mixture of catalyst particles, catalyst support if used, and an ionomer working as a binder and proton conductor, forming a complicated porous structure. Thin-film CLs usually have a dense ionomer-free structure with only a thin-film catalyst. When the CL is supported on a PEM, a CCM configuration is formed, whereas a porous transport electrode configuration is formed when the CL is supported on a PTL [36,37]. Different CL structures significantly affect their durability and degradation mechanisms.

For CCMs in PEMWEs, CLs may be mechanically damaged owing to non-uniform compression by the PTLs. As shown in Figure 9.6, microcracks form on the CCM surface when PTLs composed of titanium felt are used in PEMWEs owing to compression of the titanium fibers. The in-plane conductivity of the CL will significantly reduce owing to the distortion of the electric percolating network [38]. The in-plane conductivity of CLs can significantly affect the number of active sites and catalyst utilization, and the cracks formed reduce the number of effective reaction sites, decreasing the performance and increasing the degradation rate [39–41].

Therefore, the interface between CLs and PTLs plays a critical role in PEMWEs. CLs form void regions and cracks after long-term durability tests, as shown in

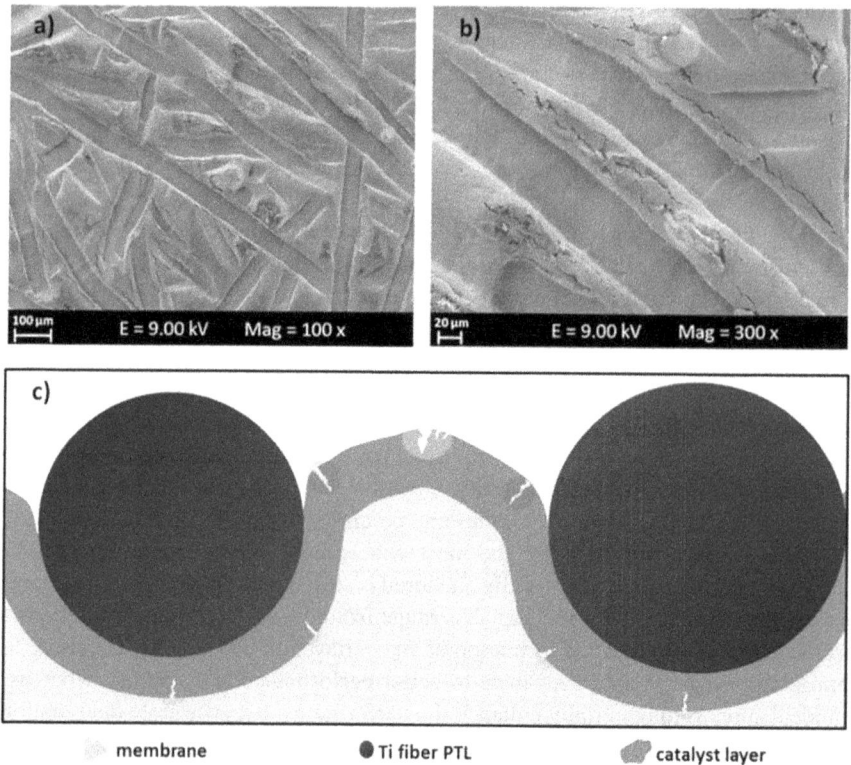

FIGURE 9.6 (a) and (b) Scanning electron microscopy (SEM) images of the surfaces of an anode CL with tension-induced microcracks and compression fractures (highlighted in red and yellow, respectively) and (c) schematic illustration of the cross-section indicating the microcracks. Reproduced from Ref. [38].

Figure 9.7 [42]. Some portions of the CLs delaminate from the PEM and attach to the titanium fiber surface, creating voids and resulting in poor CL/PEM interfaces. When bare Ti felt is used, the CL has a non-uniform temperature distribution, resulting in severe ionomer loss and CL delamination. This effect can be reduced by protecting the PTL/CL interfaces with an iridium coating.

Gas bubbles also have a dynamic effect on local CLs during bubble nucleation, growth, and detachment [43]. The number and size of gas bubbles increase significantly under high current density operation, which may lead to the leaching and detachment of catalyst particles in the CL, resulting in CL degradation. Detachment can be promoted by controlling the bubble size, which increases stability [44]. Moreover, gas bubbles may block the active sites in the CLs by occupying the void space within the porous CLs or the surface of the thin-film CLs, preventing water transport. This increases the electrochemical reaction intensity of the remaining active sites because the current passes through these regions and causes

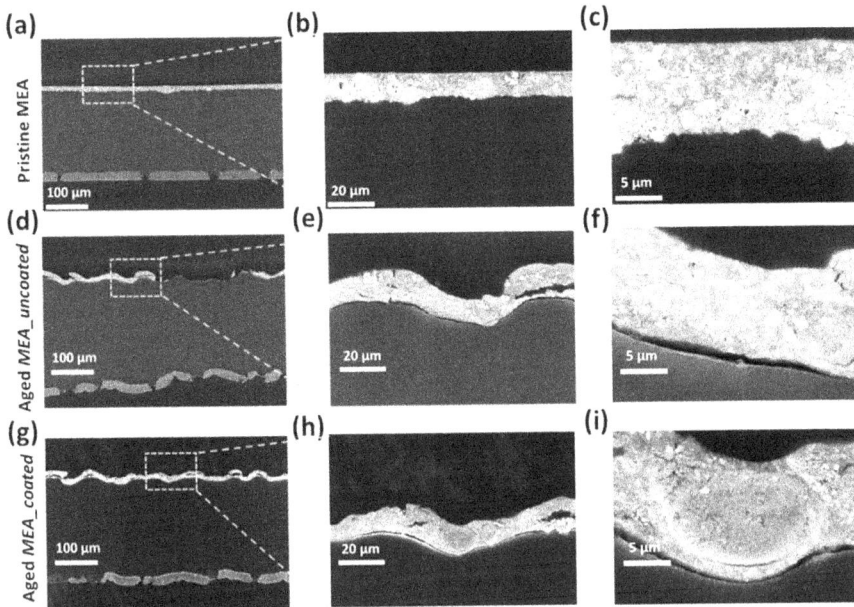

FIGURE 9.7 Cross-sectional images of a CCM: (a)–(c) pristine CCM, (d)–(f) tested CCM with uncoated anode Ti felt, and (g)–(i) tested CCM with platinum-coated anode titanium felt. Reproduced from Ref. [42].

greater degradation of the CLs. Higher water flow rates can remove gas bubbles, but the excessively high flow velocity of water may also damage CLs owing to the microscale scouring effect.

Membrane-thinning and ionomer degradation in the CLs can lower the adhesion properties of the CL during PEMWE operation, causing particle detachment and layer delamination. This effect, along with catalyst dissolution, usually causes the CLs in PEMWEs to have a high noble metal loading, which creates a buffer for catalyst material loss. CCMs with higher catalyst loadings have significantly lower degradation rates than those of CCMs with low catalyst loadings [45]. The high loading can also alleviate the effects of bubble coverage; however, it still degrades.

The external ions migrating into the CL not only impact the catalyst itself but also occupy protons in the ionomer, resulting in an increased overpotential at both the anode and cathode. Some cations form precipitates at the interface of the cathode CL and the PEM, accelerating PEMWE degradation.

9.2.3 Porous Transport Layers

PTLs are multifunctional components that significantly affect PEMWEs. PTLs are located between the CLs and BPs and facilitate the transportation of reactants and products, the conduction of electrons and heat, and the mechanical support of the CLs. Therefore, its parameters not only affect the performance of the PEMWE, but also inevitably influence its durability.

Carbon papers are widely used in PEMWEs for short-term tests in laboratories and have shown superior performance owing to their good electrical conductivity and highly efficient mass/heat transport ability; however, they rapidly corrode at the anode [46]. Therefore, metal-based PTLs are preferred because of their high corrosion resistivity. Stainless steel mesh used as anodic PTLs do not have a long life as they rapidly corrode, resulting in the release of multiple ions that may poison the catalysts and PEM, as shown in Figure 9.8 [47].

Titanium has a high corrosion resistivity and is widely used as an anode PTL material. The titanium surface during testing in a PEMWE anode easily oxidizes to form a passivation film, which acts as a conductor, semiconductor, or insulator depending on its thickness [48]. The electrical and interfacial contact resistances (ICR) of the Ti PTLs increase with an increasing passivation film thickness, resulting in a degraded performance, which in turn exacerbates the increase in the passivation film thickness. This process finally leads to failure of the PEMWE.

The passivation film not only increases the electrical resistances but also causes titanium corrosion due to the ions in the anode of the PEMWE. Fluoride ions released from the PEM can react with the titanium oxide passivation film, which deteriorates the high corrosion resistance of the Ti PTLs [49–51]. Therefore, the circulation of the passivation film and its corrosion result in gradual titanium PTL degradation. Additionally, other chemicals, such as hydrogen peroxide and Cl^-, may also corrode PTLs during PEMWE operation. The corrosion of titanium materials will initially result in an increased ICR, and the poor interfacial contact between the PTL and CL will cause the loss of electrochemically active sites within the CL, which is reflected in the degradation of the PEMWE.

FIGURE 9.8 SEM images of stainless steel mesh used as an anodic PTL and carbon paper used as a cathodic PTL in a PEMWE: (a) pristine stainless steel mesh; (b) corroded stainless steel mesh and (c) magnification of (b); (d) pristine carbon paper; (e) tested and contaminated carbon paper and (f) magnification of (e). Reproduced from Ref. [47].

Titanium PTLs are categorized as felt, sintered particle plates, and meshes. Among these, titanium felt is widely used in commercial and industrial PEMWE systems. The titanium surface of the PTLs is usually treated to enhance their anticorrosion ability, and the treatments include noble metal coating, nitridation, and other treatments [42]. Bare titanium-felt PTLs, as compared to surface-protected titanium-felt PTLs, exhibit numerous issues that can aggravate degradation. For example, passivation of the titanium fiber surface results in poor interfacial contact and electrical/thermal conductivity, which increase the ICR and exhibit an ohmic resistance dependency on the operating current densities [48]. This causes a non-uniform temperature distribution and the appearance of hot spots, which accelerate membrane and catalyst degradation.

9.2.4 BIPOLAR PLATES

BPs are critical components of PEMWEs. In a PEMWE stack, the two sides of the BPs act as the anode and cathode. Therefore, BPs should exhibit good electrical and thermal conductivities and low gas permeability. Moreover, transport of reactants and products is one of the most important functions of BPs. The flow channels of the BPs significantly affect the performance of PEMWEs and also affect the durability from several aspects, namely, surface corrosion, material oxidation, hydrogen embrittlement, and others.

BPs operate under harsh environments, particularly on the anode side. Most materials cannot withstand such a highly corrosive environment because of the high overpotential, rich oxidation substances, and possible chemicals or ions. Graphite materials are widely used in fuel cells for BPs; however, their use in PEMWEs is limited. Graphite plates are severely corroded under high overpotentials in harsh environments because carbon is easily oxidized to carbon dioxide at potentials >0.206 V vs. RHE. Figure 9.9 compares pristine and corroded graphite BPs that were only

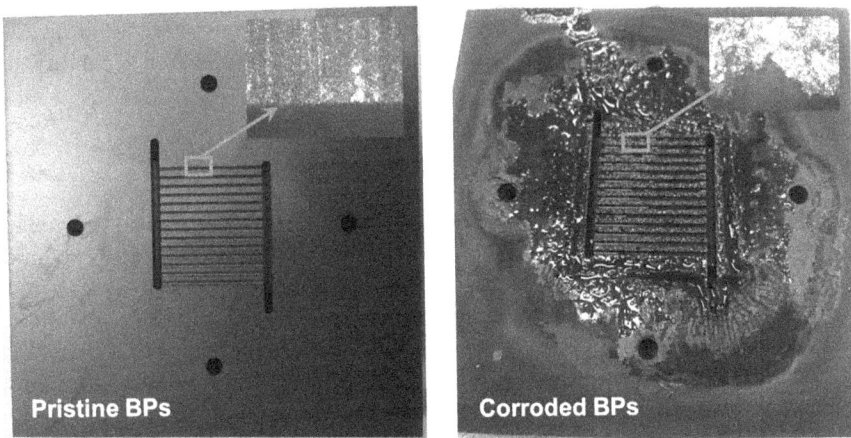

FIGURE 9.9 Photographic images comparing pristine and corroded graphite BPs.

tested at 2.3 V for 10 hours. The corroded BPs clearly exhibited a rough surface and the flow channels were damaged. Corroded BPs lead to side leakage, poor interfacial contact with adjacent components, and eventual degradation of the PEMWE. Therefore, carbon-based BPs can only be used for low-voltage and short-term testing in PEMWEs [46]. Similar to PTLs, stainless steel-based BPs have also been explored. However, they are easily corroded under such severe environments, and the ions from corrosion, such as Fe^{2+}, Ni^{2+}, and Cr^{2+}, will also contaminate the CLs, aggravating PEMWE degradation [52,53].

Titanium-based BPs are widely used in PEMWEs because of their good chemical and mechanical properties. But it also meets the situation of titanium PTLs. The surface of bare titanium BPs forms passivation layers that can be corroded by fluoride ions released from the PEM. Severe corrosion occurs when the fluoride concentration reaches 0.005 M, and this corrosion is more detrimental when the operating conditions fluctuate [49].

Hydrogen embrittlement is another factor that may accelerate the degradation of BPs at the cathode, similar to PTLs. Titanium hydrides are formed when titanium is exposed to hydrogen gas, which may cause cracks in the material and result in degradation. This is more critical when the PEMWE is operated at elevated hydrogen pressures.

9.2.5 CURRENT DISTRIBUTOR

The current distributor does not directly contact the reactant and products and only provides an electrical conduction path for the PEMWE device. Copper, iron-nickel alloys, and stainless steel are widely used for current distributors because they do not work in harsh environments like PTLs and BPs. Therefore, the current distributor is usually not a key component causing PEMWE degradation. The most critical requirements for current distributors are good electrical conductivity and interfacial contact between the current distributor and BPs. The conductivity of the current distributor is primarily determined by its bulk material. However, the interface may be affected by surface oxidation during long-term operation. Therefore, the surfaces of current distributors are usually treated to improve their antioxidation capability and reduce ICRs.

9.2.6 INSULATING LAYER AND ENDPLATE

The backside of the current distributor has an insulating layer, which is usually made of polymer chemical materials such as PTFE, PFA, and other materials. Polyimide films have recently shown excellent potential as insulating layers because of their good physical properties. This insulating layer only suffers from the influence of thermal stress and compression, which are usually not key issues related to degradation.

Endplates are used to provide uniform compression for the PEMWE assembly. Depending on the structural design, water can flow through the interior of the endplate. The endplate does not significantly affect PEMWE degradation if it does not come in contact with water. However, it may release various ions into the water if water flows through it, which may cause degradation of the CLs and PEM mentioned above. Therefore, a noncontact configuration is favored for the endplate. However, the endplates at the anode and cathode of PEMWEs are typically assembled using many bolts. To maintain compression in the PEMWE, both the endplates and bolts must have good mechanical properties to avoid assembly pressure reduction, which will cause poor internal interfacial contact in the PEMWE and degradation.

9.3 STRATEGIES FOR MITIGATING DEGRADATION

PEMWE degradation has been introduced, showing that some processes are clearly coupled with others. Therefore, PEMWE degradation mitigation strategies should consider the overall PEMWE, instead of a single component. This section discusses strategies for mitigating degradation in PEMWEs from both the materials and methods perspectives.

As mentioned in the previous section, the PEM, ionomers, and active electrochemical catalysts in CLs are very sensitive to several ions (Figure 9.10), which will lead to their degradation. Therefore, alleviating the effects of ions on the PEMWE is critical.

Ions in the water supply can be removed using high-quality water purification systems, which could provide high-purity water with a resistivity of >1 MΩ cm. By contrast, some ions leach from the PEMWE components that are in contact with water, because high-purity water (such as de-ionized water with a resistivity of approximately 18.2 MΩ cm) has a high ion adsorption ability. Ideally, the ions in water can be controlled by implementing an ion exchanger containing appropriate resins in the water line before they enter the PEMWE device. However, this is a post-process performed after the ions appear in the water, and it is more convenient and safer to prevent ions from each component from entering the water. Therefore, components such as PTLs, BPs, and tubing should be protected from ion release.

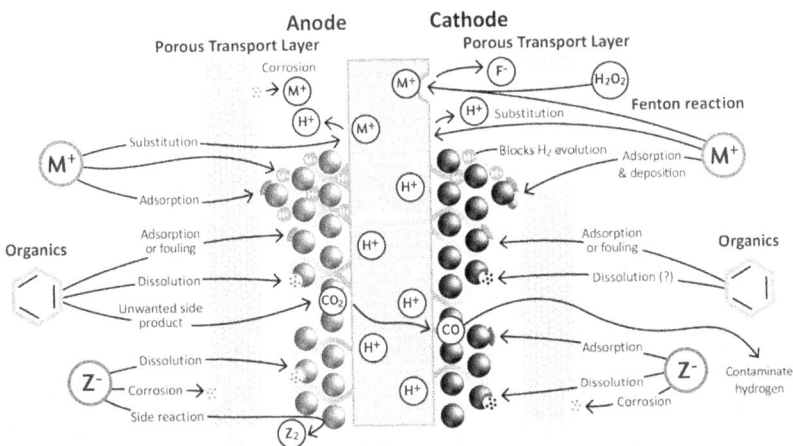

FIGURE 9.10 Schematic illustration of the impurities that may be present in PEMWEs. Reproduced from Ref. [21].

To prevent mechanical degradation of the PEM and CL, their clamping by PTLs and BPs should be carefully designed. In particular, PTLs in direct contact with CLs should provide reliable support and a small ICR. Therefore, some researchers have attempted to develop pore-graded PTLs and novel thin-tunable perforated PTLs with straight-through pores to enhance the PTL/CL interface and improve mass transport, as shown in Figure 9.11 [54–56]. The planar surface structure of perforated PTLs is one of their advantages that not only provides better interfacial contact at both the PTL/CL and PTL/FFP interfaces, but also allows for easy coating of ultrathin protective layers with significantly reduced defects. Therefore, perforated PTLs have attracted considerable academic and industrial attention. In recent years, some companies have also attempted to develop extra microporous layers on titanium-felt PTLs to obtain smaller average pore sizes in one direction facing the CLs. The integration of PTLs and BPs is another approach to reduce interfaces and enhance durability [57].

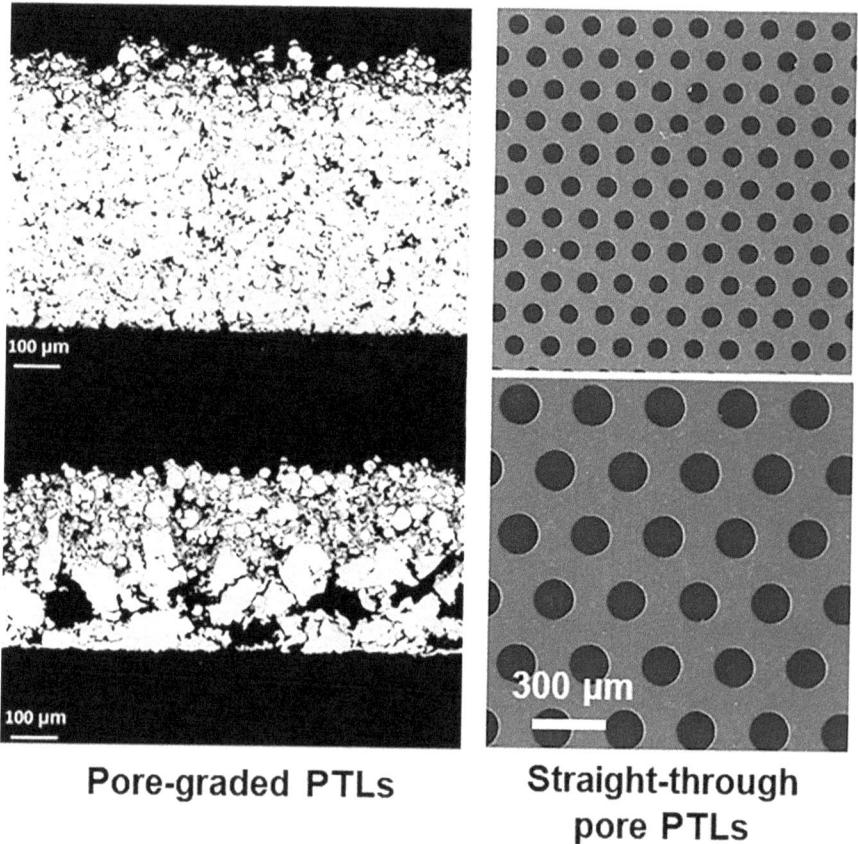

Pore-graded PTLs   **Straight-through pore PTLs**

FIGURE 9.11 Typical new titanium-based PTLs in a PEMWE. Reproduced from Refs. [40,56].

In addition, membrane structures can be redesigned and optimized, or reinforcements can be applied [58]. For example, radiation-grafted membranes can be developed by completely removing oxygen during membrane preparation to avoid the formation of weak-links, or using monomers with a protected α-position and cross-linker to mitigate degradation [59]. The temperature distribution in PEMWEs should also be considered to avoid hot spots that may cause thermal degradation.

Tuning the morphology and composition of catalyst materials significantly affects their performance and durability. Binary and ternary catalysts exhibit good stability by forming a solid solution or single-phase alloy, thereby lowering the catalyst dissolution rate. For example, tin, strontium, niobium, tantalum, and cobalt, instead of iridium and ruthenium oxides, have been used to obtain catalysts with good stability [60–62]. Catalysts with a nanotube morphology usually exhibit better stability than catalysts composed of nanoparticles. Catalyst support materials composed of inert oxides can also enhance stability [22,34,63]. For example, TaB_2 was recently used as a catalyst support material for IrO_2, achieving superior performance and good durability in PEMWEs (3.06 A/cm^2 @ 2.0 V), which meet the US DOE 2023 target [33]. Additionally, doping with transition metals, such as tantalum and tungsten, can improve the electrical conductivity of support materials, which can increase catalyst durability [64]. Furthermore, developing high-performing electrocatalyst materials can alleviate catalyst degradation. Better performance indicates a lower cell voltage or overpotential on the catalyst, which reduces stress on the active sites of the catalyst, resulting in better stability and lower degradation.

Surface treatment has been widely used to enhance the performance and durability of metal components, including PTLs and BPs [65]. Nobel metal coatings on titanium surfaces can significantly improve their durability by inhibiting the oxidation of the titanium materials, as shown in Figure 9.12 [66,67]. A titanium oxide layer underneath an iridium layer does not passivate further, as compared to an unprotected PTL, preventing the severe degradation of titanium. An ultrathin iridium or platinum layer can sustain titanium PTLs for thousands of hours [42]. Researchers have also proposed that

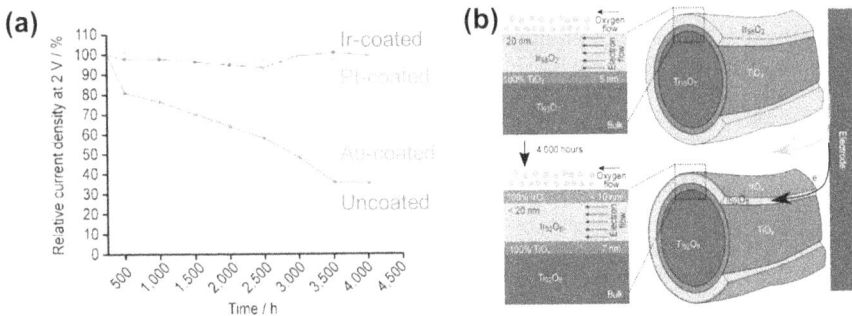

FIGURE 9.12 (a) Durability test of bare and coated titanium-felt PTLs at 2 V and (b) schematic illustrating the effects of iridium thin coatings on titanium fibers and the elemental distribution before and after durability testing. Reproduced from Ref. [66].

stainless steel or other metallic materials protected with appropriate and high-quality coatings can be used to fabricate PTLs and BPs to reduce costs [68,69].

The coatings discussed above can also feasibly and effectively prevent hydrogen embrittle. Hydrogen absorption cannot occur when hydrogen gas contains >2% moisture, except in a very acidic environment under a negative potential. Therefore, hydrogen embrittlement can be easily inhibited in an operating PEMWE owing to water diffusion from the anode to the cathode, mainly under the effects of the concentration gradient and electroosmotic drag. Therefore, hydrogen embrittlement is critical for hydrogen storage and transportation.

9.4 DIAGNOSTIC TECHNIQUES

V–i curves and electrochemical impedance spectroscopy (EIS) are the most commonly used and straightforward techniques for testing PEMWE degradation. Constant potentiostatic or galvanostatic methods are typically used for long-term stability tests, along with EIS measurements, to characterize ohmic, kinetic, and diffusion losses. Additionally, an accelerated stress test (AST) was adopted to avoid thousands of hours of continuous testing [70]. Because PEMWEs have the advantage of fast response to external electrical power, their stability and degradation under a fluctuating power input should also be investigated. This section introduces and discusses diagnostic techniques that can be used for degradation investigations.

Figure 9.13 shows examples of the degradation measurements in PEMWEs using V–i curves and EIS. V–i curves plot the changes in performance over time. However, V–i curves cannot provide valuable data for the deep analysis of degradation, as the measured voltage or current density shows the overall PEMWE characteristics. EIS is a useful tool for analyzing the losses in PEMWEs, but it can only capture irreversible degradation because reversible degradation recovers before EIS can be performed. EIS can be used to measure parameters such as the increased ohmic and charge-transfer resistances. These two methods are widely used, but they cannot clarify the degradation sources. Therefore, techniques for analyzing component degradation have been proposed.

An in situ technique for local potential characterization was recently developed and validated, as shown in Figure 9.14 [72]. The technique allows for the decoupling

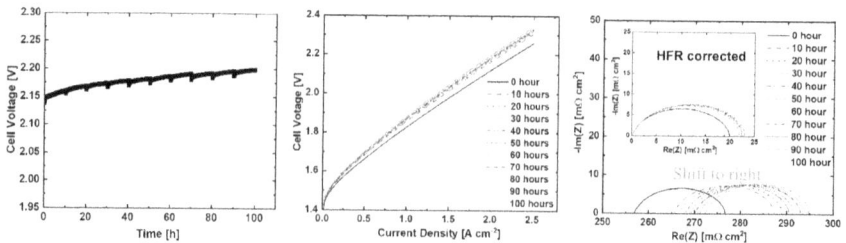

FIGURE 9.13   Examples of degradation measurements using V–i curves and EIS measurements. Reproduced from Ref. [71].

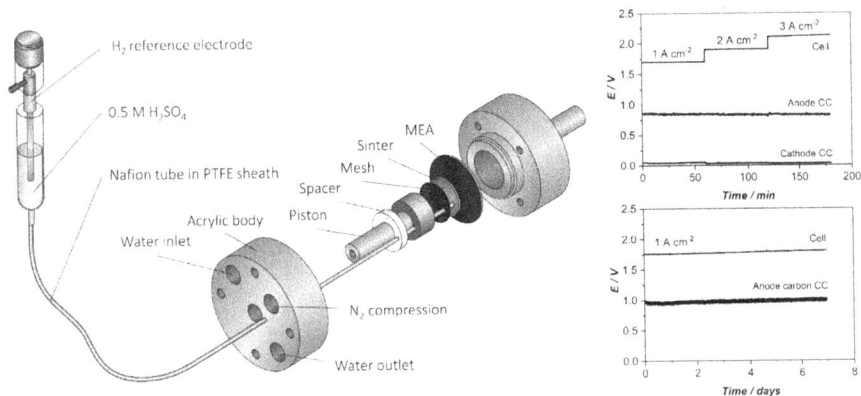

FIGURE 9.14 Schematic illustrating the local potential measurement technique and the short-term stability test results. Reproduced from Ref. [72].

of the local corrosion potential, allowing the variations in the local potentials during stability testing to be monitored.

Another simple technique was developed for sensing the internal voltage, as shown in Figure 9.15 [73]. This technique uses ultrathin voltage-sensing ribbons and helps analysis of degradation losses from individual components or interfaces. The technique also enables the direct measurement of losses through CLs, which helps clarify the degradation in PEMWEs and potentially in other electrochemical devices.

The PEM chemical degradation was in situ characterized by detecting the fluoride release rate (FRR) and sulfur emission rate of the circulated water, and was *ex situ* investigated by measuring the PEM thickness before and after testing [74]. The FRR may be a suitable parameter for estimating the change in membrane thickness [75]. Hydrogen crossover can be used to measure membrane degradation and several methods have been adopted [76–78].

Auger electron spectroscopy was used to measure the depth profiles of the oxide layers to investigate the PTL and BP degradation [79]. ICR measurements using various techniques could also be used to help understand the degradation of PTL and BP. Inductively coupled plasma (ICP) can be used to both in situ and ex situ characterize corrosion of materials.

Electrochemical methods, such as constant potentiostatic or galvanostatic and potentiodynamic or galvanodynamic methods, are the most direct and easily interpretable techniques. Parameters such as the activity stability factor (ASF) and S-number have recently been proposed to represent the instability of catalyst dissolution [80]. The ASF is defined as the ratio of oxygen generation to the equivalent dissolution current density, and the S-number is the ratio of molar oxygen production to noble metal dissolution. Therefore, measuring noble metal dissolution throughout a constant current or potential hold can be useful for assessing catalyst stability. However, this characterization method in aqueous media systems or aqueous model systems for catalyst degradation tests may not reflect the actual degradation of the catalyst

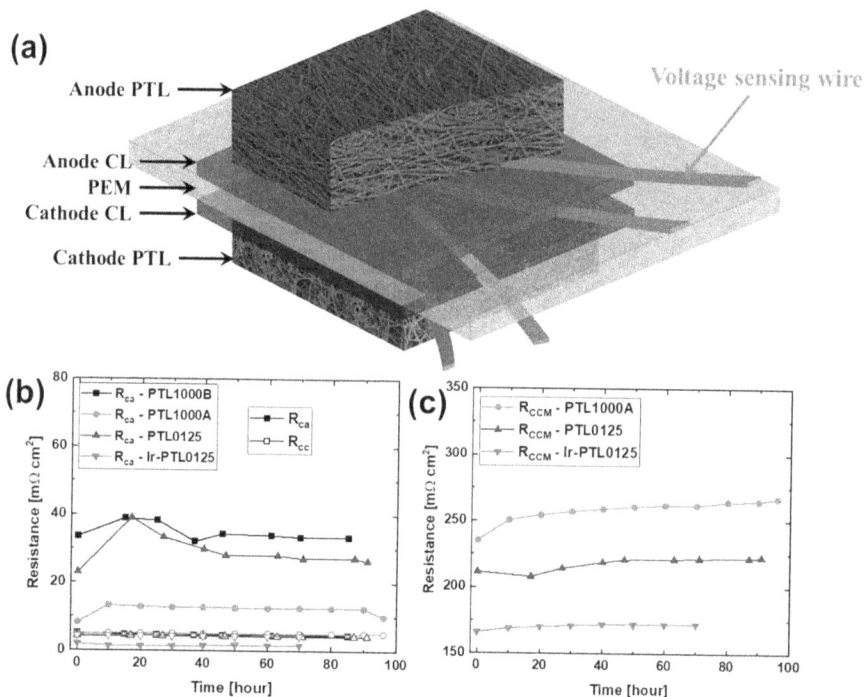

FIGURE 9.15 (a) Schematic illustration of the internal voltage characterization technique and the obtained degradation data with different anode PTLs, (b) anode and cathode ohmic resistances, and (c) CCM resistances. Reproduced form Refs. [71,73].

materials in PEMWEs. Therefore, the electrochemical surface area in PEMWE systems is more accurate for studying catalyst degradation.

9.5 SUMMARY

This chapter introduces the degradations in PEMWEs from the perspective of each component. The harsh environment in PEMWEs leads to severe corrosion of materials such as the OER catalysts, PTLs, and BPs. Catalyst degradation occurs through several mechanisms, including dissolution, leaching, agglomeration, amorphization, migration, and poisoning. PTLs and BPs that are widely fabricated from titanium or stainless steel should be protected using surface treatments or coatings. Other components, such as current distributors, are not critical components in PEMWEs and are not very important for PEMWE degradation. Diagnostic techniques, which could help understand the origins of degradation and improve the lifetime by compensating for the deficiencies of PEMWEs, were also introduced.

REFERENCES

1. Pivovar, B., Catalysts for fuel cell transportation and hydrogen related uses. *Nature Catalysis*, 2019. **2**: pp. 562–565.
2. Yang, C., et al., Size and structure tuning of FePt nanoparticles on hollow mesoporous carbon spheres as efficient catalysts for oxygen reduction reaction. *Rare Metals*, 2023. **42**(6): pp. 1865–1876
3. Ling, J., et al., Self-rechargeable energizers for sustainability. *eScience*, 2022. **2**(4): pp. 347–364.
4. Peng, X., et al., Hierarchical electrode design of highly efficient and stable unitized regenerative fuel cells (URFCs) for long-term energy storage. *Energy & Environmental Science*, 2020. **13**(12): pp. 4872–4881.
5. Lewinski, K.A., D. van der Vliet, and S.M. Luopa, NSTF advances for PEM electrolysis - the effect of alloying on activity of NSTF electrolyzer catalysts and performance of NSTF based PEM electrolyzers. *ECS Transactions*, 2015. **69**(17): p. 893.
6. Xie, H., et al., A membrane-based seawater electrolyser for hydrogen generation. *Nature*, 2022. **612**(7941): pp. 673–678.
7. Lin, C., et al., In-situ reconstructed Ru atom array on α-MnO_2 with enhanced performance for acidic water oxidation. *Nature Catalysis*, 2021. **4**: pp. 1012–1023.
8. Yu, M., E. Budiyanto, and H. Tüysüz, Principles of water electrolysis and recent progress in cobalt-, nickel-, and iron-based oxides for the oxygen evolution reaction. *Angewandte Chemie International Edition*, 2022. **61**(1): p. e202103824.
9. Office, H.a .F.C.T. *Technical Targets for Proton Exchange Membrane Electrolysis*. Available from: https://www.energy.gov/eere/fuelcells/technical-targets-proton-exchange-membrane-electrolysis (Accessed date is 07/21/2023). 2022.
10. Bessarabov, D. and P. Millet, Chapter 3 - Performance degradation, in *PEM Water Electrolysis*, D. Bessarabov and P. Millet, Editors. 2018, Academic Press. pp. 61–94.
11. Tang, H., et al., A degradation study of Nafion proton exchange membrane of PEM fuel cells. *Journal of Power Sources*, 2007. **170**(1): pp. 85–92.
12. Liu, C., et al., The impacts of membrane pinholes on PEM water electrolysis. *Journal of Power Sources*, 2023. **581**: p. 233507.
13. Mench, M.M., *Fuel Cell Engines*. 2008, Hoboken, NJ: John Wiley & Sons.
14. Millet, P., et al., Cell failure mechanisms in PEM water electrolyzers. *International Journal of Hydrogen Energy*, 2012. **37**(22): pp. 17478–17487.
15. Lapicque, F., et al., A critical review on gas diffusion micro and macroporous layers degradations for improved membrane fuel cell durability. *Journal of Power Sources*, 2016. **336**: pp. 40–53.
16. Kang, Z., et al., Performance improvement induced by membrane treatment in proton exchange membrane water electrolysis cells. *International Journal of Hydrogen Energy*, 2022. **47**(9): pp. 5807–5816.
17. Chandesris, M., et al., Membrane degradation in PEM water electrolyzer: Numerical modeling and experimental evidence of the influence of temperature and current density. *International Journal of Hydrogen Energy*, 2015. **40**(3): pp. 1353–1366.
18. Fouda-Onana, F., et al., Investigation on the degradation of MEAs for PEM water electrolysers part I: Effects of testing conditions on MEA performances and membrane properties. *International Journal of Hydrogen Energy*, 2016. **41**(38): pp. 16627–16636.
19. Grigoriev, S.A., et al., Failure of PEM water electrolysis cells: Case study involving anode dissolution and membrane thinning. *International Journal of Hydrogen Energy*, 2014. **39**(35): pp. 20440–20446.
20. Liu, H., et al., Chemical degradation: Correlations between electrolyzer and fuel cell findings, in Polymer Electrolyte Fuel Cell Durability, F.N. Büchi, M. Inaba, and T.J. Schmidt, Editors. 2009, Springer New York: New York. pp. 71–118.

21. Becker, H., et al., Impact of impurities on water electrolysis: A review. *Sustainable Energy & Fuels*, 2023. **7**: pp. 1565–1603.

22. Rozain, C., et al., Influence of iridium oxide loadings on the performance of PEM water electrolysis cells: Part I-Pure IrO₂-based anodes. *Applied Catalysis B: Environmental*, 2016. **182**: pp. 153–160.

23. Sun, S., et al., Investigations on degradation of the long-term proton exchange membrane water electrolysis stack. *Journal of Power Sources*, 2014. **267**: pp. 515–520.

24. Cherevko, S., et al., Oxygen evolution activity and stability of iridium in acidic media. Part 1. - Metallic iridium. *Journal of Electroanalytical Chemistry*, 2016. **773**: pp. 69–78.

25. Cherevko, S., et al., Oxygen evolution activity and stability of iridium in acidic media. Part 2. - Electrochemically grown hydrous iridium oxide. *Journal of Electroanalytical Chemistry*, 2016. **774**: pp. 102–110.

26. Milosevic, M., et al., In search of lost iridium: Quantification of anode catalyst layer dissolution in proton exchange membrane water electrolyzers. *ACS Energy Letters*, 2023. **8**(6): pp. 2682–2688.

27. Yu, H., et al., Microscopic insights on the degradation of a PEM water electrolyzer with ultra-low catalyst loading. *Applied Catalysis B: Environmental*, 2020. **260**: p. 118194.

28. Grigoriev, S.A., D.G. Bessarabov, and V.N. Fateev, Degradation mechanisms of MEA characteristics during water electrolysis in solid polymer electrolyte cells. *Russian Journal of Electrochemistry*, 2017. **53**(3): pp. 318–323.

29. Feng, Q., et al., A review of proton exchange membrane water electrolysis on degradation mechanisms and mitigation strategies. *Journal of Power Sources*, 2017. **366**: pp. 33–55.

30. Siracusano, S., et al., Enhanced performance and durability of low catalyst loading PEM water electrolyser based on a short-side chain perfluorosulfonic ionomer. *Applied Energy*, 2017. **192**: pp. 477–489.

31. Zhang, L., et al., The influence of sodium ion on the solid polymer electrolyte water electrolysis. *International Journal of Hydrogen Energy*, 2012. **37**(2): pp. 1321–1325.

32. Rakousky, C., et al., An analysis of degradation phenomena in polymer electrolyte membrane water electrolysis. *Journal of Power Sources*, 2016. **326**: pp. 120–128.

33. Wang, Y., et al., Nano-metal diborides-supported anode catalyst with strongly coupled TaO(x)/IrO(2) catalytic layer for low-iridium-loading proton exchange membrane electrolyzer. *Nature Communication*, 2023. **14**(1): p. 5119.

34. Puthiyapura, V.K., et al., Investigation of supported IrO₂ as electrocatalyst for the oxygen evolution reaction in proton exchange membrane water electrolyser. *International Journal of Hydrogen Energy*, 2014. **39**(5): pp. 1905–1913.

35. Fuentes, R.E., J. Farell, and J.W. Weidner, Multimetallic electrocatalysts of Pt, Ru, and Ir supported on anatase and rutile TiO₂ for oxygen evolution in an acid environment. *Electrochemical and Solid-State Letters*, 2011. **14**(3): pp. E5–E7.

36. Yu, S., et al., Tuning catalyst activation and utilization via controlled electrode patterning for low-loading and high-efficiency water electrolyzers. *Small*, 2022. **18**(14): p. 2107745.

37. Bierling, M., et al., Toward understanding catalyst layer deposition processes and distribution in anodic porous transport electrodes in proton exchange membrane water electrolyzers. *Advanced Energy Materials*, 2023. **13**(13): p. 2203636.

38. Schuler, T., T.J. Schmidt, and F.N. Büchi, Polymer electrolyte water electrolysis: Correlating performance and porous transport layer structure: Part II. Electrochemical performance analysis. *Journal of the Electrochemical Society*, 2019. **166**(10): p. F555.

39. Mo, J., et al., Discovery of true electrochemical reactions for ultrahigh catalyst mass activity in water splitting. *Science Advances*, 2016. **2**(11): p. e1600690.

40. Kang, Z., et al., Investigation of thin/well-tunable liquid/gas diffusion layers exhibiting superior multifunctional performance in low-temperature electrolytic water splitting. *Energy & Environmental Science*, 2017. **10**(1): pp. 166–175.

41. Kang, Z., et al., Performance modeling and current mapping of proton exchange membrane electrolyzer cells with novel thin/tunable liquid/gas diffusion layers. *Electrochimica Acta*, 2017. **255**: pp. 405–416.

42. Liu, C., et al., Degradation effects at the porous transport layer/catalyst layer interface in polymer electrolyte membrane water electrolyzer. *Journal of the Electrochemical Society*, 2023. **170**(3): p. 034508.

43. Li, Y., et al., In-situ investigation of bubble dynamics and two-phase flow in proton exchange membrane electrolyzer cells. *International Journal of Hydrogen Energy*, 2018. **43**(24): pp. 11223–11233.

44. Lee, H.Y., C. Barber, and A.R. Minerick, Improving electrokinetic microdevice stability by controlling electrolysis bubbles. *Electrophoresis*, 2014. **35**(12–13): pp. 1782–1789.

45. Alia, S.M., et al., Activity and durability of iridium nanoparticles in the oxygen evolution reaction. *Journal of the Electrochemical Society*, 2016. **163**(11): p. F3105.

46. Young, J.L., et al., PEM electrolyzer characterization with carbon-based hardware and material sets. *Electrochemistry Communications*, 2021. **124**: p. 106941.

47. Mo, J., et al., Electrochemical investigation of stainless steel corrosion in a proton exchange membrane electrolyzer cell. *International Journal of Hydrogen Energy*, 2015. **40**(36): pp. 12506–12511.

48. Kang, Z., et al., Effects of interfacial contact under different operating conditions in proton exchange membrane water electrolysis. *Electrochimica Acta*, 2022. **429**: p. 140942.

49. Cheng, H., et al., Effect of fluoride ion concentration and fluctuating conditions on titanium bipolar plate in PEM water electrolyser environment. *Corrosion Science*, 2023. **222**: p. 111414.

50. Wang, Z.B., et al., The effect of fluoride ions on the corrosion behavior of pure titanium in 0.05M sulfuric acid. *Electrochimica Acta*, 2014. **135**: pp. 526–535.

51. Kong, D.-S. and Y.-Y. Feng, Electrochemical anodic dissolution kinetics of titanium in fluoride-containing perchloric acid solutions at open-circuit potentials. *Journal of the Electrochemical Society*, 2009. **156**(9): p. C283.

52. Mo, J., et al., Study on corrosion migrations within catalyst-coated membranes of proton exchange membrane electrolyzer cells. *International Journal of Hydrogen Energy*, 2017. **42**(44): pp. 27343–27349.

53. Dihrab, S.S., et al., Review of the membrane and bipolar plates materials for conventional and unitized regenerative fuel cells. *Renewable and Sustainable Energy Reviews*, 2009. **13**(6): pp. 1663–1668.

54. Schuler, T., et al., Hierarchically structured porous transport layers for polymer electrolyte water electrolysis. *Advanced Energy Materials*, 2020. **10**(2): p. 1903216.

55. Lee, J.K., et al., Interfacial engineering via laser ablation for high-performing PEM water electrolysis. *Applied Energy*, 2023. **336**: p. 120853.

56. Lettenmeier, P., et al., Comprehensive investigation of novel pore-graded gas diffusion layers for high-performance and cost-effective proton exchange membrane electrolyzers. *Energy & Environmental Science*, 2017. **10**(12): pp. 2521–2533.

57. Yang, G., et al., Fully printed and integrated electrolyzer cells with additive manufacturing for high-efficiency water splitting. *Applied Energy*, 2018. **215**: pp. 202–210.

58. Kuwertz, R., et al., Influence of acid pretreatment on ionic conductivity of nafion(r) membranes. *Journal of Membrane Science*, 2016. **500**: pp. 225–235.

59. Albert, A., et al., Stability and degradation mechanisms of radiation-grafted polymer electrolyte membranes for water electrolysis. *ACS Appl Mater Interfaces*, 2016. **8**(24): pp. 15297–15306.

60. Wang, L., et al., Structurally robust honeycomb layered strontium Iridate as an oxygen evolution electrocatalyst in acid. *ACS Catalysis*, 2023. **13**(11): pp. 7322–7330.

61. Yang, L., et al., A highly active, long-lived oxygen evolution electrocatalyst derived from open-framework iridates. *Advanced Materials*, 2023. **35**(12): p. 2208539.

62. Wu, Q., et al., Advances and status of anode catalysts for proton exchange membrane water electrolysis technology. *Materials Chemistry Frontiers*, 2023. **7**(6): pp. 1025–1045.

63. Oh, H.-S., H.N. Nong, and P. Strasser, Preparation of mesoporous Sb-, F-, and in-doped SnO_2 bulk powder with high surface area for use as catalyst supports in electrolytic cells. *Advanced Functional Materials*, 2015. **25**(7): pp. 1074–1081.

64. Xiao, Y., et al., W-doped TiO_2 mesoporous electron transport layer for efficient hole transport material free perovskite solar cells employing carbon counter electrodes. *Journal of Power Sources*, 2017. **342**: pp. 489–494.

65. Wakayama, H. and K. Yamazaki, Low-cost bipolar plates of Ti_4O_7-coated Ti for water electrolysis with polymer electrolyte membranes. *ACS Omega*, 2021. **6**(6): pp. 4161–4166.

66. Liu, C., et al., Exploring the interface of skin-layered titanium fibers for electrochemical water splitting. *Advanced Energy Materials*, 2021. **11**(8): p. 2002926.

67. Kang, Z., et al., Thin film surface modifications of thin/tunable liquid/gas diffusion layers for high-efficiency proton exchange membrane electrolyzer cells. *Applied Energy*, 2017. **206**: pp. 983–990.

68. Rojas, N., et al., Coated stainless steels evaluation for bipolar plates in PEM water electrolysis conditions. *International Journal of Hydrogen Energy*, 2021. **46**(51): pp. 25929–25943.

69. Kellenberger, A., et al., Towards replacing titanium with copper in the bipolar plates for proton exchange membrane water electrolysis. *Materials (Basel)*, 2022. **15**(5): p. 1628.

70. Aßmann, P., et al., Toward developing accelerated stress tests for proton exchange membrane electrolyzers. *Current Opinion in Electrochemistry*, 2020. **21**: pp. 225–233.

71. Kang, Z., et al., In-situ and in-operando analysis of voltage losses using sense wires for proton exchange membrane water electrolyzers. *Journal of Power Sources*, 2021. **481**: p. 229012.

72. Becker, H., L. Castanheira, and G. Hinds, Local measurement of current collector potential in a polymer electrolyte membrane water electrolyser. *Journal of Power Sources*, 2020. **448**: p. 227563.

73. Kang, Z., et al., Exploring and understanding the internal voltage losses through catalyst layers in proton exchange membrane water electrolysis devices. *Applied Energy*, 2022. **317**: p. 119213.

74. Khatib, F.N., et al., Material degradation of components in polymer electrolyte membrane (PEM) electrolytic cell and mitigation mechanisms: A review. *Renewable and Sustainable Energy Reviews*, 2019. **111**: pp. 1–14.

75. Prestat, M., Corrosion of structural components of proton exchange membrane water electrolyzer anodes: A review. *Journal of Power Sources*, 2023. **556**: p. 232469.

76. Kang, Z., M. Pak, and G. Bender, Introducing a novel technique for measuring hydrogen crossover in membrane-based electrochemical cells. *International Journal of Hydrogen Energy*, 2021. **46**(29): pp. 15161–15167.

77. Ito, H., et al., Cross-permeation and consumption of hydrogen during proton exchange membrane electrolysis. *International Journal of Hydrogen Energy*, 2016. **41**(45): pp. 20439–20446.

78. Garbe, S., et al., Communication-Pt-doped thin membranes for gas crossover suppression in polymer electrolyte water electrolysis. *Journal of the Electrochemical Society*, 2019. **166**(13): p. F873.

79. Lædre, S., et al., Materials for proton exchange membrane water electrolyzer bipolar plates. *International Journal of Hydrogen Energy*, 2017. **42**(5): pp. 2713–2723.
80. Edgington, J. and L.C. Seitz, Advancing the rigor and reproducibility of electrocatalyst stability benchmarking and intrinsic material degradation analysis for water oxidation. *ACS Catalysis*, 2023. **13**(5): pp. 3379–3394.

10 Key Components and Preparation Technology for PEM Water Electrolysis

Bang Li, Qiqi Wan, and Changchun Ke

10.1 INTRODUCTION OF PEM WATER ELECTROLYSIS AND MEA PREPARATION

10.1.1 Structure of PEM Water Electrolysis (PEMWE)

Proton exchange membrane (PEM) water electrolysis is regarded as one of the most promising technologies for efficiently producing high-purity hydrogen [1]. It is especially advantageous when paired with intermittent renewable energy sources like wind and solar energy [2]. This combination facilitates a sustainable, carbon-free energy cycle. PEM water electrolysis offers advantages such as higher conversion efficiency, higher current density, reduced corrosion, minimized cross-contamination between anode and cathode gases, and compact structure [3]. PEM water electrolysis (PEMWE) cells consist of the proton exchange membrane, cathode and anode catalyst layers, cathode and anode porous transport layer (PTL, also known as the gas/liquid diffusion layer or gas diffusion layer, GDL), bipolar plates, and end plates. The proton exchange membrane plays a crucial role in facilitating proton transport, isolating gases on the cathode and anode sides, and preventing electron conduction to avoid internal short-circuits in the cell. The catalyst layer, which contains catalysts and ionic polymers, enables the conduction of electrons and protons, serving as the central site for electrochemical reactions. The PTL connects the catalyst layer to the bipolar plates, facilitating the transfer of electrons, heat, reactants, and products [4]. The bipolar plates are engraved with channels through which reactants and products enter or exit. Water can circulate only on the anode side, as shown in Figure 10.1 [5]. It is also possible for water to circulate simultaneously on both the cathode and anode sides, facilitating the release of H_2 from the cathode. Currently, most PEMWE systems only have a water inlet on the anode side, with no water circulation on the cathode [5].

DOI: 10.1201/9781003368939-10

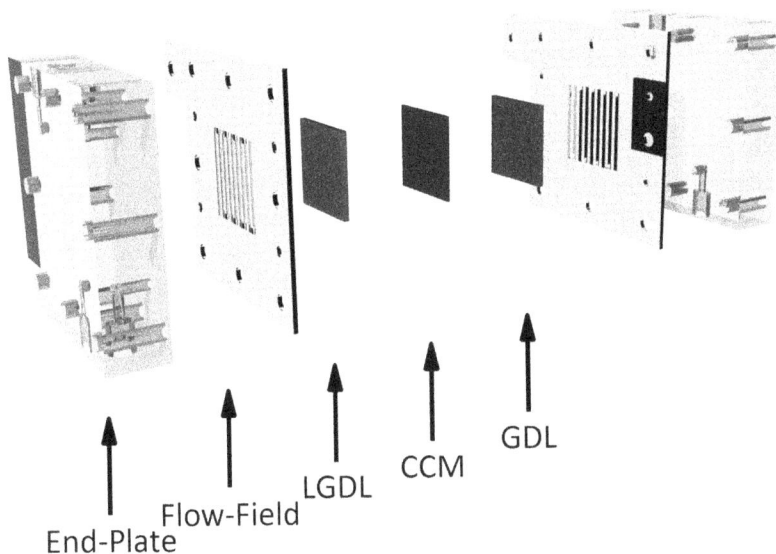

FIGURE 10.1   Structure of PEMWE.

10.1.2   Principle of PEM Water Electrolysis

10.1.2.1   Principles of PEM Water Electrolysis Reaction

Deionized water enters the flow channel and flows through the PTL. It then enters the catalyst layer where electrochemical reactions occur. In the anode catalyst layer, water molecules lose electrons and generate oxygen and protons, known as the oxygen evolution reaction (OER). The protons travel through the proton exchange membrane and reach the cathode catalyst layer, where they combine with electrons to produce H_2 through the hydrogen evolution reaction (HER). The overall reactions that occur in the electrolysis cell can be summarized as follows:

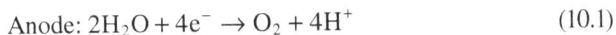

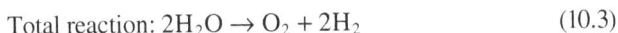

$$\text{Anode: } 2H_2O + 4e^- \rightarrow O_2 + 4H^+ \tag{10.1}$$

$$\text{Cathode: } 4H^+ + 4e^- \rightarrow 2H_2 \tag{10.2}$$

$$\text{Total reaction: } 2H_2O \rightarrow O_2 + 2H_2 \tag{10.3}$$

10.1.2.2   Thermodynamics of PEM Water Electrolysis

The water electrolysis process is an endothermic and non-spontaneous reaction, which means it requires the input of electrical and thermal energy to proceed. Therefore, when calculating the thermal equilibrium voltage for water electrolysis, it is necessary to take into consideration the external heating that is required for the reaction to occur [6]. The following formula can be used to calculate the thermal equilibrium potential difference required for the process of water electrolysis:

$$V\Delta H = \Delta H/(z \cdot F) \tag{10.4}$$

The various terms in the equation are defined as follows: $V_{\Delta H}$ is the enthalpy change voltage of 1 mole of water electrolysis, ΔH is the enthalpy change of 1 mole of water molecule electrolysis, z is the number of electrons transferred in 1 mole of water ($z = 2$), and F is the Faraday constant ($F \approx 96{,}485$ C/mol).

However, ΔH, as enthalpy change, is influenced by both thermal energy and electric energy. Indeed, when considering the minimum voltage required for water electrolysis, the Gibbs free energy change (ΔG) is commonly used. By calculating ΔG, we can determine the minimum voltage necessary to drive the electrolysis process:

$$\Delta G = \Delta H - Q = \Delta H - T \cdot \Delta S \tag{10.5}$$

The minimum voltage required for water electrolysis, also known as the reversible voltage (V_{rev}):

$$V_{rev} = \Delta G/(z \cdot F) \tag{10.6}$$

Under conditions of 298.15 K and 1 atmospheric pressure: $\Delta S = 0.163$ kJ/mol, $\Delta H = 285.84$ kJ/mol, which is used in equation (10.1) to obtain $V_{\Delta H}$ as 1.48 V. $\Delta G = 237.2$ kJ/mol, which is used in equation (10.3) to calculate V_{rev} as 1.23 V. Figure 10.2 [7] illustrates the values of ΔG, ΔH, and ΔS at different temperatures. In Figure 10.2, it can be observed that the enthalpy change (ΔH) decreases as the temperature increases before reaching 100°C. Additionally, at 100°C, there is a sudden drop in ΔH due to water vaporization. Above 100°C, ΔH shows a slight increase with further temperature escalation, but it remains relatively low. However, the Gibbs free energy change (ΔG) consistently decreases as the temperature increases, suggesting a lower requirement for electrical energy. Simultaneously, the product of temperature (T) and the entropy change (ΔS) increases with temperature, indicating a growing need for thermal energy. In general, the cost of electricity is higher than the cost of heat energy for the same amount of energy. Therefore, raising the temperature and reducing electrical losses can contribute to reducing the cost of hydrogen production [8].

10.1.2.3 PEM Water Electrolysis Kinetics

Thermodynamic principles apply to reactions or processes that approach equilibrium or reversibility. However, in the water electrolysis process, both thermodynamic and kinetic energies are consumed. When an electric current passes through the electrolytic cell, there is an irreversible loss of kinetic energy [9]. This leads to the actual water electrolysis voltage being higher than the voltage calculated based on thermodynamics. The actual voltage consists of several components, including the open circuit voltage (V_{oc}), activation energy overpotential (V_{act}), ohmic overpotential (V_{ohm}),

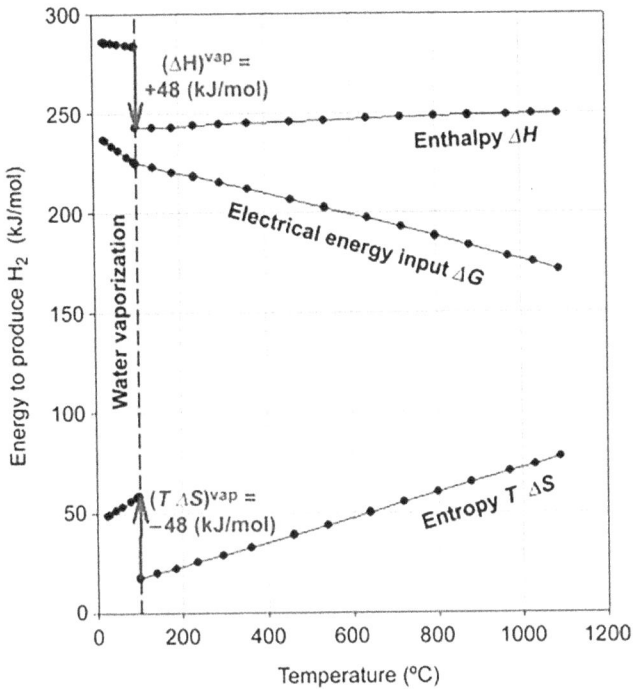

FIGURE 10.2   The value of ΔG, ΔH, and ΔS at different temperatures.

and concentration polarization overpotential (V_{con}), as described in the following equation:

$$V = V_{OC} + V_{act} + V_{ohm} + V_{con} \tag{10.7}$$

The open circuit voltage (V_{oc}) represents the thermodynamic energy that powers water electrolysis. It can be calculated using the Nernst equation (10.8) [10], which is equivalent to the thermal equilibrium voltage in the non-standard state:

$$V_{oc} = V_{rev} + \frac{RT}{2F} \ln\left(\frac{p_{H_2} \sqrt{p_{O_2}}}{a_{H_2O}} \right) \tag{10.8}$$

where V_{act} represents the activation energy needed to overcome the electrochemical reactions occurring at the interface between the electrode and water. It is the sum of the activation overpotentials at the positive and negative electrodes, typically described as a positive potential for the anode and a negative potential for the cathode. Equation (10.9) provides the expression for V_{act}, and references [5,11,12] on this topic provide more detailed information.

$$V_{act} = V_{act,a} - V_{act,c} \tag{10.9}$$

The ohmic resistance of an electrolytic cell is determined by the combination of internal resistances, which include R_{ep} (end plate resistance), R_{dp} (bipolar plate resistance), R_{lgdl} (gas–liquid diffusion layer resistance), R_{cl} (catalyst layer resistance), and R_m (proton exchange membrane resistance), along with the contact resistance at their interfaces [5,13]. The following formula incorporates the resistance of each component and the contact resistance at their interfaces:

$$V_{ohm} = IR_{ohm} = I(R_{ep} + R_{dp} + R_{lgdl} + R_{cl} + R_m) \tag{10.10}$$

The concentration overpotential is determined by the logarithm of the ratio between the concentrations of oxygen and hydrogen at the membrane under the standard reference state. This relationship is described by equation (10.11). For more in-depth information, refer to references [5,14,15].

$$V_{con} = V_{con,an} + V_{con,cat} = \frac{RT}{4F} \ln\left(\frac{c_{O_2,me}}{c_{O_2,me,0}}\right) + \frac{RT}{2F} \ln\left(\frac{c_{H_2,me}}{c_{H_2,me,0}}\right) \tag{10.11}$$

In PEMWE, the cathode overvoltage is typically lower than the anode overvoltage, as illustrated in Figure 10.3 [16]. As a result, researchers often focus on studying and reducing the anode overvoltage to enhance the performance of the electrolysis cell.

10.1.2.4 PEM Water Electrolysis Oxygen Electrocatalysis

The relationship between current and overpotential can be represented by the Butler–Volmer equation (B–V equation) [17].

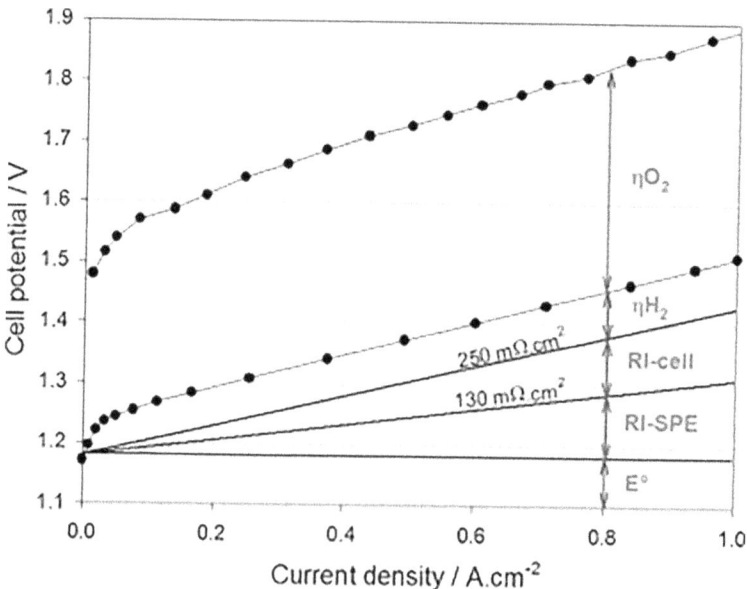

FIGURE 10.3 Polarization curves of various parts under different current densities.

$$i = i_0 \cdot \left\{ \exp\left[\frac{\alpha_a nF\eta}{RT} \right] - \exp\left[\frac{\alpha_c nF\eta}{RT} \right] \right\} \quad (10.12)$$

where i_0 is the exchange current density, representing the reaction rate; T is the temperature; F is the Faraday constant; R is the gas constant; and α is the transfer coefficient. According to the experiment of Tafel, he believed that the anode played a dominant role in the reaction process, while the cathode process could be ignored. Therefore, he simplified the B–V equation to the following equation [17]:

$$i = i_0 \cdot \exp\left[\frac{\alpha_a nF\eta}{RT} \right] \quad (10.13)$$

Take the logarithm on both sides of the equation to obtain $\eta = a + b \log j$, which is the Tafel equation [18]:

$$a = \frac{2.3RT}{\alpha F} \log i_0 \quad b = \frac{-2.3RT}{\alpha F} \quad (10.14)$$

α is related to factors such as electrode material and electrode surface state. The steady-state Tafel diagram is a commonly used method for describing charge transfer in electrocatalytic reactions. The Tafel slope (unit: mV/dec) is an important parameter for evaluating catalyst activity.

10.1.3   Preparation of Membrane Electrode Assembly (MEA)

The membrane electrode assembly (MEA) is the core of PEMWE [19], which is composed of proton exchange membrane, catalyst layers for the anode and cathode, and PTLs for the anode and cathode. Based on the function and environment of the MEA, the MEA in PEMWE should possess the following characteristics: low contact resistance, good catalyst adhesion, sufficient three-phase boundary for reactions, suitable channels for reactant and product transport, high conductivity, low ohmic impedance, good mechanical properties, effective polarity separation, and a long lifespan. In PEMWE, a slow multi-step OER process occurs at the anode [20, 21].

10.1.3.1   Membrane Treatment

Before preparing membrane electrodes, proton exchange membranes usually undergo a pretreatment process, and most references [22–24] use the following methods. First, preheat the membrane in pure water for 1 hour. Then, boil it in a 5% H_2O_2 solution for 1 hour to remove organic impurities. Next, rinse the membrane with distilled water. Then, boil the membrane in a 1 M sulfuric acid solution for 30 minutes to facilitate complete ion exchange, followed by another rinse with distilled water. Finally, boil the membrane in distilled water and store it in pure water. When preparing catalyst-coated membrane (CCM) electrodes, membrane swelling is usually caused. To overcome this problem, Chang-Soo et al. [25] initially boiled the membrane in NaOH to convert it from a protonated state to a sodium ion state. This reduces the affinity between protons and Nafion sulfonic acid groups, minimizing or even avoiding alcohol-induced swelling of the membrane. After the deposition and drying of the catalyst layer, the membrane is then converted back to a protonated state. Lagarteira et al. [26] used this method to pretreat the membrane and successfully prepared high-performance membrane electrodes.

10.1.3.2 Gas Diffusion Layer Electrode

First, the cathode and anode catalyst slurries are separately sprayed onto the cathode and anode GDLs to form the cathode and anode electrodes. Then, the cathode and anode electrodes are sandwiched on both sides of the proton exchange membrane to form the MEA. Directly coating the catalyst on the GDL offers several unique advantages. First, it can act as a protective layer [27], particularly on the anode side of the MEA. This protective layer helps to mitigate the oxidation rate of the GDL, which is subjected to a high voltage and low pH environment. Secondly, it enhances the stability of the interface between the catalyst layer and the GDL [21]. Additionally, the direct coating process is simple and well-established, facilitating the formation of porous structures. Lastly, this method avoids any swelling issues with the proton exchange membrane. The use of gas diffusion electrodes (GDEs) also poses some disadvantages. First, during the preparation of GDE, the catalyst can enter the pores of the GDL, leading to a reduced utilization rate of the catalyst. Second, the interface between the catalyst layer and the proton exchange membrane, formed through hot pressing, tends to exhibit low interfacial adhesion and increased interfacial resistance. Lastly, the hot pressing process may cause structural changes or performance loss in the catalyst layer, potentially impacting the overall efficiency of the system [28].

Buehler et al. [29] prepared GDE using the ink loading with 1.4 mg/cm^2 IrO$_2$ and 9 wt% Nafion. The ink was sprayed on sintered titanium powder and titanium fiber, respectively. This is compared with membrane electrodes prepared by CCM under the same catalyst loading. GDE exhibited better electrochemical performance at current densities above 750 mA/cm^2. Based on EIS testing, it was observed that in the high-frequency impedance (HFR) region, GDEs exhibit lower ohmic impedance compared to CCMs, indicating better material transport improvement. This could be attributed to the strong contact between the catalyst and the GDL. On the other hand, in the kinetic region, CCM demonstrates the best performance, which could be attributed to the larger contact area between the catalyst layer and the proton exchange membrane.

Magnetron sputtering is a widely employed membrane preparation technology known for its uniform particle distribution and low active catalyst loading [30]. Mo et al. [31] prepared GDE using sputtering and spraying methods, respectively. The catalyst layer obtained by spraying has a finer grain structure, thus exhibiting good electrochemical performance. However, the catalyst layer thickness prepared by sputtering method (15 nm) is much lower than that prepared by spraying method (15 μm). As a result, its overall performance is better than the former. Labou et al. [30] used magnetron sputtering technology. Iridium catalyst is sputtered from Ir target in Ar/O$_2$ plasma through DC Magnetron sputtering. The diameter of the deposited nanoparticle is 100–300 nm, and the thickness is about 500 nm.

Lee et al. [32] utilized electrodeposition to deposit IrO_2 catalyst onto carbon paper as a means to reduce the reliance on precious metal catalysts. By controlling the electrode position potential and deposition time, they were able to alter the loading amount and morphology of IrO_2. When the control voltage was set to 0.7 V for a deposition time of 10 minutes, the IrO_2 loading was determined to be 0.1 mg/cm^2. Tested at a temperature of 90°C, this configuration displayed the highest performance, achieving a current density of 1.01 A/cm^2 at 1.6 V. Choe et al. [27] employed the electrodeposition method to deposit a uniform sub-micron thick layer of IrO_2 catalyst onto an anode porous titanium mesh. The IrO_2 loading in this case was determined to be 0.4 mg/cm^2. Notably, the IrO_2 coating also serves as a protective layer for the GDL, effectively inhibiting the oxidation of the titanium mesh. The decay rate of the IrO_2 coating was measured to be 1.5 mA/cm^{2}h.

Holzapfel et al. [33] applied a direct deposition method to coat the membrane on the cathode. As shown in Figure 10.4, the electrodes prepared through this method exhibited higher activation loss when compared to those prepared using CCM. This could potentially be led to the presence of inactive catalysts within the larger pores of the anode. However, there were noticeable performance improvements in terms of ohmic impedance, mass transfer impedance, and charge transfer impedance.

10.1.3.3 Catalyst-Coated Membrane (CCM) Electrode

Inspired by fuel cell technology, it is widely accepted that the catalyst layer should be directly applied onto the membrane to form a CCM, rather than being placed on a separate substrate [22,34,35]. Various methods, such as electrodeposition, transfer printing, and spraying, have been utilized to coat the cathode and anode catalyst ink

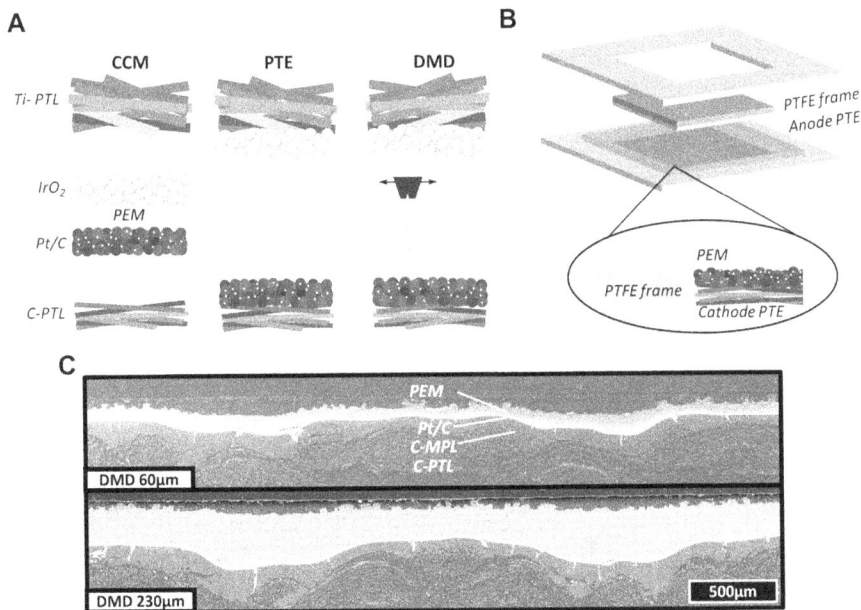

FIGURE 10.4   Preparation of membrane electrode by spray transfer printing method.

on both sides of the proton exchange membrane. The use of CCM in PEMWE offers several advantages. These include higher catalyst utilization, thanks to the direct application of the catalyst onto the membrane. The use of this method also enhances the adhesion between the catalyst layer and the proton exchange membrane, resulting in improved stability and durability. Additionally, the direct interface between the catalyst layer and the membrane ensures high proton connectivity, facilitating efficient proton transport and promoting overall cell performance [36]. While CCM offers numerous advantages, there are some associated disadvantages. These include the potential for poor adhesion between the catalyst layer and the GDL, resulting in high interfacial resistance. Additionally, the preparation process for CCM can be complex and time-consuming, requiring careful control and optimization. The use of precious metal catalysts in CCM can lead to high consumption and increased cost [37]. Moreover, the distribution of ionomers, catalysts, and pores in CCMs is typically random [38].

Mamaca et al. [39] sprayed the catalyst and Nafion onto the polytetrafluoroethylene (PTFE) substrate. As shown in Figure 10.5 [40], this method can easily transfer the catalyst layer onto the membrane. Similarly, Kus et al. [41] developed a method where they initially applied a uniform coating of conductive TiC onto a PTFE substrate. Subsequently, they hot-pressed the TiC-coated substrate onto Nafion 115 to create a supporting sublayer. Next, the catalyst was deposited onto the TiC supporting using magnetic control, establishing a three-phase interface for water electrolysis. The membrane electrode prepared through this technique possesses an exceptionally low content of precious metals, with an Ir loading of only 80 $\mu g/cm^2$. Remarkably, this membrane electrode demonstrates outstanding electrolytic performance, achieving a voltage of 1.74 V at a current density of 1 A/cm^2, operating at 80°C.

Park et al. [42] used an electrodeposited catalyst layer on the template and then transferred the catalyst layer onto the membrane to prepare the membrane electrode.

Park et al. [43] adopted the roll-to-roll coating process, to enhance the efficiency of manufacturing MEAs. However, applying a large amount of paste at once leads to longer interaction time between the dispersed medium and the membrane, resulting in increased absorption and expansion of the membrane (Figure 10.6).

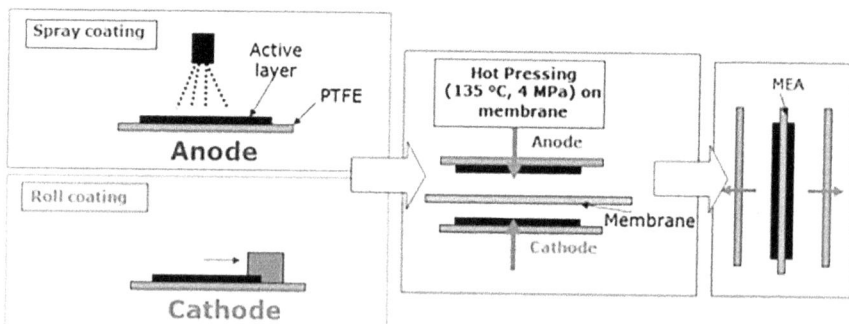

FIGURE 10.5 Roll-to-roll coating process.

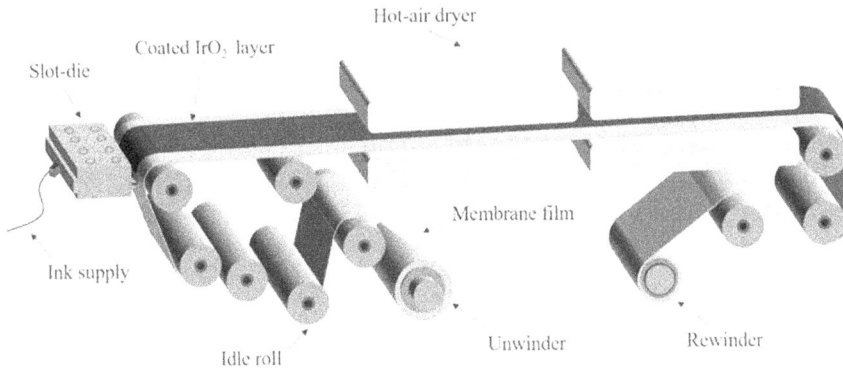

FIGURE 10.6   Directly deposit the PEM on the anode catalyst layer.

Lagarteira et al. [26] used an Aurel 900 screen printing machine equipped with Koenen Typ-10 M6 screen printing technology to prepare a highly active membrane electrode with Ir loading only 0.4 mg/cm^2.

CCM sprayed during the process may lead to swelling and deformation of the proton exchange membrane. Su et al. [44] developed a method for spray-coating electrode under illumination. In this method, the electrodes are simultaneously irradiated with infrared light during the spraying process to facilitate solvent evaporation. This significantly reduces the swelling and deformation of the electrode membrane during spraying. Due to the good contact between the cathode catalyst layer and the proton exchange membrane, as well as the uniform porous structure of the catalyst layer, the electrode membrane demonstrates excellent stability and water electrolysis performance. However, Shi et al. [45] believe that the expansion spray-coating method has the potential to prepare high-quality MEA of PEMWE. They introduced hot water as an expansion agent and heat source, which helps to prevent membrane shrinkage during the drying process of the catalyst slurry and maintain the wetting properties of the membrane. In the fuel cell, the impregnation reduction method is also used to spray catalyst on the expansion membrane [46], but the chemical process often leads to large catalyst particles, and the catalyst precursor enters the membrane during the impregnation process, resulting in low catalyst utilization [45]. Zhang et al. [47] utilized an improved CCM technique by spray-coating a layer of ionomer on the membrane surface to enhance water electrolysis performance. The improved CCM technique offers characteristics such as low catalyst loading, large three-phase interface, and low resistance between the proton exchange membrane and the catalyst layer.

10.2   POROUS TRANSPORT LAYER

The PTL is situated between the catalyst layer and bipolar plates, facilitating the uniform distribution of water, gases, electrons, and heat. Additionally, it serves to connect the catalyst layer and the bipolar plate, providing support for the catalyst layer. To fulfill these functions effectively, the PTL must establish good contact with the plates and catalyst layer, possess suitable pore size and porosity, exhibit excellent electrical conductivity and mechanical strength, and demonstrate strong corrosion resistance in the highly oxidative environment of the anode [48]. The performance of the PTL directly impacts the electrolysis efficiency as it is responsible for transporting reactants and products in MEAs. Particularly, at the interface between the PTL and the catalyst layer, it has a significant influence on ohmic contact resistance, mass transport resistance, and charge transfer resistance.

10.2.1 MICROSTRUCTURE OF POROUS TRANSPORT LAYER

10.2.1.1 Pore and Porosity

The general requirements for the PTL include high porosity, high mechanical performance, low pore size, and low thickness [49]. A high-porosity electrode facilitates the removal of gas from the electrode surface, but it also increases the ohmic resistance and contact resistance at the interface of the electrolysis cell [50]. In theory, a higher porosity may require accepting higher ohmic and thermal resistances as a trade-off to facilitate gas transport and water supply [51]. For the pore size, the smaller pore size greatly improves the utilization rate of the catalyst, improves the uniformity of thermal and electrical distribution at the CL/PTL interface, reduces the contact resistance of the interface [52], and improves the water transport capacity through capillary action [53]. The larger pore diameter in the PTL often causes the flow pattern in the flow channel to flow from dispersed bubbles to slugs [51].

Ito et al. [54] found that using titanium felt with different porosities and pore sizes as a substrate, the electrolysis performance improved with decreasing pore size when the average pore size was larger than 10 μm. The larger bubbles formed by larger pore sizes tended to become elongated bubbles, which hindered water transport. However, when the porosity exceeded 50%, the variation in porosity did not have a significant impact on water electrolysis performance. Lopata et al. [55] analyzed the polarization curves of GDLs with different pore sizes and observed a separation in the Tafel slope. They found that as the average pore size increased, the planar conductivity decreased, leading to a decrease in catalyst utilization. This separation in Tafel slope was particularly pronounced at low catalyst loading.

Grigoriev et al. [50] compared GDLs made from sintered titanium powders of various sizes. Through testing, they concluded that the optimal titanium powder size before sintering was 50–75 μm. The thickness of the GDL was 1.3 mm, with a porosity of 37% and an average pore size diameter of 10.4 μm. The optimal range for porosity was found to be 30%–50%, and within this range, the impact of pore size and permeability on GDL performance was not significant. Mo et al. [56] developed a thin titanium GDL (25 μm) with well-controlled morphology using nanofabrication technology. As shown in Figure 10.7, this new design allows for adjustability in pore size, pore shape, pore distribution, porosity, and permeability. Additionally, the flat surface of the thin titanium GDL helps to reduce contact resistance at the PTL and catalyst layer interface.

The gradient study of pore distribution in PTL has attracted increasing attention. It has been found that improving water electrolysis performance can be achieved by

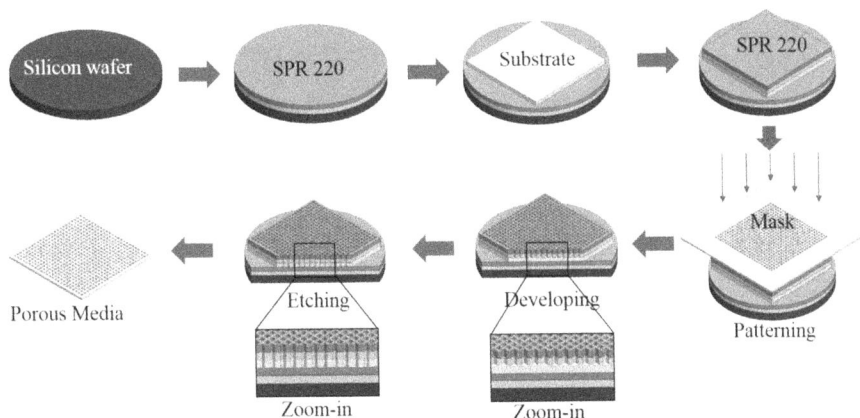

FIGURE 10.7   Typical fabrication process for titanium thin LGDLs.

increasing the porosity from the side closer to the catalyst layer to the side closer to the bipolar plate [57]. Lettenmeier et al. [58] prepared a novel PTL by using vacuum plasma spraying (VPS) to deposit a titanium layer on low carbon steel. This method allows for control of pore size, pore distribution, roughness, and thickness of the PTL by adjusting plasma parameters and titanium powder particle size. The pore size of this PTL shows a gradient distribution along the thickness, with a porosity of 20%–30%. Samples with a pore size distribution of 10 μm near the bipolar plate and 5 μm near the catalyst layer demonstrated optimal performance.

10.2.1.2   Two-Phase Flow

On the anode side, liquid water and oxygen are transported in opposite directions through the PTL. The accumulation of oxygen bubbles in the PTL can hinder the transport of liquid water toward the catalyst layer. This reduces the utilization efficiency of the catalyst and can lead to membrane electrode dehydration, potentially creating local hotspots [59,60]. Furthermore, if the bubbles are not promptly removed, they can form a barrier for mass transfer at the PTL/CL interface, leading to a decrease in water electrolysis performance [61]. In two-phase flow transport, the impedance also depends on the distance between the reaction sites and the pores in the PTL, which is determined by the average particle size of the PTL. Increasing the transport distance increases the mass transfer energy consumption and leads to an increase in gas volume in both the PTL and CL [55].

In the MEA, the formation of oxygen bubbles involves nucleation, growth, and detachment and are subsequently carried out of the electrolysis cell by liquid water [61,62]. Figure 10.8 [61] specifically illustrates these three processes. Selamet et al. [63] have visually demonstrated the two-phase flow in water electrolysis cells using neutron imaging and optical imaging techniques. They have revealed two typical behaviors of bubbles: (1) Periodic behavior, characterized by relatively rapid bubble growth, removal, and reappearance. (2) Stagnation, where bubbles cause long-term pore clogging.

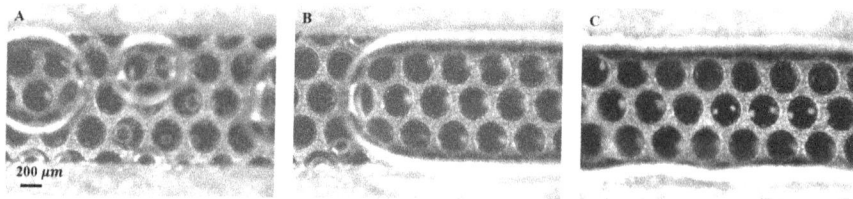

FIGURE 10.8 The formation of oxygen bubbles: (a) nucleation, (b) growth, and (c) detachment.

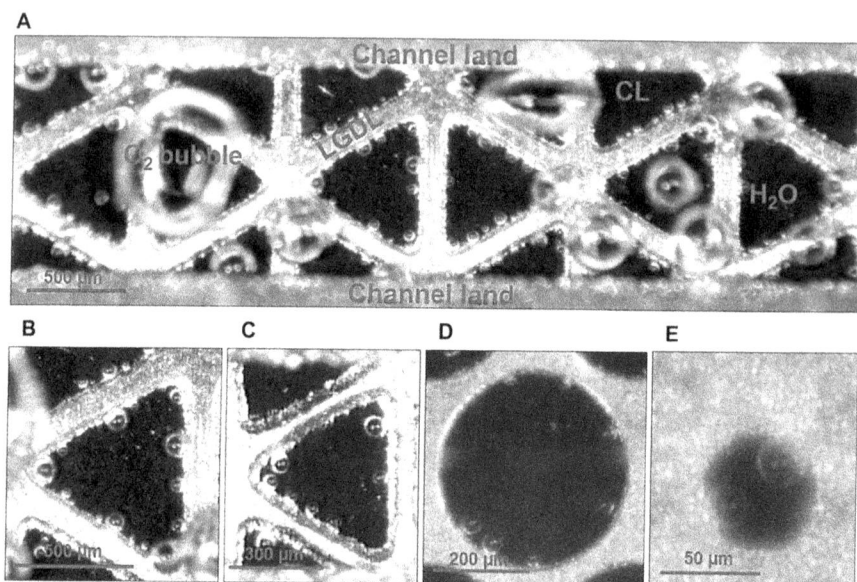

FIGURE 10.9 Microscale electrochemical reactions in PEMWE (bubbles represent the place where electrochemical reactions occur).

Mo et al. [31] observed that oxygen bubbles form on the surface of the catalyst layer (near the GDL) and exit through microchannels along with deionized water. Figure 10.9 shows the location of bubble generation. The sites where bubble generation occurs show preferences and are not uniformly distributed in the CLs. Most bubbles are generated along the edges of the pores. With an increase in current density, the number of active sites for bubble generation also increases. However, the detachment diameter of oxygen bubbles also increases, which may inhibit the OER.

It is difficult to clearly capture the transport mechanism of oxygen in PTL in experimental research of two-phase flow [64,65]. In contrast, numerical simulation can easily capture gas–liquid interfaces [66]. In recent years, the numerical simulation method has been widely used in the study of PTL two-phase flow. The commonly used methods mainly include volume of fluid (VOF) method, lattice Boltzmann method, pore grid modeling method, and phase field method [59,67–70].

However, most of the studies do not discuss the dynamic transport of oxygen in PTL, and there is a lack of systematic research on PTL structure [59]. Li et al. [59] developed a two-dimensional transient model of the anode two-phase flow in water electrolysis using the VOF method. They analyzed the transport mechanism of oxygen in the GDL and studied the effects of liquid water velocity, porosity, pore size, and contact angle on oxygen saturation and inlet pressure. In the PTL, the transport of oxygen is primarily influenced by capillary pressure. Increasing the inlet velocity facilitates bubble detachment, while increasing porosity, reducing fiber diameter, and contact angle promote oxygen transport. Arbabi et al. [71] simulated the electrolysis environment using different microfluidic chips as model structures for titanium felt, foam, and sintered powder. The results showed that even at high gas velocities, the dominant mechanism for bubble transport is capillary action. Additionally, gas is released along a path within the PTL that is independent of the water flow rate in the channels. Bubbles grow on the electrode surface until the buoyancy and shear forces acting on the bubbles exceed the adhesive forces, leading to their detachment. This implies that the growth and detachment of bubbles are influenced by the balance between these forces in the electrolysis system.

10.2.1.3 Porous Transport Layer/Catalytic Layer (PTL/CL) Interface

Understanding and improving the interface between PTL and catalyst layer is the key to improve the performance of PEMWE [72]. Mo et al. [73] used a new transparent proton exchange membrane water electrolysis, and combined with the high-speed microscale visualization system, conducted in situ research on OER. The experimental results indicate that OER only occurs at the PTL/CL interface. As the current density increases, the number of active sites at the pore edges increases.

Similar to two-phase flow, the study of PTL/CL interface is difficult to carry out under specific experimental operations, and numerical simulation is a good research method. Diaz et al. [74] first proposed the interfacial resistance of oxygen removal at the PTL/CL interface. This model considers the exchange of water between the anode and cathode, two-phase transport in a PTL, interfacial flow mechanics, gas coverage on the catalyst surface, proton conductivity in the membrane, and electrochemical reaction kinetics. They verified the experimental data reported in the literature and concluded that under the current density of 5 A/cm^2, the ohmic overpotential and the activated overpotential accounted for 52% and 38% of the voltage loss, respectively, and the oxygen component of PTL accounted for 52% of the pore space, which hindered the supply of liquid water to the catalyst layer and increased the material transport loss. The loss increased with the increase of current density. Kulkarni et al. [75] used X-ray computed tomography (CT) and modeling techniques to characterize the interface between the PTL and catalyst layer (CL) under different catalyst loadings. By quantifying the interface using X-rays and observing the oxygen transport within the channels through X-ray image, they connected the three-phase interfaces and addressed the issue of improving the existing interface. They also provided suggestions for improving the interface. For low-loading catalysts, the transport pathway of oxygen through the PTL is limited. The fibrous PTL has a higher porosity and a more uneven pore size distribution. Therefore, even with a high catalyst loading, the oxygen transport pathway is more favorable in the improved PTL. After sintering, the porosity of the PTL becomes more uniform and smaller, resulting in a more uniform distribution of oxygen in the PTL. They found that the low-porosity sintered PTL has better contact with the CL, which improves the electrolysis kinetics.

The contact of PTL/CL interface has a significant impact on the interface contact resistance and mass transport. Suermann et al. [76] modified the PTL/CL interface using femtosecond laser induction, resulting in a rough serrated surface with tips of several micrometers in diameter and depth, and covered by subtips of several hundred nanometers. By increasing the roughness of the interface, the specific surface area of PTL titanium-based fibers is increased to increase the contact area, which improves the mass transfer resistance, ohmic resistance, and contact resistance of PTL. The voltage of the electrolytic cell drops by about 30 mV after operating for 100 hours at the current density of 4 A/cm². Bernt et al. [48] believe that the conductivity between PTL and CL is also an important factor affecting PEMWE. Platinum coating on titanium PTL can minimize contact resistance and improve interfacial conductivity. In addition, IrO_x coating between PTL/CL will reduce the roughness of interface and optimize the performance of water electrolysis cell. For interface roughness, this seems to contradict the research of Suermann et al. Their purpose is to increase the contact area between interfaces and reduce contact resistance, but they adopt different methods. Peng et al. [72] have different opinions. They believe that if the contact area is too large, the PTL layer will be embedded in the catalyst layer. Although it will reduce the contact resistance, it will cause water shortage in the catalyst layer and reduce the utilization rate of the catalyst. The optimal contact area between PTL and catalyst layer should be between 29% and 40%.

10.2.2 POROUS TRANSPORT LAYER PRETREATMENT

10.2.2.1 Precious Metal Coatings

Titanium is a commonly used anode material in water electrolysis, but titanium is easy to passivate under the environment of high potential and high oxidation of anode, forming titanium oxide film, which will greatly increase the ohmic resistance of electrolytic cell. Coating a stable conductive material on the surface of PTL pretreatment can effectively protect it. Metal coatings are the most direct and effective protection, and they can be divided into three categories based on the type of coating [77]: (1) precious metal coatings, such as Au [78–81], Pt [48,79,80,82], IrO_2 [83], and Ir [84]; (2) metal oxides, such as TiO_4 [85,86]; and (3) metal nitrides, such as TiN [87].

Lin et al. [67] prepared a hydrophilic and corrosion-resistant conductive composite protective layer on the surface of titanium. They first obtained a polydopamine (PDA) film on the surface of titanium through solution oxidation, and then deposited trace amounts of Au nanoparticles on the PDA, effectively preparing an Au-PDA coating.

Fan et al. [88] coated titanium felt with a mixture of 0.43 mg/cm² TaO_x and 1 mg_{Ir+Ru}/cm² metal oxide TrO_2–RuO_2–TaO_x using pyrolysis method. This PTL has good conductivity, mass transfer performance, and OER catalytic activity, and its stability is significantly better than traditional electroplating platinum coatings. The voltage of the electrolytic cell is 1.836 V at a current density of 2 A/cm². The noble metal load of anode is 1 mg/cm^2, which is half of that of common electrolytic cell (about 2 mg/cm^2). Sung et al. [89] coated (IrO$_2$/Ta$_2$O$_5$) on the porous titanium disk as a microporous protective layer to prevent titanium from oxidation or corrosion during water electrolysis. IrO$_2$ effectively inhibits the reaction between oxygen generated by the anode and titanium, while Ta$_2$O$_5$ is an inert material that improves the stability of PTL. Within 600 hours, the voltage of the electrode was maintained between 2.33 and 2.35 V at a current density of 1 A/cm^2.

Ma et al. [90] coated the anode with Ir layer (0.5 mg/cm^2), which not only improved the performance of water electrolysis but also significantly reduced the attenuation of Nafion membrane. The EIS test found that after the membrane electrode operated at 1.6 V for 4 hours, the impedance of the uncoated membrane electrode increased twice as much as that of the coated membrane electrode. Yasutake et al. [91] deposited a layer of Ir-based particles (2–3 nm) on the surface of porous titanium sheets using arc plasma deposition (APD) and evaporation drying methods, respectively. The electrodes prepared by APD method exhibited higher OER activity. Liu et al. [84] uniformly sputtered a thin iridium layer of 20–150 nm on titanium-based PTL, preventing passivation of the PTL layer and reducing its contact resistance. Kang et al. [92] used the in situ test method to detect the change of resistance in the water electrolysis cell. They proved that the positive precious metal protective coating (Ir) of PTL can improve the contact of PTL interface, significantly reduce the anode resistance, and increase the life of the water electrolysis cell.

10.2.2.2   Acid Etching

Etching the PTL with acid can also achieve the effect of optimizing PTL performance. Bystron et al. [93] proposed a method for suppressing titanium passivation by etching. Put the pretreated titanium felt in 35% HCl at 45°C for 5 minutes. The surface roughness of the etched PTL increases, and titanium hydride compounds appear in the PTL after etching. This thin layer not only directly reduces the contact resistance at the interface but also largely inhibits the passivation of titanium, reducing the contact resistance at the interface of the electrolysis cell during the electrolysis process.

Cruz et al. [94] modified the porous titanium substrate with 0.1 M C$_2$H$_2$O$_4$. By characterizing the samples and calculating the porosity of different samples, compared with untreated titanium substrates, the pore size of the treated samples becomes smaller, which is conducive to the rapid diffusion of substances in PTL, improves the distribution rate of reactants, and avoids the accumulation of substances in the catalyst layer. According to SEM, the surface of the treated sample has high roughness, indicating that it has good adsorption on active sites. Many small particles can be observed, which may provide a higher surface area. The untreated PTL showed an increase in ohmic resistance. Combined with Energy Dispersive Spectrometer (EDS), it was found that a layer of oxide was formed on its surface, and the treated PTL had smaller contact resistance and ohmic resistance.

10.2.3 COMPOSITION OF POROUS TRANSPORT LAYER

10.2.3.1 Materials

According to the role played by the PTL in the MEA, the substrate material should have good thermal conductivity, appropriate mechanical strength, and flexibility. According to its environment, it should also have corrosion resistance and passivation resistance. In fuel cells, the GDL usually uses materials such as carbon cloth, carbon felt, and carbon fiber. However, in water electrolysis, the environment at the anode is characterized by high potential and strong oxidative conditions. So, the carbon usually has the following reaction: $C + 2H_2O \rightarrow CO_2 + 4H^+ + 4e^-$ [95]. Therefore, relatively expensive corrosion-resistant materials such as titanium are used [96]. Titanium powder sintering [49,50,57,97,98], titanium fiber [97–99], titanium felt [54,84,93,100], thin titanium [56,101], and titanium mesh [52,102] are mostly used for anode electrode. In the past decade, perforated titanium foil [56,81,103] and metal foam [104,105] have also been studied. Sintered porous titanium plates can achieve high performance [58], but limited by manufacturing processes, their thickness is generally greater than 0.6 mm and their porosity is less than 50%. The thickness of titanium felt is about 0.2 mm, and the porosity is about 80% [71]. The experiment shows that the water electrolysis performance of titanium felt is better than titanium mesh [52]. Currently, titanium felt is commonly used as the anode PTL in PEMWE.

Using carbon as the substrate material can effectively improve the conductivity of PTL. Becker et al. [96] believe that due to the low conductivity of deionized water, the corrosion area of the GDL mainly occurs at the PTL/CL interface. Protecting the interface area with a precious metal coating can effectively reduce the corrosion of the GDL, making it possible to use non-precious metals as substrate materials. Liu et al. [99] used carbon paper as the substrate and adopted an innovative method for preparing a microporous protective layer. The catalyst was first sprayed onto the membrane using CCM, followed by spraying iridium black, PTFE, and deionized water onto the carbon paper. Finally, the membrane electrode was obtained through thermal pressing. The electrode had a lifespan of 2000 hours at a current density of 1400 mA/cm². Filice et al. [106] utilized carbon fiber as a substrate and employed electrophoretic deposition to add anodic TiO_2 and cathodic Pt coatings into the PTL. Additionally, a thin Nafion hydrophilic layer was applied to the surface of the PTL to protect the catalyst nanoparticles and enhance their contact with the proton exchange membrane. The anodic TiO_2 coating exhibited positive effects on water electrolysis by increasing the hydrophilicity of the PTL, disrupting hydrogen bonding in water, facilitating PTL-membrane contact in a wet environment, and promoting the flow of O_2 within the PTL.

Stainless steel is also used as a material for PTL and bipolar plates, but it can cause severe corrosion, poisoning the membrane and catalyst layer with Fe and Ni [107–109]. Gago et al. [110] used VPS to prepare compact titanium coatings on stainless steel, and physical vapor deposition (PVD) was used to further modify the titanium coatings. It was tested in the electrolytic cell for 200 hours, and the average degradation rate was 26 μV/h. Stiber et al. [107] coated Nb/Ti coating on the GDL of stainless steel to obtain a high performance, durable, and low-cost water electrolysis tank. This coating can effectively promote the transportation of water and gas at the interface in contact with the anode. The electrochemical test of the electrolytic cell shows that the current density reaches 1.95 A/cm^2 under 2 V voltage. After testing the battery for 1000 hours, no trace of iron contamination was observed in the MEA.

10.2.3.2 Additives

Optimizing PTL by adding hydrophobic PTFE [4,111,112] has been extensively studied. Coating hydrophobic agents such as PTFE and ethylene fluoride (FEP) on the PTL can promote gas discharge, while also effectively reducing mass transport impedance and membrane hydration. However, due to the low conductivity of the hydrophobic agent, the ohmic impedance and contact impedance on the interface are improved, which has no positive impact on the improvement of MEA performance in water electrolysis.

Hwang et al. [113] studied the influence of PTFE content on the performance of GDL in the water electrolysis tank. The experiment proved that when the pore diameter of PTL is low (less than 20 µm), PTFE has no effect on water electrolysis performance. When the pore diameter of PTL is large (40–80 µm), PTFE interferes with water inflow, and water electrolysis performance decreases with the increase of PTFE content. Kang et al. [19] added PTFE to the PTL layer. Figure 10.10 shows the polarization distribution of electrolytic cells with different PTLs. From wetting to hydrophobic, the material diffusion loss, ohmic loss, and activation energy loss of the electrolytic cell increased. They further emphasized the detrimental effects of hydrophobic treatment on the PTL.

10.2.3.3 Operating Parameters of Porous Transport Layer

Under high current density conditions, mass transport losses become dominant. [114,115]. Lickert et al. [116] studied the performance variations of the PTL under different operating parameters such as water flow rate (0.2–0.8 L/min), temperature (40°C–80°C), and pressure (1–30 bar) at a current density of 5 A/cm^2. With the increase of temperature, the electrolysis voltage of water with the same density decreases. The water flow does not seem to contribute to water electrolysis, and the polarization curves obtained from the water flow of 0.2 and 0.8 L/min almost coincide. Pressure has little effect on water electrolysis under low current density

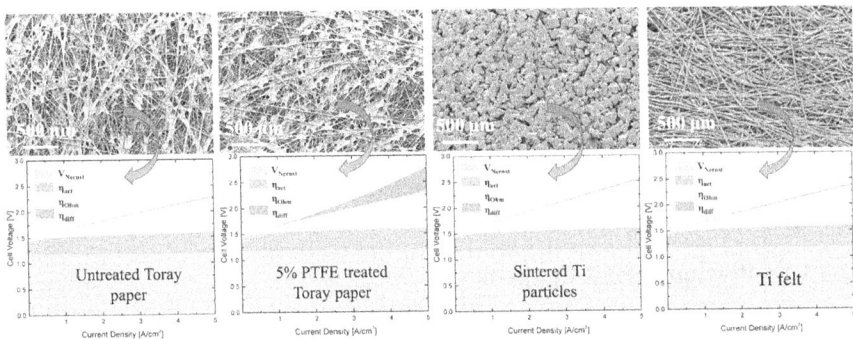

FIGURE 10.10 Polarization distribution of electrolytic cells with different PTLs.

(less than 2 A/cm²). When the current density is greater than 2 A/cm², high pressure may cause high voltage (different materials show different results), indicating that there is mass transport loss under high pressure. Finally, the author concludes that insufficient oxygen removal in anode PTL is the cause of mass transportation loss. However, with a current density within 5 A/cm² and sufficient water transportation, there will be no mass transportation loss. Generally speaking, the optimal flow rate is influenced by the size of the aperture and should increase with the increase of the average pore size [55].

10.3   CATALYTIC LAYER

The catalyst layer is a mixed porous structure consisting of catalyst particles and ionomers [117]. In PEMWE, electrochemical reactions occur only at the three-phase interface, which has electronic conductors, active catalysts, proton carriers, and pathways for mass transport [31]. In the MEA, electrons are conducted through the PTL to the catalyst layer, while protons are conducted through the polymer electrolyte membrane [118]. By optimizing the catalyst layer and improving the properties of the three-phase interface, it is possible to effectively reduce the cost of the electrolysis cell, enhance its performance, and prolong its lifespan. Currently, there are two main approaches to optimizing the catalyst layer: (1) Developing high-performance catalysts to improve their activity and the number of active sites. (2) Optimizing the structure of the catalyst layer to maximize the utilization of catalysts on the CL and reduce the impedance of the catalyst layer, including ohmic impedance, charge transfer impedance, and mass transport impedance. Generally, the conduction resistance of protons in ionomers is greater than that of electrons in the catalyst layer, so the three-phase interface is often near the proton exchange membrane side [119].

10.3.1   CATALYST

The use of high-load platinum group catalysts in the catalyst layer, such as platinum (Pt), iridium (Ir), and ruthenium (Ru), is an important reason to limit the development of PEMWE [120]. Iridium-based catalysts have high activity and stability [121,122], and they are the most widely used anode catalysts in PEMWE. The 1 MW power of the traditional PEMWE cell requires 1 kg iridium, while the global annual production of iridium is approximately 7 tons [123]. At present, people are committed to developing catalytic with low Ir load and high performance [20]. Spori et al. [124] believe that the iridium load should be less than 0.05 mg/cm², and the iridium consumption rate should be less than 0.01 g/kw. During the HER process, the required Pt catalyst content can be below 0.1 mg/cm² and will not affect the performance of the electrolytic cell [125]. However, in commercialized catalysts for OER, the loading of precious metal is typically in the range of 1–3 mg/cm² [126].

10.3.1.1   Precious Metal Catalysts

Iridium (Ir) [127,128], iridium oxide (IrO_2) [121,128–131], ruthenium oxide (RuO_2) [132,133], iridium ruthenium oxide ($IrRuO_2$) [134–139], and iridium ruthenium mixture (IrO_2–RuO_2) [140] were considered as good OER catalysts in PEMWE due to their superior catalytic activity and stability [139]. At present, the size of catalysts is almost always at the nanoscale, and nanoscale catalysts have a large specific surface area and volume, thus obtaining higher catalytic activity surface area [141]. Song et al. [35] studied the electrochemical performance and stability of Ru, Ir, RuO_2, IrO_2, and $Ru_{0.5}Ir_{0.5}O_2$ electrocatalysts through repeated cyclic voltammetry (CV), steady-state polarization curves, and stability experiments. IrO_2 shows higher stability.

10.3.1.2 Nanostructured Catalysts

Nanostructured porous metals possess unique three-dimensional porous networks, which exhibit high conductivity, a large electrochemical surface area, and highly active curved surfaces. They also have a self-supporting structure, eliminating the overpotential at the interface between the catalyst and the support [142]. However, the brittleness and structural flexibility of nanoporous metals limit their application in catalyst layers [143]. Wang et al. [144] prepared a nanowire-structured IrM (M = Ni, Co, Fe) catalyst using a eutectic oriented template. The catalyst exhibited a unique network structure of nanowires intertwined, greatly increasing the specific surface area of the catalyst. Chatterjee et al. [145] prepared nanosheet catalysts with nanopores through electrochemical dealloying using nickel iridium alloy as precursor.

10.3.1.3 Supported Catalysts

Catalyst supports (TiO_2 [24,146–148], TiC [149], SnO_2 [150–152], Ta_2O_5 [24], SiC [153], TaC [154], ATO [155], MnO_2 [156], etc.) have high specific surface area and stability, and are important aspects of catalyst research. Using supported catalysts can greatly decrease the loading of precious metals on the electrode [149]. It can also increase the number of crystallization centers during deposition, reduce the probability of catalyst agglomeration, and promote much uniform distribution within the MEA. The support and catalyst may exhibit synergistic effects, leading to enhanced catalytic activity [157]. Additionally, the presence of the support can effectively remove hydroxyl species from the catalyst surface [24]. However, the conductivity of stable support materials is often not high, as shown in Table 10.1. Rozai et al. [158] reported the applicability of small titanium particles as catalyst carriers on anodes and prepared IrO_2/Ti catalysts. They found through SEM and EIS that titanium particles on the surface of the catalyst layer can be embedded in the pores of the PTL, which is conducive to close contact between the catalyst and the PTL and reduces the ohmic resistance of the catalyst layer. The membrane electrode prepared using IrO_2/Ti catalyst has an IrO_2 loading capacity of only 0.1 mg/cm^2. The voltage of PEM electrolytic cell measured at 80°C and current density of 1 A/cm^2 is 1.72 V, and it can operate well and stably within 1000 hours, with the degradation rate of only 20 µV/h. They believe that optimizing the size of titanium particles can maximize the advantages of the catalyst.

10.3.1.4 Core–Shell Structure Catalysts

In order to further improve the utilization rate of catalysts, core–shell structured catalyst has been proposed [48]. Core–shell structure catalyst [162,163] is usually composed of a core (non-precious metal) and a shell (Ir), which reduces the Ir loading by utilizing most of the Ir atoms on the shell surface as reaction sites [164]. Tackett

TABLE 10.1
Conductivity of Common Carriers

Number	Supporter	Conductivity (S/cm)	Reference
1	ITO (In$_2$O$_3$, SnO$_2$)	>0.1	[159]
2	TiN	0.01–0.1	[159]
3	TiO$_2$	<0.01	[159]
4	Ti$_{0.7}$Ta$_{0.3}$O$_2$	0.0966	[24]
5	TixTayO$_2$	2.63×10^{-4} to 9.66×10^{-2}	[24]
6	IrO$_2$/ATO	8.36×10^{-3} to 4.8	[155]
7	ATO (antimony-doped tin oxide)	4.3×10^{-3}	[155]
8	ATO	8.2×10^{-2}	[160]
9	TO	5×10^{-4}	[160]
10	TaTO	17×10^{-4}	[160]
11	IrO$_2$	4.9	[155]
12	TaC	118	[154]
13	Si–SiC	1.8×10^{-5}	[153]
14	Sb–SnO$_2$	0.83	[150]
15	M-SnO$_2$(Sb,Nb,Ta,In)	>0.1	[161]

et al. [164] studied several transition metal (Fe, Co, Ni) nitrides as core of core–shell catalysts, and their Ir loading was reduced by about half. Pham et al. [165] prepared high-performance, low Ir loading (0.4 mg/cm^2) core–shell structures using TiO$_2$ as the core IrO$_2$@TiO$_2$ catalysts. In order to enhance the activity and stability of Ir and Ru catalysts, Shan et al. [166] prepared a synergistic Ru@IrO$_x$ catalyst. Nong et al. [163] investigated the local geometric ligand environment and electronic metal states of oxidized iridium cores in the Ni leaching catalyst IrNi@IrO$_x$. They found that the shell-layer oxidized iridium exhibited shorter Ir–O bonds and a significant number of d-band holes, which enhanced the activity of the catalyst.

10.3.2 CATALYTIC LAYER STRUCTURE

10.3.2.1 Catalytic Layer Additives

The content of ionomer (usually Nafion) has a significant impact on the performance of MEA [130,147,167], which is mainly attributed to changes in resistance between PTLs, catalyst layers, and membranes [147]. Adding Nafion ion solution in the catalyst layer can effectively improve the proton conductivity, catalytic activity, and mechanical stability of membrane electrode components [130]. Specifically, ionomers can facilitate proton conduction from the main body of the proton exchange membrane to the interior of the catalyst layer. They can also act as binders, providing a stable three-dimensional structure for the catalyst layer. Additionally, ionomers can serve as hydrophilic agents to maintain moisture in the catalyst layer [168]. However, if the ionomer content is too high, it can block mass transport channels within the catalyst layer and reduce electron conductivity. Xu et al. [168] analyzed the effect of ionomer content on membrane electrodes. At the anode, when the ionomer content increased from 5% to 25%, the area (charge capacity Qc) measured by CV gradually increased, which may be the reason for the increase in proton conductivity. On the contrary, when the ionomer increases from 25% to 40%, its area gradually decreases, which may be the reason for the decrease in electronic conductivity. Sapountzi et al. [130] sputtered IrO_2 on Ti/C and studied electrodes with different concentrations of Nafion. According to the CV, the addition of ionomers resulted in the expansion of the three-phase interface. When Nafion content is small, with the increase of Nafion load, the charge capacity (Qc) also increases, and the number of active sites at the electrode/electrolyte interface increases. The results indicate that the optimal content of Nafion is 1.5 mg/cm^2, which provides a good three-phase interface for the reactants. Bernt et al. [147] studied the influence of ionomer content on water electrolysis performance by quantifying the voltage loss caused by various resistances in the electrolytic cell. They think that the optimal content of ionomer is 11.6%, and the voltage of the electrolytic cell is 1.57 V under the current density of 1 A/cm^2. When the electrode contains lower ionomers, the proton transfer resistance dominates. When the ionomer content is higher, due to the electronic insulation of the ionomer and the filling of the electrode pores, the electron transfer and O_2 transfer resistance increase.

The catalyst layer prepared by traditional methods is mostly hydrophilic, and the most obvious drawback of this catalyst layer is the lack of oxygen transport channels, resulting in high oxygen transport resistance. Xu et al. [169] added PTFE to the traditional hydrophilic catalyst layer to prepare a hydrophilic and hydrophobic CCM electrode, which can effectively reduce the oxygen transmission resistance, and surface modified CCM can effectively improve the performance of water electrolysis.

Lagarteira et al. [26] compared cyclohexanol, propane-1,2-diol, and ethane-1,2-diol, and found that the highest porosity in the catalyst layer of cyclohexanol was 26% ± 2%, that of propane-1,2-diol was 20% ± 2%, and that of ethane-1,2-diol was 13% ± 2%. It is observed from the attached figure that cyclohexanol has the smallest adhesion area of 1280 ± 170 nm^2, indicating that its catalyst particles are most evenly distributed. Compared to cyclohexanol, the three have the smallest non-conductive region of about 82%, and its conductive network is better. The 3D morphology of the catalyst layer was observed, and the surface roughness of the catalyst layer prepared with cyclohexanol was at least 54 ± 2 nm. The current density of cyclohexanol is the highest at 1.7 V. In conclusion, it can be considered that cyclohexanol is a suitable solvent for preparing catalyst layer slurry. When preparing the catalyst layer, the thickness of the wet layer and the initial liquid phase composition will affect the cracking of the catalyst layer. Scheepers et al. [170] studied the effect of different contents of propanol on the formation of cracks in the catalyst layer. A higher thickness of the wet layer and a lower content of propanol will promote the formation of cracks in the catalyst layer. Adding propanol will reduce the surface tension of the dispersant and inhibit the formation of cracks. They believe that when the solvent mixture approaches the dynamic azeotropic point, the cracking behavior can be controlled by slight changes in the initial dispersant. Slightly changing the initial solvent content can affect the process of solvent composition during the drying process, thus having a strong impact on the formation of cracks.

10.3.2.2   Impact of Low Catalyst Loading

It is generally believed that when the catalyst load is too low (<0.1 mg/cm^2), there will be a lack of electron transport channels in the catalyst layer, which may lead to a decrease in MEA durability and catalyst utilization [117]. The interface resistance in contact with the catalyst layer will also increase [171], and the performance of water electrolysis will significantly decline [32]. However, the decrease in catalyst loading has little effect on the kinetics of electrocatalysis [172]. This may be because the electrochemical reaction only occurs at the CL/PTL interface, and a large amount of catalyst is not utilized [31]. Therefore, the decrease in catalyst loading has little effect on the electrocatalytic kinetics.

When the catalyst load decreases, the ohmic impedance will increase. Lopata et al. [55] found that as the catalyst loading decreases, the high-frequency resistance (HFR) significantly increases due to increased electron loss from the edge of the pores to the center of the pores. While this achieves a more efficient utilization of the catalyst, it also increases energy consumption. Moreover, a low catalyst loading leads to a decrease in the water permeability of the catalyst layer, making it difficult for water to access the reaction sites and for oxygen to leave the CL. Ma et al. [90] studied the influence of catalyst loading on water electrolysis with commercial iridium black (Johnson Matthey). According to V–I cycle curve, when the loading increased from 1 to 2 mg/cm^2, the electrochemical area and water electrolysis performance improved. They believed that the electronic contact area between the catalyst layer and the diffusion layer was not enough at low load, leading to high electronic resistance. When the catalyst load is too large, it will increase the ohmic resistance. When the catalyst is loaded with a high load (2.5 mg/cm^2), the electrolytic cell does not improve significantly, but the performance of water electrolysis decreases with the increase of voltage. The optimal performance of this membrane electrode is achieved when the anode is loaded with 1.5 mg/cm^2 catalyst and PTL is coated with a layer of 0.5 mg/cm^2 Ir, with a current voltage of 1.346 A/cm^2 and 1.8 V, respectively. Bernt et al. [173] analyzed the voltage loss of the water electrolysis with low platinum group load. Due to the extremely fast hydrogen evolution kinetics (HER) of cathode Pt, the reduction of cathode Pt/C catalyst load from 0.3 to 0.025 mg/cm^2 has little effect on the performance of water electrolysis cell. The optimal loading capacity of anode Ir is 1–2 mg/cm^2, and the thickness of the anode catalyst layer is about 4–8 μm. When the catalyst loading is too high and the thickness of the catalyst layer exceeds 10 μm, the ohmic impedance of the electrolysis cell increases. This is due to the increased water transport resistance through the thick catalyst layer, which leads to a decrease in the water content at the anode interface and a reduction in the membrane conductivity. On the other hand, when the catalyst loading is less than 0.5 mg/cm^2 and the catalyst layer thickness is less than 2 μm, the electrolysis cell performance drastically declines due to the unevenness and discontinuity of such thin catalyst layers. This can be improved by modifying the GDL with a microporous layer (MPL) to enhance the performance of the electrolysis cell.

Rozain et al. [172] indirectly reacted with the electrochemical surface area (ESCA) of the catalyst using the number of voltammetric charges (Q*). When the load is less than 1 mg/cm^2, Q* linearly increases with the increase of load, and the optimal load of the catalyst is between 1.5 and 2 mg/cm². Through SEM, they found that the pores in the catalyst layer were filled with ionic conductive polymers, and protons could not contact the internal active site of the catalyst, so the catalyst load had a threshold. Subsequently, they established a model to explain the relationship between catalyst load and Q*. They believed that there was a threshold of 0.5 mg/cm² for catalyst load, below which catalyst performance would significantly decrease. Above this threshold, battery voltage would not depend on catalyst load. When the current density of water electrolysis is 1 A/cm², the voltage is 1.72 V under IrO₂ load of 0.5 mg/cm². They believe that if we want to continue reducing the amount of catalyst, it is necessary to add a large area of conductive carrier to make all catalyst sites have electrochemical activity.

10.3.2.3 Conductivity of Catalytic Layer

An increase in electrode conductivity can improve OER kinetics, reduce battery ohmic resistance, and obtain more reaction sites [31,174]. Mo et al. conducted experimental observations using transparent visualized PEMWE and found that electrochemical reactions did not occur uniformly in CL, and most tree-phase boundaries (TPBs) do not play a role. The reason is that electrochemical reactions require not only TPB but also good electron conduction. They embedded conductive and non-conductive wires with a diameter of 50 μm into the catalyst layer and observed a significant generation of bubbles only around the conductive wires. This phenomenon was observed at different locations, leading them to believe that the conductivity of the catalyst layer acts as a threshold for electrochemical reactions.

Yang et al. [175] inserted the Au mesh between the catalyst layer and the membrane as the electron layer, which increased the interface conductivity by 4000 times and reduced the anode ohmic resistance to 1/3 of the conventional resistance. At the same time, carbon nanotube (CNT) coatings were also analyzed. CNT coatings have lower conductivity than Au coatings, but exhibit better electrochemical performance. The reason is that the CNT coating has a larger porosity and specific surface area, which increases the reaction sites and reaction area, while also improving the transportation performance of bubbles This indicates that conductivity and nanopore structure are important factors affecting conductivity. Hegge et al. [176] used electrospinning (equipment: IME Technologies) IrOₓ nanofiber technology to apply an intermediate layer of IrOₓ nanofiber with high conductivity on the traditional catalyst layer of IrOₓ nanoparticles, as shown in Figure 10.11. This novel design not only increases the performance and durability of the electrode, but also reduces the load of precious metals, with an iridium load of only 0.2 mg/cm².

However, the high conductivity of the catalyst layer may have adverse effects on proton transport. Kang et al. [119] used four wire sensing technology to obtain the internal voltage loss in PEMWE, especially analyzing the catalyst layer resistance. The results indicate that the catalyst layer resistance of the cathode and anode exhibits different values and trends, mainly due to the difference in conductivity between the two layers. The cathode uses Pt/C catalyst, and its conductivity is much higher than that of the anode. The OER active center in the anode increases with the increase of current density, making it more conducive to electron transport and inhibiting proton transport. Moreover, its active center tends to move toward the PTL layer, which is

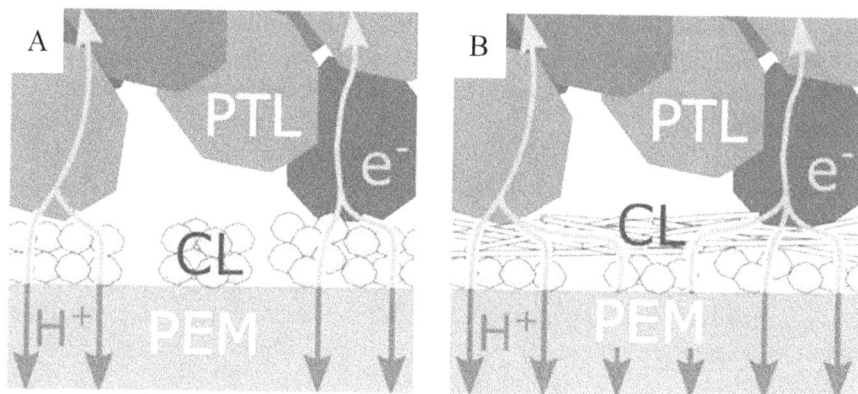

FIGURE 10.11 IrO_x nanofiber intermediate layer.

more unfavorable for proton transport. The current in the catalyst layer is the sum of the current provided by electron and proton transport. The suppression of proton transport at high current density results in an exponential increase in the ohmic resistance of the anode catalyst layer with current density.

10.3.3 PREPARATION OF CATALYTIC LAYER

10.3.3.1 Ordered Catalytic Layer

Ordered membrane electrode is the arrangement of mass transfer channels in the GDL in a regular manner, reducing the curvature of the transfer channels. This allows water, gas, and electrons to pass more directly through the GDL. Debe et al. [177] employed the technique of electron-beam evaporation of electrically insulating carriers onto an organic membrane under ultra-high vacuum conditions, followed by deposition of catalyst onto the carrier to form a continuous conductive film. This successful preparation method led to the development of ordered nanostructured thin film electrodes. The straight pores and ultrathin thickness of this film electrode facilitate efficient material transport.

Xu et al. [178] utilized polystyrene spheres as colloidal templates and employed tin-doped indium oxide (ITO) and proton-conductive phosphoric acid as raw materials to synthesize three-dimensional ordered array structure carriers with mixed proton and electron conduction capabilities. Initially, the precursor solution is poured onto the polystyrene template, followed by the addition of tin-doped indium oxide (ITO) to form 3-DOM ITO. Finally, proton-conductive phosphoric acid is added to create 3-DOM TIP-ITO. The resulting supported catalyst exhibits a high specific surface area (180 cm²/g), approximately a fivefold enhancement in OER activity with IrO_2 loading, and high stability. It can sustain stable operation for 1150 hours at a current density of 0.35 A/cm². Liu et al. [150] also used a similar method to prepare antimony-doped tin oxide nanowire support structure by using an expandable electrospinning method. The prepared Sb-SnO_2 NW structure support makes the catalyst

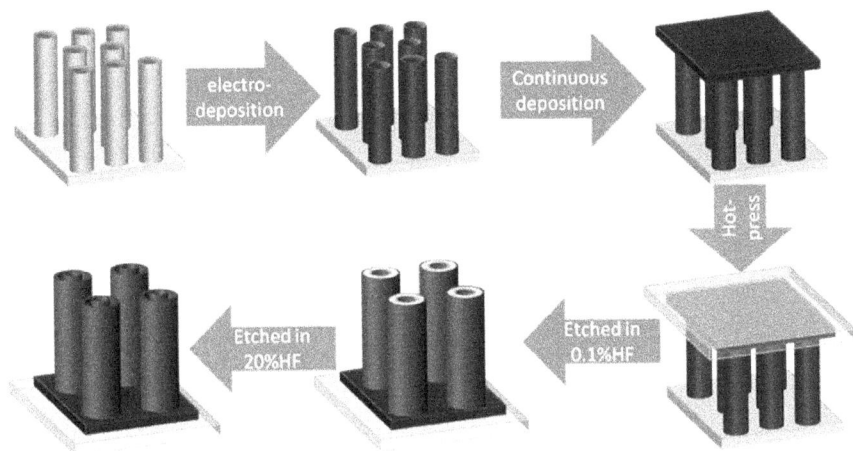

FIGURE 10.12 Preparation of catalyst ordered membrane electrode.

layer have a more uniform pore size distribution, improves the charge transfer rate, and has its catalyst activity nearly three times higher than that of pure IrO_2.

Lu et al. [179] synthesized vertically arranged IrO_x nanoarrays using titanium nanotemplates as substrates through electrodeposition, including three processes: electrodeposition, transfer, and template removal, as shown in Figure 10.12. The ordered membrane electrode of this catalyst greatly improves the ion transport capacity. Zeng et al. [38] constructed an efficient nanoporous ultrathin membrane electrode by embedding IrO_2 nanoparticles into nanoporous gold through simple dealloying and thermal decomposition methods. The electrode has three-dimensional interconnected nanopores and ultrathin thickness. The nanoporous ultrathin structure has no binder, which is conducive to improving the electrochemical surface area, enhancing the mass transfer, and promoting the release of oxygen in the water electrolysis process. This electrode has an extremely low noble metal loading, with IrO_2 and Au loading capacities of 0.086 and 0.1 mg/cm^2, respectively. The electrode has an electrolytic voltage of 1.728 V at a current density of 2 A/cm^2 at 80°C.

10.3.3.2 Non-Uniform Catalytic Layer

At present, almost all the studies believe that homogeneous catalyst layers should be prepared on the membrane or on the GDL substrate. Mo et al. [31] believe that electrochemical reactions do not occur uniformly in CL, and through experiments, it has been observed that most bubbles are generated along the edge of pores, so CL can only deposit at the edge of PTL pores. Kang et al. [118] argue that the uniformity of catalyst in PEMWE may not have a significant impact in PEMWE. They demonstrated a two-dimensional patterned electrode with edge effects, which indicate that when protons reach the membrane, they can migrate to more distant locations. This is shown in Figure 10.13. In terms of proton transport, their findings differ from Hegge et al. Through experimentation, it has been discovered that partially covered catalyst or patterned electrode structures offer several advantages, such as

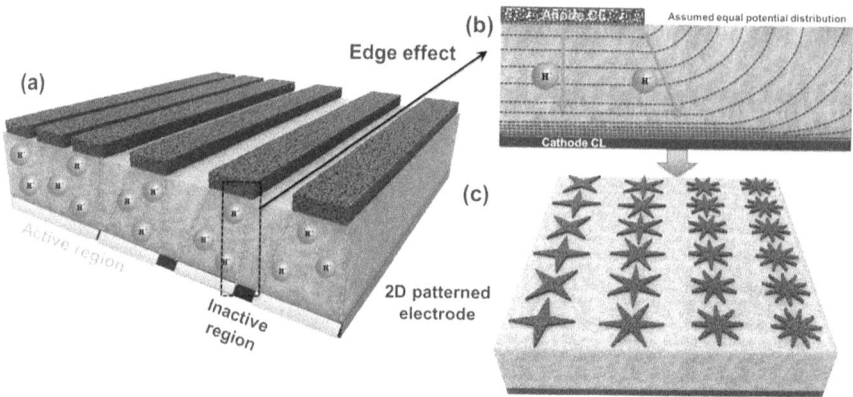

FIGURE 10.13   Membrane electrode with edge effects.

reducing catalyst loading, improving transport, adjusting the three-phase boundary, and enhancing catalyst utilization.

10.3.3.3   Alternate Catalytic Layer

Toshiba Corporation in Japan has developed an Alternating Catalyst layer Structure (ACLS) in proton exchange membrane fuel cells. This catalyst layer consists of multiple alternating platinum thin layers and gas gap layers. Due to the absence of carbon, this catalyst layer exhibits higher durability, with almost no voltage increase even after 7000 hours of operation [123]. Yoshinaga et al. [123] applied this technology to PEMWE and prepared a catalyst layer with alternating iridium thin films and gas gap layers. By forming porous metal targets and iridium targets, both materials are separately sputtered onto the membrane. This process is repeated for the desired number of sputtered layers, followed by removal of the porous metal to obtain the ACLS catalyst layer. The catalyst layer has extremely low iridium loading (approximately $0.2 mg/cm^2$) and noticeable gaps between iridium films, resulting in a higher porosity. When measured at a current density of $2 A/cm^2$, traditional MEAs prepared with low iridium content experience a sharp increase in voltage, while the ACLS MEA maintains stable voltage. This is because, with reduced catalyst amount, the active surface area decreases and the activation overpotential of the catalyst increases, but its overall effect on voltage is minimal. At this point, the concentration overpotential dominates, and due to its higher porosity, ACLS demonstrates better catalytic performance. Moreover, the electrode in the preparation process does not undergo noble metal treatment, reducing the usage of precious metals.

10.4   MEMBRANE ELECTRODE ATTENUATION

The expected life of commercial PEMWE should be more than 50,000 h [180]. Long-time operation will lead to attenuation of membrane electrode. The performance of a MEA is controlled by different degradation mechanisms. At a given current density, different losses can result in an increase in voltage, consequently affecting the performance of the MEA. The performance degradation of the MEA can mainly be attributed to several factors: (1) passivation or corrosion of the PTL on the anode side, (2) poisoning of the MEA by impurity cations, and (3) changes in the structure or dissolution of the catalyst layer. These factors can alter the morphology and internal structure of the MEA, leading to increase in internal ohmic resistance and decrease in electrochemical active surface area [181,182].

10.4.1 Porous Transport Layer Attenuation

10.4.1.1   Ti Passivation in Porous Transport Layers

In the process of water electrolysis, the anode high potential accelerates the passivation of Ti-based PTL [109], and the PTL passivation leads to the increase of ohmic resistance, so the anode needs a higher overpotential. At higher operating temperatures, titanium passivation significantly increased. Fouda Onana et al. [183] compared the passivation of PTLs at 60°C and 80°C. At 60°C, the titanium layer underwent passivation, but at a relatively slow rate. Furthermore, membrane thinning occurred during operation, leading to an overall decrease in overpotential. However, at 80°C, the passivation of the titanium layer accelerated. Frensch et al. [182] found that at different temperatures, the membrane thickness decreases with increasing operating time. However, they observed an increase in total ohmic resistance, leading to a decline in overall performance. They attributed this phenomenon to the passivation of the Ti plate. Rozai et al. [158] conducted detailed analysis on the performance degradation of membrane electrodes through aging experiments. They identified three main factors contributing to the degradation of membrane electrodes: (1) an increase in the ohmic resistance of the electrolyzer, (2) an increase in the anode charge transfer resistance, and (3) an increase in the capacitance value measured by electrochemical impedance spectroscopy. At the temperature of 60°C, the ohmic resistance remained relatively stable or even decreased, which might be attributed to membrane thinning. However, as the temperature increased, both the ohmic resistance and the anode charge transfer resistance increased, potentially due to titanium oxidation. When the voltage exceeded 2 V, the resistance increased at a higher rate.

10.4.1.2   Porous Transport Layer Corrosion

Rakousky et al. [181] conducted observations on samples using XRD and TEM. They found that there were no changes observed in the anode catalyst particles before and after operating the electrolysis. They also performed EDS analysis on the elements in the membrane electrode and did not detect impurity ions such as Ca, Na, and Fe. However, Ti, Ir, and Pt elements were found to be widely distributed in the membrane electrode. The difference was that Ir and Pt only diffused at their respective electrodes, while Ti diffused from the anode to the cathode. Therefore, they concluded that the corrosion of titanium resulted in performance degradation, where the mechanism involved a decrease in proton conductivity during the Ti diffusion process, leading to a significant decrease in the anode exchange current density. Subsequently, they coated Pt on the PTL surface and observed a significant reduction in degradation in the electrolyzer, indicating that Ti in the PTL was the source of degradation in the membrane electrode.

Steen et al. [184] observed a significant amount of iron oxide contaminants on the cathode electrode using EDX analysis. As the anode GDL was made of 316L stainless steel mesh and other components did not contain iron elements, iron was likely to originate from the anode. Testing of the membrane showed no signs of leakage, which indicates that iron elements migrated through the membrane and attached to the cathode GDL. Mo et al. [109] used 316 stainless steel mesh as the GDL. They discovered iron ions in the catalyst layer and membrane of the cathode and anode, and traced the path of iron ions from the anode to the cathode through the membrane using EDS. The results showed that iron ions mainly adhere to sulfonic functional groups on the Nafion electrolyte, which hinders proton transport and reduces electrolysis efficiency.

10.4.2 CATALYTIC LAYER ATTENUATION

10.4.2.1 Dissolution of Catalytic Layer

The dissolution of catalysts in the catalyst layer is slow, therefore it does not lead to sudden failure of the MEA [117]. The stability of a catalyst can be represented by changes in catalyst activity or by directly measuring the physicochemical properties of catalyst, such as dissolution or aggregation. Although the decline in catalyst activity is often associated with stability, it does not always reflect the actual stability of the catalyst. For example, if catalyst dissolution leads to the formation of a roughened surface that exposes more active sites, the decline in activity may not be observed. In such cases, stability loss can be masked. Additionally, if the initial catalyst dissolution yields a more active catalyst structure, the instability of the initial composition may manifest as an increase in activity [185]. Although the dissolution of catalyst may be beneficial to water electrolysis reaction in a short time [117], the catalyst layer will be destroyed if the dissolution process of catalyst layer continues. Spori et al. [185] provided a detailed review on the stability and degradation mechanism of OER catalysts in acidic environments. Since OER occurs through cyclic transitions between Ir^{4+}/Ir^{3+} states, it is currently widely accepted that electrochemical reactions and dissolution have a common intermediate (Ir^{3+}), and therefore, both processes occur simultaneously [186,187].

Cherevko et al. [188] compared Tafel slopes and dissolution rates. All studied metals exhibited both transient and steady-state dissolution, with transient dissolution occurring during oxide formation and reduction processes. Metals in oxide form, such as Ru and Au, which involve oxygen in the OER, showed lower Tafel slopes and higher dissolution rates. In contrast, metals such as Pt and Pd, which directly evolve oxygen from adsorbed water, exhibited higher Tafel slopes and lower dissolution rates.

In addition to the catalyst, the ionomer also degrades during the operation of the electrolytic cell. Park et al. [189] found in the fuel cell that applying voltage would accelerate the degradation of the ionomer and eventually lead to the collapse of the catalyst layer. In general, higher temperatures tend to lead to more severe degradation. The degradation of ionomers in a short period of time may also have a positive effect. Lettenmeier et al. [190] conducted AFM (atomic force microscopy) tests on the

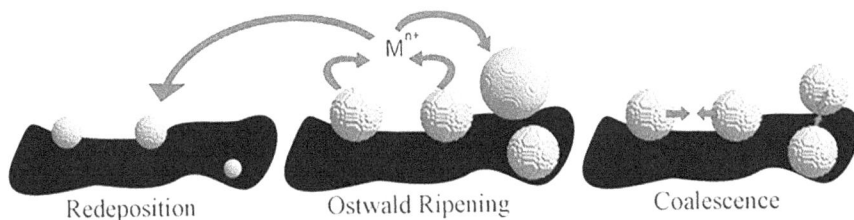

FIGURE 10.14   Three modes of particle growth.

surface of the catalyst layer and found that the conductive area of the anode increased by about 50%. According to image analysis, the catalyst is initially uniformly distributed but has a large non-conductive area. After operation, the catalyst agglomerates, but the ionomer loses and the conductivity of the catalyst layer increases.

10.4.2.2   Catalyst Agglomeration

Due to size reasons, nanoscale catalyst particles are inherently unstable. There are three main particle growth modes, as shown in Figure 10.14 [185]: (1) Ostwald growing, (2) redeposition, and (3) fusion. They can occur independently or simultaneously. The first two methods are based on dissolution. In the first method, dissolved ions re-deposit onto the existing particles, leading to an increase in the average particle size. The second method is similar to the first, with the difference being that aggregates form at different sites (not on existing particles), resulting in a wider size distribution and larger differences in the sizes of formed aggregates. The third method involves the coalescence of nearby particles, leading to an increase in the average particle size [185].

Alia et al. [191] found that at moderate voltages (1.5–1.6 V), only a small amount of Ir was observed in the electrolyte, indicating that particle aggregation may lead to a loss of electrochemical surface area. However, at higher potentials, a significant amount of Ir dissolved in the electrolyte. Claudel et al. [192] analyzed the degradation mechanism of catalysts using transmission electron microscopy, X-ray photoelectron spectroscopy, and electrochemical measurements. Cycling the catalysts (IrO_x) in the range of 1.2–1.6 V for 30,000 cycles resulted in a decrease in OER activity. Additionally, the coverage of hydroxyl groups and water increased during this cycling process. However, under the catalyst of IrO_2, the OER activity remained constant, and the coverage of hydroxyl groups and water remained unchanged. This suggests that the decrease in activity is a result of increased oxidation state of Ir, leading to an increase in the coverage of hydroxyl groups and water. TEM observations revealed significant Ir dissolution, redistribution and IrOx nanoparticle migration, aggregation, separation, which also contributed to the decline in OER activity.

10.4.2.3   Catalytic Layer Poisoning

In the process of water electrolysis, the catalyst layer may be polluted by cations of impurities (such as Ca^{2+}, Fe^{3+}, and Cu^{2+}), which may come from water supply, pipes and the preparation process of the stack [193]. The entry of impurity ions will occupy the catalytic sites and transport sites of protons, which not only increases the ohmic resistance of the electrolytic cell, but also affects the transport of protons, leading to an increase in the charge transfer resistance and a decrease in exchange current density [194]. In addition, the migration of impurity cations to the cathode can cause changes in the catalyst electrolyte interface, leading to an increase in the double layer capacitance, which is related to interface roughness, crystallinity, and anion adsorption [16]. The presence of certain impurities (such as fluorides) in the electrolyte or battery can also poison the catalyst layer, leading to deactivation or destruction of the catalyst support [186].

Sun et al. [195] conducted electrolyzer tests for 7800 hours, with an average decay rate of 35.5 µV/h. They investigated the impact of impurity cations on the MEA by comparing CV curves before and after operation. The post-operation CV curve exhibited expansion, and the total integrated charge increased. They believed that this was attributed to the oxidation of active Ir atoms in the catalyst layer, resulting in an increase in active sites for iridium oxidation. This indicates that the impurity ions did not occupy the active sites in the catalyst layer, and the performance decline was attributed to their occupation of the proton exchange sites in the catalyst layer. Through electron probe analysis, impurity cations, mainly Ca, Fe, and Cu (from other parts of the pipeline and stack), were found to occupy the proton exchange sites in the catalyst layer and proton exchange membrane, resulting in increased proton transfer resistance and ohmic resistance. Babic et al. [196] reached a similar conclusion, where an increase in impurity Fe^{3+} content in the cathode catalyst layer resulted in a doubling of the proton transport resistance. Additionally, impurity ions also caused a decrease in H^+ activity, leading to an increase in hydrogen evolution overpotential.

10.4.2.4 Catalytic Layer Detachment and Deformation

The swelling of membranes and the generation of bubbles can cause catalyst particles to detach from the catalyst layer, resulting in thinning of the catalyst layer [197]. Zeradjanin et al. [198] suggest that particle loss or detachment of the catalyst layer is caused by mechanical damage due to bubble formation or stress generated from blocked active sites that cannot participate in the reaction. Chanderis et al. [199] studied the membrane thinning process induced by oxygen crossover and temperature degradation. They observed that when the electrolysis cell operated at temperatures of 80°C or higher, the membrane underwent significant thinning (50% loss after 10,000 hours). The degradation and thinning of the membrane can decrease the adhesion of the catalyst layer, thereby inducing the detachment of the catalyst layer. Panchenko et al. [197] demonstrated a visual identification method for bubble formation within the catalyst layer. They observed that the detachment of iridium from the catalyst layer originated at the interface between the catalyst layer and the membrane, where intense gas evolution occurred, leading to catalyst detachment. Later, they also discovered oxygen bubbles at the interface between the catalyst layer and the GDL, and these oxygen bubbles became the driving force for catalyst separation. After 2 hours of operation, the catalyst loss rate in the CCM exceeded 60%, with most of the catalyst separating from the catalyst layer and no longer participating in electrochemical reactions. Additionally, as the operating time increased, the electrode surface blocked by bubbles did not participate in the reactions, resulting in an increase in current through the remaining active sites of the catalyst, causing higher stress on the active regions.

After a period of operation, catalyst particles in the electrolysis cell can diffuse toward the membrane and undergo reduction within the polymer [190]. Yu et al. [200] conducted a study on membrane electrodes with ultralow catalyst loading (0.3 mg/cm^2 Pt, 0.08 mg/cm^2 Ir). After 4800 hours of testing, only 30% of the iridium content remained in the anode. They discovered oxide stripes formed by iridium dissolution at the anode/membrane interface and observed iridium deposition on the cathode catalyst layer. Lettenmeier et al. [190] observed sample cross-sections using SEM and found a 5 μm interlayer between the catalyst layer and the membrane. They also observed small particles within the proton exchange membrane, which were not present in the unused MEA. Based on these findings, they concluded that this interlayer and the presence of small particles were likely caused by the diffusion of dissolved iridium.

The deformation of the anode catalyst layer is primarily caused by compression from the PEM electrolyte cell [117]. If the contact pressure is uneven, it can lead to mechanical damage of the catalyst layer, resulting in CL thinning and intrusion of polymer electrolyte into the porous layer of the PTL. This can increase the overpotential of the electrochemical reaction and impact the kinetics of the reaction [201]. When analyzing the catalyst layer at different temperatures, it was observed that as the temperature increases, the membrane's water uptake increases, leading to increased compression of the CL against the PTL. The deformation of the catalyst layer interrupts the proton transport pathway, resulting in increased proton transport resistance at the anode and higher interface resistance. Additionally, higher temperatures can cause cracks in the CL, further adding to the interface resistance [202].

10.5   SUMMARY AND OUTLOOK

MEA is the core component of PEMWE, and in-depth study of MEA is conducive to promoting the practical application of PEMWE. This review mainly summarizes the research and optimization of the PTL and catalyst layer in the anode of the membrane electrode, as well as the attenuation of the GDL and catalyst layer in the membrane electrode. Based on the review content, the following suggestions are proposed for the development of MEA:

1. The mass transport in PTL has an impact on the ohmic overpotential, concentration overpotential, and activated overpotential of the membrane electrode. By optimizing the microstructure of PTL, the mass transport capacity and interface contact of the membrane electrode can be effectively improved, as well as the performance of the membrane electrode.
2. Titanium passivation in PTL can increase the ohmic impedance, mass transport impedance, and activation impedance in MEA, which has a negative impact on the performance of MEA. It is necessary to strengthen research on coatings in order to avoid titanium passivation by applying stable and high-performance coatings.

3. Currently, the load of precious metal catalysts in membrane electrodes is relatively high. On the one hand, improving catalyst activity or introducing non-precious metal catalysts can reduce the content of precious metals. On the other hand, optimizing the catalyst layer structure can improve catalyst utilization.

4. After running the electrolytic cell for a period of time, the catalyst will detach from the catalyst layer and no longer function. Therefore, it is necessary to improve the adhesion of the catalyst layer on the membrane electrode, reduce catalyst detachment, and increase the lifespan of the membrane electrode.

REFERENCES

1. Kang, Z., et al., Investigation of thin/well-tunable liquid/gas diffusion layers exhibiting superior multifunctional performance in low-temperature electrolytic water splitting. *Energy & Environmental Science*, 2017. **10**(1): pp. 166–175.

2. Xu, Q., et al., Integrated reference electrodes in anion-exchange-membrane electrolyzers: Impact of stainless-steel gas-diffusion layers and internal mechanical pressure. *ACS Energy Letters*, 2020. **6**(2): pp. 305–312.

3. Majasan, J.O., et al., Two-phase flow behaviour and performance of polymer electrolyte membrane electrolysers: Electrochemical and optical characterisation. *International Journal of Hydrogen Energy*, 2018. **43**(33): pp. 15659–15672.

4. Eom, K., et al., Optimization of GDLs for high-performance PEMFC employing stainless steel bipolar plates. *International Journal of Hydrogen Energy*, 2013. **38**(14): pp. 6249–6260.

5. Maier, M., et al., Mass transport in PEM water electrolysers: A review. *International Journal of Hydrogen Energy*, 2022. **47**(1): pp. 30–56.

6. Schalenbach, M., et al., Pressurized PEM water electrolysis: Efficiency and gas crossover. *International Journal of Hydrogen Energy*, 2013. **38**(35): pp. 14921–14933.

7. Millet, P., *Hydrogen production by polymer electrolyte membrane water electrolysis*, *Compendium of Hydrogen Energy*. Woodhead Publishing, 2015. pp. 255–286.

8. Bessarabov, D. and P. Millet, *Fundamentals of water electrolysis*, In PEM Water Electrolysis. Academic Press, 2018. pp. 43–73.

9. Aouali, F.Z., et al., Analytical modelling and experimental validation of proton exchange membrane electrolyser for hydrogen production. *International Journal of Hydrogen Energy*, 2017. pp. 1366–1374.

10. Han, B., et al., Modeling of two-phase transport in proton exchange membrane electrolyzer cells for hydrogen energy. *International Journal of Hydrogen Energy*, 2017. **42**(7): pp. 4478–4489.

11. García-Valverde, R., N. Espinosa, and A. Urbina, Simple PEM water electrolyser model and experimental validation. *International Journal of Hydrogen Energy*, 2012. **37**(2): pp. 1927–1938.

12. Toghyani, S., et al., Thermal and electrochemical analysis of different flow field patterns in a PEM electrolyzer. *Electrochimica Acta*, 2018. **267**: pp. 234–245.

13. Yang, G., et al., Fully printed and integrated electrolyzer cells with additive manufacturing for high-efficiency water splitting. *Applied Energy*, 2018. **215**: pp. 202–210.

14. Aouali, F.Z., et al., Analytical modelling and experimental validation of proton exchange membrane electrolyser for hydrogen production. *International Journal of Hydrogen Energy*, 2017. **42**(2): pp. 1366–1374.

15. Toghyani, S., et al., Optimization of operating parameters of a polymer exchange membrane electrolyzer. *International Journal of Hydrogen Energy*, 2019. **44**(13): pp. 6403–6414.
16. Rozain, C. and P. Millet, Electrochemical characterization of polymer electrolyte membrane water electrolysis cells. *Electrochimica Acta*, 2014. **131**: pp. 160–167.
17. Rossmeisl, J., A. Logadottir, and J.K. Nørskov, Electrolysis of water on (oxidized) metal surfaces. *Chemical Physics*, 2005. **319**(1–3): pp. 178–184.
18. Aikens and A.J.J.o .C.E. D., Electrochemical methods, fundamentals and applications. *Journal of Chemical Education*, 2004. **60**(1): pp. 669–676.
19. Kang, Z., et al., Effects of various parameters of different porous transport layers in proton exchange membrane water electrolysis. *Electrochimica Acta*, 2020. **354**: p. 136641.
20. Saveleva, V.A., et al., Operando evidence for a universal oxygen evolution mechanism on thermal and electrochemical iridium oxides. *Journal of Physical Chemistry Letters*, 2018. **9**(11): pp. 3154–3160.
21. Buehler, M., et al., From catalyst coated membranes to porous transport electrode based configurations in PEM water electrolyzers. *Journal of the Electrochemical Society*, 2019. **166**(14): pp. F1070–F1078.
22. Rasten, E., G. Hagen, and R. Tunold, Electrocatalysis in water electrolysis with solid polymer electrolyte. *Electrochimica acta*, 2003. **48**(25–26): pp. 3945–3952.
23. Selvarani, G., et al., A phenyl-sulfonic acid anchored carbon-supported platinum catalyst for polymer electrolyte fuel cell electrodes. *Electrochimica Acta*, 2007. **52**(15): pp. 4871–4877.
24. Lv, H., et al., Activity of IrO_2 supported on tantalum-doped TiO_2 electrocatalyst for solid polymer electrolyte water electrolyzer. *RSC Advances*, 2017. **7**(64): pp. 40427–40436.
25. Chang-Soo, K., et al., Method for fabricating membrane and electrode assembly for polymer electrolyte membrane fuel cells. United States Patent US6180276. 2001, US.
26. Lagarteira, T., et al., Highly active screen-printed $Ir-Ti_4O_7$ anodes for proton exchange membrane electrolyzers. *International Journal of Hydrogen Energy*, 2018. **43**(35): pp. 16824–16833.
27. Choe, S., et al., Electrodeposited IrO_2/Ti electrodes as durable and cost-effective anodes in high-temperature polymer-membrane-electrolyte water electrolyzers. *Applied Catalysis B-Environmental*, 2018. **226**: pp. 289–294.
28. Siracusano, S., et al., Degradation issues of PEM electrolysis MEAs. *Renewable Energy*, 2018. **123**: pp. 52–57.
29. Buehler, M., et al., Optimization of anodic porous transport electrodes for proton exchange membrane water electrolyzers. *Journal of Materials Chemistry A*, 2019. **7**(47): pp. 26984–26995.
30. Labou, D., et al., Performance of laboratory polymer electrolyte membrane hydrogen generator with sputtered iridium oxide anode. *Journal of Power Sources*, 2008. **185**(2): pp. 1073–1078.
31. Mo, J., et al., Discovery of true electrochemical reactions for ultrahigh catalyst mass activity in water splitting. *Science Advances*, 2016. **2**(11): pp.2375–2548.
32. Lee, B.-S., et al., Development of electrodeposited IrO_2 electrodes as anodes in polymer electrolyte membrane water electrolysis. *Applied Catalysis B: Environmental*, 2015. **179**: pp. 285–291.
33. Holzapfel, P., et al., Directly coated membrane electrode assemblies for proton exchange membrane water electrolysis. *Electrochemistry Communications*, 2020. **110**: p. 106640.
34. Marshall, A., et al., Electrochemical characterisation of $Ir_xSn_{1-x}O_2$ powders as oxygen evolution electrocatalysts. *Electrochim Acta*, 2006. **51**(15): pp. 3161–3167.

35. Song, S., et al., Electrochemical investigation of electrocatalysts for the oxygen evolution reaction in PEM water electrolyzers. *International Journal of Hydrogen Energy*, 2008. **33**(19): pp. 4955–4961.

36. Leonard, E., et al., Interfacial analysis of a PEM electrolyzer using X-ray computed tomography. *Sustainable Energy & Fuels*, 2020. **4**(2): pp. 921–931.

37. Carmo, M., et al., A comprehensive review on PEM water electrolysis. *International Journal of Hydrogen Energy*, 2013. **38**(12): pp. 4901–4934.

38. Zeng, Y., et al., A cost-effective nanoporous ultrathin film electrode based on nanoporous gold/IrO_2 composite for proton exchange membrane water electrolysis. *Journal of Power Sources*, 2017. **342**: pp. 947–955.

39. Mamaca, N., et al., Electrochemical activity of ruthenium and iridium based catalysts for oxygen evolution reaction. *Applied Catalysis B-Environmental*, 2012. **111**: pp. 376–380.

40. Kokoh, K.B., et al., Efficient multi-metallic anode catalysts in a PEM water electrolyzer. *International Journal of Hydrogen Energy*, 2014. **39**(5): pp. 1924–1931.

41. Kus, P., et al., Magnetron sputtered Ir thin film on TiC-based support sublayer as low-loading anode catalyst for proton exchange membrane water electrolysis. *International Journal of Hydrogen Energy*, 2016. **41**(34): pp. 15124–15132.

42. Park, J.E., et al., Ultra-low loading of IrO_2 with an inverse-opal structure in a polymer-exchange membrane water electrolysis. *Nano Energy*, 2019. **58**: pp. 158–166.

43. Park, J., et al., Roll-to-roll production of catalyst coated membranes for low-temperature electrolyzers. *Journal of Power Sources*, 2020. **479**: p. 228819.

44. Su, H., et al., Study of catalyst sprayed membrane under irradiation method to prepare high performance membrane electrode assemblies for solid polymer electrolyte water electrolysis. *International Journal of Hydrogen Energy*, 2011. **36**(23): pp. 15081–15088.

45. Shi, Y., et al., Fabrication of membrane electrode assemblies by direct spray catalyst on water swollen Nafion membrane for PEM water electrolysis. *International Journal of Hydrogen Energy*, 2017. **42**(42): pp. 26183–26191.

46. Pethaiah, S.S., et al., Evaluation of platinum catalyzed MEAs for PEM fuel cell applications. *Solid State Ionics*, 2011. **190**(1): pp. 88–92.

47. Zhang, Y., et al., Study on a novel manufacturing process of membrane electrode assemblies for solid polymer electrolyte water electrolysis. *Electrochemistry Communications*, 2007. **9**(4): pp. 667–670.

48. Bernt, M., et al., Effect of the IrOx conductivity on the anode electrode/porous transport layer interfacial resistance in PEM water electrolyzers. *Journal of the Electrochemical Society*, 2021. **168**(8): acleb4.

49. Borgardt, E., et al., Mechanical characterization and durability of sintered porous transport layers for polymer electrolyte membrane electrolysis. *Journal of Power Sources*, 2018. **374**: pp. 84–91.

50. Grigoriev, S.A., et al., Optimization of porous current collectors for PEM water electrolysers. *International Journal of Hydrogen Energy*, 2009. **34**(11): pp. 4968–4973.

51. Babic, U., et al., Critical review-identifying critical gaps for polymer electrolyte water electrolysis development. *Journal of the Electrochemical Society*, 2017. **164**(4): pp. F387–F399.

52. Steen, S.M., III, et al., Investigation of titanium liquid/gas diffusion layers in proton exchange membrane electrolyzer cells. *International Journal of Green Energy*, 2017. **14**(2): pp. 162–170.

53. Majasan, J.O., et al. *Effect of Microstructure of Porous Transport Layer on Performance in Polymer Electrolyte Membrane Water Electrolyser.* In *3rd Annual Conference on Energy Storage and its Applications (CDT-ESA-AC).* 2018. Univ Sheffield, Engn & Phys Sci Res Council Ctr Doctoral Training Energy S, Sheffield, England.

54. Ito, H., et al., Experimental study on porous current collectors of PEM electrolyzers. *International Journal of Hydrogen Energy*, 2012. **37**(9): pp. 7418–7428.
55. Lopata, J., et al., Effects of the transport/catalyst layer interface and catalyst loading on mass and charge transport phenomena in polymer electrolyte membrane water electrolysis devices. *Journal of the Electrochemical Society*, 2020. **167**(6): p. 064507.
56. Mo, J.K., et al., Thin liquid/gas diffusion layers for high-efficiency hydrogen production from water splitting. *Applied Energy*, 2016. **177**: pp. 817–822.
57. Omrani, R. and B. Shabani, Gas diffusion layer modifications and treatments for improving the performance of proton exchange membrane fuel cells and electrolysers: A review. *International Journal of Hydrogen Energy*, 2017. **42**(47): pp. 28515–28536.
58. Lettenmeier, P., et al., Comprehensive investigation of novel pore-graded gas diffusion layers for high-performance and cost-effective proton exchange membrane electrolyzers. *Energy & Environmental Science*, 2017. **10**(12): pp. 2521–2533.
59. Li, Q., et al., Two-dimensional numerical pore-scale investigation of oxygen evolution in proton exchange membrane electrolysis cells. *International Journal of Hydrogen Energy*, 2022. **47**(37): pp. 16335–16346.
60. Millet, P., et al., Cell failure mechanisms in PEM water electrolyzers. *International Journal of Hydrogen Energy*, 2012. **37**(22): pp. 17478–17487.
61. Li, Y., et al., In-situ investigation of bubble dynamics and two-phase flow in proton exchange membrane electrolyzer cells. *International Journal of Hydrogen Energy*, 2018. **43**(24): pp. 11223–11233.
62. Wüthrich, R., C. Comninellis, and H. Bleuler, Bubble evolution on vertical electrodes under extreme current densities. *Electrochimica Acta*, 2005. **50**(25): pp. 5242–5246.
63. Selamet, O.F., et al., Two-phase flow in a proton exchange membrane electrolyzer visualized in situ by simultaneous neutron radiography and optical imaging. *International Journal of Hydrogen Energy*, 2013. **38**(14): pp. 5823–5835.
64. Seweryn, J., et al., Communication-neutron radiography of the water/gas distribution in the porous layers of an operating electrolyser. *Journal of the Electrochemical Society*, 2016. **163**(11): pp. F3009–F3011.
65. Park, J., et al., Neutron imaging investigation of liquid water distribution in and the performance of a PEM fuel cell. *International Journal of Hydrogen Energy*, 2008. **33**(13): pp. 3373–3384.
66. Iulianelli, A., K. Ghasemzadeh, and A. Basile, Progress in methanol steam reforming modelling via membrane reactors technology. *Membranes*, 2018. **8**(3): p. 65.
67. Lin, N., S. Feng, and J. Wang, Multiphysics modeling of proton exchange membrane water electrolysis: From steady to dynamic behavior. *Aiche Journal*, 2022. **68**(8): p. e17742.
68. Han, B., et al., Effects of membrane electrode assembly properties on two-phase transport and performance in proton exchange membrane electrolyzer cells. *Electrochimica Acta*, 2016. **188**: pp. 317–326.
69. Lafmejani, S.S., A.C. Olesen, and S.K. Kaer, VOF modelling of gas-liquid flow in PEM water electrolysis cell micro-channels. *International Journal of Hydrogen Energy*, 2017. **42**(26): pp. 16333–16344.
70. Nouri-Khorasani, A., et al., Model of oxygen bubbles and performance impact in the porous transport layer of PEM water electrolysis cells. *International Journal of Hydrogen Energy*, 2017. **42**(8): pp. 28665–28680.
71. Arbabi, F., et al., Feasibility study of using microfluidic platforms for visualizing bubble flows in electrolyzer gas diffusion layers. *Journal of Power Sources*, 2014. **258**: pp. 142–149.
72. Peng, X., et al., Insights into interfacial and bulk transport phenomena affecting proton exchange membrane water electrolyzer performance at ultra-low iridium loadings. *Advanced Science (Weinheim)*, 2021. **8**(21): p. e2102950.

73. Mo, J., et al., In situ investigation on ultrafast oxygen evolution reactions of water splitting in proton exchange membrane electrolyzer cells. *Journal of Materials Chemistry A*, 2017. **5**(35): pp. 18469–18475.
74. Diaz, D.F.R., E. Valenzuela, and Y. Wang, A component-level model of polymer electrolyte membrane electrolysis cells for hydrogen production. *Applied Energy*, 2022. **321**: p. 119398.
75. Kulkarni, D., et al., Elucidating effects of catalyst loadings and porous transport layer morphologies on operation of proton exchange membrane water electrolyzers. *Applied Catalysis B-Environmental*, 2022. **308**: p. 121213.
76. Suermann, M., et al., Femtosecond laser-induced surface structuring of the porous transport layers in proton exchange membrane water electrolysis. *Journal of Materials Chemistry A*, 2020. **8**(9): pp. 4898–4910.
77. Liu, Y., et al., Novel Au nanoparticles-inlaid titanium paper for PEM water electrolysis with enhanced interfacial electrical conductivity. *International Journal of Minerals Metallurgy and Materials*, 2022. **29**(5): pp. 1090–1098.
78. Jung, H.-Y., et al., Performance of gold-coated titanium bipolar plates in unitized regenerative fuel cell operation. *Journal of Power Sources*, 2009. **194**(2): pp. 972–975.
79. Wang, S.-H., J. Peng, and W.-B. Lui, Surface modification and development of titanium bipolar plates for PEM fuel cells. *Journal of Power Sources*, 2006. **160**(1): pp. 485–489.
80. Hwang, M.-J., et al., Characterization of passive layers formed on Ti-10wt% (Ag, Au, Pd, or Pt) binary alloys and their effects on galvanic corrosion. *Corrosion Science*, 2015. **96**: pp. 152–159.
81. Kang, Z., et al., Thin film surface modifications of thin/tunable liquid/gas diffusion layers for high-efficiency proton exchange membrane electrolyzer cells. *Applied Energy*, 2017. **206**: pp. 983–990.
82. He, Z., et al., Electrode materials for vanadium redox flow batteries: Intrinsic treatment and introducing catalyst. *Chemical Engineering Journal (Lausanne, Switzerland: 1996)*, 2022. **427**: p. 131680.
83. Wang, S.-H., et al., Performance of the iridium oxide (IrO$_2$)-modified titanium bipolar plates for the light weight proton exchange membrane fuel cells. *Journal of Fuel Cell Science and Technology*, 2013. **10**(4): p. 041002.
84. Liu, C., et al., Performance enhancement of PEM electrolyzers through iridium-coated titanium porous transport layers. *Electrochemistry Communications*, 2018. **97**: pp. 96–99.
85. Wakayama, H. and K. Yamazaki, Low-cost bipolar plates of Ti$_4$O$_7$-coated Ti for water electrolysis with polymer electrolyte membranes. *ACS Omega*, 2021. **6**(6): pp. 4161–4166.
86. Chen, Y.-z., et al., Anodized metal oxide nanostructures for photoelectrochemical water splitting. *International Journal of Minerals, Metallurgy and Materials*, 2020. **27**(5): pp. 584–601.
87. Toops, T.J., et al., Evaluation of nitrided titanium separator plates for proton exchange membrane electrolyzer cells. *Journal of Power Sources*, 2014. **272**: pp. 954–960.
88. Fan, Z., et al., Low precious metal loading porous transport layer coating and anode catalyst layer for proton exchange membrane water electrolysis. *International Journal of Hydrogen Energy*, 2022. **47**(44): pp. 18963–18971.
89. Sung, C.-C. and C.-Y. Liu, A novel micro protective layer applied on a simplified PEM water electrolyser. *International Journal of Hydrogen Energy*, 2013. **38**(24): pp. 10063–10067.
90. Ma, L., S. Sui, and Y. Zhai, Investigations on high performance proton exchange membrane water electrolyzer. *International Journal of Hydrogen Energy*, 2009. **34**(2): pp. 678–684.

91. Yasutake, M., et al., GDL-integrated electrodes with Ir-based electrocatalysts for polymer electrolyte membrane water electrolysis. *ECS Transactions*, 2019. **92**(8): pp. 833–843.

92. Kang, Z., et al., In-situ and in-operando analysis of voltage losses using sense wires for proton exchange membrane water electrolyzers. *Journal of Power Sources*, 2021. **481**: pp. 229012.

93. Bystron, T., et al., Enhancing PEM water electrolysis efficiency by reducing the extent of Ti gas diffusion layer passivation. *Journal of Applied Electrochemistry*, 2018. **48**(6): pp. 713–723.

94. Cruz, J.C., et al., Electrochemical and microstructural analysis of a modified gas diffusion layer for a PEM water electrolyzer. *International Journal of Electrochemical Science*, 2020. **15**(6): pp. 5571–5584.

95. Doan, T.L., et al., A review of the porous transport layer in polymer electrolyte membrane water electrolysis. *International Journal of Energy Research*, 2021. **45**(10): pp. 14207–14220.

96. Becker, H., L. Castanheira, and G. Hinds, Local measurement of current collector potential in a polymer electrolyte membrane water electrolyser. *Journal of Power Sources*, 2020. **448**: p. 227563.

97. Panchenko, O., et al., In-situ two-phase flow investigation of different porous transport layer for a polymer electrolyte membrane (PEM) electrolyzer with neutron spectroscopy. *Journal of Power Sources*, 2018. **390**: pp. 108–115.

98. Schuler, T., et al., Polymer electrolyte water electrolysis: correlating porous transport layer structural properties and performance: Part I. Tomographic analysis of morphology and topology. *Journal of the Electrochemical Society*, 2019. **166**(4): pp. F270–F281.

99. Liu, C.-Y., L.-H. Hu, and C.-C. Sung, Micro-protective layer for lifetime extension of solid polymer electrolyte water electrolysis. *Journal of Power Sources*, 2012. **207**: pp. 81–85.

100. Ito, H., et al., Influence of pore structural properties of current collectors on the performance of proton exchange membrane electrolyzer. *Electrochimica Acta*, 2013. **100**: pp. 242–248.

101. Mo, J., et al., Mask-patterned wet etching of thin titanium liquid/gas diffusion layers for a PEMEC. *ECS Transactions*, 2015. **66**(24): pp. 3–10.

102. Li, H., et al., Optimum structural properties for an anode current collector used in a polymer electrolyte membrane water electrolyzer operated at the boiling point of water. *Journal of Power Sources*, 2016. **332**: pp. 16–23.

103. Kang, Z., et al. Micro/nano manufacturing of novel multifunctional layers for hydrogen production from water splitting. In *2017 IEEE 12th International Conference on Nano/Micro Engineered and Molecular Systems (NEMS)*. 2017.

104. Baumann, N., et al., Membrane electrode assemblies for water electrolysis using WO_3-supported $Ir_xRu_{1-x}O_2$ catalysts. *Energy Technology*, 2016. **4**(1): pp. 212–220.

105. Baroutaji, A., et al., Application of open pore cellular foam for air breathing PEM fuel cell. *International Journal of Hydrogen Energy*, 2017. **42**(40): pp. 25630–25638.

106. Filice, S., et al., Applicability of a new sulfonated pentablock copolymer membrane and modified gas diffusion layers for low-cost water splitting processes. *Energies*, 2019. **12**(11): p. 2064.

107. Stiber, S., et al., A high-performance, durable and low-cost proton exchange membrane electrolyser with stainless steel components. *Energy & Environmental Science*, 2022. **15**(1): pp. 109–122.

108. Mo, J., et al., Electrochemical investigation of stainless steel corrosion in a proton exchange membrane electrolyzer cell. *International Journal of Hydrogen Energy*, 2015. **40**(36): pp. 12506–12511.

109. Mo, J., et al., Study on corrosion migrations within catalyst-coated membranes of proton exchange membrane electrolyzer cells. *International Journal of Hydrogen Energy*, 2017. **42**(44): pp. 27343–27349.

110. Gago, A.S., et al., Protective coatings on stainless steel bipolar plates for proton exchange membrane (PEM) electrolysers. *Journal of Power Sources*, 2016. **307**: pp. 815–825.

111. Mendoza, A.J., et al., Raman spectroscopic mapping of the carbon and PTFE distribution in gas diffusion layers. *Fuel Cells*, 2011. **11**(2): pp. 248–254.

112. Chun, J.H., et al., Numerical modeling and experimental study of the influence of GDL properties on performance in a PEMFC. *International Journal of Hydrogen Energy*, 2011. **36**(2): pp. 1837–1845.

113. Hwang, C.M., et al., Influence of properties of gas diffusion layers on the performance of polymer electrolyte-based unitized reversible fuel cells. *International Journal of Hydrogen Energy*, 2011. **36**(2): pp. 1740–1753.

114. Suermann, M., T.J. Schmidt, and F.N. Büchi, Investigation of mass transport losses in polymer electrolyte electrolysis cells. *ECS Transactions*, 2015. **69**(17): pp. 1141–1148.

115. Suermann, M., T.J. Schmidt, and F.N. Büchi, Cell performance determining parameters in high pressure water electrolysis. *Electrochimica Acta*, 2016. **211**: pp. 989–997.

116. Lickert, T., et al., On the influence of the anodic porous transport layer on PEM electrolysis performance at high current densities. *International Journal of Hydrogen Energy*, 2020. **45**(11): pp. 6047–6058.

117. Feng, Q., et al., A review of proton exchange membrane water electrolysis on degradation mechanisms and mitigation strategies. *Journal of Power Sources*, 2017. **366**: pp. 33–55.

118. Kang, Z., et al., Discovering and demonstrating a novel high-performing 2D-patterned electrode for proton-exchange membrane water electrolysis devices. *ACS Applied Materials & Interfaces*, 2022. **14**(1): pp. 2335–2342.

119. Kang, Z., et al., Exploring and understanding the internal voltage losses through catalyst layers in proton exchange membrane water electrolysis devices. *Applied Energy*, 2022. **317**: p. 119213.

120. Kang, Z., M. Pak, and G. Bender, Introducing a novel technique for measuring hydrogen crossover in membrane-based electrochemical cells. *International Journal of Hydrogen Energy*, 2021. **46**(29): pp. 15161–15167.

121. Zhao, S., et al., Determining the electrochemically active area of IrOx powder catalysts in an operating proton exchange membrane electrolyzer. *ECS Transactions*, 2015. **69**(17): pp. 877–881.

122. Reier, T., M. Oezaslan, and P. Strasser, Electrocatalytic oxygen evolution reaction (OER) on Ru, Ir, and Pt Catalysts: A comparative study of nanoparticles and bulk materials. *ACS Catalysis*, 2012. **2**(8): pp. 1765–1772.

123. Yoshinaga, N., et al., Development of ACLS electrodes for a water electrolysis cell. *ECS Transactions*, 2019. **92**(8): pp. 749–755.

124. Spori, C., et al., Molecular analysis of the unusual stability of an IrNbOx catalyst for the electrochemical water oxidation to molecular oxygen (OER). *ACS Applied Materials & Interfaces*, 2021. **13**(3): pp. 3748–3761.

125. Neyerlin, K.C., et al., Study of the exchange current density for the hydrogen oxidation and evolution reactions. *Journal of the Electrochemical Society*, 2007. **154**(7): p. B631.

126. Ayers, K., et al., Perspectives on low-temperature electrolysis and potential for renewable hydrogen at scale. *Annual Review of Chemical and Biomolecular Engineering*, 2019. **10**(1): pp. 219–239.

127. Fu, L., et al., Ultrasmall Ir nanoparticles for efficient acidic electrochemical water splitting. *Inorganic Chemistry Frontiers*, 2018. **5**(5): pp. 1121–1125.

128. Taie, Z., et al., Pathway to complete energy sector decarbonization with available iridium resources using ultralow loaded water electrolyzers. *ACS Applied Materials & Interfaces*, 2020. **12**(47): pp. 52701–52712.

129. Li, Q., et al., Ultrafine-grained porous Ir-based catalysts for high-performance overall water splitting in acidic media. *ACS Applied Energy Materials*, 2020. **3**(4): pp. 3736–3744.

130. Sapountzi, F.M., et al., The role of Nafion content in sputtered IrO_2 based anodes for low temperature PEM water electrolysis. *Journal of Electroanalytical Chemistry*, 2011. **662**(1): pp. 116–122.

131. Chourashiya, M.G. and A. Urakawa, Solution combustion synthesis of highly dispersible and dispersed iridium oxide as an anode catalyst in PEM water electrolysis. *Journal of Materials Chemistry A*, 2017. **5**(10): pp. 4774–4778.

132. Jirkovský, J., et al., Particle size dependence of the electrocatalytic activity of nanocrystalline RuO[sub 2] electrodes. *Journal of the Electrochemical Society*, 2006. **153**(6): p. E111.

133. Feng, Q., et al., Influence of surface oxygen vacancies and ruthenium valence state on the catalysis of pyrochlore oxides. *ACS Applied Materials & Interfaces*, 2020. **12**(4): pp. 4520–4530.

134. Ye, F., et al., A RuO_2IrO_2 electrocatalyst with an optimal composition and novel microstructure for oxygen evolving in the single cell. *Korean Journal of Chemical Engineering*, 2022. **39**(3): pp. 596–604.

135. Xu, C., et al., Synthesis and characterization of novel high-performance composite electrocatalysts for the oxygen evolution in solid polymer electrolyte (SPE) water electrolysis. *International Journal of Hydrogen Energy*, 2012. **37**(4): pp. 2985–2992.

136. Corona-Guinto, J.L., et al., Performance of a PEM electrolyzer using RuIrCoOx electrocatalysts for the oxygen evolution electrode. *International Journal of Hydrogen Energy*, 2013. **38**(28): pp. 12667–12673.

137. Hong Hanh, P., et al., Nanosized $IrxRu1-xO_2$ electrocatalysts for oxygen evolution reaction in proton exchange membrane water electrolyzer. *Advances in Natural Sciences-Nanoscience and Nanotechnology*, 2015. **6**(2): p. 025015.

138. Siracusano, S., et al., Enhanced performance and durability of low catalyst loading PEM water electrolyser based on a short-side chain perfluorosulfonic ionomer. *Applied Energy*, 2017. **192**: pp. 477–489.

139. Wang, L., et al., Highly active anode electrocatalysts derived from electrochemical leaching of Ru from metallic $Ir_{0.7}Ru_{0.3}$ for proton exchange membrane electrolyzers. *Nano Energy*, 2017. **34**: pp. 385–391.

140. Tunold, R., et al. *Materials for Electrocatalysis of Oxygen Evolution Process in PEM Water Electrolysis Cells.* In *Symposium on Interfacial Electrochemistry in Honor of Brian E. Conway held during the 216th Meeting of the Electrochemical-Society (ECS)*. 2009. Vienna, Austria.

141. Campelo, J.M., et al., Sustainable preparation of supported metal nanoparticles and their applications in catalysis. *ChemSusChem*, 2009. **2**(1): pp. 18–45.

142. Luc, W. and F. Jiao, Nanoporous metals as electrocatalysts: State-of-the-art, opportunities, and challenges. *ACS Catalysis*, 2017. **7**(9): pp. 5856–5861.

143. Biener, J., A.M. Hodge, and A.V. Hamza, *Applied Physics Letters*, 2005. **87**(12).

144. Wang, Y., et al., Nanoporous iridium -based alloy nanowires as highly efficient electrocatalysts toward acidic oxygen evolution reaction. *ACS Applied Materials & Interfaces*, 2019. **11**(43): pp. 39728–39736.

145. Chatterjee, S., et al., Nanoporous iridium nanosheets for polymer electrolyte membrane electrolysis. *Advanced Energy Materials*, 2021. **11**(34): p. 2101438.

146. Moeckl, M., et al., Durability testing of low-iridium PEM water electrolysis membrane electrode assemblies. *Journal of the Electrochemical Society*, 2022. **169**(6): p. 064505.

147. Bernt, M. and H.A. Gasteiger, Influence of ionomer content in IrO_2/TiO_2 electrodes on PEM water electrolyzer performance. *Journal of the Electrochemical Society*, 2016. **163**(11): pp. F3179–F3189.

148. Cha, J.I., et al., Improved utilization of IrOx on Ti_4O_7 supports in membrane electrode assembly for polymer electrolyte membrane water electrolyzer. *Catalysis Today*, 2022: pp. 19–27.

149. Sui, S., L. Ma, and Y. Zhai, Investigation on the proton exchange membrane water electrolyzer using supported anode catalyst. *Asia-Pacific Journal of Chemical Engineering*, 2009. **4**(1): pp. 8–11.

150. Liu, G., et al., An oxygen evolution catalyst on an antimony doped tin oxide nanowire structured support for proton exchange membrane liquid water electrolysis. *Journal of Materials Chemistry A*, 2015. **3**(41): pp. 20791–20800.

151. Lim, J.Y., et al., Highly stable RuO_2/SnO_2 nanocomposites as anode electrocatalysts in a PEM water electrolysis cell. *International Journal of Energy Research*, 2014. **38**(7): pp. 875–883.

152. Liu, G., et al., Nanosphere-structured composites consisting of Cs-substituted phosphotungstates and antimony doped tin oxides as catalyst supports for proton exchange membrane liquid water electrolysis. *International Journal of Hydrogen Energy*, 2014. **39**(5): pp. 1914–1923.

153. Nikiforov, A.V., et al., Preparation and study of IrO_2/SiC-Si supported anode catalyst for high temperature PEM steam electrolysers. *International Journal of Hydrogen Energy*, 2011. **36**(10): pp. 5797–5805.

154. Polonský, J., et al., Tantalum carbide as a novel support material for anode electrocatalysts in polymer electrolyte membrane water electrolysers. *International Journal of Hydrogen Energy*, 2012. **37**(3): pp. 2173–2181.

155. Puthiyapura, V.K., et al., Physical and electrochemical evaluation of ATO supported IrO2 catalyst for proton exchange membrane water electrolyser. *Journal of Power Sources*, 2014. **269**: pp. 451–460.

156. Sun, W., et al., In situ formation of grain boundaries on a supported hybrid to boost water oxidation activity of iridium oxide. *Nanoscale*, 2021. **13**(32): pp. 13845–13857.

157. Mazur, P., et al., Non-conductive TiO_2 as the anode catalyst support for PEM water electrolysis. *International Journal of Hydrogen Energy*, 2012. **37**(17): pp. 12081–12088.

158. Rozain, C., et al., Influence of iridium oxide loadings on the performance of PEM water electrolysis cells: Part II - Advanced oxygen electrodes. *Applied Catalysis B: Environmental*, 2016. **182**: pp. 123–131.

159. Zhao, S., et al., Highly active, durable dispersed iridium nanocatalysts for PEM water electrolyzers. *Journal of the Electrochemical Society*, 2018. **165**(2): pp. F82–F89.

160. Sola-Hernandez, L., et al., Doped tin oxide aerogels as oxygen evolution reaction catalyst supports. *International Journal of Hydrogen Energy*, 2019. **44**(45): pp. 24331–24341.

161. Ohno, H., et al., Effect of electronic conductivities of iridium oxide/doped SnO_2 oxygen-Evolving catalysts on the polarization properties in proton exchange membrane water electrolysis. *Catalysts*, 2019. **9**(1): p. 74.

162. Zheng, Y., et al., CO induced phase-segregation to construct robust and efficient IrRux@Ir core-shell electrocatalyst towards acidic oxygen evolution. *Journal of Power Sources*, 2022. **528**: p. 231189.

163. Nong, H.N., et al., A unique oxygen ligand environment facilitates water oxidation in hole-doped IrNiOx core-shell electrocatalysts. *Nature Catalysis*, 2018. **1**(11): pp. 841–851.

164. Tackett, B.M., et al., Reducing iridium loading in oxygen evolution reaction electrocatalysts using core-shell particles with nitride cores. *ACS Catalysis*, 2018. **8**(3): pp. 2615–2621.

165. Pham, C.V., et al., IrO$_2$ coated TiO$_2$ core-shell microparticles advance performance of low loading proton exchange membrane water electrolyzers. *Applied Catalysis B: Environmental*, 2020. **269**: p. 118762.

166. Shan, J., et al., Charge-redistribution-enhanced nanocrystalline Ru@IrOx electrocatalysts for oxygen evolution in acidic media. *Chem*, 2019. **5**(2): pp. 445–459.

167. Su, H., V. Linkov, and B.J. Bladergroen, Membrane electrode assemblies with low noble metal loadings for hydrogen production from solid polymer electrolyte water electrolysis. *International Journal of Hydrogen Energy*, 2013. **38**(23): pp. 9601–9608.

168. Xu, W. and K. Scott, The effects of ionomer content on PEM water electrolyser membrane electrode assembly performance. *International Journal of Hydrogen Energy*, 2010. **35**(21): pp. 12029–12037.

169. Xu, J., et al., A novel catalyst layer with hydrophilic-hydrophobic meshwork and pore structure for solid polymer electrolyte water electrolysis. *Electrochemistry Communications*, 2011. **13**(5): pp. 437–439.

170. Scheepers, F., et al., Layer formation from polymer carbon-black dispersions. *Coatings*, 2018. **8**(12): p. 450.

171. Rozain, C., et al., Influence of iridium oxide loadings on the performance of PEM water electrolysis cells: Part I-Pure IrO$_2$-based anodes. *Applied Catalysis B: Environmental*, 2016. **182**: pp. 153–160.

172. Alia, S.M., et al., Activity and durability of iridium nanoparticles in the oxygen evolution reaction. *Journal of the Electrochemical Society*, 2016. **163**(11): pp. F3105–F3112.

173. Bernt, M., A. Siebel, and H.A. Gasteiger, Analysis of voltage losses in PEM water electrolyzers with low platinum group metal loadings. *Journal of the Electrochemical Society*, 2018. **165**(5): pp. F305–F314.

174. Kang, Z., et al., Performance improvement of proton exchange membrane electrolyzer cells by introducing in-plane transport enhancement layers. *Electrochimica Acta*, 2019. **316**: pp. 43–51.

175. Yang, G., et al., Favorable morphology and electronic conductivity of functional sublayers for highly efficient water splitting electrodes. *Journal of Energy Storage*, 2021. **36**: p. 102342.

176. Hegge, F., et al., Efficient and stable low iridium loaded anodes for PEM water electrolysis made possible by nanofiber interlayers. *ACS Applied Energy Materials*, 2020. **3**(9): pp. 8276–8284.

177. Debe, M.K. and R.J. Poirier, Postdeposition growth of a uniquely nanostructured organic film by vacuum annealing. *Journal of Vacuum Science & Technology A: Vacuum, Surfaces, and Films*, 1994. **12**(4): pp. 2017–2022.

178. Xu, J., et al., Oxygen evolution catalysts on supports with a 3-D ordered array structure and intrinsic proton conductivity for proton exchange membrane steam etectrolysis. *Energy & Environmental Science*, 2014. **7**(2): pp. 820–830.

179. Lu, Z.-X., et al., Electrochemical fabrication of IrOx nanoarrays with tunable length and morphology for solid polymer electrolyte water electrolysis. *Electrochimica Acta*, 2020. **348**: p. 136302.

180. Pham, C.V., et al., Essentials of high performance water electrolyzers - From catalyst layer materials to electrode engineering. *Advanced Energy Materials*, 2021. **11**(44): p. 2101998.

181. Rakousky, C., et al., An analysis of degradation phenomena in polymer electrolyte membrane water electrolysis. *Journal of Power Sources*, 2016. **326**: pp. 120–128.

182. Frensch, S.H., et al., Influence of the operation mode on PEM water electrolysis degradation. *International Journal of Hydrogen Energy*, 2019. **44**(57): pp. 29889–29898.

183. Fouda-Onana, F., et al., Investigation on the degradation of MEAs for PEM water electrolysers part I: Effects of testing conditions on MEA performances and membrane properties. *International Journal of Hydrogen Energy*, 2016. **41**(38): pp. 16627–16636.

184. Steen, S.M., III, F.Y. Zhang, and Iop. *Energy Dispersive X-Ray and Electrochemical Impedance Spectroscopies for Performance and Corrosion Analysis of PEMWEs*. In *22nd International Conference on Spectral Line Shapes (ICSLS)*. 2014. Tullahoma, TN.

185. Spori, C., et al., The stability challenges of oxygen evolving catalysts: Towards a common fundamental understanding and mitigation of catalyst degradation. *Angewandte Chemie International Edition - England*, 2017. **56**(22): pp. 5994–6021.

186. Martelli, G.N., R. Ornelas, and G. Faita, Deactivation mechanisms of oxygen evolving anodes at high current densities. *Electrochimica Acta*, 1994. **39**(11): pp. 1551–1558.

187. Cherevko, S., et al., Oxygen evolution activity and stability of iridium in acidic media. Part 2 - Electrochemically grown hydrous iridium oxide. *Journal of Electroanalytical Chemistry*, 2016. **774**: pp. 102–110.

188. Cherevko, S., et al., Dissolution of noble metals during oxygen evolution in acidic media. *ChemCatChem*, 2014. **6**(8): pp. 2219–2223.

189. Park, S., et al., Degradation of the ionic pathway in a PEM fuel cell cathode. *The Journal of Physical Chemistry C*, 2011. **115**(45): pp. 22633–22639.

190. Lettenmeier, P., et al., Durable membrane electrode assemblies for proton exchange membrane electrolyzer systems operating at high current densities. *Electrochimica Acta*, 2016. **210**: pp. 502–511.

191. Alia, S.M., et al., Activity and durability of iridium nanoparticles in the oxygen evolution reaction. *ECS Transactions*, 2015. **69**(17): pp. 883–892.

192. Claudel, F., et al., Degradation mechanisms of oxygen evolution reaction electrocatalysts: A combined identical-location transmission electron microscopy and X-ray photoelectron spectroscopy study. *ACS Catalysis*, 2019. **9**(5): pp. 4688–4698.

193. Wang, X., et al., The influence of ferric ion contamination on the solid polymer electrolyte water electrolysis performance. *Electrochimica Acta*, 2015. **158**: pp. 253–257.

194. Jie, X., et al., The influence of sodium ion as a potential fuel impurity on the direct methanol fuel cells. *Electrochimica Acta*, 2010. **55**(16): pp. 4783–4788.

195. Sun, S., et al., Investigations on degradation of the long-term proton exchange membrane water electrolysis stack. *Journal of Power Sources*, 2014. **267**: pp. 515–520.

196. Babic, U., et al., CO_2-Assisted regeneration of a polymer electrolyte water electrolyzer contaminated with metal ion impurities. *Journal of the Electrochemical Society*, 2019. **166**(10): pp. F610–F619.

197. Panchenko, O., et al., Non-destructive in-operando investigation of catalyst layer degradation for water electrolyzers using synchrotron radiography. *Materials Today Energy*, 2020. **16**: p. 100394.

198. Zeradjanin, A.R., et al., Rational design of the electrode morphology for oxygen evolution - enhancing the performance for catalytic water oxidation. *RSC Advances*, 2014. **4**(19): pp. 9579–9587.

199. Chandesris, M., et al., Membrane degradation in PEM water electrolyzer: Numerical modeling and experimental evidence of the influence of temperature and current density. *International Journal of Hydrogen Energy*, 2015. **40**(3): pp. 1353–1366.

200. Yu, H., et al., Microscopic insights on the degradation of a PEM water electrolyzer with ultra-low catalyst loading. *Applied Catalysis B-Environmental*, 2020. **260**: p. 118194.

201. Babic, U., et al., Understanding the effects of material properties and operating conditions on component aging in polymer electrolyte water electrolyzers. *Journal of Power Sources*, 2020. **451**: p. 227778.

202. Garbe, S., et al., Understanding degradation effects of elevated temperature operating conditions in polymer electrolyte water electrolyzers. *Journal of the Electrochemical Society*, 2021. **168**(4): p. 044515.

11 Review of the State-of-the-Art Anion Exchange Membranes

Haolin Tang, Yucong Liao, and Letian Wang

11.1 ANION EXCHANGE MEMBRANE WATER ELECTROLYSIS (AEMWE)

As renewable energy consumption continues to increase, the intermittent nature of energy sources has made energy storage an increasingly vital problem to solve. Hydrogen energy stands out as a promising solution due to several inherent benefits. First, hydrogen electricity can be effectively exchanged using a proton exchange membrane (PEM) [1,2]. Second, hydrogen has a high energy density and is easily stored. Third, hydrogen conversion to electricity has the potential for large-scale application, making it a valuable resource in the transition to greener energy. Among various methods being investigated for hydrogen extraction, water electrolysis is particularly promising [3]. Currently, among electrolytic water hydrogen production technologies, alkaline water electrolysis (AWE) hydrogen production and proton exchange membrane water electrolysis (PEMWE) hydrogen production have been gradually industrialized. Although AWE is mature and cost-effective, its efficiency and performance are relatively low, posing limitations to its viability for large-scale hydrogen production. PEMWE has a higher performance than AWE [4]. However, PEMWE technology is expensive due to the use of costly materials such as Nafion membranes and noble metal electrocatalysts. Anion exchange membrane water electrolysis (AEMWE) technology integrates the benefits of both traditional alkaline liquid electrolyte electrolysis and PEM electrolysis. The use of an AEM and proton exchange membrane offers similar advantages, but the alkaline system avoids the heavy usage of precious metals, resulting in significantly reduced equipment costs compared to PEM hydroelectric solution pools. Non-precious metal accelerators like Ni, Co, and Fe can be utilized in an alkaline medium, thereby eliminating the need for alkaline liquid and reducing product gas pollution. However, the AEM's suboptimal performance poses a significant challenge to the development of AEM electrolysis technology [5]. The membrane's poor thermal and chemical stability, as well as limited anion conduction energy, limits the life and electrolytic performance of AEM electrolytic cells. Despite these challenges, efforts are underway to overcome these limitations and optimize AEM electrolysis technology for efficient and sustainable hydrogen production.

DOI: 10.1201/9781003368939-11

11.1.1   STRUCTURE OF ANION EXCHANGE MEMBRANE WATER ELECTROLYZER

In Figure 11.1, a typical AEM electrolyzer is presented. AEM electrolyzers consist of a membrane electrode assembly (MEA) that comprises a hydrocarbon AEM and two transition metal catalyst-based electrodes, where the electrochemical reactions take place [6]. In addition, the electrolyzers include gas diffusion layers (GDLs), bipolar plates, and end plates. Unlike traditional alkaline electrolysis which uses concentrated KOH solution as an electrolyte, AEMWE cells can utilize distilled water or low-concentration alkaline solution as the electrolyte. This approach combines the advantages of both PEM and alkaline electrolysis. The AEM is a core component of AEM electrolysis systems, transferring hydroxyl ions from the cathode chamber to the anode chamber while preventing gas crossover and electron transmission during electrochemical operation. AEMs consist of a hydrocarbon polymer backbone as the main chain, with anion exchange functional groups forming a side chain. Typically, polysulfone (PSF) or polystyrene (PS) connects divinyl benzenes (DVB) to form the polymeric backbone. The ion exchange groups in AEMs typically contain ammonium ($-NH_3^+$, $-RNH_2^+$, $-RN^+$, $=R_2N^+$) or phosphonium ($-R_3P^+$) groups [7,8]. For an ion exchange membrane to be considered effective, it must demonstrate specific characteristics such as high perm-selectivity, excellent ionic conductivity, strong thermal and mechanical durability, and exceptional chemical stability [9]. As AEM electrolysis technology is still in the developmental phase, further research and improvements are necessary to achieve commercially viable hydrogen production. Specifically, the power efficiency, membrane stability, robustness, ease of handling, and cost reduction of AEM electrolysis must be investigated and optimized to realize its full potential.

11.1.2   BASIC PRINCIPLE OF ANION EXCHANGE MEMBRANE ELECTROLYSIS

AEM electrolysis is an electrochemical process that uses an AEM to split water into hydrogen and oxygen, as shown in Figure 11.2 [10]. An external DC power supply is

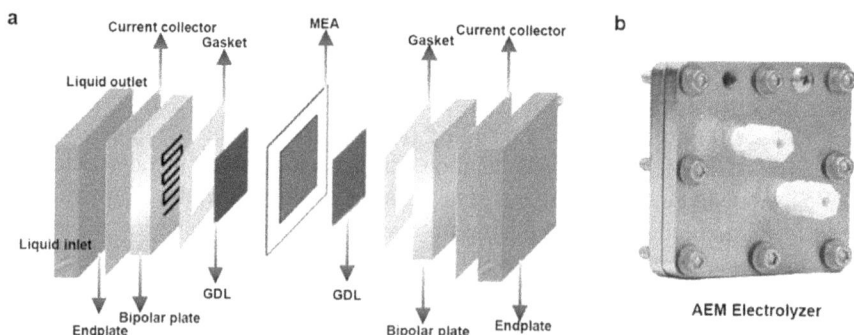

FIGURE 11.1 Configuration and actual images of typical AEM electrolyzers. (a) Schematic diagram of an AEM electrolyzer. (b) Typical square AEM electrolyzer. Reproduced with permission from Yan et al. [6]. Copyright (2021), Nature.

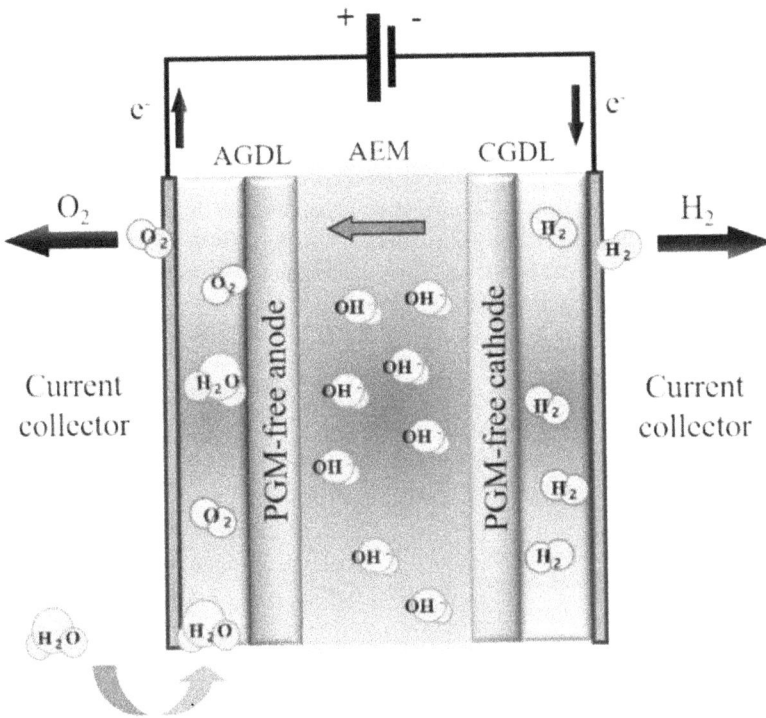

FIGURE 11.2 The Schematic demonstration for the operating process of AEMWE. Reproduced with permission from Vincent et al. [10]. Copyright (2021), *Nature*.

connected to the anode and cathode to initiate the hydrogen evolution reaction (HER) and the oxygen evolution reaction (OER), which are two half-cell reactions. In the anode chamber, water is reduced to form hydrogen and hydroxyl ions by accepting two electrons. The hydroxyl ions are attracted to the anode and diffuse through the AEM, while the electrons are transported through the external circuit to the anode. The hydroxyl ions recombine with electrons in the anode chamber to form water and oxygen, which are then released as bubbles from the surface of the anode. Catalytic activity is required in both half-cell reactions to form and release the respective gases from the electrode surfaces. To optimize AEM electrolysis for efficient hydrogen production, it is essential to focus on improving the catalytic activity, membrane stability, and overall system efficiency. In AEMWE devices, the two half-cell reactions, as well as the overall cell reaction, can be described as follows [11]:

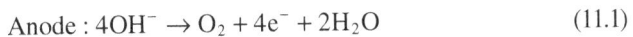

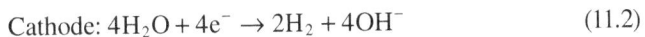

$$\text{Anode}: 4\text{OH}^- \rightarrow \text{O}_2 + 4\text{e}^- + 2\text{H}_2\text{O} \tag{11.1}$$

$$\text{Cathode}: 4\text{H}_2\text{O} + 4\text{e}^- \rightarrow 2\text{H}_2 + 4\text{OH}^- \tag{11.2}$$

$$\text{Overall}: 2\text{H}_2\text{O} \rightarrow 2\text{H}_2 + \text{O}_2 \tag{11.3}$$

To split water into hydrogen and oxygen in AEMWE devices, the overall reaction requires a theoretical thermodynamic cell voltage of 1.23 V at 25°C. This voltage is determined by the free Gibbs energy (ΔG) of water splitting under standard conditions. However, in practical applications, an additional voltage is necessary to overcome both the kinetics and ohmic resistance of the electrolyte and components of the electrolyzer, resulting in a higher operational cell voltage. For instance, alkaline and PEM electrolysis typically require operational cell voltages of 1.85–2.05 V and 1.75 V at 70°C–90°C, respectively [12]. To achieve highly efficient AEMWE devices, it is crucial to develop AEMs with high anion conductivity and electrodes with high electrocatalytic activity to minimize overpotentials.

11.2 KEY PARAMETER OF ANION EXCHANGE MEMBRANE

The primary focus of AEM characterization methods is to examine various aspects of the membrane, including chemical homogeneity, structure, stability, and mechanical properties [13]. To accomplish this, there are several analytical techniques available, such as microscopy, including scanning electron microscopy (SEM), and spectroscopy, including energy-dispersive X-ray (EDX), nuclear magnetic resonance (NMR), Fourier-transform infrared (FTIR), and small-angle X-ray scattering (SAXS). These methods help to determine the molecular composition of the membrane, such as the uniform distribution of head groups and the formation of ion clusters, as well as the structure of the membrane surface, such as pore structure and surface smoothness. In cases where asymmetrical membranes are synthesized, comparisons can be made between both membrane surfaces to better understand the impact of surface differences on membrane properties [14,15].

To evaluate the performance of AEM, certain parameters are typically measured. These include the IEC (ion exchange capacity), swelling ratio, water uptake (WU), water content, contact angle, conductivity, and alkaline stability. These measurements help to assess how well the AEM performs and how stable it is in different conditions [16,17].

11.2.1 Ion Exchange Capacity (IEC)

The IEC is a measure of the number of ions that can be exchanged per unit of the dry weight of the membrane. The IEC is typically calculated using either a traditional acid–base titration technique or Mohr's method. The AEM is first submerged in a strong base solution to convert it to the OH⁻ form before determining the membrane IEC using the acid–base titration technique in the Cl⁻ form. The AEM is then transferred to a precise volume and concentration of a strong acid solution to convert it to the Cl⁻ form [18]. Finally, the AEM is removed and washed with deionized water, and the standardized base solution is titrated to a phenolphthalein endpoint with a diluted acid solution. This approach provides precise IEC measurement, which is critical for maximizing AEM performance for specific applications. The computation of IEC is shown in the following equation:

$$IEC = \frac{(V_{acid} * C_{acid}) - (V_{base} * C_{base})}{m_d} \qquad (11.4)$$

The variables V_{acid}/V_{base} and C_{acid}/C_{base} represent the volume and molar concentration of the acid/base, respectively, while MD denotes the mass of the dried membrane after titration. However, the acid/base titration method has a drawback of OH^- group CO_2 poisoning. When the AEM is in the OH^- form, it can react with the environment containing CO_2 and convert to the HCO_3^- form, which can affect the IEC value. To mitigate any discrepancies in IEC measurement due to pH variation during titration, it is recommended to measure IEC in the Cl^- form (unit: mmol Cl^-/g). First, the membrane is soaked in $NaNO_3$ and acidified with HNO_3 before measuring the IEC. Second, the solution containing Cl^- ions is titrated with $AgNO_3$ using Ag-tetrodes until all Cl^- has converted to AgCl.

$$IEC = \frac{V_{AgNO_3} * C_{AgNO_3}}{m_d} \qquad (11.5)$$

The variable md refers to the mass of the dried membrane. Mohr's method employs the aforementioned equation to calculate the IEC. First, the AEM is immersed in a salt solution to transform it into the Cl^- form. Second, the AEM is brought to equilibrium with a Na_2SO_4 solution to release Cl^-. The AEM in the Na_2SO_4 solution is titrated until the K_2CrO_4 indicator endpoint.

$$2Ag^+ + CrO_4^{2-} \rightarrow Ag_2CrO_{4(s)} \qquad (11.6)$$

11.2.2 SWELLING RATIO (SR)

The swelling ratio (SR) is a measure of the linear expansion of the membrane upon exposure to water. It is expressed as a percentage difference between the lengths of the wet and dry membranes, and can be calculated using the following equation:

$$SR = \frac{L_w - L_d}{L_d} * 100\% \qquad (11.7)$$

where L_w and L_d are the lengths of the wet and dry membranes, respectively.

11.2.3 WATER UPTAKE (WU)

WU is a parameter that measures how the mass of the membrane changes when it comes into contact with water. It is expressed as a percentage difference between the masses of the wet and dry membranes. The "dry" membrane state is defined as mentioned earlier for IEC.

$$WU = \frac{m_w - m_d}{m_d} * 100\% \qquad (11.8)$$

where m_w and m_d are the mass of wet and dry membranes, respectively.

11.2.4 MEMBRANE WATER CONTENT (γ)

The water content of a membrane provides an estimate of the number of water molecules per mobile anion and can be calculated by dividing the water absorption by the molecular weight of water and the IEC. However, because the WU is expressed as a percentage and the IEC is expressed in mmol/g, the WU value must be multiplied by 10. This adjustment ensures that the units are consistent and the calculation is accurate.

$$\gamma = \frac{10*WU}{MW_{H_2O}*IEC}$$ (11.9)

Note that the WU is multiplied by 10 to account for the WU being reported in percent and the IEC being reported in mmol/g.

11.2.5 WATER CONTACT ANGLE (θ)

The water contact angle (θ) is an indicator of the degree of wetness of the surface of a membrane, with a large contact angle suggesting a highly hydrophobic surface. This parameter can be determined using the sessile-drop technique.

11.2.6 HYDROXIDE CONDUCTIVITY (σ)

Hydrogen conductivity in an AEM can be measured using electrochemical impedance spectroscopy (EIS) with a two- or four-electrode testing cell. To do so, the AEM is soaked in DI water overnight and then fixed in the testing cell to collect impedance data using changing AC currents nonlinear least squares regression analysis, the membrane ionic resistance (Rm) can be determined, and the conductivity (σ) can be calculated using a specific equation [18]:

$$\sigma = \frac{L}{R_m * A} = \frac{L}{ASR(\Omega cm^2)}$$ (11.10)

where σ is the conductivity (S/cm), L is the distance between electrodes (cm), A is the cross area of the membrane (cm^2), and R_m is the measured resistance (Ω) of the membrane. The area-specific resistance (ASR) is expressed as a product of Rm and area (A). It should be noted that the presence of CO_2 makes it difficult to measure the true hydroxide conductivity because of the formation of carbonates (CO_3^{2-}) and bicarbonates (HCO_3^-) as follows:

$$OH^- + CO_2 \rightleftharpoons HCO_3^-$$ (11.11)

$$OH^- + HCO_3^- \rightleftharpoons CO_3^{2-} + H_2O$$ (11.12)

Therefore, it is recommended to release CO_2 in advance for the measurement of OH$^-$ conductivities of AEMs. Ziv and Dekel reported the practical method for measuring the OH$^-$ conductivity of AEM [19]. They applied the constant direct current to the membrane, and membranes were exposed to continuous N_2 flow. The in-plane conductivity was calculated by the following equation:

$$\sigma(\text{in-plane}) = \frac{L}{R_m * w * t_w} \quad (11.13)$$

where w is the width of the membrane and t_w is the thickness of the membrane under wet conditions. Similarly, the Henkensmeier group measured the in-plane conductivity in DI water by excluding carbonates by injecting N_2 in water and applying the voltage until reaching a constant value [40].

The transportation of hydroxyl ions in AEMs is intricately linked to both polymer dynamics and interactions with water molecules, as depicted in Figure 11.3 [20]. The two primary transport mechanisms are vehicular and Grotthuss mechanisms, but vehicular diffusion (standard diffusion) is the dominant mode of transport in AEMs, with a higher activation energy barrier for OH⁻ diffusion. Therefore, to enhance anion conductivity, it is crucial to operating at higher temperatures. Additionally, the pathway of OH⁻ transport can be influenced by the hydration state of the membrane. For instance, Foglia et al [21] employed quasi-elastic neutron scattering across a broad range of times-cales (10^0–10^3 ps) to disentangle the dynamics of water, polymer relaxation, and hydroxyl ion diffusion under different hydration conditions. In commercial AEMs like Fumatech FAD-55, vehicular diffusion of hydroxyl ions and Grotthuss proton exchange are contingent on the hydration state.

Under low hydration conditions, the transportation of hydroxyl ions necessitated the presence of one water molecule (equivalent to two slow protons), leading to a relatively sluggish vehicular mechanism. At moderate hydration levels, diffusivity was predominantly influenced by the interaction between water mobility and polymer.

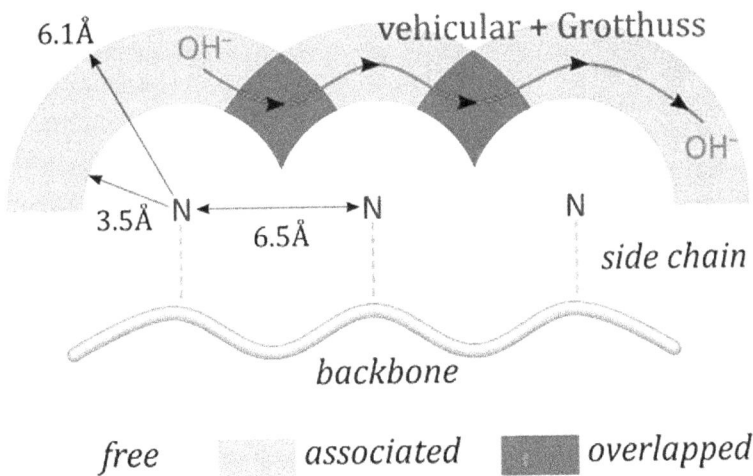

FIGURE 11.3 Different transport modes of OH⁻ in AEM. Reproduced with permission from Chen et al. [20]. Copyright (2016), American Chemical Society.

By contrast, bulk water dynamics were the main driver of conductivity at high hydration levels. The faster Grotthuss proton transport could be achieved at both moderate and high hydration, emphasizing the significance of maintaining a certain level of hydration to ensure anionic hopping and, in turn, minimize the operating resistance of AEMWE devices.

Improving the conductivity of AEMs plays a vital role in reducing energy loss caused by potential drops within the membrane, thereby enhancing the overall effectiveness of AEM water electrolysis devices. To achieve better ion transport efficiency, it is crucial to gain a deeper understanding of the kinetics of anion transport and to design the polymer backbone with cationic groups in a well-considered manner. While increasing the IEC is a direct approach to enhancing conductivity, it may also result in high WU and swelling of the AEM, which could further undermine its mechanical stability.

11.2.7 Alkaline Stability

Alkaline stability is a fundamental parameter that characterizes the ability of an AEM to maintain its performance under high-pH conditions over an extended period [22]. While the testing conditions may vary depending on the specific experimental setup, the general methodology involves immersing the AEM in a high-pH solution, typically 1–10 M KOH, for an extended duration, either at room temperature or at an elevated temperature. The membrane's IEC is then periodically monitored to evaluate the extent and nature of any changes that occur over time, providing insights into the membrane's long-term stability and suitability for various applications. Inconsistencies in the conditions used for testing alkaline stability can pose significant challenges, as studies have demonstrated that the hydration level of the nucleophile (OH^-) can impact the stability of alkaline solutions [23]. Specifically, decreasing hydration levels has been shown to reduce alkaline stability, while higher hydration levels, where water molecules surround the OH^- ion, can act as a protective shield, reducing its nucleophilic character and improving alkaline stability. Ex situ testing of alkaline stability has commonly been conducted using KOH or NaOH solutions of up to 10 M, which corresponds to a water content of approximately 5 ($\lambda = 5$) [24,25]. However, it is important to note that higher concentrations of KOH may lead to lower water contents, which can affect the hydration level of the nucleophile and, in turn, the measured alkaline stability. Additionally, the increased viscosity of high-concentration KOH solutions can negatively impact OH^- diffusivity, which can further impact the accuracy of measured alkaline stability. In alkaline exchange membrane fuel cells (AEMFCs), high current densities can cause the cathode to become water-depleted, resulting in ultralow hydration levels ($\lambda = 0$) in the AEM. Previous work by Dekel et al. showed that the stability of quaternary ammonium (QA) groups in AEMs was excellent at $\lambda = 4$, but significantly decreased at $\lambda = 0$ due to changes in S_N2 reaction energies, which was the primary degradation mechanism for the QA group studied. As the hydration level declined, so did OH nucleophilicity, resulting in reduced activation and reaction energy. This implies that the existing ex situ alkalinity stability testing with aqueous solutions may exaggerate alkalinity stability results, which may not accurately reflect in situ alkaline stability in AEMFCs. Dekel et al. presented an alternative ex situ alkalinity stability testing approach based on NMR and water-free hydroxide (crown ether/KOH) solutions, where the water/OH$^-$ ratio (c) may be manipulated to assess alkaline stability at different hydration levels. Thermal stability and tensile strength are routinely assessed to evaluate the mechanical characteristics of AEMs. Because AEM fuel cells (up to 200°C) and water electrolyzers (usually 50°C–70°C) operate at high temperatures, assessing the thermal stability of the membrane is crucial. This can be accomplished using techniques such as thermogravimetric analysis (TGA) and differential scanning calorimetry (DSC) [26,27]. TGA provides information about the thermal stability of the membrane by monitoring changes in weight as a function of temperature, which can be attributed to water loss, head group decomposition, and/or polymer decomposition. DSC, on the other hand, can be used to evaluate the glass transition temperature, as well as the effects of thermal cycling, changes in polymer crystallinity, and cross-linking [28,29]. Additionally, the tensile properties of the membrane, such as tensile strength, stress–strain curves, and elongation at break, can be determined by stretching membrane samples in a universal testing machine.

11.3 ADVANCEMENT OF WATER ELECTROLYSIS BY ANION EXCHANGE MEMBRANE

So far, the main issues that have prevented the widespread commercial development of AEMWEs are the weaker ionic conductivity of AEMs compared to PEMs and the drawback of poor durability under alkaline environments. AEMWEs have developed as a result of recent studies that addressed the two problems mentioned above and put forth fresh ideas and concepts to promote the development of AEMs in terms of increased ionic conductivity and alkaline stability. Anion exchange polymers (AEPs) have cationic headgroups linked to their polymeric backbones and are used to create AEMs [30].

11.3.1 METHODS TO IMPROVE CRITICAL AEM PROPERTIES

11.3.1.1 Cross-Linking

The hydrophilic cationic functional groups that are necessary for hydroxide ion conduction are supported by the hydrophobic and mechanically robust polymer backbone. Ordinarily, for polymers with more cations or a higher IEC (mmol/g). The material's ability to absorb water also rises with an increase in IEC, which might result in mechanical failure and catalyst layer flaking because of polymer swelling [31,32]. Each polymer/cation system's IEC and mechanical properties must be balanced to maximize APE performance. Cross-linking can assist APE maintain its mechanical qualities even at high IEC if the membrane exhibits significant swelling [33].

To reduce swelling, cross-linking forms chemical links between molecules found in ion-conducting polymers. Optimal cross-linking also preserves the strong ionic conductivity of AEP and AEM [30]. Physical cross-linking introduces ion–ion or van der Waals interactions between molecules. In particular, Lee et al. [34] reported a series of additive-induced physically cross-linked systems for use as AEMs were produced by mixing a dodecyl-substituted and quaternary ammonium-functionalized PPO (poly(2,6-dimethyl-1,4-phenylene oxide)) with long alkyl and QA groups. As for the chemical cross-linking, Zhang et al. [35] used PPO as the main chain and performed the cross-linking reaction through "thiol-ene" chemistry. By chemical cross-linking, the possible entanglement in the polymer chains has increased the tensile strength to 53.2 MPa, while the water absorption and swelling coefficients have been well controlled. The cross-linked AEM had an IEC of 1.27 meq/g and WU of 12.8 wt% when hydrophilic dithiol was used as the cross-linking agent, which was nearly twice as high as that of the membrane with the hydrophobic cross-link AEM (WU of 6.6 wt% and IEC = 1.28 meq/g) (Figure 11.4).

In physical cross-linking, ion–ion interactions [36] or intermolecular forces [34] (van der Waals forces) are generally used, while in chemical cross-linking, in addition to the thiol-ene chemistry mentioned in the appeal [35,37,38], there are Menshutkin reactions between halogenated methylated polymers and commercially available diamines [39], ring-opening complex decomposition polymerization [40], olefin complex decomposition [41], and thermal cross-linking [42]. After cross-linking, the rate of swelling and water absorption will be somewhat slower, which will lessen the impact of OH⁻ on the higher functional groups and the main chain of AEP. After cross-linking, the rate of swelling and water absorption will be somewhat slower, which will lessen the impact of OH⁻ on the higher functional groups and the main chain of AEP. However, it has been demonstrated that cross-linking techniques using polycationic side chains or end groups are efficient. For example, Wang et al. [43] offered d_3, a brand-new degludec alcohol cross-linker that has two QA groups and a diphenyl ether group. The QA group could serve as an anion exchange site in addition to acting as a cross-linker to reduce swelling and increase the mechanical stability of the AEM during fuel cell operation. The norbornene derivative and Grubbs III catalyst are combined, and ROMP copolymerizes norbornene (NB) and d_3 to create the cross-linked AEM. Wu et al. [42] found a new and facile way to synthesize self-cross-linked AEMs without any cross-linker or catalyst is reported. Soluble copolymers with olefin side chains were synthesized by the Menshutkin

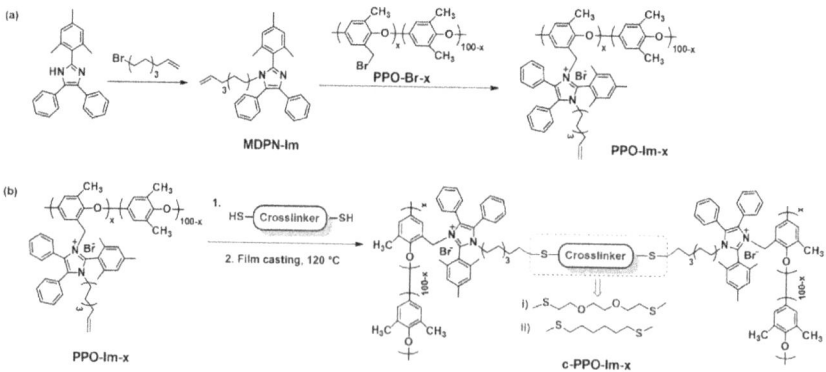

FIGURE 11.4 (a) Synthesis of cross-linkable PPO-Im-x precursor and (b) the preparation of cross-linked AEM c-PPO-Im-x via the "thiol-ene" click reaction [35].

reaction. The cross-linked derivatives were then prepared by thermal cross-linking of unsaturated side chains during membrane formation. The strategy produced an AEM that yielded a peak power density of 42 mW/cm^2 in an H/O fuel cell at 60°C. For instance, Chen et al. [44] described an array of polycationic cross-linked membranes with good dimensional and alkali stability and high OH$^-$ conductivity (155.80 S/cm at 0°C). End-group cross-linked PSF membranes were created by Lee et al. [45] by adding benzyl groups to the ends of PSF polymer chains. Ionic conductivity (11.80 S/cm at 0°C) and stability of dimension were both increased by cross-linking. AEM's performance can be significantly impacted by inappropriate cross-linking, even though cross-linking can increase the mechanical strength of AEM. For instance, using lengthy chains of cross-linkers can cause AEP to crystallize, which alters AEM's hydrophilicity, impacts the OH$^-$ transport route, and influences the ionic conductivity further. Sung et al. [46] reported a series of the usage of oxygen-containing cross-linkers that may result in higher crystallinity because of the production of hydrogen bonds by oxygen atoms and interactions between oxygen atoms, according to several studies that were done to prove this. By doing this, the polymeric membrane produced more bound water overall. Additionally, it was discovered that as the polymeric membrane's ability to retain water rose, so did its ionic conductivity and alkali stability. But too much oxygen causes crystallization, which lowers conductivity and alkali stability. It is feasible to enhance not only the fundamental characteristics of AEMs but also their cell performance, which is essential for these AEMs to be useful. This may be accomplished by using molecules that lessen the hydrophobicity of the cross-linker between each ionic conductivity group. The cross-linking process will also significantly increase the complexity of the reaction, which will hurt the film formation. If the cross-linking is excessive, it will also affect the original structural stability, which will affect the brittleness of the mechanical properties [47].

The term "IPN" (Interpenetrating Polymer Network) refers to a network structure made up of two or more co-blended polymers having at least one polymer molecular chain cross-linked chemically. The challenge of exceeding the mechanical strength of the original APE while preserving the original ionic conductivity is resolved by IPN. Ion-conductive networks are created by IPN AEMs, but mechanical stability is preserved by non-ion-conductive networks. Without using covalent bonds, these networks are cross-linked at the molecular level. For instance, IPN AEM cross-linked quaternized poly(epichlorohydrin)/poly(tetrafluoroethylene) (PTFE) and IPN AEM polyvinyl alcohol/polyethyleneimine based. To overcome the restriction of chain flexibility caused by chemical cross-linking in IPN, Lin et al. [48] created a new allyl imidazole monomer through an electrophilic substitution reaction between chloropropene and commercially available 1,2-dimethylimidazole. They also prepared AEM using thermoplastic interpenetrating polymer networks (TIPN).

11.3.1.2 Microphase Separation

Ionic conducting polymers can contain a wetting part (usually an ionic conducting part) and a non-wetting part. When moderately compatible polymers are blended, phase separation occurs because it is not a completely thermodynamically stable system, but the two polymers are again compatible, so that this phase separation is microscopic or submicroscopic in size and homogeneous in appearance, with no macroscopic delamination visible (as distinguished from macroscopic phase separation), and when such polymers come into contact with a liquid such as water or an electrolyte, the polymer molecules can be reoriented in such a way as to bring the wetted part of the polymer into contact with the liquid, leading to the formation of liquid clusters. The wetted and non-wetted portions will produce two phases if the molecules have specific spatial properties. This effect is called microphase separation [49]. Constructing hydrophilic/hydrophobic microphase separation structures, also known as constructing ion channels, provides quicker and smoother conduction domains for ion conduction, which is how the necessary phase separation structure in AEMs is often achieved.

Controlling the number of maximally permeable ion domains to enable the creation of additional high-quality ion channels is one of the key considerations when building APEs for microphase separation. It is also preferable to have a uniform and continuous distribution of ion-conducting domains. Controlling the ensuing microphase separation's shape improves the mechanical characteristics of the AEM and the ionic conductivity. To achieve the effect of microphase separation, current research typically focuses on the relationship between the cationic head group and the backbone network. This is done by varying the length and position of side chains, functional groups, and main chains to find the best match, which alters the ionic conductivity and water absorption [50]. Designing polymer architectures logically to create well-developed microphase-separated structures is essential. Examples include functionalized polyelectrolytes with well-defined side-linked branches that resemble combs, ionic clusters, cross-linked structures, and/or macroblock structures [51] (Figure 11.5).

The placement of the wetting ion-conducting section in the side chain or multi-block copolymer including alternate wetting and non-wetting portions is a key tactic for achieving effective phase separation. Zhou et al. [52] developed comb-like polyvinylidene-indole piperidine copolymers with various side chain lengths to create the polymer structures with the most effective ion transport channels. The influence of side chain length on AEM performance and alkaline fuel cell performance was examined. Hydrophobic alkyl side chains were constructed directly on the aryl ether-free main chain. With a higher hydroxide conductivity of 5.80 mS/cm at 134°C than the control PITP-Q85 membrane without a side chain, the PITP-C85Q10 membrane with a C10 side chain contributed to a bigger ion domain size. In the above-mentioned study of side-chain-type AEP, the researchers concluded that the length of the chain or the position of the end groups affects the performance of AEP. The hydrophilicity and flexibility of ionic side chains have been shown to play a critical role in the manufacture of high-performance AEMs. Yang et al. [53] reported that an acid-catalyzed hydroxylation reaction was used to create the poly aryl piperidinium backbone, poly biphenyl N-methyl piperidine (PBP), and then QA cations with different kinds of substituent side chain tether chains were used to create the AEMs resins. AEM structure and performance were compared, and the side chains included hydrophobic alkyl chains, hydrophilic PEG chains, and multi-PEG chains. PBP-PEG demonstrated the best performance because well-defined, continuous water channels formed in the membrane; in contrast, PBP-TPEG showed poor performance because

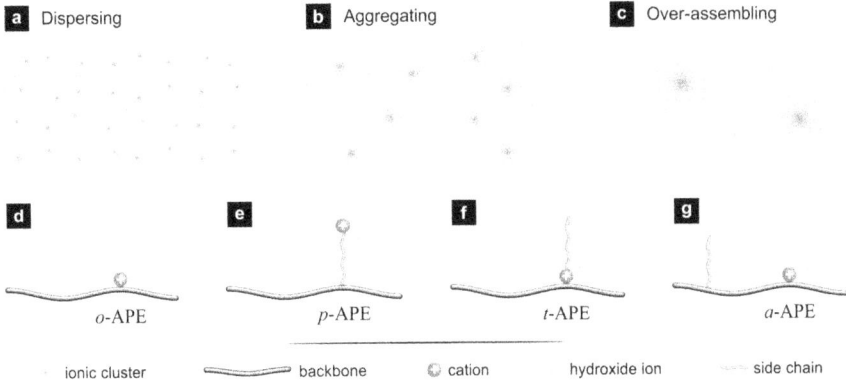

FIGURE 11.5 Schematic illustrations of the strategy and methodology for constructing an ionic highway in alkaline polymer electrolytes (APEs). (a) Ordinarily, small ionic clusters (grey) are dispersed in a hydrophobic matrix (white) in APEs. (b) By introducing additional hydrophobic structure, the ionic clusters can be driven aggregated, which facilitates the formation of interconnected, broad ionic channels. (c) But inappropriate designs may cause an over-assembly of ionic clusters, resulting in partitioned domains in the APE. (d) The original structure of APE (o-APE), where the cation is attached closely to the backbone and the anion (OH$^-$) is dissociated in the aqueous phase. Additional hydrophobic structures can be incorporated in three styles: (e) pendant-type APE (p-APE), where a hydrophobic side chain links the backbone and the cation; (f) tadpole-type APE (t-APE), where the hydrophobic side chain is attached to the cation; and (g) a new style of APE, where the cation and the hydrophobic side chain are separately attached to the backbone. This design (denoted as a-APE) turns out to be more efficient in forming the ion-aggregating structure [51].

the many PEG substituents decreased the cation's alkalinity and increased water absorption (Figure 11.6).

Since surface site jumping of hydroxides is more effective than surface site diffusion and migration of hydroxides, the authors attribute this to the increased conductivity brought about by the introduction of PEG groups as well as the hydrophilic PEG groups that can form a hydrogen bonding network with water and then act as additional jump sites to facilitate conduction. PBP-TPEG with three hydrophilic side chains has a substantially lower conductivity, though. This is because the many substituents surrounding the QA group partially prevent the hydroxide ion from dissociating, hence lowering the cation's alkalinity.

Atomic force microscopy (AFM) and transmission electron microscopy (TEM) are typically used to assess the degree of microphase separation of AEP and disclose the morphology–property link, and the degree of microphase separation can also be demonstrated by SAXS. The effect of microphase separation is obtained by the contrast and position distribution of light and dark phases. SAXS testing necessitates substantially less sample preparation than AFM and TEM. Important SAXS model parameters include the appearance and location of the ionophore peaks. The formula $d = 2\pi/q$ can be used to determine the characteristic separation size of ionic clusters (microphase separation), and the bigger the value of d, the more prominent the microphase separation in the membrane [54] (Figure 11.7).

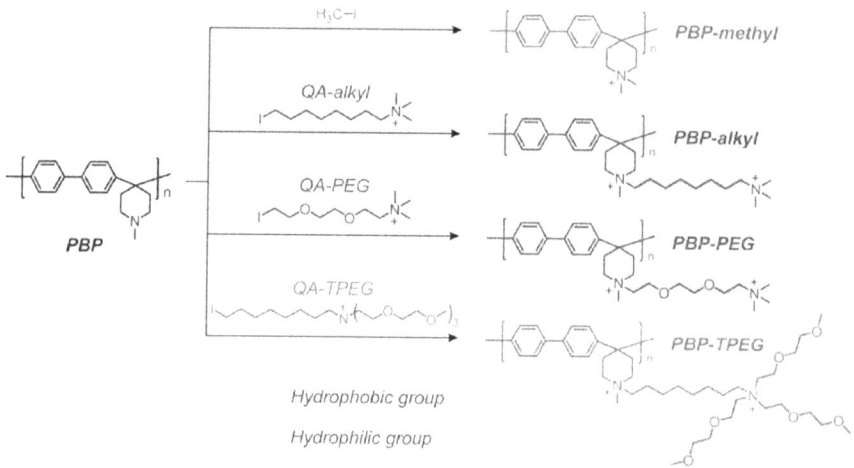

FIGURE 11.6 Synthesis of poly(aryl piperidinium)-based AEMs tethered with quaternary ammonium cations [53].

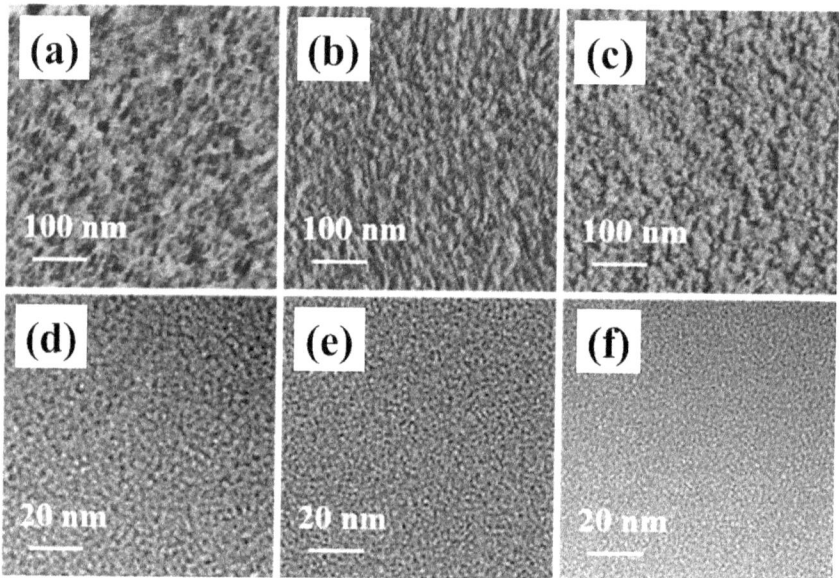

FIGURE 11.7 (a–c) AFM and (d–f) TEM images of QABNP (left), QPBNP (middle), and QAQPP (right) [55].

There are three advantages to the hydrophilic/hydrophobic microphase separation structure that forms in AEM. First, the development of ion transport channels in the membrane can significantly boost OH⁻ conductivity. Second, the hydrophobic segment restricts the membrane's ability to expand in size while submerged in water and gives the AEM a strong mechanical foundation. Third, because the hydrophobic phase inhibits the nucleophilic assault of OH^- to the backbone, the alkaline stability of AEM can be improved. The electrical conductivity and alkaline stability of AEM have significantly increased as a result of recent advances [56].

11.3.1.3 Organic/Inorganic Composite AEMs

Another tactic to increase AEM performance is to use organic/inorganic composites. In general, there are two methods for creating composite films: the first is to embed inorganic nanoparticles directly into organic AEP and the second is to put AEP into inert porous carriers [57]. Many different types of inorganic nano-ions can be embedded, including functionalized carbon nanotubes, metal oxides, silica, metal ions, and graphene oxide. These ions are being studied more and more due to their relatively easy modification and capacity to simultaneously change the superiority or inferiority of properties. Particles are typically added to composite membranes or porous support membranes to prevent membrane water absorption or to provide cation loading sites to increase ion conduction efficiency. In composite AEM, the particles and porous support membranes are typically non-ionic and prevent the uptake of water, but the polycations offer significant ionic loading and encourage ionic conductivity. Composite AEMs have been used in earlier research because they exhibit improved ionic conductivity as well as thermal, chemical, and mechanical stability while minimizing WU. For example, Vijayakumar Elumalai et al. [58] reported that a series of hydrothermal preparation of TNTs with tubular form was verified by TEM and XRD measurements. The investigations indicated above revealed that 5 wt% QTNT composite membrane had the best electrochemical characteristics, with an OCV of 0.92 V and a maximum power density of 285 mW/cm^2 at an operating temperature of 60°C. Lee et al. [59] then constructed quaternized polyvinylidene ether ketone (QPAEK)/f-TiO_2 composite membranes by first preparing TiO_2 and then modifying it, using PTMA precursors to anchor it uniformly on its surface, and then adding the functionalized TiO_2 to the polyvinylidene ether ketone to form composite AEMs (Figure 11.8).

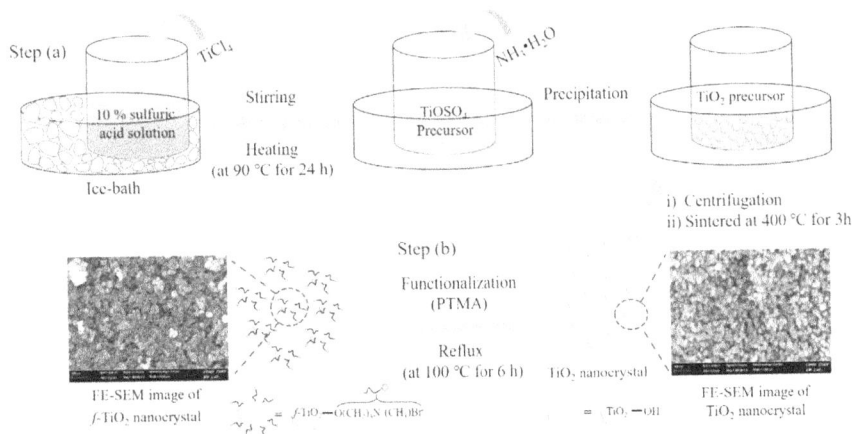

FIGURE 11.8 The synthetic steps of (a) TiO_2 and (b) f-TiO_2 nanocrystals [59].

Inorganic elements are typically added directly into the polymer solution when utilizing AEM. The inorganic material's dispersion in the polymer matrix, however, restricts the employment of inorganic materials in AEM. Even 10% can hurt the ionic conductivity of AEM. The amount of inorganic material doping should not be more than 20% of the mass of the polymer [60]. Zirconia (ZrO_2) is a preferable option when it comes to inorganic nanoparticles since it considerably enhances the anion exchange capabilities and fuel cell efficiency of imidazole-functionalized polysulfone (ImPS) inclusion. The membranes' thermal stability, mechanical strength, and water absorption capacity were all increased by the combination of ZrO_2 nanoparticles. IEC, ionic conductivity, and alkaline stability are all very performant due to the strong and efficient interaction between ZrO_2 and ImPS. ImPS/ZrO_2 co-blended membranes have shown promise in fuel cell performance experiments, particularly ImPS membranes containing 10% ZrO_2, which have higher OCP and maximum power density composites than pure ImPS by 35% and 39%, respectively [61]. Chen et al. [62] changed the traditional way of doping inorganic materials in AEM by demonstrating porous sandwich-structure AEMs with composite films exhibiting higher hydroxide conductivity, base stability, and fuel cell performance, as well as better mechanical properties and dimensional stability. The TC-QAPPO membrane's surface is covered with a high alkali-stabilized QA-LDH layer that serves as a barrier against OH^- erosion, extending the membrane's useful life. The QA-LDH/TC-QAPPO composite membrane's highest hydroxide conductivity was 122 mS/cm, and its maximum power density was 267 mW/cm² at a current density of 554 mA/cm². This illustrates how QA-LDH/TC-QAPPO membranes can be used with actual machinery [62] (Figure 11.9).

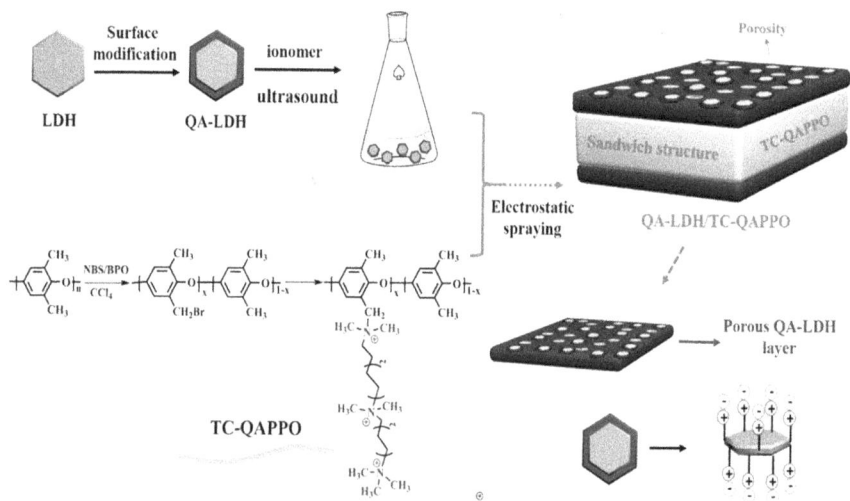

FIGURE 11.9   Fabrication of sandwich-structure AEMs [62].

11.3.1.4 Commercial Membranes for Anion Exchange Membrane Water Electrolysis

The utilization of commercial AEMs to produce high-performance AEMWEs, particularly inside MEA components, has been the subject of several research as interest in MEA has increased. There have been various commercial AEMs created and launched in recent years [63]. Few of these examples includeFumasep®FAA3 (Fumatech Co., Germany), A201® (Tokuyama Co., Japan), Aemion™ (Ionomr Innovation Co., Canada), SUSTAINION® (Dioxide Materials Co., USA), and Orion TM1™ (Orion Polymer, USA) [64] (Table 11.1). Early on in the development of the A201®, it was discovered that it performed insufficiently for AEMWE (350–550 mA/cm²) [65] at 1.9 Vcell [66,67]. As a result, the product was put on hold. When examining the fabrication method, MEA parameters, and operating conditions, Park et al. [68,69] found that Fumapem® membranes improved the performance of AEMWE to 1500 mA/cm² at 1.9 Vcell. However, when used as AEM, A201® and Fumapem® are reported to have low durability. Using FASe50® (Fumatech, Germany), Liu et al. [70,71] quadrupled the AEMWE performance in their report on the performance and durability of Sustainion® membranes created by Dioxide Materials Co. At 180 hours, FASe50®'s voltage dramatically changed, although it was still operational at 1950 hours. The AEMWE with Aemion™ membrane was conducted with 2000 mA/cm² at 1.82 Vcell and 60°C, according to Fortin et al. [72].

TABLE 11.1
Commercial AEMs and Their Reported Properties [64]

Brand Name	Company	Product Code	IEC (meq/g)	Ion Conductivity (mS/cm)	Tensile Strength (MPa)	Elongation at Break (%)
Fumasep®	Fumatech	FAA-3–30	1.7–2.1	4–7 (Cl)	25–40	20–40
Fumasep®		FAA-3–50	1.85	As above	As above	As above
Fumasep®		FAA-3-PK-75	1.39 (Cl)	>2.5 (Cl)	20–45	30–50
A201	Tokuyama	A201	1.8	42 (OH)	96 (dry, Cl)	62 (dry, Cl)
AEMION™	Ionomr	AF1-HNN8–50-X	2.1–2.5	>80	60 (dry, I)	85–110 (dry, I)
AEMION™		AF1-HNN8–25-X	2.1–2.5	>80	60 (dry, I)	85–110 (dry, I)
AEMION™		AF1-HNN5–50-X	1.4–1.7	15–25	60 (dry, I)	85–110 (dry, I)
AEMION™		AF1-HNN5–25-X	1.4–1.7	15–25	60 (dry, I)	85–110 (dry, I)
SUSTAINION®	Dioxide Materials	Sustaining 37–50	NA	80 (1 M KOH, 30°C)	Cracks when dry	Cracks when dry
Orion TM1	Orion Polymer	Pure material m-TPN1	2.19(OH)	19(Cl) 54(OH) >60	30	35

Additionally, the degradation rate was 3.21 mV/h at 50°C. AEMWE testing has been conducted with all AEMs except Orion TM1™ [63].

Wang et al. [73] reported a series of poly(aryl piperidine) AEMs (PAP HEMs) with high IEC and basic compatible cations. With 2,2,2-trifluoro acetophenone in place of various piperidones, they were able to achieve good conductivity, chemical stability, and mechanical strength. PAP HEMS continued to be flexible and had good ionic conductivity after 2,000 hours in 1 M KOH solution at 100 °C. They also have excellent electrical conductivity (ranging from 78 mS/cm at 20°C to 193 mS/cm at 95°C) and prototype PEM can absorb water similarly to commercial Nafion. It is also the prototype of the PiperION AEM (Versogen Co., USA), which is one of the best overall performance membranes on the market today. Wang et al. [73] illustratively demonstrate hydroxide exchange membranes and hydroxide exchange ionomers based on poly(aryl piperidinium) (PAP), which simultaneously exhibit appropriate ionic conductivity, chemical stability, mechanical robustness, gas separation, and selective solubility. These characteristics result from the interaction of the stiff aryl backbone with the piperidinium cation. With H_2/CO_2-free air at 95°C, a low-Pt membrane electrode assembly with an Ag-based cathode demonstrated a good peak power density of 920 mW/cm² and operated steadily at a steady current density of 500 mA/cm² for 300 hours (Figure 11.10).

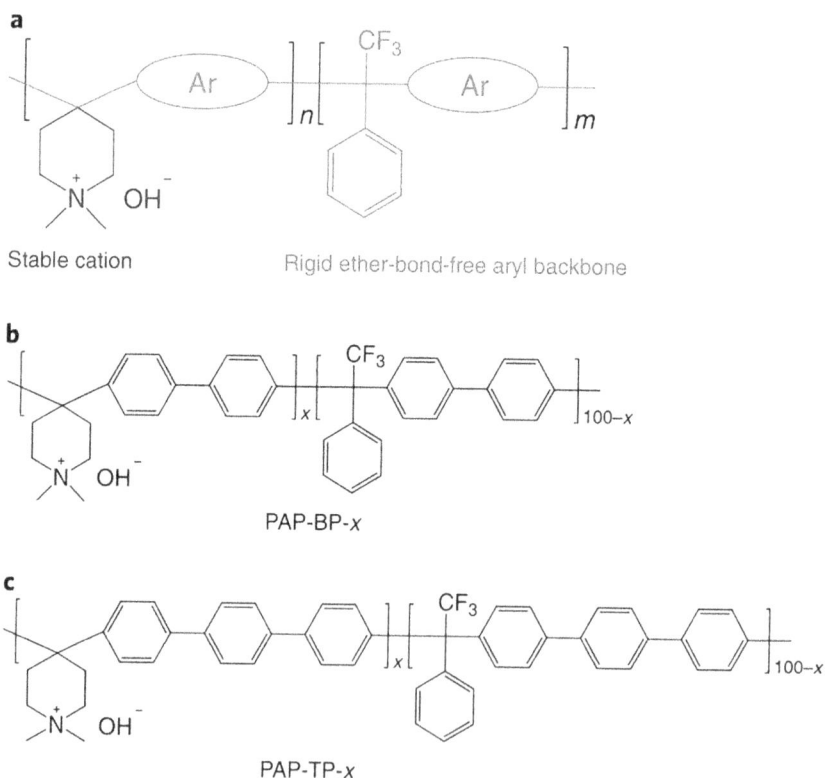

FIGURE 11.10 Chemical structure of the PAP HeM family: (a) General chemical structure, (b) PAP-BP-x, and (c) PAP-TP-x [73].

PAP-BP-x and PAP-TP-x in the aforementioned illustration stand for PAP based on biphenyl and triphenyl, respectively, and x represents the molar ratio (in percent) between N-methyl-4-piperidone and aryl monomers, with the remaining (100x) being 2,2,2-trifluoro acetophenone. Despite the stiff aryl chain segments on the PAP-N backbone, high relative molecular masses (molecular weight >60 kg/mol) were achieved, and the polymers were soluble in common solvents such as dichloromethane (CH_2Cl_2) and trichloromethane ($CHCl_3$) [73]. Due to its stiff, ether-bond-free aromatic backbone, this commercial resin solution offers high chemical durability and stability over a pH range of 1–14. Additionally, it offers durable performance over a broad temperature range. Anode and cathode catalyst films with superionic conductivity and high mechanical strength are produced.

11.4 DEGRADATION FACTORS IMPACT ANION EXCHANGE MEMBRANES

The target for AEMs before 2010 was to reach an ionic conductivity of 0.1 S/cm at 60°C–80°C. Since then, some AEMs have exceeded 0.2 S/cm, and the target has expanded tenfold. AEMWE has the potential to be a high-performance and inexpensive device for (large-scale) hydrogen production for a sustainable energy future based on a green hydrogen economy by combining the benefits of AWE and PEMWE. AEMWE development is currently in its early phases, but in the years to come, more focus and work will be put into developing such electrolyzers. Most current research on AEMs focuses on enhancing mechanical and chemical stability, as well as pushing the boundaries of working temperature. To improve the stability under alkaline conditions, the degradation pathway of the main chain under alkaline conditions should be clarified first.

11.4.1 HYDROXIDE ANIONS ATTACK

Alkaline stability has been the primary focus of AEM research due to the fragility of their functional groups, which can deteriorate when attacked by nucleophilic OH⁻. However, the pure polymer backbone of AEMs remains stable under the same conditions. QA cations also decay under alkaline circumstances, with the rate of breakdown increasing with temperature. Many QA groups have been studied as potential cationic groups for AEMs, with some exhibiting exceptional alkaline stability toward hydroxide in lab-scale tests. The nucleophilic activity, basicity, and flexibility of amine groups are all factors that can affect stability. Figure 11.11 illustrates the four main degradation mechanisms via hydroxide ion attack: S_N2 benzyl substitution (main, nucleophilic substitution), S_N2 methyl substitution (minor, nucleophilic substitution), b-elimination substitution (Hofmann elimination) and nucleophilic substitution, and ylide-intermediated rearrangements [74–76]. Hofmann elimination is an E2 reaction that occurs quickly when OH⁻ ions attack the neighboring b-hydrogen, resulting in the elimination of a tertiary amine. If b-hydrogen is not present, nucleophilic attack, specifically the S_N2 reaction, can occur on both the backbone and functional group, as OH⁻ is a strong nucleophile [77]. Groups that are close to OH⁻ are more vulnerable to nucleophilic substitution. For instance, benzyl trimethyl ammonium hydroxide

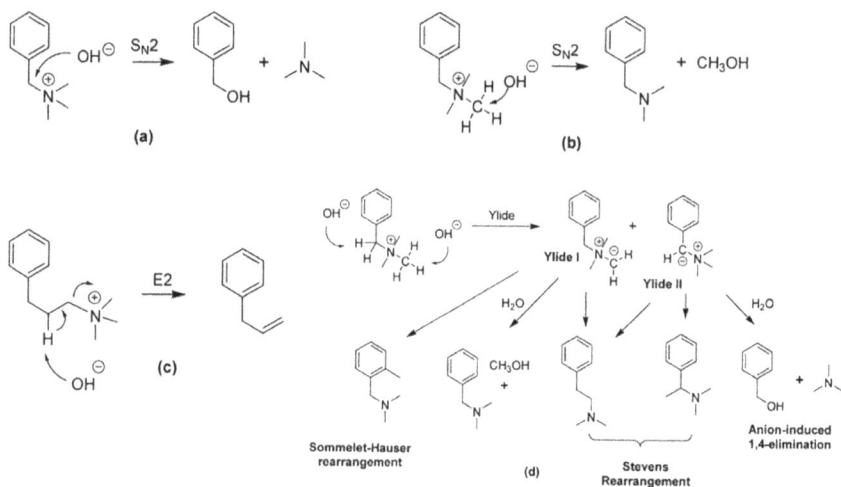

FIGURE 11.11   Possible degradation pathways for quaternary ammonium groups: (a) S_N2 benzyl substitution (main, nucleophilic substitution), (b) S_N2 methyl substitution (minor, nucleophilic substitution), (c) β-elimination substitution (Hofmann elimination), and (d) ylide-intermediated rearrangements [79].

can undergo substitution on both the a-carbon, which converts the QA group to a tertiary amine and produces alcohol as a minor pathway and the benzylic carbon atom, resulting in the loss of a tertiary amine as the main pathway [78].

The ylide production pathway involves the removal of a proton from the methyl group by OH⁻ ions, resulting in the formation of a nitrogen line intermediate. This intermediate is then converted to tertiary amine and water through the Stevens or Sommelet–Hauser rearrangements, which can be influenced by factors such as temperature, nucleophile strength, and concentration. However, some studies suggest that this process is reversible and does not lead to the significant overall deterioration [80,81]. Im-functionalized AEMs exhibit superior chemical stability in high-pH environments compared to QA-functionalized AEMs due to their ring-shaped structure. However, as illustrated in Figure 11.12a–c, Im groups are still susceptible to ring-opening, S_N2 methyl substitution, and heterocycle deprotonation. Two effective methods for preventing functional group degradation in AEMs are the use of electron-donating groups and steric hindrance strategies. The electron donation approach adds electron-donating groups, such as the C2-methyl-substitution of the imidazolium group, to stabilize the electron distribution in the functional group and protect against nucleophilic attack. The steric hindrance approach shields the functional group with a bulky structure to prevent degradation. The alkaline stability improves when the C2 position is occupied by electron-donating groups, such as 1,2-dimethylimidazole (DmIm) [82]. Long spacers, such as DmIm, 1-butyl imidazole (BuIm), 1-methyl imidazole (EtIm), and 1-aminoethyl-2,3-dimethyl imidazolium (AeIm) [83], were tethered to the imidazolium to mitigate the degradation of functional groups. However, this conclusion is still debated, as others have reported that

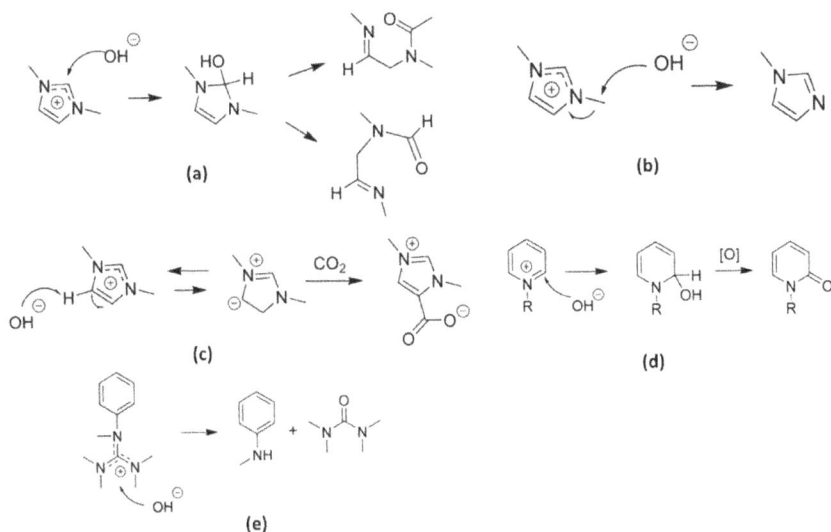

FIGURE 11.12 (a) Ring opening (imidazolium), (b) S_N2 methyl substitution (imidazolium), (c) heterocycle deprotonation (imidazolium), (d) S_N2 and ring opening (piperidinium, pyrrolidinium, and morpholinium), and (e) nucleophilic degradation (guanidinium) [79].

the increased length of a long alkyl affixed to the N−3 position in Im decreases the alkaline stability of functionalized membranes [84]. In addition to aliphatic substitution, benzyl substitution, such as 1-benzimidazole (BZM), provides better alkaline stability due to steric hindrance. Besides QA and Im, other cationic groups also suffer from OH⁻, as shown in Figure 11.12d and e, with pyridinium suffering from nucleophilic addition and displacement, and guanidinium suffering from nucleophilic degradation [79].

Including electron-donating groups or reducing electron-withdrawing substituents can slow the degradation of OH⁻. For instance, when subjected to alkaline conditions, the DMH cation was more effective at inhibiting PPO backbone hydrolysis in PPO-based AEMs than the TMA cation. However, the mitigation effect was not observed with the PSU backbone due to the presence of the electron-withdrawing sulfone group [85]. Additionally, MOF can enhance the alkaline stability of AEMs by reducing the interaction between strongly polar nitrile groups, such as benzonitrile, and side-chain functional cations, such as QA. The energy levels of the lowest unoccupied molecular orbital (LUMO) energies of benzyl imidazolium, benzyl morpholinium, benzyl imidazolium/benzonitrile, and benzyl morpholinium/benzonitrile were calculated to be −4.64, −4.40, −4.07, and −4.18 eV, respectively [86]. The LUMO energy of benzyl morpholinium was observed to be higher than that of benzyl imidazolium. The introduction of benzonitrile groups increased the LUMO energies of both imidazolium and morpholinium, due to the interaction between benzonitrile and functional groups. The high LUMO energy indicates reduced electron stabilization and acidity, which weakens the cationic head groups and reduces

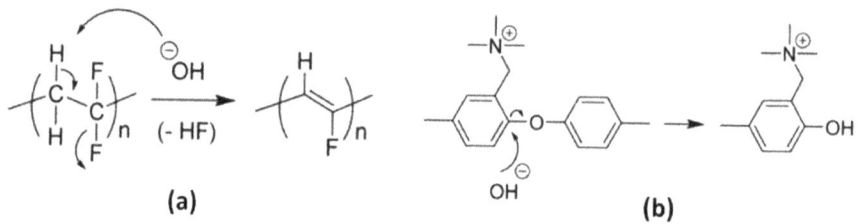

FIGURE 11.13 The degradation mechanism of polymer backbones: (a) dehydrochlorination and (b) $S_N Ar$ aryl ether cleavage [75].

their interaction with OH⁻ ions. Beyond attacking the cationic group, OH⁻ ions can also cause quaternary carbon and ether hydrolysis, which damages the backbone of AEMs in alkaline environments [87]. The degradation mechanism of the backbone under high-pH conditions is illustrated in Figure 11.13 [75]. Functionalization modifies the stability of the backbone by introducing functional groups, which depend on the charge distribution in the modified polymer. Consequently, the ether group in the backbone, as seen in PPO and PSF, becomes more vulnerable to OH⁻ ion attack after functionalization [88].

The chemical stability of AEMs is a crucial factor that affects their long-term durability, and it is determined by their alkaline and oxidative stability. The stability of the membrane is heavily influenced by the functional groups present in the head group. To enhance the alkaline stability of the head group, various strategies have been employed. These include the use of spacers between the head group and backbone, bulkier head groups, or electron-donating groups adjacent to the quaternized head group to reduce its acidity and make it less susceptible to nucleophilic attack [89]. While some strategies have been effective, they often lead to reduced ionic conductivity. For instance, incorporating a long spacer chain between functional groups and the polymer backbone can enhance membrane stability, with alkyl imidazolium proving more stable than benzyl imidazolium. The addition of a lengthy spacer chain can significantly reduce the chances of $S_N 2$ nucleophilic substitution reactions by OH⁻ on functional groups by stabilizing the attack reaction transition state, thereby improving the alkaline stability of the AEM. However, the effectiveness of this strategy is dependent on the specific AEM backbone and head group used, as it was not observed when applied to other AEM backbones, such as PPO [90].

Moreover, the density of functional groups can have an impact on the conductivity of the membrane, with densely packed functional groups exhibiting better conductivity than loosely packed ones. It has also been observed that di-quaternized membranes tend to exhibit larger ionic clusters than monoquaternized membranes. Increasing steric hindrance around the functional groups can protect against nucleophilic attack by OH⁻, as demonstrated by the incorporation of adjacent bulky groups near the reactive C2 position of the benzimidazolium group. This steric crowding hinders the attack, providing improved stability. Additionally, positioning the QA groups on the side chain instead of the main chain reduces steric hindrance in the aromatic backbone, promoting phase separation. However, this may also lead to a decrease in overall ionic conductivity [91,92].

11.4.2 Hydroxide Anions Attack

In addition to OH⁻ attacks, radicals like hydroxyl (OH) and peroxyl (•OOH) can be generated through electrode electrochemical reactions, such as those at the cathode of a fuel cell or the anode of an electrolyzer, or through chemical reactions involving oxygen or the OH⁻ counterion and cationic head group [93]. These radicals, along with hydroxide ions, can cause severe and irreversible degradation of the membrane. Given the harsh working conditions that AEMs in fuel cell systems endure, including high temperature, humidity, and exposure to oxidizing chemicals, oxidative stability is a critical factor. If an AEM lacks oxidative stability, it can deteriorate chemically over time, leading to a decline in performance, mechanical damage, and ultimately failure of the fuel cell system. Therefore, selecting AEMs with high oxidative stability is essential to ensure the long-term dependability and performance of the cell system [94,95]. Although alkaline stability is a critical factor, it is not enough to guarantee membrane durability in deionized water-fed AEMFC and AEMWE. Under these conditions, oxidative stability becomes crucial, as the presence of free radicals, such as hydroxyl free radicals (•OH) and superoxide anion radicals ($O2^{\bullet-}$), can cause polymer electrolyte chain scission, leading to a significant reduction in mechanical strength [96]. Visible changes in membrane color and mechanical properties loss after stability testing indicated that membrane electrolytes could degrade at an accelerated rate. Oxygen was found to increase the degradation rate by four times more than nitrogen, suggesting its role in accelerating the process. Chemical analysis revealed the loss of functional groups and removal of the benzene ring connecting functional groups and backbones as responsible for membrane degradation, with the rate of degradation closely linked to oxidant concentration. These findings highlight the importance of oxidative stability for AEMs and the need to improve oxidation resistance. Figure 11.14 illustrates the three main steps in the formation of hydroxyl (•OH) and superoxide anion radicals ($O_2^{\bullet-}$) [78]. Carbanions are produced through deprotonation in alkaline conditions (Step 1). In Step 2, the carbanion reduces dioxygen, generating the superoxide anion radical and an organic free radical. Step 3 involves the hydroxide ion donating an electron to the organic free radical, renewing the carbanion, and generating the highly reactive hydroxyl free radical. At high pH levels, the hydroperoxyl radical ($HO_2^{\bullet}$) is deprotonated to form $O_2^{\bullet-}$.

The most widely used method for evaluating oxidative stability involves measuring weight loss, IEC loss, or reduced ionic conductivity after immersing the material

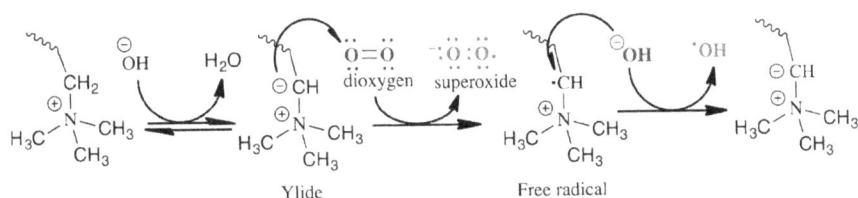

FIGURE 11.14 The mechanism for generating reactive oxygen species by one-electron reduction of dioxygen under alkaline conditions. Reproduced with permission from Parrondo et al. [78].

in oxygenated deionized water for extended periods or in Fenton's solution (3 wt% H_2O_2 plus 2 ppm $FeSO_4$) for shorter durations [97]. Reduction of OH^- can lead to the formation of oxidizing species, such as OH and $\cdot OOH/H_2O_2$. Additionally, oxygen can be reduced to produce superoxide ions. The reduction of OH^- and oxygen can be catalyzed by the head group ylide and the presence of an electrocatalyst. During AEM oxidative stability tests, $\cdot OH$ and $O_2 \cdot^-$ were detected in oxygen. When oxygen molecules acquire the ylide electron, they give rise to QA radicals that subsequently break down into tertiary amines and ethylene [74]. Wierzbicki et al. utilized electron paramagnetic resonance (EPR) spectroscopy to detect and identify a radical both during and after extended AEMFC operation. The study examined four types of AEMs, including hydrocarbon backbone membranes with QA groups, radiation-grafted membranes composed of low-density polyethylene (LDPE) with covalently bonded benzyl trimethylammonium head groups, a $2,2'',4,4'',6,6''$-hexamethylpterphenylene membrane, and N-methylated poly(benzimidazolium) (PMBI) and commercial FAA-3-PK-130 membranes marked FAA3. The primary adducts detected during micro-AEMFC operation were DMPOOOH and DMPO-OH on the cathode side and DMPO-H on the anode side.

To enhance oxidative stability, free radical scavengers such as ceria and sulfide groups can be incorporated into the membrane structure. These scavengers have proven effective in sister technology to AEMs, such as PEMFCs. For example, Bu et al. developed a 1,2,4-Triazole-functionalized poly(arylene ether ketone) (PAEK)-based PEM with improved oxidative stability by introducing sulfide groups into the membrane. These sulfide-containing membranes exhibited excellent resistance to oxidative degradation and maintained their shape for over 50 hours in Fenton's solution. However, these strategies are yet to be tested in AEMs. Although phenolic groups in the solution can potentially enhance the oxidative stability of AEMs, they are not regenerated and therefore provide only limited sacrificial protection. While aryl imidazolium demonstrated stability in 3 M KOH for 10,000 hours, its conductivity and electrolysis performance were poor even at high concentrations of KOH.

The ionic conductivity of AEMs is determined by their IEC, hydration level, and micro-morphology, with higher IEC typically leading to greater water absorption and ionic conductivity. However, to achieve optimal membrane performance, a balance must be struck between ionic conductivity and water absorption.

11.4.3 Measures to Slow Degradation

Poly(arylene ether), poly(arylene ether ketone), and poly(arylene ether sulfone) are examples of APEs with heteroatomic linkages that should be avoided because they are prone to hydroxide attack via nucleophilic aromatic substitution (S_NAr) at the ether linkage, which can result in a loss of molecular weight [98]. AEM cross-linking may result via hydroxide assault on fluorinated phenyl groups via S_NAr, which in this case results in the production of phenol or aryl ethers [99]. When utilized as ionomers, APEs containing aromatic groups can adsorb onto electrocatalysts and "poison the catalyst," resulting in performance loss and deterioration of the ionomer via phenyl oxidation, resulting in further performance loss [100–102]. Because the AEMWE anode's working potential is substantially higher than the AEMFC cathode's, encouraging the oxidation of phenyl in ionomers, this general degradation route is more common in AEMWE than in AEMFC [88]. The adsorption energy of the electrocatalyst can be dramatically changed by changes in the structure of aromatic molecules [100]. As a result, employing aromatic groups with lower adsorption energies can increase the functionality and stability of aromatic APEs [103]. Therefore, to prevent degradation through S_NAr and phenyl oxidation, we advise using completely aliphatic APEs, such as poly(norbornene) and polyethylene, which are more stable under alkaline and oxidizing environments [104] (Figure 11.15).

An intensive study conducted over the previous 10 years has shown growing tendencies in AEP design. Due to its strong ionic conductivity and off-site endurance, QA (quaternary amine group) is by far the most researched cationic group among them. Aryl ether-free AEPs have been demonstrated to perform better for polymer backbones than aryl ether AEPs [105]. Since they are widely accessible, have a stable AEP backbone, and have well-designed side chains that prevent Hoffman elimination and counterion condensation, AEPs with acyclic QAs continue to be of interest. Due to their high alkaline solution stability, heterocyclic aliphatic QAs might garner more attention [30]. Aliphatic heterocyclic QA cations with spirocyclic structures are a specific subclass of this type. Due to the high transition state of the spiro ring structure, which resists degrading processes, this form of AEP demonstrates exceptional alkaline stability [106]. Examples of QA salt incorporation into polymer backbones, direct aliphatic or aromatic polymer attachment, and introduction as cross-linkers to create networks are included in the presented investigations [107]. To learn more about how the cationic structure and the base stability of AEP are related, in situ AEMWE cell testing is necessary [108] (Figure 11.16).

The ether-free backbone is favored for structural design when it comes to AEPs because polybenzimidazole (PBI), polyphenylene, and polyolefin-type AEMs have all been thoroughly investigated. AEPs of the PBI and polyphenylene types are highly thermally and chemically stable. A high density of electronegative pyridine nitrogen (N) is provided by the special benzimidazole repeating unit in the main chain, which can also establish hydrogen bonds to conduct OH^-. However, their ionic conductivity and processability are insufficient for AEMWE [109,110]. In the meantime, poly aryl piperidine structures are utilized frequently in modern times due to their great conductivity balance and superior base stability.

AEMWE has the potential to take the place of AWE and PEMWE for large-scale applications, as evidenced by its significantly enhanced performance. Future academic and industrial research on AEMWE should follow a roadmap that includes the following steps: (1) designing and synthesizing APE with high IEC and mechanical stability to close the gap between dilute alkaline and pure water AEMWE; (2) testing electrocatalysts, standardization of performance metrics, and complete elimination of PGM electrocatalysts in AEMWE; and (3) testing dilute alkaline and pure water AEMWE systems to optimize electrolyzers. It is necessary to provide standardized accelerated stress test methodologies for AEMWE [104].

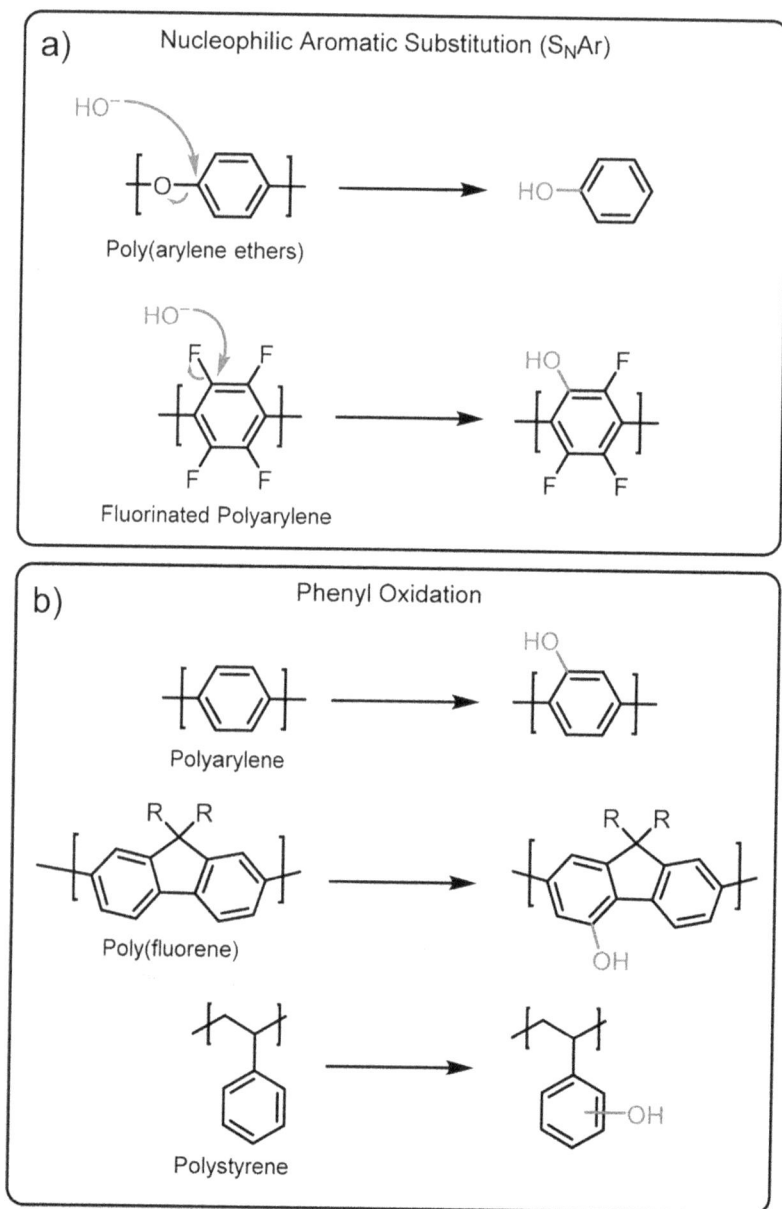

FIGURE 11.15 Polymer backbone degradation mechanism under alkaline and oxidative conditions. (a) Nucleophilic aromatic substitution (S$_N$Ar) degradation mechanism observed for poly(arylene ether), poly(arylene ether ketone), poly(arylene ether sulfone), and fluorinated poly(arylenes). (b) Phenyl oxidation was observed in APEs containing aromatic moieties [104].

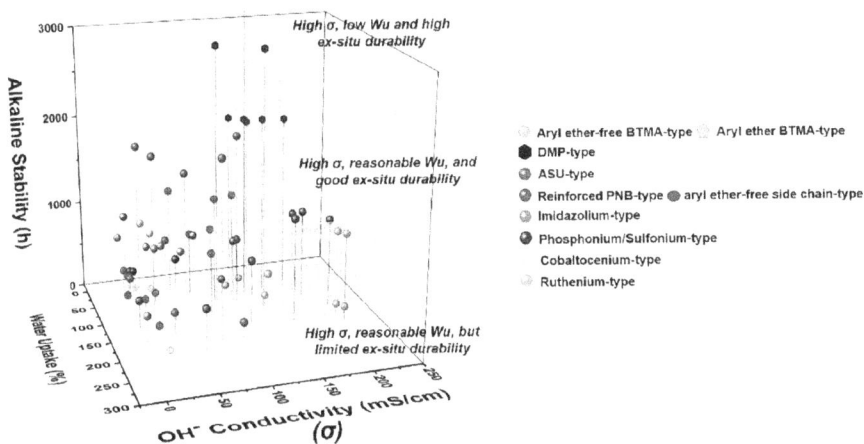

FIGURE 11.16 Comparison between the water uptake, OH⁻ conductivity (σ), and ex situ stability of typical BTMA-, DMP-, ASU-, side-chain-, imidazolium-, phosphonium/sulfonium-, cobaltocenium-, and ruthenium-type AEPs. The water uptake (WU) corresponds to the σ value at the same temperature (most AEPs are recorded at 80°C, but some for the side-chain-, imidazolium-, sulfonium-, and ruthenium-type AEPs are plotted at room temperature and 60°C due to insufficient information). The alkaline stability was recorded based on the temporal stability of AEPs in 1 M NaOH or KOH at 80°C with degradation <10% and some of the stable AEPs were evaluated at harsher conditions [105].

11.5 CHALLENGES AND PROSPECTS

Nevertheless, AEM electrolysis faces several challenges in achieving low-cost hydrogen generation, and several issues must be resolved in this regard. Current AEM electrolysis technology has several flaws that require attention, including the following:

1. It is essential to develop conductive polymers that possess high stability, conductivity, and minimal gas crossover.
2. Understanding the interaction between hydroxyl ions, QA functional groups, and polymer backbones is crucial in preventing nucleophilic attack and Hoffmann elimination.
3. To enhance the performance of AEMWEs, hydrophilic QA functionalized styrene copolymer and hydrophobic hydrophilic block cross-linked AEMs can be employed.
4. Examining polarization, gas crossover, electrochemical impedance, and chronoamperometry is necessary to describe the attributes of AEMs under AEM water electrolysis circumstances.
5. Membranes containing multi-cations on the side chains may be a useful material for enhancing ionic conductivity by effectively utilizing water due to the structure-induced segregation of microphases.

6. Increasing the content of imidazole groups can lead to improved IEC and water absorption.

7. The use of nanoparticle fillers and blends has the potential to enhance hydroxyl ion conductivity. However, many ionomers are toxic, and the growth of ionomers can be slow. Therefore, creating a chemically stable, highly conductive, and thermally stable ionomer should be pursued for AEM electrolysis.

Functional groups are a critical component of membrane chemical stability, and various techniques are utilized to enhance the alkaline stability of the head group. These approaches include incorporating spacers between the head group and backbone, using bulkier head groups, and introducing adjacent electron-donating groups to decrease the acidity of quaternized head groups and their susceptibility to nucleophilic attack. While these strategies can be effective in certain situations, they often result in a reduction in ionic conductivity and depend on the specific backbone and head group employed. For instance, membrane stability can be improved by inserting a long spacer chain between functional groups and the polymer main chain. Alkylimidazolium has been discovered to be more stable than benzyl imidazolium due to the long spacer chain's ability to stabilize the transition state of the attack reaction and reduce the S_N2 nucleophilic substitution attack of OH^- on the functional group, thereby enhancing AEM alkaline stability. However, this effect was not observed when used in other AEM backbones, such as PPO. Additionally, dense functional groups may exhibit better conductivity than loose ones, with the di-quaternized membrane demonstrating a larger scattering for ionic clusters than the mono-quaternized membrane. Increasing steric hindrance can also protect functional groups, such as incorporating adjacent bulky groups near the reactive C2 position of the benzimidazolium group to hinder nucleophilic attack by OH^- due to steric crowding. Furthermore, incorporating QA groups on the side chain rather than the main chain of the membrane decreases the aromatic backbone's steric hindrance and promotes phase separation. However, this may reduce ionic conductivity accordingly. The degree of phase separation is determined by the intensity of ion aggregation, and the resulting polymer's properties are largely determined by the synergism of dense functional groups per segment with the size effect of the phase blocks. Hydrophilic features enhance hydroxide ion mobility and provide wide ion transport channels due to the strong-field effects of the dense functional clusters, while hydrophobic segments, such as fluorinated hydrophobic moieties, decrease the possibility of hydroxide attack and improve backbone stability. As a result, selecting appropriate functional groups and materials with improved stability in alkaline media and considering the stability, degradation, and modification of functional groups are important factors to consider when designing effective AEMs for ion transfer in AEMWE devices.

During the operation of AEMWE devices, it is expected that the local pH at the electrode reaction region and the ion conduction channel within the AEM will increase. This can lead to the degradation of the molecular structure of AEM due to the high reactivity of hydroxide anions with functional groups in a high-pH environment, resulting in a deterioration of anion conductivity. This degradation process occurs when hydroxide anions directly attack the cationic sites of the ionomeric membrane. To address this issue, the search for appropriate functional groups and novel cationic materials with improved stability in alkaline conditions is a promising solution. In AEM, functional groups play a crucial role in ion transfer and should be carefully considered in the design of AEMs. Their stability, deterioration, and modification processes are critical factors to consider when designing effective AEMs.

REFERENCES

1. Yu, Z.Y., et al., Clean and affordable hydrogen fuel from alkaline water splitting: Past, recent progress, and future prospects. *Adv Mater*, 2021. **33**(31): p. e2007100.
2. Yang, Y., et al., Electrocatalysis in alkaline media and alkaline membrane-based energy technologies. *Chem Rev*, 2022. **122**(6): pp. 6117–6321.
3. Du, N., et al., Anion-exchange membrane water electrolyzers. *Chem Rev*, 2022. **122**(13): pp. 11830–11895.
4. Xu, Q., et al., Anion exchange membrane water electrolyzer: Electrode design, lab-scaled testing system and performance evaluation. *EnergyChem*, 2022. 4(5): p. 100087.
5. Chand, K. and O. Paladino, Recent developments of membranes and electrocatalysts for the hydrogen production by anion exchange membrane water electrolysers: A review. *Arab J Chem*, 2023. **16**(2): p. 104451.
6. Yan, X., et al., A membrane-free flow electrolyzer operating at high current density using earth-abundant catalysts for water splitting. *Nat Commun*, 2021. **12**(1): p. 4143.
7. Pavel, C.C., et al., Highly efficient platinum group metal free based membrane-electrode assembly for anion exchange membrane water electrolysis. *Angew Chem Int Ed Engl*, 2014. **53**(5): pp. 1378–1381.
8. Merle, G., M. Wessling, and K. Nijmeijer, Anion exchange membranes for alkaline fuel cells: A review. *J Membrane Sci*, 2011. **377**(1–2): pp. 1–35.
9. Li, C. and J.-B. Baek, The promise of hydrogen production from alkaline anion exchange membrane electrolyzers. *Nano Energy*, 2021. **87**: p. 106162.
10. Vincent, I., E.C. Lee, and H.M. Kim, Comprehensive impedance investigation of low-cost anion exchange membrane electrolysis for large-scale hydrogen production. *Sci Rep*, 2021. **11**(1): p. 293.
11. Leng, Y., et al., Solid-state water electrolysis with an alkaline membrane. *J Am Chem Soc*, 2012. **134**(22): pp. 9054–9057.
12. Zoulias, E., *A Review on Water Electrolysis*. Tcjst, 2004.
13. Berezina, N.P., et al., Characterization of ion-exchange membrane materials: Properties vs structure. *Adv Colloid Interface Sci*, 2008. **139**(1–2): pp. 3–28.
14. Pan, J., et al., Constructing ionic highway in alkaline polymer electrolytes. *Energy Environ Sci*, 2014. **7**(1): pp. 354–360.
15. Jiang, S. and B.P. Ladewig, High ion-exchange capacity semihomogeneous cation exchange membranes prepared via a novel polymerization and sulfonation approach in porous polypropylene. *ACS Appl Mater Interfaces*, 2017. **9**(44): pp. 38612–38620.
16. Carmo, M., et al., A comprehensive review on PEM water electrolysis. *Int J Hydrogen Energy*, 2013. **38**(12): pp. 4901–4934.
17. Hagesteijn, K.F.L., S. Jiang, and B.P. Ladewig, A review of the synthesis and characterization of anion exchange membranes. *J Mater Sci*, 2018. **53**(16): pp. 11131–11150.
18. Najibah, M., et al., PBI nanofiber mat-reinforced anion exchange membranes with covalently linked interfaces for use in water electrolysers. *J Membrane Sci*, 2021. **640**: p. 119832.

19. Ziv, N. and D.R. Dekel, A practical method for measuring the true hydroxide conductivity of anion exchange membranes. *Electrochem Commun*, 2018. **88**: pp. 109–113.

20. Chen, C., et al., Hydroxide solvation and transport in anion exchange membranes. *J Am Chem Soc*, 2016. **138**(3): pp. 991–1000.

21. Foglia, F., et al., Disentangling water, ion and polymer dynamics in an anion exchange membrane. *Nat Mater*, 2022. **21**(5): pp. 555–563.

22. Thomas, O.D., et al., A stable hydroxide-conducting polymer. *J Am Chem Soc*, 2012. **134**(26): pp. 10753–10756.

23. Dekel, D.R., et al., Effect of water on the stability of quaternary ammonium groups for anion exchange membrane fuel cell applications. *Chem Mater*, 2017. **29**(10): pp. 4425–4431.

24. Lin, B., et al., Cross-linked alkaline ionic liquid-based polymer electrolytes for alkaline fuel cell applications. *Chem Mater*, 2010. **22**(24): pp. 6718–6725.

25. Fang, J., et al., Cross-linked, ETFE-derived and radiation grafted membranes for anion exchange membrane fuel cell applications. *Int J Hydrogen Energy*, 2012. **37**(1): pp. 594–602.

26. Li, Z., et al., Enhancing hydroxide conductivity and stability of anion exchange membrane by blending quaternary ammonium functionalized polymers. *Electrochim Acta*, 2017. **240**: pp. 486–494.

27. Stokes, K.K., J.A. Orlicki, and F.L. Beyer, RAFT polymerization and thermal behavior of trimethylphosphonium polystyrenes for anion exchange membranes. *Polym Chem*, 2011. **2**(1): pp. 80–82.

28. Faraj, M., et al., New LDPE based anion-exchange membranes for alkaline solid polymeric electrolyte water electrolysis. *Int J Hydrogen Energy*, 2012. **37**(20): pp. 14992–15002.

29. Mabrouk, W., et al., Ion exchange membranes based upon crosslinked sulfonated polyethersulfone for electrochemical applications. *J Membrane Sci*, 2014. **452**: pp. 263–270.

30. Du, N., et al., Anion-exchange membrane water electrolyzers. *Chem Rev*, 2022. **122**(13): pp. 11830–11895.

31. Mandal, M., G. Huang, and P.A. Kohl, Highly conductive anion-exchange membranes based on cross-linked poly (norbornene): Vinyl addition polymerization. *ACS Appl Energy Mater*, 2019. **2**(4): pp. 2447–2457.

32. You, W., et al., Expeditious synthesis of aromatic-free piperidinium-functionalized polyethylene as alkaline anion exchange membranes. *Chem Sci*, 2021. **12**(11): pp. 3898–3910.

33. You, W., et al., Highly conductive and chemically stable alkaline anion exchange membranes via ROMP of trans-cyclooctene derivatives. *Proc Natl Acad Sci*, 2019. **116**(20): pp. 9729–9734.

34. Lee, B., et al., Physically-crosslinked anion exchange membranes by blending ionic additive into alkyl-substituted quaternized PPO. *J Membrane Sci*, 2019. **574**: pp. 33–43.

35. Zhang, X., et al., Enhancement of the mechanical properties of anion exchange membranes with bulky imidazolium by "thiol-ene" crosslinking. *J Membrane Sci*, 2020. **596**: p. 117700.

36. Kwon, S., B. Lee, and T.-H. Kim, High performance blend membranes based on densely sulfonated poly (fluorenyl ether sulfone) block copolymer and imidazolium-functionalized poly (ether sulfone). *Int J Hydrogen Energy*, 2017. **42**(31): pp. 20176–20186.

37. Pan, J., et al., Self-crosslinked alkaline polymer electrolyte exceptionally stable at 90C. *Chem Commun*, 2010. **46**(45): pp. 8597–8599.

38. Tibbits, A.C., et al., A single-step monomeric photo-polymerization and crosslinking via thiol-ene reaction for hydroxide exchange membrane fabrication. *Journal of The Electrochemical Society*, 2015. **162**(10): p. F1206.

39. Park, J.-S., et al., Performance of solid alkaline fuel cells employing anion-exchange membranes. *J Power Sources*, 2008. **178**(2): pp. 620–626.

40. Robertson, N.J., et al., Tunable high performance cross-linked alkaline anion exchange membranes for fuel cell applications. *J Am Chem Soc*, 2010. **132**(10): pp. 3400–3404.

41. Wang, L. and M.A. Hickner, Low-temperature crosslinking of anion exchange membranes. *Polymer Chem*, 2014. **5**(8): pp. 2928–2935.

42. Wu, L., et al., Thermal crosslinking of an alkaline anion exchange membrane bearing unsaturated side chains. *J Membrane Sci*, 2015. **490**: pp. 1–8.

43. Wang, C., et al., Controllable cross-linking anion exchange membranes with excellent mechanical and thermal properties. *Macromol Mater Eng*, 2018. **303**(3): p. 1700462.

44. Chen, N., et al., Tunable multi-cations-crosslinked poly (arylene piperidinium)-based alkaline membranes with high ion conductivity and durability. *J Membrane Sci*, 2019. **588**: p. 117120.

45. Lee, K.H., et al., Highly conductive and durable poly (arylene ether sulfone) anion exchange membrane with end-group cross-linking. *Energy Environ Sci*, 2017. **10**(1): pp. 275–285.

46. Sung, S., et al., Crosslinked PPO-based anion exchange membranes: The effect of crystallinity versus hydrophilicity by oxygen-containing crosslinker chain length. *J Membrane Sci*, 2021. **619**: p. 118774.

47. Hagesteijn, K.F., S. Jiang, and B.P. Ladewig, A review of the synthesis and characterization of anion exchange membranes. *J Mater Sci*, 2018. **53**(16): pp. 11131–11150.

48. Lin, J., et al., Thermoplastic interpenetrating polymer networks based on polybenzimidazole and poly (1, 2-dimethy-3-allylimidazolium) for anion exchange membranes. *Electrochim Acta*, 2017. **257**: pp. 9–19.

49. Arges, C.G., et al., Interconnected ionic domains enhance conductivity in microphase separated block copolymer electrolytes. *J Mater Chem A*, 2017. **5**(11): pp. 5619–5629.

50. Heitner-Wirguin, C., Recent advances in perfluorinated ionomer membranes: Structure, properties and applications. *J Membrane Sci*, 1996. **120**(1): pp. 1–33.

51. Pan, J., et al., Constructing ionic highway in alkaline polymer electrolytes. *Energy Environ Sci*, 2014. **7**(1): pp. 354–360.

52. Zhou, X., et al., Rational design of comb-shaped poly(arylene indole piperidinium) to enhance hydroxide ion transport for H_2/O_2 fuel cell. *J Membrane Sci*, 2021. **631**: p. 119335.

53. Yang, L., et al., Poly(aryl piperidinium) anion exchange membranes with cationic extender sidechain for fuel cells. *J Membrane Sci*, 2022. **653**: p. 120448.

54. Niu, M., et al., Pendent piperidinium-functionalized blend anion exchange membrane for fuel cell application. *Int J Hydrogen Energy*, 2019. **44**(29): pp. 15482–15493.

55. Gao, W.T., et al., High-performance tetracyclic aromatic anion exchange membranes containing twisted binaphthyl for fuel cells. *J Membrane Sci*, 2022. **655**: p. 120578.

56. Xu, F., Y. Su, and B. Lin, Progress of alkaline anion exchange membranes for fuel cells: The effects of micro-phase separation. *Front Mater*, 2020. **7**: p. 4.

57. Ran, J., et al., Ion exchange membranes: New developments and applications. *J Membrane Sci*, 2017. **522**: pp. 267–291.

58. Elumalai, V. and D. Sangeetha, Preparation of anion exchangeable titanate nanotubes and their effect on anion exchange membrane fuel cell. *Mater Design*, 2018. **154**: pp. 63–72.

59. Lee, K.H., et al., Functionalized TiO_2 mediated organic-inorganic composite membranes based on quaternized poly(arylene ether ketone) with enhanced ionic conductivity and alkaline stability for alkaline fuel cells. *J Membrane Sci*, 2021. **634**: p. 119435.

60. Li, J., et al., Enhanced hydroxide conductivity of imidazolium functionalized polysulfone anion exchange membrane by doping imidazolium surface-functionalized nanocomposites. *RSC Adv*, 2016. **6**(63): pp. 58380–58386.

61. Rambabu, K., et al., ZrO$_2$ incorporated polysulfone anion exchange membranes for fuel cell applications. *Int J Hydrogen Energy*, 2020. **45**(54): pp. 29668–29680.
62. Chen, N., et al., High-performance layered double hydroxide/poly(2,6-dimethyl-1,4-phenylene oxide) membrane with porous sandwich structure for anion exchange membrane fuel cell applications. *J Membrane Sci*, 2018. **552**: pp. 51–60.
63. Chand, K. and O. Paladino, Recent developments of membranes and electrocatalysts for the hydrogen production by anion exchange membrane water electrolysers: A review. *Arab J Chem*, 2023. **16**(2): p. 104451.
64. Henkensmeier, D., et al., Overview: State-of-the art commercial membranes for anion exchange membrane water electrolysis. *J Electrochem Energy Convers Storage*, 2021. **18**(2): p. 024001.
65. Pavel, C.C., et al., Highly efficient platinum group metal free based membrane-electrode assembly for anion exchange membrane water electrolysis. *Angew Chem Int Ed*, 2014. **53**(5): pp. 1378–1381.
66. Cho, M.K., et al., Factors in electrode fabrication for performance enhancement of anion exchange membrane water electrolysis. *J Power Sources*, 2017. **347**: pp. 283–290.
67. Vincent, I., A. Kruger, and D. Bessarabov, Development of efficient membrane electrode assembly for low cost hydrogen production by anion exchange membrane electrolysis. *Int J Hydrogen Energy*, 2017. **42**(16): pp. 10752–10761.
68. Park, J.E., et al., High-performance anion-exchange membrane water electrolysis. *Electrochim Acta*, 2019. **295**: pp. 99–106.
69. Park, E.J., et al., How does a small structural change of anode ionomer make a big difference in alkaline membrane fuel cell performance? *J Mater Chem A*, 2019. **7**(43): pp. 25040–25046.
70. Liu, Z., et al., An alkaline water electrolyzer with sustainion(tm) membranes: 1 A/cm² at 1.9 V with base metal catalysts. *ECS Trans*, 2017. **77**(9): p. 71.
71. Liu, Z., et al., The effect of membrane on an alkaline water electrolyzer. *Int J Hydrogen Energy*, 2017. **42**(50): pp. 29661–29665.
72. Fortin, P., et al., High-performance alkaline water electrolysis using Aemion(tm) anion exchange membranes. *J Power Sources*, 2020. **451**: p. 227814.
73. Wang, J., et al., Poly (aryl piperidinium) membranes and ionomers for hydroxide exchange membrane fuel cells. *Nat Energy*, 2019. **4**(5): pp. 392–398.
74. Espiritu, R., et al., Degradation of radiation grafted anion exchange membranes tethered with different amine functional groups via removal of vinylbenzyl trimethylammonium hydroxide. *J Power Sources*, 2018. **375**: pp. 373–386.
75. Mustain, W.E., et al., Durability challenges of anion exchange membrane fuel cells. *Energy Environ Sci*, 2020. **13**(9): pp. 2805–2838.
76. Espiritu, R., et al., Degradation of radiation grafted hydroxide anion exchange membrane immersed in neutral pH: Removal of vinylbenzyl trimethylammonium hydroxide due to oxidation. *J Mater Chem A*, 2017. **5**(3): pp. 1248–1267.
77. Chen, J., et al., A general strategy to enhance the alkaline stability of anion exchange membranes. *J Mater Chem A*, 2017. **5**(13): pp. 6318–6327.
78. Parrondo, J., et al., Reactive oxygen species accelerate degradation of anion exchange membranes based on polyphenylene oxide in alkaline environments. *Phys Chem Phys*, 2016. **18**(29): pp. 19705–19712.
79. Elwan, H.A., M. Mamlouk, and K. Scott, A review of proton exchange membranes based on protic ionic liquid/polymer blends for polymer electrolyte membrane fuel cells. *J Power Sources*, 2021. 484: p. 229197.
80. Lu, W., et al., Preparation and characterization of imidazolium-functionalized poly (ether sulfone) as anion exchange membrane and ionomer for fuel cell application. *Int J Hydrogen Energy*, 2013. **38**(22): pp. 9285–9296.

81. Hossain, M.A., et al., Novel hydroxide conducting sulfonium-based anion exchange membrane for alkaline fuel cell applications. *Int J Hydrogen Energy*, 2016. **41**(24): pp. 10458–10465.

82. Yan, X., et al., Long-spacer-chain imidazolium functionalized poly(ether ketone) as hydroxide exchange membrane for fuel cell. *Int J Hydrogen Energy*, 2016. **41**(33): pp. 14982–14990.

83. Yan, X., et al., Imidazolium-functionalized polysulfone hydroxide exchange membranes for potential applications in alkaline membrane direct alcohol fuel cells. *Int J Hydrogen Energy*, 2012. **37**(6): pp. 5216–5224.

84. Hibbs, M.R., Alkaline stability of poly(phenylene)-based anion exchange membranes with various cations. *J Polym Sci Part B: Polym Phy*, 2013. **51**(24): pp. 1736–1742.

85. Arges, C.G., et al., Mechanically stable poly(arylene ether) anion exchange membranes prepared from commercially available polymers for alkaline electrochemical devices. *J Electrochem Soc*, 2015. **162**(7): pp. F686–F693.

86. Yan, X., et al., Improvement of alkaline stability for hydroxide exchange membranes by the interactions between strongly polar nitrile groups and functional cations. *J Membrane Sci*, 2017. **533**: pp. 121–129.

87. Arges, C.G. and V. Ramani, Two-dimensional NMR spectroscopy reveals cation-triggered backbone degradation in polysulfone-based anion exchange membranes. *Proc Natl Acad Sci U S A*, 2013. **110**(7): pp. 2490–2495.

88. Li, D., et al., Durability of anion exchange membrane water electrolyzers. *Energy Environ Sci*, 2021. **14**(6): pp. 3393–3419.

89. Lu, W., et al., Preparation of anion exchange membranes by an efficient chloromethylation method and homogeneous quaternization/crosslinking strategy. *Solid State Ionics*, 2013. **245–246**: pp. 8–18.

90. Xiao Lin, C., et al., Quaternized triblock polymer anion exchange membranes with enhanced alkaline stability. *J Membrane Sci*, 2017. **541**: pp. 358–366.

91. Li, T., et al., Surface microstructure and performance of TiN monolayer film on titanium bipolar plate for PEMFC. *Int J Hydrogen Energy*, 2021. **46**(61): pp. 31382–31390.

92. Wang, X.Q., et al., Anion exchange membranes from hydroxyl-bearing poly(ether sulfone)s with flexible spacers via ring-opening grafting for fuel cells. *Int J Hydrogen Energy*, 2017. **42**(30): pp. 19044–19055.

93. Maurya, S., et al., Stability of composite anion exchange membranes with various functional groups and their performance for energy conversion. *J Membrane Sci*, 2013. **443**: pp. 28–35.

94. Zhang, P., B. Shen, and H. Pu, Robust, dimensional stable, and self-healable anion exchange membranes via quadruple hydrogen bonds. *Polymer*, 2022. 245: p. 124698.

95. Sung, S., et al., Preparation of crosslinker-free anion exchange membranes with excellent physicochemical and electrochemical properties based on crosslinked PPO-SEBS. *J Mater Chem A*, 2021. **9**(2): pp. 1062–1079.

96. Wierzbicki, S., et al., Are radicals formed during anion-exchange membrane fuel cell operation? *J Phys Chem Lett*, 2020. **11**(18): pp. 7630–7636.

97. Manohar, M., et al., Efficient and stable anion exchange membrane: Tuned membrane permeability and charge density for molecular/ionic separation. *J Membrane Sci*, 2015. **496**: pp. 250–258.

98. Mohanty, A.D., et al., Systematic alkaline stability study of polymer backbones for anion exchange membrane applications. *Macromolecules*, 2016. **49**(9): pp. 3361–3372.

99. Miyanishi, S. and T. Yamaguchi, Highly conductive mechanically robust high M w polyfluorene anion exchange membrane for alkaline fuel cell and water electrolysis application. *Polym Chem*, 2020. **11**(23): pp. 3812–3820.

100. Matanovic, I., et al., Adsorption of polyaromatic backbone impacts the performance of anion exchange membrane fuel cells. *Chem Mater*, 2019. **31**(11): pp. 4195–4204.

101. Maurya, S., et al., On the origin of permanent performance loss of anion exchange membrane fuel cells: Electrochemical oxidation of phenyl group. *J Power Sources*, 2019. **436**: p. 226866.

102. Li, D., et al., Phenyl oxidation impacts the durability of alkaline membrane water electrolyzer. *ACS Appl Mater Interfaces*, 2019. **11**(10): pp. 9696–9701.

103. Li, D., et al., Highly quaternized polystyrene ionomers for high performance anion exchange membrane water electrolysers. *Nat Energy*, 2020. **5**(5): pp. 378–385.

104. Li, Q., et al., Anion exchange membrane water electrolysis: The future of green hydrogen. *J Phys Chem C*, 2023. **127**(17): pp. 7901–7912.

105. Chen, N. and Y.M. Lee, Anion exchange polyelectrolytes for membranes and ionomers. *Progress Polym Sci*, 2021. **113**: p. 101345.

106. Gu, L., et al., Spirocyclic quaternary ammonium cations for alkaline anion exchange membrane applications: An experimental and theoretical study. *RSC Adv*, 2016. **6**(97): pp. 94387–94398.

107. Olsson, J.S., T.H. Pham, and P. Jannasch, Poly (N, N-diallylazacycloalkane) s for anion-exchange membranes functionalized with N-spirocyclic quaternary ammonium cations. *Macromolecules*, 2017. **50**(7): pp. 2784–2793.

108. Meek, K.M., et al., High-throughput anion exchange membrane characterization at NREL. *ECS Trans*, 2019. **92**(8): p. 723.

109. Savagado, O. and B. Xing, Hydrogen/oxygen polymer electrolyte membrane fuel cell (PEMFC) based on acid-doped polybenzimidazole (PBI). *J New Mater Electrochem Syst*, 2000. 3(4): pp. 343–347.

110. Zeng, L., et al., A high-performance sandwiched-porous polybenzimidazole membrane with enhanced alkaline retention for anion exchange membrane fuel cells. *Energy Environ Sci*, 2015. **8**(9): pp. 2768–2774.

12 Advanced Electrocatalysts for AEMWE

Jiacheng Wang and Xunlu Wang

12.1 INTRODUCTION

As a clean energy carrier, hydrogen has the potential to solve the energy crisis [1]. The future development direction of hydrogen energy should depend on green hydrogen production technology using renewable electricity [2–4]. China considered that the amount of hydrogen produced from renewable energy could reach 100,000–200,000 tons/year by 2025. The production of "green hydrogen" by electrolyzing water from renewable energy sources is a zero-carbon and environmentally friendly way [5]. With the rapid increase of new energy installed capacity, the cost of green hydrogen could be lower than that of coal power in the future, which is expected to achieve parity in 2030. As the key equipment for the preparation of green hydrogen, electrolytic cells have attracted much attention in the industry [6].

At present, three kinds of low-temperature water electrolysis technologies have been developed, including alkaline water electrolysis (AWE), proton exchange membrane water electrolysis (PEMWE), and anion exchange membrane water electrolysis (AEMWE) [7,8]. In principle, AEMWE technology combines the advantages of the other two technologies [9,10]. First, AEMWE employs an anion exchange membrane as a separator to provide an alkaline interfacial environment, which allows for the utilization of the cost-efficient catalysts and hardware [9]. Second, AEMWE can also run for a long time at high current density to produce high-purity hydrogen. However, the performance of AEMWE is not ideal as expected due to the low conductivity of AEM and the slow kinetics of the catalytic reaction [11]. Therefore, there is still need to develop more efforts based on material design, component optimization, and performance evaluation to enhance the performance of AEMWE and increase their commercial competitiveness.

Advanced electrocatalyst and electrode design is always the core technique for developing AEMWE [12]. The current HER electrocatalysts at the industrial level are highly dependent on the noble metal catalysts (e.g., Pt and Ru). However, the large-scale commercialization of the noble metal-based OER electrocatalysts is greatly hindered by their high cost, low crustal reserve, and relatively poor stability [13,14]. In addition, noble metals are mostly subjected to dissolution, agglomeration, and poor tolerance for poisoning during electrocatalysis [15]. Recently, some non-noble metal-based OER electrocatalysts, such as Fe-, Co-, and Ni-based catalysts, have demonstrated

DOI: 10.1201/9781003368939-12

291

superior catalytic performance to noble metal catalysts. Similarly, for anodic OER catalysts, in addition to the excellent noble metals IrO_2- and RuO_2-based catalysts, transition metal and perovskite catalysts also show excellent catalytic activity [16,17]. However, it is necessary to consider the issues of damage to catalyst during long-term operation under harsh operating conditions. Especially, deactivation of a catalyst due to severe changes of surface atomic structure and composition results in catalyst dissolution and agglomeration, as well as lower WE efficiency. Therefore, it is necessary to fully understand the internal catalytic mechanism of various catalysts and develop catalysts with high activity and long-term durability to meet the needs of AEMWE.

In this chapter, we start with the introduction of AEMWE, including the basic structure, operation mechanism, and advantages and disadvantages. In order to further improve the performance of AEMWE, the electrocatalysts as the core technology of AEMWE are very important, which are discussed in detail. Then, we classify and summarize recent advances in noble metal-based and non-noble metal-based electrocatalysts for AEMWE. We mainly introduce the existing problems and improvement strategies of various catalysts in detail, as well as obtain in-depth understanding of the internal mechanism of various catalysts from a theoretical perspective. Finally, it is concluded with the current states and challenges, along with some possible solutions and future directions in this chapter. We hope that this chapter can provide systematic insights for the development of highly active and stable electrocatalysts and promote the research and development of AEMWE related technologies.

12.2   ANION EXCHANGE MEMBRANE WATER ELECTROLYSIS

AEMWE has been considered as an emerging third-generation technology, which combines the merits of AWE and PEMWE. Therefore, AEMWE with low-cost and high-performance electrodes is expected to be promising in green hydrogen production. However, AEMWE is at an early stage of development. And more efforts in material design, component optimization, and performance evaluation are needed to increase its commercial competitiveness [18].

As shown in Figure 12.1a, AEMWEs are composed of bipolar plates, gas diffusion layers (GDLs), catalyst layers (CLs), and a polymer anion exchange membrane.

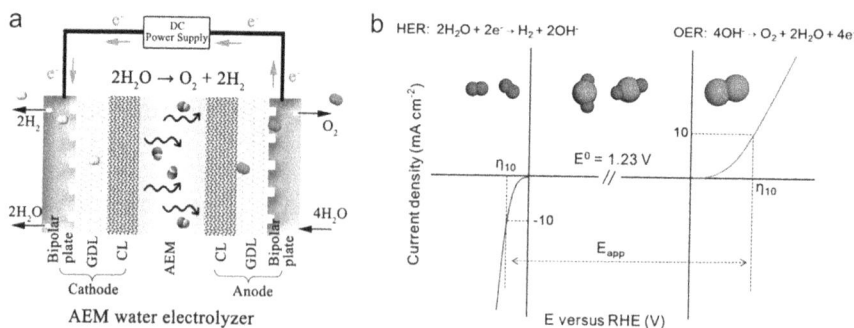

FIGURE 12.1   (a) Schematic illustration of the major components and working principles of anion exchange membrane (AEM) water electrolyzers. (b) The typical HER and OER polarization curves.

The membrane electrode assembly (MEA) is normally fabricated by pressing the GDL, CL, and AEM together under high temperature and pressure. In the AEMWEs, water is reduced and H_2 and OH^- are generated at the cathode. Then, OH^- moves through the AEM to the anode region, where they are oxidized to form H_2O and O_2 [12]. The detailed electrode reaction equations in AEMWEs are presented as follows:

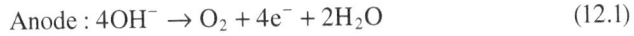

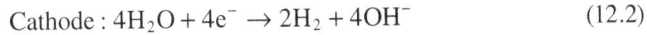

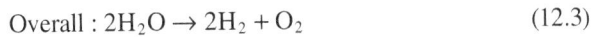

$$Anode : 4OH^- \rightarrow O_2 + 4e^- + 2H_2O \tag{12.1}$$

$$Cathode : 4H_2O + 4e^- \rightarrow 2H_2 + 4OH^- \tag{12.2}$$

$$Overall : 2H_2O \rightarrow 2H_2 + O_2 \tag{12.3}$$

Theoretically, the thermodynamic water splitting voltage is 1.23 V at 298 K as shown in Figure 12.1b [19]. However, an overpotential is required to drive the water decomposition reaction due to the factors such as sluggish electrode dynamics and ohmic resistance of electrolytes and other components. Therefore, it is of great significance to find efficient and stable catalysts to reduce the overpotential of water decomposition and improve the efficiency of water decomposition.

12.3 HER CATALYSTS

Water electrolysis with renewable energy provides a promising approach of producing "green hydrogen" and has been considered as the core of the future carbon-neutral energy systems. Hydrogen evolution reaction (HER) is a crucial half-reaction in water electrolysis. And the efficient electrocatalysts are required to accelerate the reaction kinetics, which are especially sluggish in alkaline media [20,21]. However, the kinetics of cathodic HER reaction in alkaline medium is still not suitable for practical application. Even the activity of the most desirable Pt catalysts decreases by several orders of magnitude when the pH of the electrolyte changes from acidic to basic [20,22,23]. Therefore, efficient alkaline HER catalysts need to be developed urgently. The kinetic pathway of the HER generally follows the Volmer–Heyrovskey or Volmer–Tafel mechanism as shown in Figure 12.2a [21,24]. Both consist of water adsorption, followed by water dissociation (Volmer step, equation 12.4), and then either hydrogen dissociation via chemical desorption (Tafel step, equation 12.5) or electrochemical desorption (Heyrovsky step, equation 12.6) to form H_2 [25].

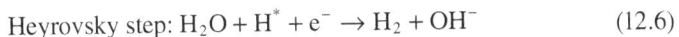

$$Volmer\ step : 2H_2O + 2e^- \rightarrow 2H^* + 2OH^- \tag{12.4}$$

$$Tafel\ step : 2H^* \rightarrow H_2 \tag{12.5}$$

$$Heyrovsky\ step: H_2O + H^* + e^- \rightarrow H_2 + OH^- \tag{12.6}$$

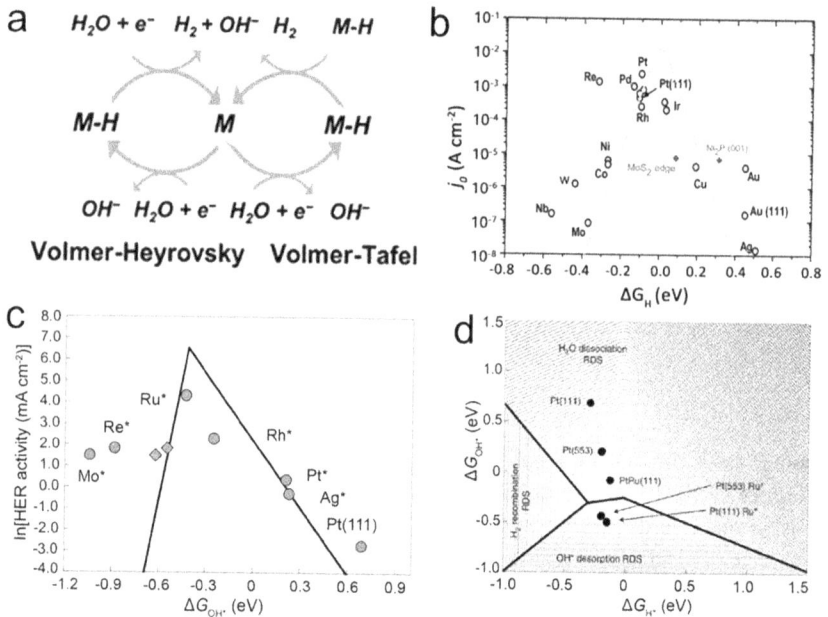

FIGURE 12.2 (a) HER reaction pathways [25]. (b) Volcano plot of exchange current density against the activity descriptor of ΔGH in acidic electrolytes [28]. (c) The plot of the simulated and experimentally measured HER rates on Pt(553), Pt(553) with Mo*, Re*, Ru*, Rh*, and Ag* adsorbed at the step and Pt(111) vs. ΔG_{OH*}. (d) Logarithm of the HER rate (contours) as a function of HBE and OHBE [29].

Relevant studies suggested that the dissociative adsorption energy of water on catalysts and the adsorption energies of H* and OH* species are the three key and correlative factors that influence the alkaline HER process [26,27]. Correspondingly, in order to overcome the sluggish alkaline HER kinetics, an efficient HER electrocatalyst should be designed according to the following principles: low water dissociation energy barrier and appropriate H-binding and OH-binding energies (HBEs and OHBEs) with catalysts (Figure 12.2b–d) [28–30]. However, these intrinsically functional requirements are hard to be met simultaneously on a single component electrocatalyst due to the scaling relations. In contrast, the composite electrocatalysts consisting of different functional components and structures could provide potential possibilities for fine-tuning the water dissociation kinetics, adsorption and desorption abilities of H* and OH* species simultaneously, which are favorable for the alkaline HER process.

12.3.1 PLATINUM GROUP METAL-BASED CATALYSTS

12.3.1.1 Pt-Based Catalysts

Even for the state-of-the-art Pt catalysts, the kinetics of the cathodic HER in alkaline media remain unsatisfactory for practical applications due to high water decomposition energy barrier and inappropriate H^* or OH^* adsorption energy [31,32]. Therefore, a handful of strategies such as the construction of heterogeneous interfaces, alloying, and the creation of local acidic environments, have been explored to promote the HER activity of Pt-based catalysts in alkaline media.

Wang et al. developed a facile approach to synthesize two-dimensional C_{60} nanosheets and constructed a Pt/C_{60} heterostructure [33]. As shown in Figure 12.3a, theoretical results illustrate that the metal-support electronic interactions lead to the transfer of 1.90 electrons from the Pt cluster to the C_{60} plane, which is also consistent with the experimental results. The pronounced charge redistributions at the diverse interface of Pt/C_{60} alter the binding energy of key intermediates at the Pt

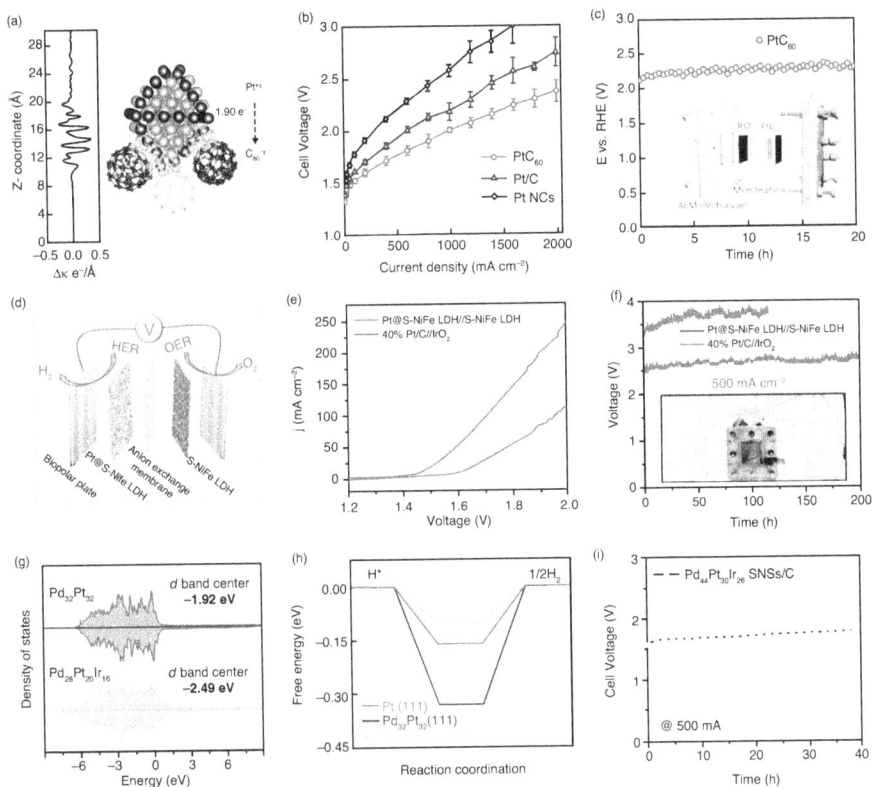

FIGURE 12.3 (a) Electronic density difference due to the interactions between Pt and C_{60} as an integrated 1D profile and 3D-isosurface plotted at 0.02 $e/Å^3$ value. (b) LSV curves of the AEM electrolyzers using $PtC_{60}\|IrO_2$, $PtC\|IrO_2$, and Pt NCs$\|IrO_2$, respectively. (c) Durability test of the AEM electrolyzer at a current density of 1 A/cm² [33]. (d) Schematic illustration of an alkaline AEM cell. (e) Polarization curves for water electrolysis. (f) Chronopotentiometric curve at 500 mA/cm² and the photograph of AEM cell [34]. (g) Calculated PDOS spectra of the $Pd_{32}Pt_{32}(111)$ and $Pd_{28}Pt_{20}Ir_{16}(111)$ surfaces. (h) Free-energy diagrams for the HER at pH=14 on Pt(111), $Pd_{32}Pt_{32}(111)$, and $Pd_{28}Pt_{20}Ir_{16}(111)$ surfaces, respectively. (i) Chronopotentiometry tests of the $Pd_{44}Pt_{30}Ir_{26}$ ASNSs/C catalysts for continuous H_2 production at 0.5 A using the AEM electrolyzer carried out at 80°C [36].

site, accelerating all elementary steps in alkaline HER. Figure 12.3b displays that the inherent activity of alkaline HER of Pt/C_{60} is significantly improved (12 times) compared to the most advanced Pt/C catalyst. In addition, Figure 12.3c shows that an AEM electrolyzer assembled with Pt/C_{60} composites achieved 1 A/cm² with minimal activity loss under a 20-hour-long stability test. In addition to carbon matrix, transition metal-based materials can also be used as the substrate of metal Pt to change the redistribution of electrons at the heterogeneous interface and the energy barrier of H–OH bond cleavage. Mai et al. designed Pt-quantum-dot-modified sulfur-doped NiFe-layered double hydroxides (Pt@S-NiFe LDHs), which can greatly facilitate interfacial electron transport between the Pt active sites and the substrate [34]. The heterogeneous interface of Pt@S-NiFe LDHs promotes the desorption of H_{ads} and accelerates the dissociation of water, with HER performance in alkaline environments exceeding 40% of commercially available Pt/C and most reported Pt-based electrocatalysts. Furthermore, the strong metal–substrate interaction and a lower metal dissolution rate significantly enhanced the stability of Pt@S-iFe LDH. As shown in Figure 12.3d–f, an AEM cell based on Pt@S-NiFe-LDH//S–NiFe-LDH exhibits smaller overpotentials than that of Pt/C//IrO_2 and can be stable for 200 hours at a current of 500 mA/cm², which is a promising candidate for large-scale hydrogen production. Hui et al. developed an alkaline HER catalyst consisting of dense Pt nanoparticles (NPs) immobilized in oxygen vacancy-rich NiO_x heterojunctions (Pt/NiO_x-O_V) [35]. A combined theoretical and experimental studies manifest that anchoring Pt NPs on NiO_x–O_V leads to electron-rich Pt species with altered density of states (DOS) distribution, which can efficiently optimize the d-band center and the adsorption of reaction intermediates as well as enhance the water dissociation ability. The assembled alkaline electrolyzer based on Pt/NiO_x-O_V achieved 1000 mA/cm² under an extremely low voltage of 1.776 V and can operate stably for more than 400 hours.

Alloying of heterogeneous elements is also an effective means to effectively enhance HER in the alkaline environment of Pt-based catalysts. Specifically, the cooperative coupling of surface atomically heterogeneous active sites can effectively regulate the electronic structure of active sites and the interaction with reactants/intermediates from the point of view of catalyst design. Xie et al. synthesized ternary $Pd_{44}Pt_{30}Ir_{26}$ assembled supernanosheets (ASNSs) with abundant parallel subnanometer interlayer spacings [36]. Heterogeneous elements in $Pd_{44}Pt_{30}Ir_{26}$ ASNSs can effectively regulate the electronic structure of platinum. As shown in Figure 12.3g, the d-band center of $Pd_{28}Pt_{20}Ir_{26}$ (111) negatively shifts from −1.92 to −2.49 eV after Ir incorporation. Besides, Figure 12.3h shows that the H_2O molecule is easier to dissociate into H* and OH* on $Pd_{28}Pt_{20}Ir_{16}$(111) and H* can be further reduced to H_2 easily with the incorporation of Ir. In alkaline electrolyte, $Pd_{44}Pt_{30}Ir_{26}$ ASNSs can achieve a mass activity of 8.86 mA/µg at an overpotential of 70 mV, which is 11.8 times that of commercial Pt/C. Significantly, as shown in Figure 12.3i, $Pd_{44}Pt_{30}Ir_{26}$ ASNSs were applied as both the anode and cathode catalysts in an AEM electrolyzer, which can continuously catalyze the electrolysis of water for more than 40 hours at a large current of 500 mA, achieving efficient and stable hydrogen production.

In addition to metallic heteroelements, the coordination effect between non-metallic heteroelements and Pt can also effectively regulate the chemical environment around the active center, thus enhancing the HER performance of Pt-based catalysts in alkaline environment. Wang et al. develop a facile procedure of irradiation-impregnation to precisely adjust the axial ligand (–F, –Cl, –Br, –I, and –OH) on the Pt single sites, based on which chemical-environment/HER-activity relationship can be determined clearly [37]. The coordination of the axial ligand to the Pt single sites is shown in Figure 12.4a. Electrochemical measurement displayed that the HER activity follows the order of Cl-Pt/LDH > F-Pt/LDH > HO-Pt/LDH > Br-Pt/LDH > I-Pt/LDH, confirming the significant axial ligand effect on HER activity. Both experiment

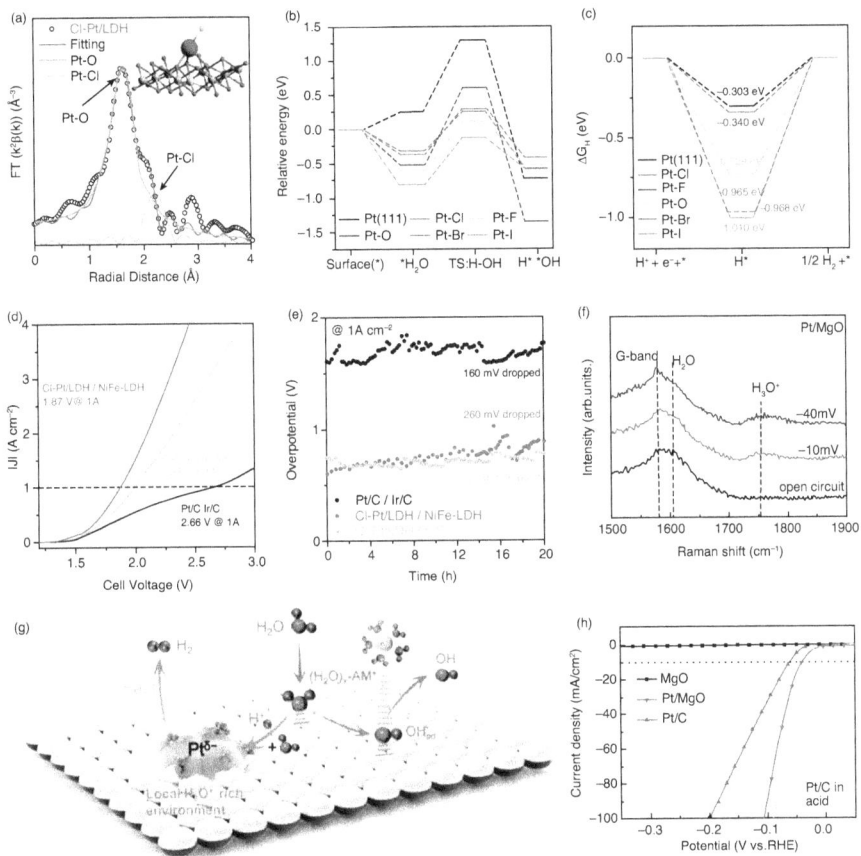

FIGURE 12.4 (a) Representative EXAFS fitting curve (the inset is the magnified local structure) of Cl-Pt/LDH. (b) Calculated energy barriers of water dissociation kinetics and (c) adsorption free energies of H* on the surfaces of the Pt-SACs, and the Pt (111) slab as a reference. (d) LSV curves of the MEA reactors using Cl-Pt/LDH/NiFe-LDH, Cl-Pt/LDH/Ir/C, and commercial Pt/C/Ir/C. (e) Stability tests of the MEA water electrolyzers at a current density of 1 A/cm^2 at 60°C [37]. (f) The operando Raman spectra of Pt/MgO in 1 M KOH. (g) Schematic representation of water dissociation, formation of H$_3$O$^+$ intermediates, and subsequent formation of H$_2$ as well as OH$^-$ desorption on the Pt/MgO nanosheets. (h) Linear sweep voltammetry (LSV) curves of Pt/MgO, Pt/C, and MgO in 1 M KOH, with that of Pt/C in H$_2$SO$_4$ obtained for comparison [38].

and calculation results prove that halogen coordination can effectively regulate the chemical environment of Pt sites. As shown in Figure 12.4b and c, Cl-coordinated Pt sites exhibit most optimized water dissociation energy and H-binding energy, consequently facilitating the sluggish Volmer step and accelerating kinetics for the conversion of H* to H_2. Cl-Pt/LDH was also evaluated in a MEA-based alkaline water electrolyzer at industrially relevant reaction rates. As shown in Figure 12.4d, the electrolyzer based on Cl-Pt/LDH/NiFe-LDH exhibits a much lower cell voltage (1.87 V) than those based on Cl-Pt/LDH/Ir/C (1.99 V) and the Pt/C/Ir/C (2.66 V) at a current density of 1.0 A/cm². Figure 12.4e shows that the Cl-Pt/LDH/Ir/C electrolyzer exhibits good stability with negligible overpotential loss.

The general consensus is that during the alkaline HER, the sluggish Volmer step could affect the rate-determining step, but this step is unnecessary in an acidic solution. Therefore, creating a local acid-like environment for HER in alkaline medium can fundamentally solve the disadvantages of alkaline HER. Yan et al. selected Pt/ MgO as the prototypical example to construct an acid-like reaction environment in an alkaline medium by virtue of multiple physicochemical interactions between the substrate, metal active site, and reaction intermediate [38]. As shown in Figure 12.4f, the operando Raman spectra confirmed that massive amounts of H_3O^+ intermediates are produced on the surface of MgO, which accumulate around the negatively charged Pt ($Pt^{\delta-}$). In alkaline media, $Pt^{\delta-}$ accelerates the migration of H_3O^+, forming an acid-like environment around $Pt^{\delta-}$, thereby increasing HER in alkaline media. This acid-like environment provides Pt with a favorable reaction condition for the HER in the alkaline electrolyte. Based on the above experimental and theoretical results, an HER mechanism was proposed for the $Pt^{\delta-}$ nanoparticles in an alkaline electrolyte as shown in Figure 12.4g. As a result, the Pt/MgO catalyst exhibits an overpotential of 39 mV at a current density of 10 mA/cm², which is significantly lower than 20 wt% Pt/C in an alkaline medium and close to the acidic HER behavior of Pt/C (33 mV) (Figure 12.4h). This study provides insight into tailoring the local reaction environment to design high-performance electrocatalysts in a more rational and precise way.

12.3.1.2 Ru-Based Catalysts

Ru is another PGM that is attracting attention as a potential HER catalyst for alkaline electrolytes. The ~65 kcal/mol H-bonding energy of Ru is similar to that of Pt [39]. Therefore, Ru has become a viable candidate for large-scale AEMWE catalysts if Ru-based catalysts achieve high HER activities and long-term stability.

Chen et al. dispersed Ru clusters into the CuO matrix, leading to a strong Ru–O– Cu bond construction at the interface [40]. Both experimental and theoretical results show that Ru–O–Cu centers enhance the electron coupling between Ru and CuO, which is conducive to water adsorption. This could reduce the energy barrier for water dissociation and weaken the hydrogen adsorption for easier H_2 desorption as shown in Figure 12.5a and b. Therefore, Ru-CuO-SA catalysts have a higher HER activity in alkaline environments than most reported Ru-based materials and commercial Pt/C catalysts due to the accelerated Volmer and Tafel steps. To further test the potential of Ru-CuO-SA catalyst in practical applications, the authors assembled an AEM electrolyzer based on the catalyst. As shown in Figure 12.5c, the Ru-CuO-SA electrolyzer exhibits a current density of 216 mA/cm² at the cell voltage

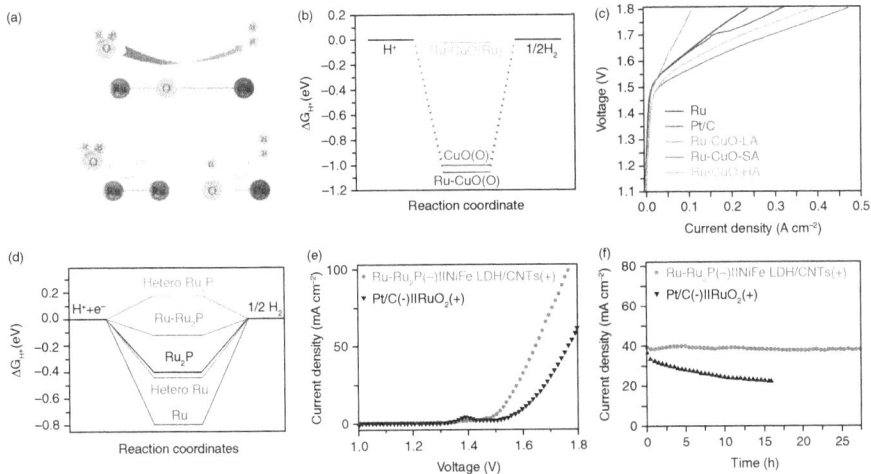

FIGURE 12.5 (a) Illustrations of alkaline HER proceeding on Ru–O–Cu, Ru–Ru, and O–Cu bonds. (b) HER free-energy diagrams. (c) Polarization curves achieved on these catalysts in the AEM electrolyzer [40]. (d) Gibbs free-energy diagram of HER for different models. (e) LSV curves and (f) *I–T* curves of the alkaline electrolyzer using Ru–Ru$_2$P (or 20 wt% Pt/C) as the cathode and NiFe-LDH/CNTs (or commercial RuO$_2$) as the anode [41].

of 1.65 V and achieves a good long-term durability. Zhou et al. designed Ru-Ru$_2$P heterogeneous nanoparticles via in situ green phosphating strategy [41]. XPS and XAFS results mean that electrons at the heterogeneous interface are transferred from Ru to Ru$_2$P. DFT calculation results determined that strong electronic redistribution occurs at the heterointerface of Ru–Ru$_2$P, which effectively modulated the electronic structure at the interface to achieve an optimized hydrogen adsorption strength as shown in Figure 12.5d. In addition, AEM electrolyzer of (−) Ru-Ru$_2$P‖NiFe-LDH/CNTs (+) exhibits a low cell voltage of 1.53 V at a current density of 10 mA/cm^2. Meanwhile, the long-term stability of water splitting for (−) Ru-Ru$_2$P‖NiFe-LDH/CNTs (+) was also evaluated as shown in Figure 12.5e–f, which demonstrates the excellent stability of (−) Ru-Ru$_2$P‖NiFe-LDH/CNTs (+), which was much better than that of (−) 20 wt% Pt/C‖RuO$_2$ (+).

In alkaline electrolyte, water dissociation as a source of hydrogen is generally regarded as the rate-determining step for HER. The orientation and polarization of interfacial water could affect the hydrolytic dissociation and HER activity. Therefore, in addition to the traditional structurally regulated strategies, engineering the interfacial microenvironment is emerging as an alternative and powerful way to regulate the electrocatalytic kinetics through the non-covalent interaction among intermediates/species in an electrical double layer (EDL). Huang et al. built a nanocone-assembled Ru$_3$Ni (NA-Ru$_3$Ni) catalyst [42]. Both finite element simulations and experimental results exhibit that the local electric field is enhanced by the nano-tip of Ru$_3$Ni catalyst, which greatly increases the interface K$^+$ concentration. Further mechanistic studies reveal that the locally increased hydrated K$^+$ concentration enhanced the polarization of the H–OH bond of interfacial water, resulting in the reduced energy barrier for water dissociation and thus accounting for greatly improved alkaline HER activity as shown in Figure 12.6a. Furthermore, the authors evaluated the performance of NA-Ru$_3$Ni/C as a cathode in a practical AEM electrolyzer. As shown in Figure 12.6b, the polarization of NA-Ru$_3$Ni/C||NA-Ru$_3$Ni/C curve exhibits a small cell voltage (2.048 V) at 1 A/cm^2 for water splitting, which is superior to commercial Pt/C||IrO$_2$/C. The assembled AEM electrolytic cell can have no change for 2,000 hours at a current density of 1.0 A/cm^2, which is the record high performance of the location to date and is expected to be commercialized.

Liu et al. constructed the IrRu dizygotic single-atom sites (IrRu DSACs) with an atomically asymmetric local electric field, regulating the adsorption configuration and orientation of H$_2$O and thus optimizing its dissociation process [43]. The integral differential phase contrast STEM (iDPC-STEM) was performed to study the local electric field of IrRu DSACs. As shown in Figure 12.6c, the IrRu DSACs possess a high average charge density variation of about 4.00 e/Å^2, corresponding to an electric field intensity of 4×10^{10} N/C, which is higher than Ir and Ru SACs. In situ Raman and molecular dynamics simulations revealed that the local electric field affected the reorientation of the "H-down" adsorption configuration of interfacial water and promoted the dissociation of water, thus obtaining higher HER activity. The MEA electrolyzer was assembled based on IrRu DSACs catalyst to evaluate its practical applications. As shown in Figure 12.6d, the MEA electrolyzer can maintain high stability for more than 300 hours at 1 A/cm^2, demonstrating that IrRu DSACs can be a promising catalyst for the practical alkaline HER.

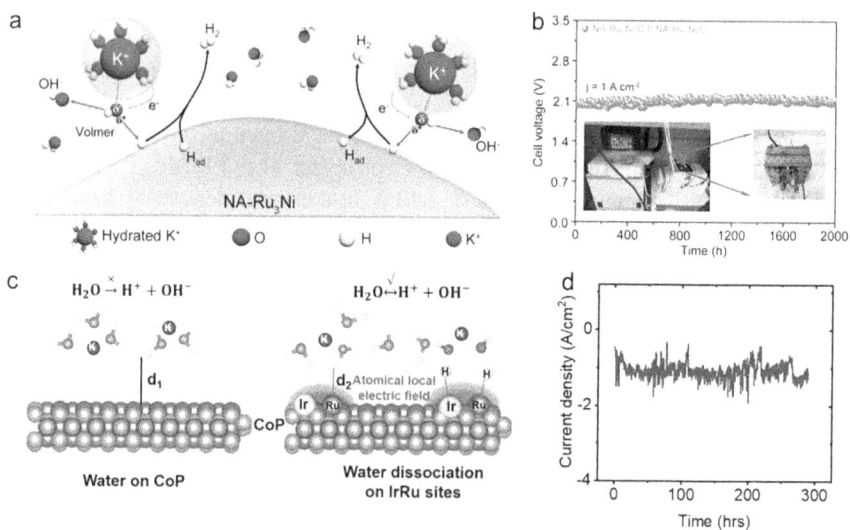

FIGURE 12.6 (a) A schematic showing how K$^+$ promotes interfacial water dissociation on the NA-Ru$_3$Ni surface. (b) Chronopotentiometry curve for AEM electrolysis using NA-Ru$_3$Ni/C as cathode and anode catalysts operating at 1 A/cm^2. The inset shows the photographs of the AEM electrolyzer [41]. (c) Schematic of interface H$_2$O reorientation induced by atomic electric field. H$_2$O adsorbed on CoP and IrRu DSACs. (d) The stability test of IrRu DSACs in MEA at 1 A/cm^2 [43].

12.3.2 Ni-Based Catalysts

Ni is an abundant metal that is used in traditional WE electrolyzers as an HER and OER catalyst, thus making it a candidate of high interest to replace Pt- or Ru-based catalysts for alkaline conditions [44]. Ni possesses good water adsorption, but high Ni–H bond energy results in lower HER activity of pure nickel catalyst than Pt/C catalyst [45].

Alloying of Ni has been shown to alter HER activity, and attempts to improve the activity and stability of pure Ni catalysts include forming Ni binary and ternary alloys with different elements such as Mo, Co, Fe, Ce, Zn, and Cu [46]. Li et al. prepared Ni_4Mo alloy nanoparticles. As shown in Figure 12.7a–c, the HER activity of Ni_4Mo alloy was significantly improved when compared with Ni metal catalyst [47]. The experimental results combined with theoretical calculations reveal that the alloying

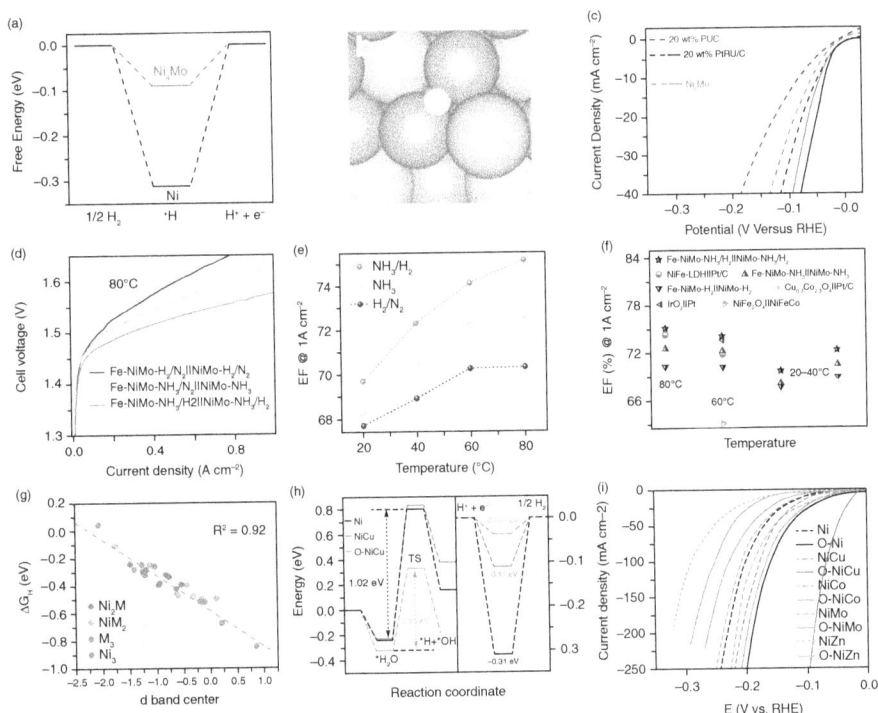

FIGURE 12.7 (a) Gibbs free-energy profiles for hydrogen adsorption (ΔG_{H*}) on Ni_4Mo and Ni. (b) Optimized hydrogen adsorption configuration on Ni_4Mo. (c) HER polarization curves of Ni_4Mo, Ni, 20 wt% Pt/C and 20 wt% PtRu/C in (dotted) 0.1 M KOH and (solid) 1 M KOH [47]. (d) The performance of MEAs employing Fe-NiMo-NH_3/H_2||NiMo-NH_3/ H_2, Fe-NiMo-NH_3||NiMo-NH_3, and Fe-NiMo-N_2/H_2||NiMo-N_2/H_2 pairs at 80°C. (e) The energy conversion efficiencies of MEAs employing different catalysts from 20°C to 80°C. (f) Comparison of efficiencies of the MEAs for Fe-NiMo-NH_3/H_2||NiMo-NH_3/H_2 pair with state-of-the-art examples [48]. (g) Dependence of ΔG_H on the improved d-band center. (h) Corresponding energy differences for the water dissociation process and the diagram of ΔG_H for Ni, NiCu. (i) LSV of NiM [49].

of Ni and Mo significantly weakens the H* adsorption, reduces the reaction barrier formed by water, and contributes to the dissociation of H*. Hu et al. reported a series of NiMo alloy HER catalysts and obtained NiMo-NH$_3$/H$_2$ with the highest activity of HER by adjusting the reduction temperature and atmosphere [48]. The AEM electrolyzer integrated by NiMo-NH$_3$/H$_2$ cathode delivers 500 mA/cm^2 at an overpotential of 244 mV and achieves energy conversion efficiency as high as 75% as shown in Figure 12.7d and e. Furthermore, Figure 12.7f shows that the MEA incorporating Fe-NiMo-NH$_3$/H$_2$‖NiMo-NH$_3$/H$_2$ catalysts outperforms all previous examples. Xing et al. designed Ni–M (M=Ti, V, Cr, Mn, Fe, Co, Cu, Zn, Mo, W) bimetallic electrocatalysts by introducing a series of common non-noble metal heteroatoms into Ni alloys [49]. As shown in Figure 12.7g, the authors propose an improved d-band center theory to establish the relationship between ΔG_H and the electronic structure and predict that Cu elements can weaken excessive ΔG_H, thereby balancing the *H adsorption/desorption process. Figure 12.7h displays that *H adsorbed on the NiCu surface could recombine to form H$_2$ quickly, thus achieving fast HER kinetics in alkaline media, which also proves the above conjecture. Therefore, introducing Cu remarkably improves the HER activity as shown in Figure 12.7i.

In addition to Ni alloy catalysts, nickel-based oxides, sulfur compounds, and phosphates have high catalytic activity in the cathode HER. Li et al. designed a Ni$_3$Sn$_2$–NiSnO$_x$ electrocatalyst with functional components (alloy and alloy oxide) [50]. Theoretical calculation revealed that Ni$_3$Sn$_2$ has an ideal hydrogen adsorption capacity and a weak hydroxyl adsorption capacity, while NiSnO$_x$ promotes the process of hydrolytic ionization and hydroxyl transfer as shown in Figure 12.8a. As a result, the combination of these two components accelerates the efficient dissociation of water and promotes the desorption of OH$^-$, thus greatly improving HER kinetics in the composite (Figure 12.8b–d). Zhao et al. developed a synergistic hybrid Ni$_3$S$_2$/Cr$_2$S$_3$ site where hydrolytic dissociation/hydrogen generation using hydrogen overflow bridging occurs (Figure 12.8e) [51]. It can eliminate the inhibition of high hydrogen coverage at hydrolytic dissociation sites induced by high current density while promoting the Volmer/Tafel process. It is worth noting that the catalyst can be synthesized in large quantities and has strong processability, making it suitable for industrial-scale electrode manufacturing. As illustrated in Figure 12.8f, the design concept and manufacturing method of the catalyst proposed in this work can also be used to prepare the cathode suitable for industrial-scale alkaline water electrolyzer. The composite electrocatalyst composed of different functional components can simultaneously regulate the hydrolytic dissociation kinetics, as well as the adsorption and desorption capacity of H* and OH$^-$, which is conducive to the alkaline HER process.

12.4   OER CATALYSTS

The mechanism of OER at anodes is complicated and still controversial. Traditional OER mechanism in alkaline electrolytes involves several electron–proton coupling reactions in which OH$^-$ is oxidized into oxygen molecules and water molecules, also known as adsorbent evolution mechanism (AEM) as shown in Figure 12.9a [52]. During the OER process, the active site first adsorbs OH$^-$ and then deprotonates

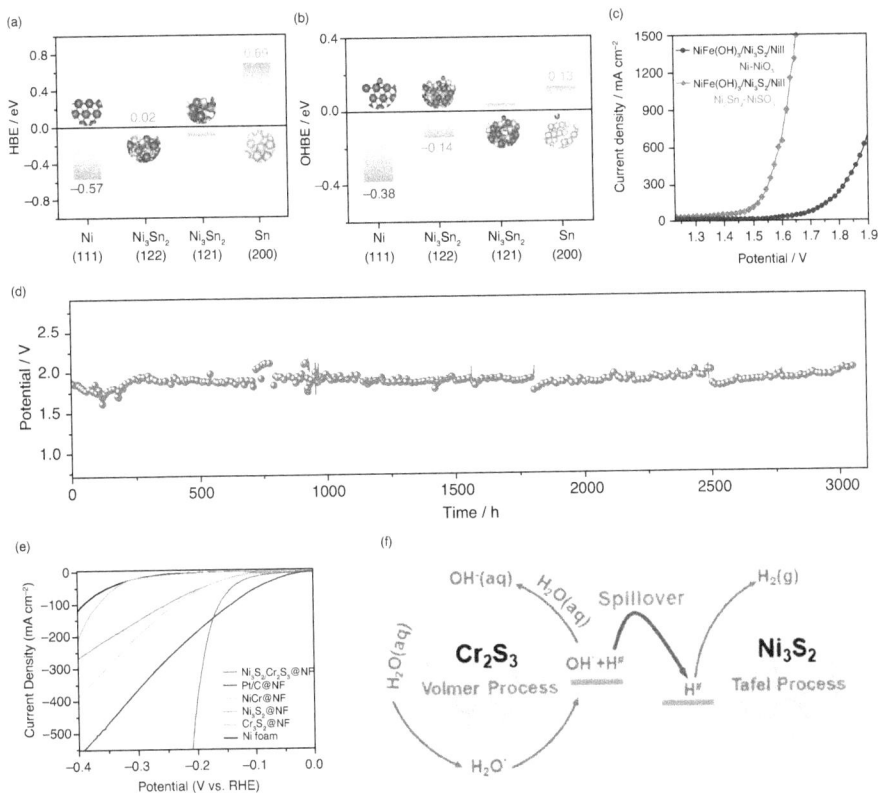

FIGURE 12.8 (a) Calculated H-binding energies (HBEs, ΔE_H) and (b) OH-binding energies (OHBEs, ΔE_{OH}) for Ni (111), Ni_3Sn_2 (122), Ni_3Sn_2 (121), and Sn (200) surfaces. (c) LSV curves of the two-electrode systems for water electrolysis in 1 M KOH at 25 °C. (d) Chronopotentiometric curve at 1000 mA/cm² for the NiFe(OH)x/Ni_3S_2/Ni||Ni_3Sn_2-NiSnOx electrolyzer toward water splitting in 30% KOH at 80 °C [50]. (e) iR$_{initial}$-compensated LSV curves of Ni_3S_2/Cr_2S_3@NF, NiCr@NF, Ni_3S_2@NF, Cr_2S_3@NF, Pt/C@NF, and NF at a scan rate of 1 mV/s in 1.0 M KOH. (f) Schematic illustrating the proposed reaction mechanism on Ni_3S_2/Cr_2S_3@NF [51].

to form *O intermediates. Then, the O–O coupling produces *OOH intermediates through a nucleophilic attack of OH⁻. Finally, the catalyst surface desorbs O_2 molecules through secondary deprotonation and release active metal sites [53]. Thermodynamic perspective indicates that the ideal adsorption energy difference between *OOH and *OH is 2.46 eV. However, the actual energy difference between *OOH and *OH is about 3.2 eV, thus still requiring a high overpotential (η) to drive the reaction, with a calculated theoretical limit of about 370 mV [54]. Therefore, in alkaline environment, anode OER involved with complex four-electron oxidation process and slow kinetics leads to large overpotential η, resulting in a large loss of overall water decomposition efficiency. A fundamental understanding of potential OER active sites and the association of OER activity with specific descriptors is essential for the search for new catalysts with high OER activity.

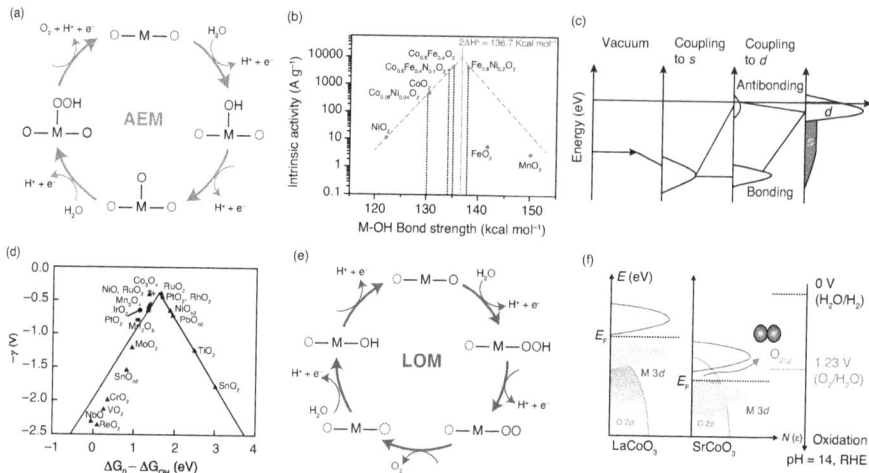

FIGURE 12.9 (a) A conventional OER mechanism based on adsorbate evolution mechanism [53]. (b) Volcano plot of the OER intrinsic activities against the descriptor of M–OH bond strength [62]. (c) Schematic illustration of bond formation at a transition metal surface. The lower the d states are in energy relative to E_f, the more filled the antibonding states and the weaker the adsorption bond [57]. (d) Volcano plot of the calculated η against the descriptor of the value of $(\Delta G_O - \Delta G_{OH})$ [60]. (e) A new proposed OER mechanism based on lattice oxygen mechanism [53]. (f) Schematic rigid band diagrams of $LaCoO_3$ and $SrCoO_3$ [61].

As early as 1955, Rüetschi and Delahay first correlated the OER activity with the M–OH bond energy and they found that η decreases in an approximate linear correlation with the increasing bond energy of M–OH [55]. Hu and colleagues synthesized a series of transition metal oxygen (hydroxide) thin film catalysts by electrodeposition [56]. As shown in Figure 12.9b, OER activities of catalysts are closely related to M–OH bond strength. Notably, the ternary $NiCoFeO_x$ and binary $NiFeO_x$ and $CoFeO_x$ have a more ideal M–OH bond strength than the unary NiO_x, CoO_x, FeO_x, and MnO_x.

Another descriptor used to predict OER activity is the d-band center of the metal site. In the 1990s, Hammer and Nørskov first used the d-band theory to establish the relationship between the electronic structure of transition metals and their catalytic activity [57]. The antibonding states are above the d states, and its filling governs the bond strength in terms of the distance from the band center to the Fermi level (E_f) as shown in Figure 12.9c. Recently, a series of studies have shown that the highest OER activity can be achieved by adjusting the chemical environment of the catalyst active site to obtain the optimal Ed level [58]. This can be explained by the fact that M–OH bond strength decreases with the increase of the d-band center, resulting in an increase in OER activity [59].

Recent advances in DFT calculations have shown that OER activity depends primarily on the surface adsorption energy of intermediates (*OH, *O, and *OOH) (Figure 12.4a). Rosssmeisl et al. found that the difference between *OOH and *OH adsorption energies ($\Delta G_{OOH} - \Delta G_{OH}$) of the catalyst was consistently around 3.2 eV [60]. Figure 12.9d shows the volcanic-type relationship between the descriptor

(ΔG_O–ΔG_{OH}) and the OER activity. ΔG_O–ΔG_{OH} has become a widely accepted descriptor for predicting OER activity considering the multi-step reaction process during OER process.

In addition to the conventional AEM, the lattice oxygen mechanism (LOM) has been identified as a new OER reaction pathway (Figure 12.9e) [53]. LOM directly forms O–O by coupling between lattice oxygen and *O, thus bypassing the limitations associated with AEM mechanisms. First, the two *OH groups at the metal site undergo a deprotonation reaction to form two metal-oxo compounds. Second, these two adjacent oxo species combine directly to form O–O bonds. Finally, O_2 was released, and the two empty metal centers are occupied by OH⁻. Shao-Horn and colleagues suggested that oxygen molecules produced during the OER process of some highly active perovskite oxides come from lattice oxygen [61]. An increase in Co-O covalency from $LaCoO_3$ to $SrCoO_3$ can move the oxide Fermi level below the O_2/H_2O redox energy (Figure 12.9f), which can activate the surface oxygen of $SrCoO_3$. Therefore, the lattice oxygen mechanism has a lower reaction barrier, which can bypass the theoretical η upper limit limited by the scaling relationship in the traditional mechanism. However, the OER process involving lattice oxygen can weaken the stability of the catalyst, which is a major challenge for the application of these catalysts [62].

OER electrocatalysts for industrial electrocatalysis require low OER overpotential and long-term stability at high current densities. The design and preparation of highly efficient and durable OER electrocatalysts is of great significance for the realization of high efficiency electrolyzer. Currently, commonly used OER catalysts can be divided into precious metal catalysts, transition metal-based catalysts, and perovskite catalysts. Based on the above analysis, OER is a four-electron step reaction that faces adsorption and analytical interactions with multiple intermediates, resulting in high overpotential and low hydrogen production efficiency. Therefore, in order to obtain satisfactory electrolytic water performance, it is necessary to further improve the activity and stability of OER by modifying the catalysts.

12.4.1 Precious Metal-Based OER Catalyst

At present, the OER catalysts used in water electrolysis are mainly Ir- and Ru-based catalysts [63,64]. In addition to the high overpotential and slow reaction kinetics analyzed above, the precious metal catalyst also faces the problem of dissolution and agglomeration at high reaction potential [65]. To solve these problems, different catalyst design strategies have been proposed, such as interface engineering and doping or alloying, as well as designing catalysts with reasonable coordination structures. The catalyst can be effectively endowed with high activity and durability by optimizing the chemical environment of the catalyst active site. Sun et al. reported hollow Co-based N-doped porous carbon spheres decorated with ultrafine Ru nanoclusters (HS-RuCo/NC) as an efficient OER electrocatalyst [66]. As shown in Figure 12.10a, the coupling of RuO_2 and Co_3O_4 can optimize the electron configuration of the RuO_2/Co_3O_4 heterostructure and reduce the energy barrier in the OER process. The resultant HS-RuCo/NC exhibits superior catalytic activity, rapid reaction kinetics,

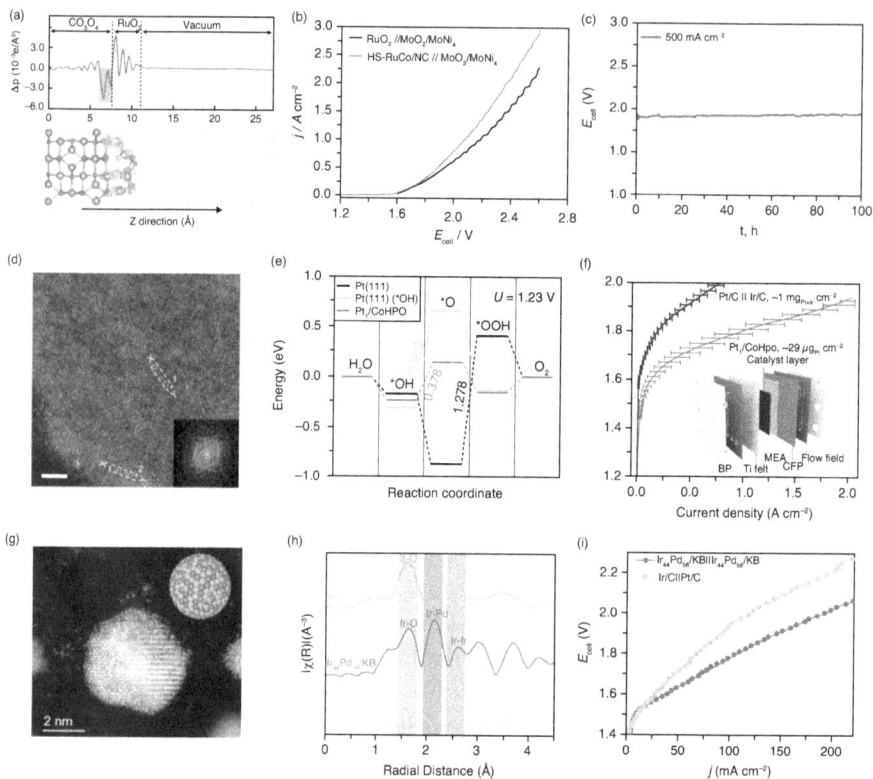

FIGURE 12.10 (a) Planar-averaged electron density difference and charge density difference for RuO_2/Co_3O_4 heterostructure. (b) LSV curves of AEM water electrolyzers for the $RuO_2//MoO_2/MoNi_4$ and HS-RuCo/NC//$MoO_2/MoNi_4$ cells without iR compensation at room temperature in 1 M KOH. (c) Stability test of AEM water electrolyzers for the HS-RuCo/NC//$MoO_2/MoNi_4$ cells at a current density of 500 mA/cm² [66]. (d) Representative magnified HAADF-STEM image, showing that only Pt single atoms are present in the CoHPO support. Inset, the FFT image. Scale bar, 2 nm. (e) The OER free-energy diagram for Pt (111), OH-covered Pt (111), and $Pt_1/CoHPO$ surfaces at equilibrium potential of 1.23 V. (f) Electrocatalytic water splitting properties of the $Pt_1/CoHPO$ and the benchmark Pt/C + Ir/C measured in an alkaline AEMWE setup operating at 80°C. Inset: a typical single AEMWE setup comprising a membrane electrode assembly (MEA) and bipolar plates (BP) with a flow field is presented, wherein the MEA comprises a gas diffusion layers with a Ti felt and a carbon fiber paper (CFP) at the anodic and cathodic sides, respectively, anodic and cathodic catalyst layers and an anion exchange membrane (AEM) [67]. (g) HAADF-STEM images of $Ir_{44}Pd_{56}/KB$. (h) Fourier transformation of the EXAFS spectra of $Ir_{44}Pd_{56}/KB$, metallic Ir, and IrO_2. (i) I–V curves of AEM electrolyzers using $Ir_{44}Pd_{56}/KB$ as anodic and cathodic catalyst, and commercial Ir/C as anodic and commercial Pt/C as the cathodic catalyst, at room temperature and ambient pressure. No cell voltages were iR compensated [69].

and excellent long-term stability in alkaline medium. In addition, the presence of Co_3O_4 can effectively inhibit the excessive oxidation of RuO_2 and make the catalyst have high stability. Figure 12.10b and c illustrates that AEMWE integrated with HS-RuCo/NC exhibits a cell voltage of 2.07 V at a current density of 1 A/cm² and has good long-term stability at 500 mA/cm^2, which is better than the commercial RuO$_2$-based AEMWE (2.19 V).

Guo et al. designed an efficient and solubility-resistant Pt-based OER catalyst by carefully controlling the local coordination environment and electronic structure of Pt [67]. Specifically, an atomically dispersed Pt site is inserted into a cobalt hydrogen phosphate (CoHPO) carriers to obtain OER catalysts with unique Pt(OH)(O$_3$)/Co(P) coordination as shown in Figure 12.10d. The experimental results and theoretical analysis show that the enhanced surface activity of the catalyst is originated from the unique coordination of the Pt(OH)(O$_3$)/Co(P) site and the strong electron coupling between the isolated Pt atom and the surrounding Co atom. As illustrated in Figure 12.10e, the resultant charge redistribution not only optimizes the binding energies of oxygenated intermediates, but also lowers the OER energy barriers. Furthermore, the strong electron coupling between Pt and adjacent Co atoms can inhibit the formation of soluble Pt$^{x>4}$ substances, thus enhancing the durability of the catalyst. AEM electrolyzer assembled with Pt(OH)(O$_3$)/Co(P) catalyst realizes an industrial-level current density of 1 A/cm^2 at 1.8 V with a high durability (Figure 12.10f). Tan et al. adopted a facile self-reconstruction strategy for single-atom Ir catalysts with controllable deposition of isolated Ir atoms on free-standing nanoporous (Ni$_{0.74}$Fe$_{0.26}$)$_3$P (denoted as np-Ir/NiFeP) [68]. Experimental characterization and theoretical calculation analysis show that the d-electron domination of surface Ni and Fe atoms was optimized by the introduction of single-atom Ir, resulting in the enrichment of surface effective charges on np-Ir/NiFeO. The coupling effect between Ir and Ni(Fe) oxyhydroxides greatly reduces the rate-controlled step energy barrier of Ir, Ni, and Fe sites, thereby increasing the OER activity of the catalyst. Meanwhile, the contraction of the Ni–O bond and other bonds makes the Ir atoms stable without aggregation in the OER process, which gives the catalyst overall excellent activity and stability.

Shuang Li et al. reported the preparation of IrPd alloy catalyst by introducing Pd to modulate the electronic state of Ir as shown in Figure 12.10g and h [69]. Spectral results show that Ir loses electrons and Pd gains electrons in Ir$_{44}$Pd$_{56}$/KB, and there is a clear electron transfer between Pd and Ir sites. The Pd in Ir$_{44}$Pd$_{56}$/KB can optimize the chemical environment of Ir to achieve the best binding energy with oxygen intermediates, thus obtaining the best OER activity. AEM electrolyzer shows a good stability of >20 hours at a current of 250 mA/cm^2 when the Ir$_{44}$Pd$_{56}$/KB catalyst was applied to the anode and cathode electrodes, indicating its practical application value in hydrogen production (Figure 12.10i).

12.4.2   3D Transition Metal-Based OER Catalyst

The high price and scarcity of precious metals limit their wide application. In contrast, in AEM electrolyzers, hydroxides, selenides, nitrides, and phosphates of the transition metals Ni, Co, and Fe can provide a cheaper alternative to highly efficient OER electrocatalysts [70–73].

12.4.2.1   3*d* Transition Metal (Oxygen) Hydroxides (MOxHy)

3*d* Transition metal (oxygen) hydroxides are effective OER electrocatalysts due to their tunable electronic structure, which can optimize the energetics of OER intermediates [74]. Due to the energetics change between the high-valence dopants and 3*d* metal, high-valence dopants have a stronger effect on modifying electronic structures on adjacent 3*d* metal sites, such as Fe, Co, and Ni. Liu et al. showed that high-valence dopants can regulate the valence state of bimetallic NiFe sites, thereby enhancing the adsorption of O intermediates, reducing the reaction energy of potential limiting steps (*OH → *O), and increasing OER activity (Figure 12.11a) [75]. Based on the

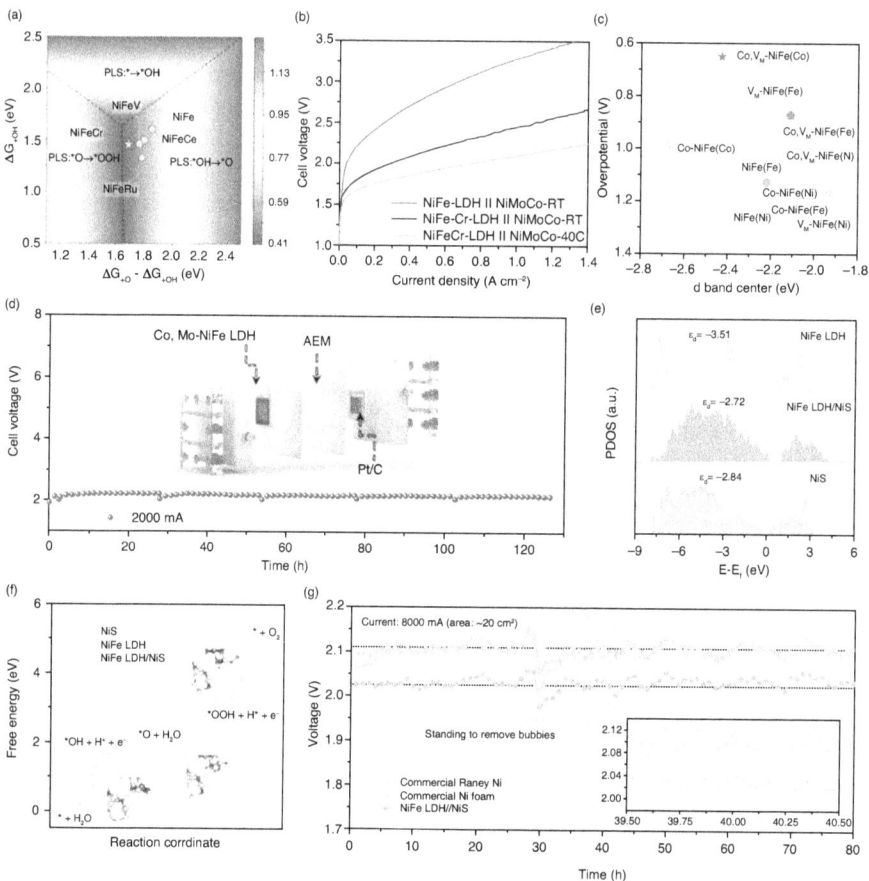

FIGURE 12.11   (a) Oxygen evolution reaction (OER) activity volcano diagram, in which the overpotential is plotted as a function of ΔG_{*OH} and $\Delta G_{*O}-\Delta G_{*OH}$. (b) Cell voltages of NiFe-LDH ‖ NiMoCo alloy at room temperature (RT), NiFeCr-LDH ‖ NiMoCo alloy at RT, and NiFeCr-LDH ‖ NiMoCo alloy at 40°C at different current densities for AWE without iR compensation. (c) The correlation of overpotentials and d-band center (ε_d). (d) Chronopotentiometry curves of the AEM water electrolyzer [80]. (e) The projected density on d orbital for NiFe-LDH, NiS, and NiFe-LDH/NiS. (f) Free-energy diagram. (g) The cell voltages of the electrolyzer at constant catalytic current of 8000 mA (area: ≈20 cm^2) for 80 hours at 80°C–85°C [75].

above theoretical results, the authors prepared a series of high-priced metal-doped OER catalysts, in which NiFeCr-LDH showed high OER performance, exceeding that of IrO_2 and binary controls. The NiFeCr-LDH and NiMoCo alloy coupled AEM electrolyzers can maintain a high current density at a lower voltage, which also proves their feasibility as anode catalysts in AEM electrolyzers (Figure 12.11b).

The high-valence doping strategy could provide a new way for the rational design of efficient OER catalysts. Vacancy engineering, especially for metal cation vacancies with multiple electron configurations and orbitals, can improve OER activity by manipulating band structures, carrier concentrations, and spin states [76]. However, the increase of vacancy concentration tends to destroy the structure of the catalyst [77,78]. Therefore, it is powerless to achieve high activity and stability of OER catalysts via a single vacancy engineering [79]. Yu et al. achieved the simultaneous introduction of co-doping and metal vacancy in NiFe-oxidized hydroxides (Co, V_M-NiFeOOH) by in situ leaching of Mo atoms in an electrochemical process [80]. As shown in Figure 12.11c, DFT results show that Co doping and the introduction of metal vacancy can lead to local charge transfer, which can effectively regulate the d-band center of the metal active site, thus optimizing the adsorption energy of oxidation intermediates in the OER process. At the same time, in situ experimental characterization showed that the co-action of cationic vacancy and Co doping effectively promoted the oxidation of metal sites, thus reducing the overpotential to 255 mV at 100 mA/cm^2. Co, V_M-NiFe OOH electrodes applied to industrial water decomposition electrolytic cells can run stably for 100 hours at 8 A current (Figure 12.11d). This work elucidates the positive effects of metal doping and cationic defects on enhancing the OER activity of catalysts, which could help advance the development of industrial OER catalysts.

The construction of heterojunctions by transition metal (oxygen) hydroxides with metal-like transition metal compounds (such as chalcogenides and phosphates) can induce strong electronic interactions at the heterogeneous interface, which often results in significant electrocatalytic properties [81]. Wen et al. constructed a strongly coupled NiFe-LDH/NiS Schottky heterojunction [75]. As shown in Figure 12.11e and f, the d-band center of Ni(Fe) atoms in NiFe-LDH/NiS is well-tuned through interfacial charge transfer, suggesting the optimized intermediate adsorption for obtaining an accelerated OER kinetics. Furthermore, the NiS nanosheet arrays promote electrolyte penetration and O_2 release even under a large current. As illustrated in Figure 12.11g, in an industrial-grade AEMWE device, the scaled-up NiFe-LDH/NiS electrode can maintain a stable cell voltage of 2.01 V under an ultra-high current of 8 A for 80 hours.

12.4.2.2 3d Transition Metal Oxides, Selenides, Nitrides, and Phosphates

Earth-abundant metal oxides, selenides, nitrides, and phosphates derived from the transition metals Ni, Co, and Fe could provide a cheaper option for highly efficient OER electrocatalysts [62,82–84]. Sargent et al. prepared nanocrystalline Ni–Co–Se by low-temperature ball milling [85]. The operando X-ray absorption spectra showed that the Ni–Co–Se structure was anodized during the OER process, Se was leached from the original structure, and water molecules hydrated the defective parts of Ni and Co to further evolve into active Ni–Co oxyhydroxide. Figure 12.12a clearly

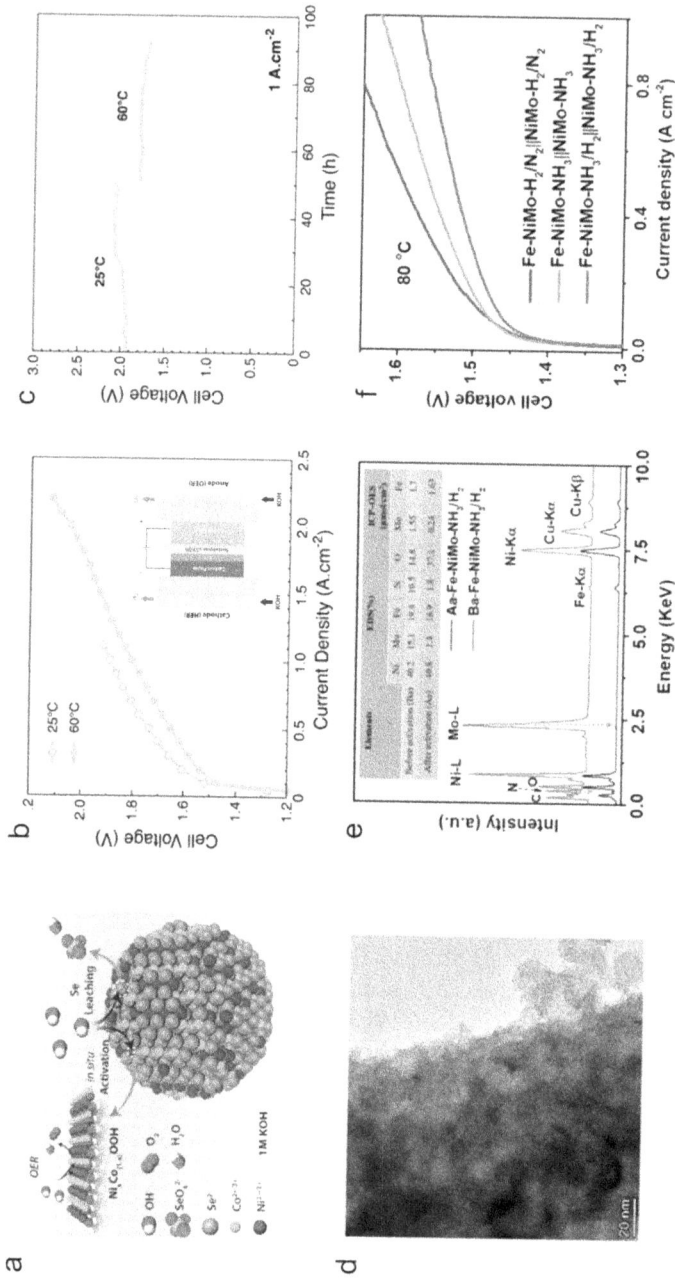

FIGURE 12.12 (a) Schematic illustration of the activation process for $(NiCo)_3Se_4$ $(NiCo)_3Se_4$ in AEMWE cell [85]. (d) HRTEM image of the Fe-NiMo-NH$_3$/H$_2$ catalyst after anodic activation. (e) The EDS spectrum and corresponding element contents (EDS and ICP-OES) of Fe-NiMo-NH$_3$/H$_2$ catalyst before and after anodic activation. (f) The performance of MEAs employing Fe-NiMo-NH$_3$/H$_2$||NiMo-NH$_3$/H$_2$, Fe-NiMo-NH$_3$||NiMo-NH$_3$, and Fe-NiMo-N$_2$/H$_2$||NiMo-N$_2$/H$_2$ pairs at 80°C [48].

describes the reconstruction process of Ni–Co–Se precatalyst, which has also been verified by theoretical calculation. The reconstructed electrocatalyst exhibits an overpotential of 329 mV at 1 A/cm^2 and lasts for 500 hours without performance degradation. The resulting catalyst is applied to an AEM electrolyzer, yielding a current density of 1 A/cm^2 at a low voltage of 1.75 V and can run for 95 hours without decay as shown in Figure 12.12b and c. Moreover, Chen et al. synthesized NiMoN$_x$ precatalyst through a one-step annealing process and then achieved Fe-modified NiMoN$_x$ precatalysts through anodic oxidation [48]. As shown in Figure 12.12 d and e, EDS and ICP-MS results showed that the original NiMoN$_x$ precatalyst underwent a phase transition in which the reconstructed amorphous surface proved to be FeOOH and NiOOH materials. AEM electrolyzer integrated with these catalysts delivered 1.0 A/cm^2 at 1.57 V at 80°C, outperforming the commercial electrolyzer (Figure 12.12f). It has been confirmed that these catalysts undergo reconfiguration during OER process and form an amorphous metal oxide hydroxide phase on the surface [86–88]. The enhanced OER activity could be attributed to the increased active area, the formed defects, disordered nanostructure, and unusual amorphous phase of in situ generated metal oxide/hydroxides. However, the mechanism of reconfiguration and the actual active site are not fully understood, and the role of S, Se, Te, P, or N anions remains unclear. In order to fundamentally understand the structure, degradation, and active sites of catalysts, further ex situ and in situ characterization techniques are required.

12.4.3 PEROVSKITES

The general structure of perovskite is ABO$_3$, which is composed of rare and alkaline earth metals at site A and 3d TM at site B. Due to the characteristics of flexible composition/structure and rich functionality, perovskites are another class of materials studied extensively as catalysts for the OER in alkaline media [89–91]. Recently, the research of perovskite OER catalysts has made great progress, and LaFeO$_3$ and SrCoO$_3$ with high activity have been reported. But these highly reactive materials are reported to be unstable under oxidation conditions in AEMWEs. Therefore, in order to realize the application of perovskite in AEM electrolyzer, the activity and stability of the perovskite should be further improved.

As shown in Figure 12.13a, Shao et al. proposed an A/B co-doping strategy to regulate the crystal structure and electronic structure, realizing the efficient and stable operation of Fe-based perovskite catalyst [92]. Compared to the SrFeO$_{3-\delta}$ parent oxide, Sr$_{0.95}$Ce$_{0.05}$Fe$_{0.9}$Ni$_{0.1}$O$_{3-\delta}$ (SCFN) perovskite oxide with minor Ce/Ni co-doping in A/B sites shows significant enhancement in OER activity as shown in Figure 12.13b. Detailed investigations revealed that doping Ce at the A site contributes to the formation of three-dimensional strongly connected cubic structure. Furthermore, doping Ni at the B site is conducive to the formation of strong interaction between Fe–Ni active sites. The three-dimensional strongly connected cubic structure and the strong interaction between Fe–Ni active sites play a decisive role in maintaining the efficient and stable operation of the catalyst. More importantly, this strategy is universal and provides a new idea for rational and simple design of efficient and stable Fe-based catalysts.

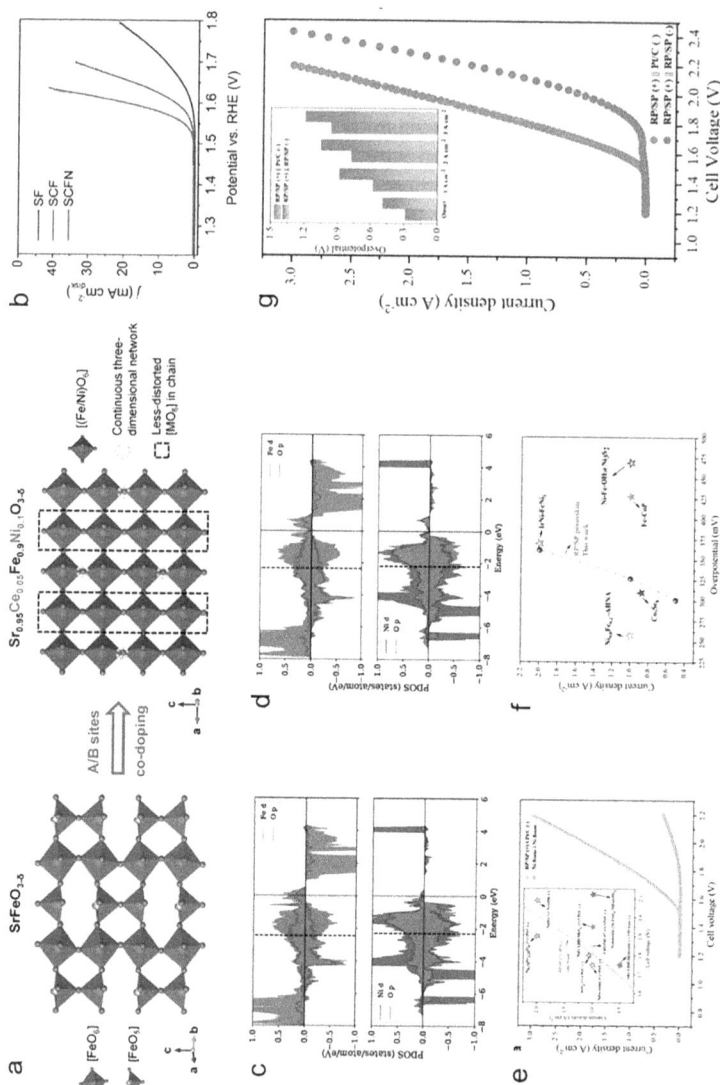

FIGURE 12.13 (a) Schematic illustration of the structural evolution from pristine SF to Ce/Ni-co-doped SCFN. (b) Polarization curves for SF, SCF, and SCFN electrocatalysts in O_2-saturated 0.1 M KOH [93]. Local spin-polarized PDOS for (c) LaFe$_{0.75}$Ni$_{0.25}$O$_3$ and (d) La$_{0.625}$Ca$_{0.375}$Fe$_{0.75}$Ni$_{0.25}$O$_3$ for O $2p$ and Fe and Ni $3d$ orbitals [93]. (e) Polarization curves of overall water splitting in an AEMEC fabricated with RP/SP anode and Pt/C cathode. The inset shows a comparison of our RP/SP (+) ‖ Pt/C (−) AEMEC with the representative AEM electrolyzer cells or water electrolyzers reported in the literature. (f) Overpotentials for the OER at high current densities achieved by our RP/SP catalyst, in comparison with some of the reported representative electrocatalysts operated at high current densities. (g) Polarization curves of overall water splitting in the symmetric AEMEC (RP/SP (+) ‖ PR/SP (−)) compared with the asymmetric counterpart (RP/SP (+) ‖ Pt/C (−)). The inset presents a comparison of the overpotentials required for reaching certain current densities in symmetric and asymmetric AEMECs [94].

Abakumov et al. synthesized $La_{0.6}Ca_{0.4}Fe_{0.7}Ni_{0.3}O_{2.9}$ perovskite using a modified ultrasonic spray pyrolysis technique. Compared with $LaFe_{0.7}Ni_{0.3}O_3$, Ca-doped $La_{0.6}Ca_{0.4}Fe_{0.7}Ni_{0.3}O_{2.9}$ has a more rational electronic structure [93]. Figure 12.13c and d displays that the doping of Ca increases the oxidation states of Ni and Fe, improves the covalency of Ni/Fe-O bonds, moves the center of O $2p$ band toward Fermi level, reduces the formation energy of oxygen vacancy, activates the lattice oxygen mechanism of OER, and improves the catalytic activity. The experimental result shows that $La_{0.6}Ca_{0.4}Fe_{0.7}Ni_{0.3}O_{2.9}$ demonstrates a mass activity of ~ 400 A/g_{oxide} at 1.61 V vs. RHE and a low Tafel slope of 52 ± 2.6 mv/dec, which proves that doping Ca^{2+} ions can promote the activity of the catalyst.

Recently, Shao et al. demonstrated for the first time that perovskites without precious metals can be used in AEM water decomposition devices operating at high current densities [94]. Specifically, Ruddlesden–Popper (RP) and single perovskite (SP) composite (RP/SP) were used as anode materials for AEM electrolyzer. The strong interfacial interaction between the two phases of the electrode material plays an important role in promoting lattice oxygen participation in water oxidation, thus improving the performance of the electrolyzer. Figure 12.13e and f shows that such an electrolyzer with the RP/SP perovskite anode and Pt/C cathode reached an overall water splitting current density as high as 2.01 A/cm^2 at a cell voltage of 2.00 V, which is among the best reported performance of AEMECs based on precious metal free electrodes in the literature. Inspired by the bifunctional catalytic properties of perovskites, the authors successfully applied RP/SP perovskites as anode and cathode electrocatalysts to symmetric AEMECs as shown in Figure 12.13g. Therefore, this work effectively ensures that perovskite oxides can be used as highly efficient and stable electrocatalysts for practical high current density operations.

12.5 CONCLUSIONS AND PERSPECTIVES

Although much progress has been made in the application of catalysts in AEM electrolyzers, developing catalysts with low cost, high activity, and practical stability applications remains a great challenge. Herein, we summarize several important alkaline OER and HER electrocatalysts developed in recent years and highlight the regulation strategies of various electrocatalysts, which can help us understand the sensible structure–activity and stability correlations. Although great progress has been made in the development of AWE, some challenges still remain, and there is a long way ahead for clean, affordable, and sustainable hydrogen production.

Standard evaluation protocols of the OER performance are lacking. Even for catalysts with the same composition and structure, the catalytic performance reported by different research groups varies greatly. However, that catalyst performance is affected by many factors, including electrolyte, catalyst loadings, supporting substrate, material morphology, and electrochemical measurement methods. Therefore, in order to screen out electrocatalysts with excellent performance, it is necessary to compare the catalysts of different researchers in a fair manner.

Industrial-scale preparation of catalysts should be enhanced. In order to meet the requirements of industrial-scale preparation, a facile and large-scale synthesis method is urgently needed. Among previously reported catalysts in alkaline electrolytes, metal oxides/hydroxides and Ni-based alloys often can be easily prepared under mild conditions, thus making them very promising for large-scale application. However, there are still some catalysts, such as metal nitrides and metal carbides, that require dangerous and harsh conditions for preparation. Therefore, more efforts are still needed to develop methods for synthesizing catalysts on a large scale.

In most studies, laboratory-scale electrolysis is usually performed at room temperature, with low electrolyte concentrations and short operating times. However, the operation of MEA electrolysis systems requires a greater current density, higher temperature, higher electrolyte concentration, and longer operating time, which may be overlooked in laboratory-scale testing [95,96]. Therefore, in order to bridge the gap between laboratory-scale and large-scale performance, it is necessary to further consider the reaction environment of the catalyst to rationally design the catalyst to achieve excellent activity and stability.

The research on the mechanism of catalytic reaction can provide reasonable guidance for the development of catalysts with excellent performance. However, the mechanism of anode and cathode reaction and the factors affecting the activity and stability of catalysts are still controversial. The studies have shown that the high activity of some catalysts is often at the cost of stability. For example, catalysts that undergo reconstruction during catalysis may be structurally damaged during the reconfiguration process. In addition, the complex composition of the reconstructed catalysts introduces additional difficulty in the exploration of catalytic mechanisms and the recognition of real catalytic sites. Therefore, it is very important to construct the relationship between catalyst structure and catalytic activity and stability. More accurate in situ characterization techniques are needed to provide experimental evidence for the identification of key intermediates, catalyst reconstruction processes, actual catalytic sites, and reaction pathways [97–100]. In addition, the reconstruction process and the function of each component can be understood through in situ characterization techniques to guide effective catalyst design. Therefore, in situ characterization technology can provide experimental evidence for the relationship between catalyst structure and activity and stability, and then effectively guide the design of catalysts.

REFERENCES

1. Chu, S. and A. Majumdar, Opportunities and challenges for a sustainable energy future. *Nature*, 2012. **488**(7411): pp. 294–303.
2. Faber, M.S. and S. Jin, Earth-abundant inorganic electrocatalysts and their nanostructures for energy conversion applications. *Energy Environ Sci*, 2014. **7**(11): pp. 3519–3542.
3. Ren, X., et al., Cobalt-borate nanowire array as a high-performance catalyst for oxygen evolution reaction in near-neutral media. *J Mater Chem A*, 2017. **5**(16): pp. 7291–7294.
4. Abe, J.O., et al., Hydrogen energy, economy and storage: Review and recommendation. *Int J Hydrogen Energy*, 2019. **44**(29): pp. 15072–15086.
5. Chatenet, M., et al., Water electrolysis: From textbook knowledge to the latest scientific strategies and industrial developments. *Chem Soc Rev*, 2022. **51**(11): pp. 4583–4762.
6. Xu, Q., et al., Anion exchange membrane water electrolyzer: Electrode design, lab-scaled testing system and performance evaluation. *EnergyChem*, 2022. 4(5): p. 100087.

7. Du, N., et al., Anion-exchange membrane water electrolyzers. *Chem Rev*, 2022. **122**(13): pp. 11830–11895.

8. Phillips, R. and Charles W. Dunnill, Zero gap alkaline electrolysis cell design for renewable energy storage as hydrogen gas. *RSC Adv*, 2016. **6**(102): pp. 100643–100651.

9. Abbasi, R., et al., A roadmap to low-cost hydrogen with hydroxide exchange membrane electrolyzers. *Adv Mater*, 2019. **31**(31): p. e1805876.

10. Liu, Y., et al., Shining light on anion-mixed nanocatalysts for efficient water electrolysis: Fundamentals, progress, and perspectives. *Nanomicro Lett*, 2022. **14**(1): p. 43.

11. Li, D., et al., Highly quaternized polystyrene ionomers for high performance anion exchange membrane water electrolysers. *Nature Energy*, 2020. 5: pp. 378–385.

12. Sun, L., et al., Material libraries for electrocatalytic overall water splitting. *Coord Chem Rev*, 2021. 444: p. 214049.

13. Marini, S., et al., Advanced alkaline water electrolysis. *Electrochim Acta*, 2012. **82**: pp. 384–391.

14. Wu, H., et al., Metal-organic frameworks and their derived materials for electrochemical energy storage and conversion: Promises and challenges. *Sci Adv* 3: p. eaap9252.

15. Wang, W., B. Lei, and S. Guo, Engineering multimetallic nanocrystals for highly efficient oxygen reduction catalysts. *Adv Energy Mater*, 2016. **6**(17): p. 1600236.

16. Lu, X.F., et al., Interfacing manganese oxide and cobalt in porous graphitic carbon polyhedrons boosts oxygen electrocatalysis for Zn-air batteries. *Adv Mater*, 2019. **31**(39): p. e1902339.

17. Lu, X.F., et al., Bimetal-organic framework derived CoFe(2) O(4)/C porous hybrid nanorod arrays as high-performance electrocatalysts for oxygen evolution reaction. *Adv Mater*, 2017. 29(3): p. 1604437.

18. Sun, H., et al., Electrochemical water splitting: bridging the gaps between fundamental research and industrial applications. *Energy Environ Mater*, 2023. 6: p. e12441.

19. Lu, F., et al., First-row transition metal based catalysts for the oxygen evolution reaction under alkaline conditions: Basic principles and recent advances. *Small*, 2017. 13(45): pp. 15441–15451.

20. Jiao, Y., et al., Design of electrocatalysts for oxygen- and hydrogen-involving energy conversion reactions. *Chem Soc Rev*, 2015. **44**(8): pp. 2060–2086.

21. Zheng, Y., et al., The hydrogen evolution reaction in alkaline solution: from theory, single crystal models, to practical electrocatalysts. *Angew Chem Int Ed Engl*, 2018. **57**(26): pp. 7568–7579.

22. Sheng, W., et al., Hydrogen oxidation and evolution reaction kinetics on platinum: Acid vs alkaline electrolytes. *J. Electrochem. Soc.* **157**: p. B1529.

23. Stamenkovic, V.R., et al., Energy and fuels from electrochemical interfaces. *Nat Mater*, 2016. **16**(1): pp. 57–69.

24. Lei, C., et al., Efficient alkaline hydrogen evolution on atomically dispersed Ni-Nx Species anchored porous carbon with embedded Ni nanoparticles by accelerating water dissociation kinetics. *Energy Environ Sci*, 2019. 12: pp. 149–156.

25. Jin, H., et al., Nanocatalyst design for long-term operation of proton/anion exchange membrane water electrolysis. *Adv. Energy Mater.*, 2020. 11(4): p. 2003188.

26. Jie Zheng, W.S., Zhongbin Z., Bingjun X., Yushan Y, Universal dependence of hydrogen oxidation and evolution reaction activity of platinum-group metals on pH and hydrogen binding energy. *Sci Adv.*, 2: p. e1501602.

27. Sheng, W., et al., Correlating hydrogen oxidation and evolution activity on platinum at different pH with measured hydrogen binding energy. *Nat Commun*, 2015. **6**: p. 5848.

28. Morales-Guio, C.G., L.A. Stern, and X. Hu, Nanostructured hydrotreating catalysts for electrochemical hydrogen evolution. *Chem Soc Rev*, 2014. **43**(18): pp. 6555–6569.

29. McCrum, I.T. and M.T.M. Koper, The role of adsorbed hydroxide in hydrogen evolution reaction kinetics on modified platinum. *Nature Energy*, 2020. **5**(11): pp. 891–899.

30. Medford, A.J., et al., From the Sabatier principle to a predictive theory of transition-metal heterogeneous catalysis. *J Catal*, 2015. **328**: pp. 36–42.

31. Mahmood, J., et al., An efficient and pH-universal ruthenium-based catalyst for the hydrogen evolution reaction. *Nat Nanotechnol*, 2017. **12**(5): pp. 441–446.

32. Li, M.F., et al., Single-atom tailoring of platinum nanocatalysts for high-performance multifunctional electrocatalysis. *Nat. Catal.*, 2019. **2**(6): pp. 495–503.

33. Chen, J., et al., Diversity of platinum-sites at platinum/fullerene interface accelerates alkaline hydrogen evolution. *Nat Commun*, 2023. **14**(1): p. 1711.

34. Lei, H., et al., Pt-quantum-dot-modified sulfur-doped NiFe layered double hydroxide for high-current-density alkaline water splitting at industrial temperature. *Adv Mater*, 2023. **35**(15): p. e2208209.

35. Wang, K., et al., Dense platinum/nickel oxide heterointerfaces with abundant oxygen vacancies enable ampere-level current density ultrastable hydrogen evolution in alkaline. *Adv Functional Mater*, 2022. 33(8): p. 2211273.

36. Lyu, Z., et al., Two-dimensionally assembled Pd-Pt-Ir supernanosheets with subnanometer interlayer spacings toward high-efficiency and durable water splitting. *ACS Catal*, 2022. **12**(9): pp. 5305–5315.

37. Zhang, T., et al., Pinpointing the axial ligand effect on platinum single-atom-catalyst towards efficient alkaline hydrogen evolution reaction. *Nat Commun*, 2022. **13**(1): p. 6875.

38. Tan, H., et al., Engineering a local acid-like environment in alkaline medium for efficient hydrogen evolution reaction. *Nat Commun*, 2022. **13**(1): p. 2024.

39. Abbas, S.A., et al., Synergistic effect of nano-Pt and Ni spine for HER in alkaline solution: hydrogen spillover from nano-Pt to Ni spine. *Sci Rep*, 2018. **8**(1): p. 2986.

40. Xu, J., C. Chen, and X. Kong, Ru-O-Cu center constructed by catalytic growth of Ru for efficient hydrogen evolution. *Nano Energy*, 2023. 111: p. 108403.

41. Yu, Q., et al., Hydroxyapatite-derived heterogeneous Ru-Ru(2) P electrocatalyst and environmentally-friendly membrane electrode toward efficient alkaline electrolyzer. *Small*, 2023. **19**(25): p. e2208045.

42. Gao, L., et al., Engineering a local potassium cation concentrated microenvironment toward the ampere-level current density hydrogen evolution reaction. *Energy Environ Sci*, 2023. **16**(1): pp. 285–294.

43. Cai, C., et al., Atomically local electric field induced interface water reorientation for alkaline hydrogen evolution reaction. *Angew Chem Int Ed Engl*, 2023. **62**(26): p. e202300873.

44. Colli, A.N., H.H. Girault, and A. Battistel, Non-precious electrodes for practical alkaline water electrolysis. *Materials (Basel)*, 2019. 12(8): p. 1336.

45. Quaino, P., et al., Volcano plots in hydrogen electrocatalysis - uses and abuses. *Beilstein J Nanotechnol*, 2014. **5**: pp. 846–854.

46. Barati Darband, G., M. Aliofkhazraei, and A.S. Rouhaghdam, Facile electrodeposition of ternary Ni-Fe-Co alloy nanostructure as a binder free, cost-effective and durable electrocatalyst for high-performance overall water splitting. *J Colloid Interface Sci*, 2019. **547**: pp. 407–420.

47. Wang, M., et al., Alloying nickel with molybdenum significantly accelerates alkaline hydrogen electrocatalysis. *Angew Chem Int Ed Engl*, 2021. **60**(11): pp. 5771–5777.

48. Chen, P. and X. Hu, High-efficiency anion exchange membrane water electrolysis employing non-noble metal catalysts. *Adv Energy Mater*, 2020. 10(39): 2002285.

49. Wang, J., et al., Manipulating the water dissociation electrocatalytic sites of bimetallic nickel-based alloys for highly efficient alkaline hydrogen evolution. *Angew Chem Int Ed Engl*, 2022. **61**(30): p. e202202518.

50. Wang, X., et al., Rationally modulating the functions of Ni(3) Sn(2) -NiSnO(x) nano-composite electrocatalysts towards enhanced hydrogen evolution reaction. *Angew Chem Int Ed Engl*, 2023. **62**(19): p. e202301562.

51. Fu, H.Q., et al., Hydrogen spillover-bridged volmer/tafel processes enabling ampere-level current density alkaline hydrogen evolution reaction under low overpotential. *J Am Chem Soc*, 2022. **144**(13): pp. 6028–6039.

52. Suen, N.T., et al., Electrocatalysis for the oxygen evolution reaction: recent development and future perspectives. *Chem Soc Rev*, 2017. **46**(2): pp. 337–365.

53. Chen, F.-Y., et al., Stability challenges of electrocatalytic oxygen evolution reaction: From mechanistic understanding to reactor design. *Joule*, 2021. **5**(7): pp. 1704–1731.

54. Gao, Z.W., et al., Engineering NiO/NiFe LDH intersection to bypass scaling relation-ship for oxygen evolution reaction via dynamic tridimensional adsorption of intermedi-ates. *Adv Mater*, 2019. **31**(11): p. e1804769.

55. Rüetschi, P. and P. Delahay, Influence of electrode material on oxygen overvoltage: A theoretical analysis. *J Chem Phys*, 1955. **23**(3): pp. 556–560.

56. Morales-Guio, C.G., L. Liardet, and X. Hu, Oxidatively electrodeposited thin-film transition metal (Oxy)hydroxides as oxygen evolution catalysts. *J Am Chem Soc*, 2016. **138**(28): pp. 8946–8957.

57. Norskov, J.K., et al., Density functional theory in surface chemistry and catalysis. *Proc Natl Acad Sci U S A*, 2011. **108**(3): pp. 937–943.

58. Zhang, K. and R. Zou, Advanced transition metal-based OER electrocatalysts: current status, opportunities, and challenges. *Small*, 2021. **17**(37): p. e2100129.

59. Norskov, J.K., et al., Towards the computational design of solid catalysts. *Nat Chem*, 2009. **1**(1): pp. 37–46.

60. Man, I.C., et al., Universality in oxygen evolution electrocatalysis on oxide surfaces. *ChemCatChem*, 2011. **3**(7): pp. 1159–1165.

61. Grimaud, A., et al., Activating lattice oxygen redox reactions in metal oxides to catalyse oxygen evolution. *Nat Chem*, 2017. **9**(5): pp. 457–465.

62. Yu, Z.Y., et al., Clean and affordable hydrogen fuel from alkaline water splitting: Past, recent progress, and future prospects. *Adv Mater*, 2021. **33**(31): p. e2007100.

63. Frydendal, R., et al., Benchmarking the stability of oxygen evolution reaction cata-lysts: The importance of monitoring mass losses. *ChemElectroChem*, 2014. **1**(12): pp. 2075–2081.

64. Lee, Y., et al., Synthesis and activities of rutile IrO$_2$ and RuO$_2$ nanoparticles for oxygen evolution in acid and alkaline solutions. *J Phys Chem Lett*, 2012. **3**(3): pp. 399–404.

65. Spori, C., et al., The stability challenges of oxygen evolving catalysts: Towards a com-mon fundamental understanding and mitigation of catalyst degradation. *Angew Chem Int Ed Engl*, 2017. **56**(22): pp. 5994–6021.

66. Du, J., et al., Highly stable and efficient oxygen evolution electrocatalyst based on Co oxides decorated with ultrafine Ru nanoclusters. *Small*, 2023. **19**(28): p. e2207611.

67. Zeng, L., et al., Anti-dissolution Pt single site with Pt(OH)(O$_3$)/Co(P) coordination for efficient alkaline water splitting electrolyzer. *Nat Commun*, 2022. **13**(1): p. 3822.

68. Jiang, K., et al., Dynamic active-site generation of atomic iridium stabilized on nanopo-rous metal phosphides for water oxidation. *Nat Commun*, 2020. **11**(1): p. 2701.

69. Yang, X., et al., IrPd nanoalloy-structured bifunctional electrocatalyst for efficient and pH-universal water splitting. *Small*, 2023. **19**(27): p. e2208261.

70. Gong, M. and H. Dai, A mini review of NiFe-based materials as highly active oxygen evolution reaction electrocatalysts. *Nano Res*, 2014. **8**(1): pp. 23–39.

71. Wu, H., et al., Identification of facet-governing reactivity in hematite for oxygen evolu-tion. *Adv Mater*, 2018. **30**(52): p. e1804341.

72. Zhang, B., et al., Homogeneously dispersed multimetal oxygen-evolving catalysts. *Sci Adv*, 352 (6283): p. 333.

73. Yang, H., et al., Preparation of nickel-iron hydroxides by microorganism corrosion for efficient oxygen evolution. *Nat Commun*, 2020. **11**(1): p. 5075.

74. Wu, Z.P., et al., Non-noble-metal-based electrocatalysts toward the oxygen evolution reaction. *Adv Functional Mater*, 2020. 30(15): p. 1910274.

75. Wen, Q., et al., Schottky heterojunction nanosheet array achieving high-current-density oxygen evolution for industrial water splitting electrolyzers. *Adv Energy Mater*, 2021. **11**(46): p. 2102353.

76. Dou, Y., et al., Approaching the activity limit of CoSe(2) for oxygen evolution via Fe doping and Co vacancy. *Nat Commun*, 2020. **11**(1): p. 1664.

77. Yang, C., et al., Cation insertion to break the activity/stability relationship for highly active oxygen evolution reaction catalyst. *Nat Commun*, 2020. **11**(1): p. 1378.

78. Li, G., G.R. Blake, and T.T. Palstra, Vacancies in functional materials for clean energy storage and harvesting: the perfect imperfection. *Chem Soc Rev*, 2017. **46**(6): pp. 1693–1706.

79. Li, S., et al., Oxygen-evolving catalytic atoms on metal carbides. *Nat Mater*, 2021. **20**: pp. 1240–1247.

80. Zhao, Y., et al., Operando reconstruction toward dual-cation-defects co-containing NiFe oxyhydroxide for ultralow energy consumption industrial water splitting electrolyzer. *Adv Energy Mater*, 2023. 13(10): p. 2203595.

81. Xue, Z., et al., Interfacial electronic structure modulation of NiTe nanoarrays with NiS nanodots facilitates electrocatalytic oxygen evolution. *Adv Mater*, 2019: p. e1900430.

82. Stern, L.A., et al., Ni2P as a Janus catalyst for water splitting: the oxygen evolution activity of Ni2P nanoparticles. *Energy Environ Sci*, 2015. **8**(8): pp. 2347–2351.

83. Chen, P., et al., Metallic Co4N porous nanowire arrays activated by surface oxidation as electrocatalysts for the oxygen evolution reaction. *Angew Chem Int Ed Engl*, 2015. **54**(49): pp. 14710–14714.

84. Peng, X., et al., Recent progress of transition metal nitrides for efficient electrocatalytic water splitting. *Sustain Energy Fuels*, 2019. **3**(2): pp. 366–381.

85. Abed, J., et al., In situ formation of nano Ni-Co oxyhydroxide enables water oxidation electrocatalysts durable at high current densities. *Adv Mater*, 2021. **33**(45): p. e2103812.

86. Chen, W., et al., In situ electrochemical oxidation tuning of transition metal disulfides to oxides for enhanced water oxidation. *ACS Cent Sci*, 2015. **1**(5): pp. 244–251.

87. Ryu, J., et al., In situ transformation of hydrogen-evolving CoP nanoparticles: toward efficient oxygen evolution catalysts bearing dispersed morphologies with co-oxo/hydroxo molecular units. *ACS Catal*, 2015. **5**(7): pp. 4066–4074.

88. Mabayoje, O., et al., The role of anions in metal chalcogenide oxygen evolution catalysis: Electrodeposited thin films of nickel sulfide as "pre-catalysts". *ACS Energy Lett*, 2016. **1**(1): pp. 195–201.

89. Grimaud, A., et al., Double perovskites as a family of highly active catalysts for oxygen evolution in alkaline solution. *Nat Commun*, 2013. **4**: p. 2439.

90. Jin Suntivich, K.J.M., Hubert A. Gasteiger, John B. Goodenough, and Yang Shao-Horn, A perovskite oxide optimized for oxygen evolution catalysis from molecular orbital principles. *Science*, 2011. **334**: pp. 1383–1385.

91. May, K.J., et al., Influence of oxygen evolution during water oxidation on the surface of perovskite oxide catalysts. *J Phys Chem Lett*, 2012. **3**(22): pp. 3264–3270.

92. She, S., et al., Realizing high and stable electrocatalytic oxygen evolution for iron-based perovskites by Co-doping-induced structural and electronic modulation. *Adv Funct Mater*, 2021. **32**(15): p. 2111091.

93. Porokhin, S.V., et al., Mixed-cation perovskite $La_{0.6}Ca_{0.4}Fe_{0.7}Ni_{0.3}O_{2.9}$ as a stable and efficient catalyst for the oxygen evolution reaction. *ACS Cataly*, 2021. **11**(13): pp. 8338–8348.

94. Tang, J., et al., Perovskite-based electrocatalysts for cost-effective ultrahigh-current-density water splitting in anion exchange membrane electrolyzer cell. *Small Methods*, 2022. **6**(11): p. e2201099.

95. Kuang, Y., et al., Solar-driven, highly sustained splitting of seawater into hydrogen and oxygen fuels. *Proc Natl Acad Sci U S A*, 2019. **116**(14): pp. 6624–6629.

96. Zeng, K. and D. Zhang, Recent progress in alkaline water electrolysis for hydrogen production and applications. *Prog Energy Combust Sci*, 2010. **36**(3): pp. 307–326.

97. Cheng, W., et al., Lattice-strained metal-organic-framework arrays for bifunctional oxygen electrocatalysis. *Nat Energy*, 2019. **4**: pp. 155–122.

98. Duan, Y., et al., Scaled-up synthesis of amorphous NiFeMo oxides and their rapid surface reconstruction for superior oxygen evolution catalysis. *Angew Chem Int Ed Engl*, 2019. **58**(44): pp. 15772–15777.

99. Yan, Z., et al., Anion insertion enhanced electrodeposition of robust metal hydroxide/oxide electrodes for oxygen evolution. *Nat Commun*, 2018. **9**(1): p. 2373.

100. Cao, L., et al., Identification of single-atom active sites in carbon-based cobalt catalysts during electrocatalytic hydrogen evolution. *Nat Catal*, 2018. **2**: pp. 134–141.

13 CO_2 Electrolysis in Solid Oxide Electrolysis Cells

Key Materials, Application, and Challenges

Xiang Wang, Liang Hu, Min Yang,
Zongying Han, and Zhibin Yang

13.1 INTRODUCTION

In the modern world, the overuse of fossil fuels has resulted in a surge in atmospheric CO_2 concentration, which has further induced a wide range of environmental issues such as global warming and the rise of sea level [1]. Meanwhile, as fossil fuels become scarcer, renewable energy sources such as wind, solar, and geothermal are being developed. However, compared to traditional thermal power generation, the electricity generated by these renewable energy sources is generally intermittent and fluctuating, which may lead to remarkable energy wastage [2]. To address this issue, one of the primary methods of storing these renewable energy that cannot be consumed in time is to convert them into chemical energy in the form of H_2 or CO [3]. Therefore, carbon dioxide capture and utilization (CCU) technologies are urgently needed to be developed [4]. However, efficient conversion of carbon dioxide into chemicals with high added value was difficult due to its extremely stable C=O bond. Currently, several methods have been developed for the electrochemical reduction of CO_2, such as H-type electrolysis cells, flow electrolysis cells, molten salt electrolysis cells, and solid oxide electrolysis cells (SOEC) [5]. However, due to their slower reaction kinetics, H-type electrolysis cells, flow electrolysis cells, and molten salt electrolysis cells were unable to attain high energy efficiency and stability [6]. On the other hand, SOEC, as a unique and extremely efficient energy conversion device, can electrolyze carbon dioxide at high temperatures (>600°C) to generate CO, with a substantially greater system efficiency than the other electrolysis methods [7]. Notably, SOEC can not only convert electrical energy into chemical energy but also produce high-purity oxygen, which can be used as rocket propellant or essential substance for sustaining life in the outer space. Despite the advantages of high efficiency, environmental friendliness, and high energy density, numerous hurdles still limit the promotion of SOEC technology in various applications. In this chapter, we comprehensively reviewed the development of key materials involved in SOEC, summarized the main applications of SOEC to CO_2 electrolysis, and systematically investigated the challenges facing SOEC technology in CO_2 electrolysis. It

DOI: 10.1201/9781003368939-13

is expected that the information collected in this chapter will be a valuable information source for researchers and provide direction and guidance for future research in SOEC technology.

13.2   WORKING PRINCIPLE OF CO$_2$ ELECTROLYSIS IN SOEC

13.2.1   THE REACTION MECHANISM

SOEC is an all-solid-state structure energy conversion device, which mainly consists of three parts: dense electrolyte layer, porous cathode and anode layers (Figure 13.1) [8]. During operation, direct current is applied to both sides of the cathode and anode, providing driving force for CO$_2$ electrolysis. After being charged into the cathode chamber, the CO$_2$ is first adsorbed on the surface of the cathode and then decomposed into CO and O^{2-} after receiving electrons [7]. CO gas can be collected from the exhaust gas of cathode. Simultaneously, oxygen ions (O^{2-}) are delivered to the anode through the solid electrolyte layer and then converted into O$_2$ by losing electrons.

In this process, CO$_2$ reduction reaction (CO$_2$RR, equation 13.1) occurs at the cathode, while the following oxygen evolution reaction (OER, equation 13.2) occurs at the anode side. The total reaction can be described as shown in equation (13.3) [9].

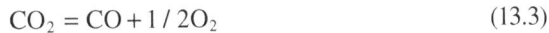

$$CO_2 + 2e^- = CO + O_2 \tag{13.1}$$

$$O^{2-} = 1/2O_2 + 2e^- \tag{13.2}$$

$$CO_2 = CO + 1/2O_2 \tag{13.3}$$

In the SOEC system, when using an electrolyte that was a pure oxygen ionic conductor, there was no current through the system and its open circuit voltage was approximately equal to the theoretical electric potential [10]. For some electrolytes with oxygen ion migration number less than 1, the existence of electron conductance

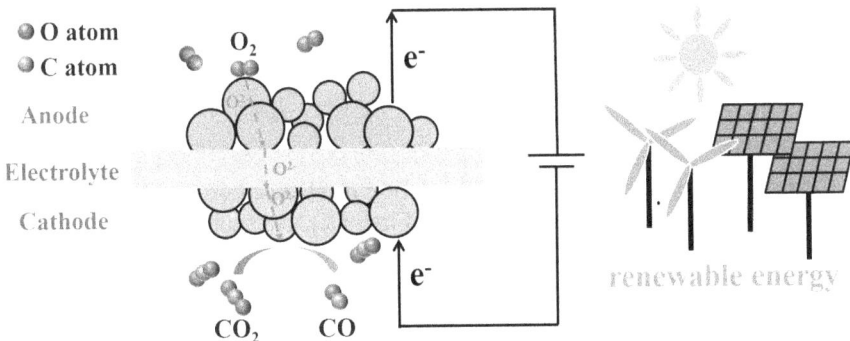

FIGURE 13.1   Schematic diagram of SOEC working principle.

led to short circuit in the circuit, and the open circuit voltage was less than the theoretical electromotive force. Three types of polarization typically occurred in SOEC, namely ohmic polarization, activation polarization, and concentration polarization [11]. Ohmic polarization was mainly due to the polarization phenomenon caused by the contact of the components within the cell, resulting in ionic and electronic conduction obstruction. Activation polarization was attributed to the phenomenon caused by electrocatalytic activity of the electrode that was insufficient to limit the overall rate of electrochemical reaction, whereas concentration polarization was due to the charge transfer rate that was too fast, and the electrode reaction of the active substance was not transferred to the active site in a timely manner, resulting in the concentration of reactive substances deviated from the concentration of the body of the phenomenon caused by the phenomenon.

The main direction of research in recent years was to reduce the working temperature to medium temperature ($500°C–700°C$) [12]. However, the lower operating temperature will cause an increase in the electrode polarization impedance, especially on the anode, and the increase in the electrode polarization impedance will result in a decrease in the CO_2RR and OER activity of electrode as well as the oxygen transfer rate in SOEC [13].

The detailed reaction mechanism of the CO_2RR process varies depending on the cathode materials. Yang et al. investigated the reaction mechanism of CO_2RR occurring on the surface of $La_xSr_{1-x}FeO_{3-\delta}$ (LSF) cathode by Raman spectroscopy and density functional theory (DFT) calculations [14]. As shown in reactions (13.4)–(13.7), CO_2 molecules initially interact with surface lattice oxygen ions to create adsorbed carbonate species, which were then dissolved by successive electron acceptance. Finally, the produced CO molecules desorb off the cathode surface into the exhaust gas.

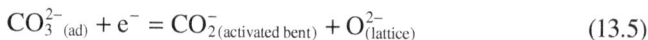

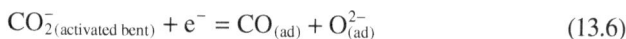

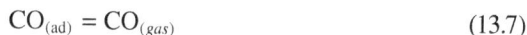

$$CO_{2(g)} + O^{2-}_{(lattice)} = CO^{2-}_{3\,(ad)} \tag{13.4}$$

$$CO^{2-}_{3\,(ad)} + e^- = CO^-_{2\,(activated\ bent)} + O^{2-}_{(lattice)} \tag{13.5}$$

$$CO^-_{2\,(activated\ bent)} + e^- = CO_{(ad)} + O^{2-}_{(ad)} \tag{13.6}$$

$$CO_{(ad)} = CO_{(gas)} \tag{13.7}$$

In the study of Opitz et al., a different reaction mechanism of the CO_2RR process on perovskite cathodes was proposed [15]. First, CO_2 molecules react with oxygen vacancies at the cathode surface and electrons given by an external circuit to form $(CO_2)^-_{(ad)}$ (reaction 13.8). Then, the monodentate carbonate reacts with oxygen ions at the surface to generate bidentate carbonate (reactions 13.9). Finally, the bidentate carbonate is decomposed into a CO molecule and two oxygen ions with the assistance of another electron (reactions 13.10).

$$CO_{2(g)} + Vac + e^- = (CO_2)^-_{(ad)} \tag{13.8}$$

$$(CO_2)^-_{(ad)} + O^{2-} = CO_3^{2-}_{(ad)} \tag{13.9}$$

$$CO_3^{2-}_{(ad)} + e^- = CO_{(g)} + 2O^{2-} \tag{13.10}$$

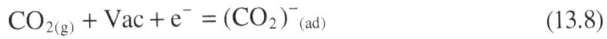

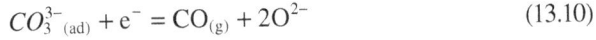

For cerium oxide-based cathodes, CO$_2$RR is mainly mediated by the redox process of trivalent cerium ions (Ce^{3+}) and tetravalent cerium ions (Ce^{4+}), as illustrated in reactions (13.11–13.13) [15]. First, CO$_2$ combines with the surface oxygen ions to produce carbonate (CO$_3^{2-}$), then Ce^{3+} is oxidized to Ce^{4+} and electrons are separated, and finally carbonate combines with two electrons to disintegrate into CO and oxygen ions.

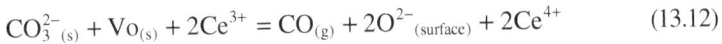

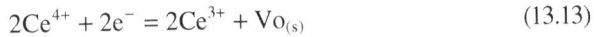

$$CO_{2(g)} + O^{2-}_{(surface)} = CO_3^{2-}_{(surface)} \tag{13.11}$$

$$CO_3^{2-}_{(s)} + Vo_{(s)} + 2Ce^{3+} = CO_{(g)} + 2O^{2-}_{(surface)} + 2Ce^{4+} \tag{13.12}$$

$$2Ce^{4+} + 2e^- = 2Ce^{3+} + Vo_{(s)} \tag{13.13}$$

Feng et al. proposed that additional chemical step occurred on the CeO$_2$-based cathode surface, as demonstrated in reactions (13.14–13.15) [16]. The CO$_2$RR process on the surface of the Sm$_{0.2}$Ce$_{0.8}$O$_2$ cathode is divided into three steps. First, Ce^{4+} is reduced to Ce^{3+}, then CO$_2$ reacts with surface oxygen ions and surface Ce^{3+} to form carbonate species, and finally, carbonate species combine with Ce^{3+} to form CO molecules and two oxygen ions.

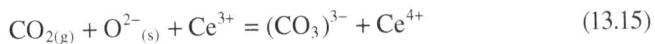

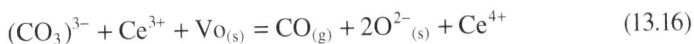

$$2Ce^{4+} + 2e^- = 2Ce^{3+} \tag{13.14}$$

$$CO_{2(g)} + O^{2-}_{(s)} + Ce^{3+} = (CO_3)^{3-} + Ce^{4+} \tag{13.15}$$

$$(CO_3)^{3-} + Ce^{3+} + Vo_{(s)} = CO_{(g)} + 2O^{2-}_{(s)} + Ce^{4+} \tag{13.16}$$

13.2.2 THERMODYNAMIC ANALYSIS

During the operation of SOEC, the energy consumption of the reaction can be calculated according to equation (13.17). Figure 13.2 shows the calculated energy demands

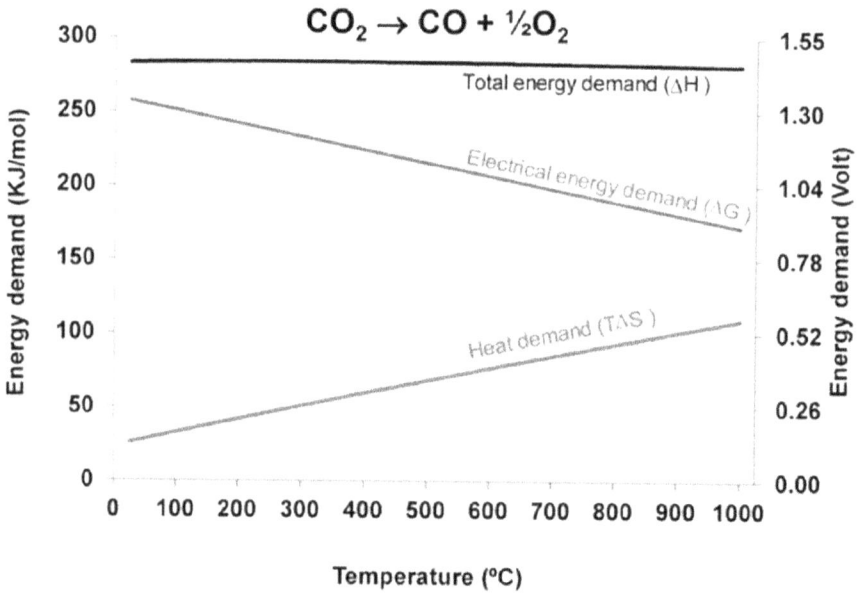

FIGURE 13.2   Energy demand for CO_2 electrolysis [7].

for CO_2 electrolysis. ΔHr represents the total energy consumption for the electrolysis of CO_2, which contains the applied electrical energy (the Gibbs free energy of the reaction, ΔGr) and heat requirements for the reaction process (the product of the temperature (T) and the entropy change of the reaction (ΔSr)) [9].

$$\Delta Hr = \Delta Gr + T\Delta Sr \qquad (13.17)$$

In the standard state, assuming that the electrolysis of carbon dioxide was a reversible process, its electrical energy can be calculated as shown in equation (13.18). The variables n, $E\ominus$, and F are the number of electrons moved throughout the reaction process, the standard reversible electrolysis voltage, and the Faraday constant, respectively [17]. Since each carbon dioxide molecule's breakdown during electrolysis results in the transfer of two electrons, n is equal to 2. The standard reversible electrolysis voltage for carbon dioxide is calculated to be 1.33 V. As the temperature increased, the accompanying E decreased since the values of both ΔH_r and ΔGr depend on the temperature. Due to different polarizations, the actual voltage during electrolysis was substantially higher than the value of E. By altering the cell component materials and optimizing the cell configuration and cell microstructure, it is possible to partially eliminate various polarizations present in the system.

$$\Delta Gr^\circ = nEr^\circ F \qquad (13.18)$$

In an adiabatic environment, the thermal neutral voltage (E_{tn}) represents the lowest voltage necessary for reactant decomposition. Equation (13.19) demonstrates the relationship between the thermal neutral voltage and the overall energy of the electrolysis reaction.

$$\Delta Hf = nEtnF \qquad (13.19)$$

Under the standard condition, the thermal neutral voltage for CO$_2$ decomposition is 1.48 V. When the applied voltage is higher than the reversible electrolysis voltage and lower than the thermo-neutral voltage, external heat supply is required to keep the electrochemical reaction going. When the applied voltage is higher than the thermal neutral voltage, the heat released by the system itself is sufficient to sustain the electrochemical reaction.

13.2.3 STRUCTURE TYPES OF SOEC

Two basic forms of SOEC structures are currently in use, flat plate and tubular constructions, depending on the various shapes [2]. The SOEC with a flat plate structure stood out among the others for its straightforward structural design, ease of processing, shorter electron transmission channel, higher energy density per unit area, and other benefits. However, there are still a lot of challenging issues facing the flat plate-type SOCE due to the larger exposed area, such as more difficult encapsulation, thermal stress, lower thermal cycling efficiency, and the larger performance degradation rate.

Furthermore, SOEC can also be categorized into cathode-supported type, electrolyte-supported type, and metal-supported type depending on the support body [3]. The electrolyte-support type is the first structure adopted for SOEC. However, the wide promotion of the electrolyte-support type SOEC is limited by the high ohmic impedance induced by the thick electrolyte support. The ohmic impedance and the operating temperature can be efficiently reduced by using electrodes or metal as support and thinning electrolyte layer.

13.3 KEY MATERIALS

13.3.1 ELECTROLYTE

Electrolyte candidates include typical ZrO$_2$-, LaGaO$_3$-, and CeO$_2$-based electrolytes. However, specific considerations are also needed for electrolyte materials, especially regarding impurity tolerance and chemical stability.

13.3.1.1 Stabilized Zirconia

ZrO$_2$-based electrolytes are the first kind of electrolyte materials to be explored and are still one of the most widely used electrolytes in SOEC systems nowadays. However, ZrO$_2$ has a monoclinic crystal at ambient temperature that can only be transformed into stable cubic crystal structure at temperature of above 2300°C [18]. As a result, pure ZrO$_2$ cannot be directly employed as electrolyte materials. As shown in Figure 13.3a, the stabilized cubic crystal structure could be obtained by doping ZrO$_2$ with low-valent metal elements [19]. Simultaneously, a large number

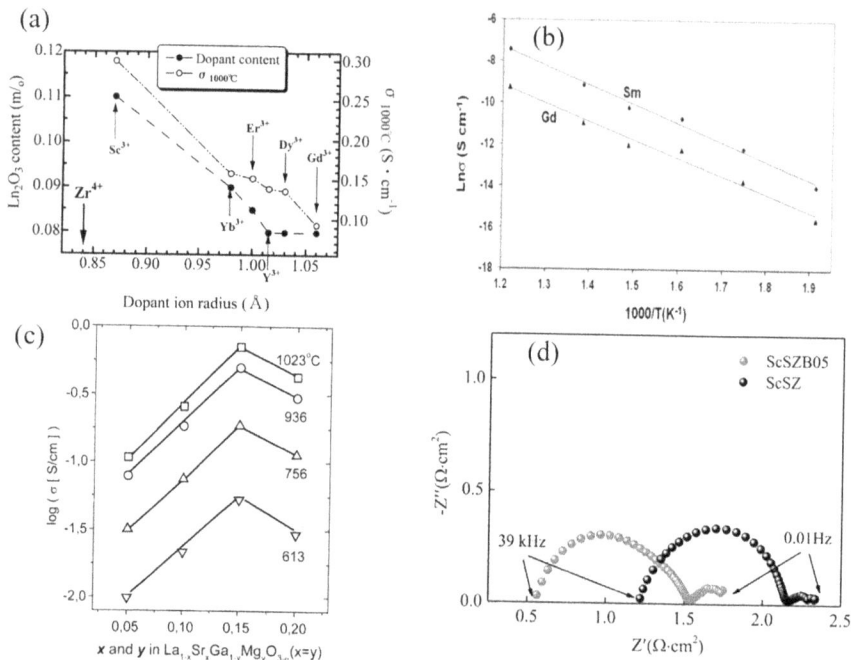

FIGURE 13.3 (a) Dopant content showing the highest conductivity at 1000°C vs. dopant ion radius curves for ZrO_2-based electrolyte materials [19], (b) Arrhenius plots of SDC and GDC samples [22], (c) electrical conductivity of $La_{1-x}Sr_xGa_{1-y}Mg_yO_{3-\delta}$ at different temperatures as a function of doping with equal quantities of Sr and Mg ($x=y$) [23], and (d) the impedance spectra of the hydrogen SOFCs measured under open circuit at 800°C for ScSZ and 5 mol% Bi_2O_3-ScSZ [24].

of oxygen vacancies are introduced to improve the ionic conductivity of ZrO_2-based materials, making them high-performance electrolyte materials. It has been recognized that a strong relationship exists between the concentration of oxygen vacancy defects and the conductivity of ZrO_2-based material systems [2]. Generally, oxygen vacancies and conductivities initially increase with increasing degree of doping. However, after a specific quantity of doping was added, the increase in doping concentration caused a drop in conductivity and oxygen vacancy activity characteristics. The Y_2O_3-stabilized ZrO_2 (YSZ) electrolyte has been regarded as one of the most promising electrolyte materials. YSZ exhibits excellent chemical and thermal stability, as well as outstanding mechanical properties, in both oxidizing and reducing atmospheres. At 1000°C, 8 mol%Y_2O_3-stabilized ZrO_2 (8YSZ) has the highest conductivity (0.13 S/cm) with good stability and compatibility to other cell components [18]. Sc_2O_3-stabilized ZrO_2 (ScSZ) possesses the highest conductivity among ZrO_2-based materials at medium–low temperatures (<800°C) [20]. However, ScSZ is prone to phase transition from a cubic phase with high conductivity to a rhombic phase with low conductivity when SOEC was operated at low–medium temperatures. The stability of the cubic phase of ScSZ can be improved by doping with CeO_2, which was accompanied by slight increase in conductivity [21]. The high cost of

Sc element and the complexity of preparing ScSZ materials hamper their commercialization. Therefore, research on the SOEC with 8YSZ electrolyte remained the primary research focus in practical applications.

13.3.1.2 Doped Ceria

Although ZrO$_2$-based materials exhibit admirable performance, they are excessively affected by temperature. Unlike ZrO$_2$-based materials, pure cerium oxide (CeO$_2$) has a cubic fluorite structure at room temperature and inherently has a large ionic conductivity [25]. As shown in Figure 13.3b, the ionic conductivity can also be considerably improved after doping with low-valent metal elements [22].

Under the same temperature and pressure circumstances, the conductivity of Gd-doped CeO$_2$ (GDC) materials reaches several times or even tens of times that of 8YSZ [26]. However, under low oxygen partial pressure or higher temperature conditions, Ce^{4+} ions in CeO$_2$-based materials are partially reduced to Ce^{3+} ions, resulting in the appearance of electronic conductivity and lattice expansion of the electrolyte materials, both of which have significant adverse effect on their electrochemical and physical properties [16]. The 20 mol% Sm-doped CeO$_2$ (SDC) material has single-phase cubic fluorite structures and electrical conductivity of up to 4.5×10^{-2} S/cm [27,28]. Enabling CeO$_2$-based materials to have higher oxygen ionic conductivity and lower electronic conductivity at medium temperature conditions has been one of the current key research directions. In summary, CeO$_2$-based materials show great promise as electrolyte materials for medium- and low-temperature SOEC toward CO$_2$ electrolysis.

13.3.1.3 Doped LaGaO$_3$

Perovskite oxides (ABO$_3$) exhibit stable crystal structure and superb oxygen ion conductivity at low and medium temperatures. The (Sr, Mg)-doped LaGaO$_3$-based materials (LSGM) are the most frequently used electrolyte materials with perovskite oxide structure applied in SOEC systems [29,30]. The LSGM electrolyte has high chemical stability and ionic conductivity that is significantly higher than that of 8YSZ. As shown in Figure 13.3c, the conductivity of LSGM increased with the addition of Sr and Mg due to the formation of oxygen vacancies [23]. The increase in the doping amount of Sr ions decreases the conductivity activation energy (E$_a$), while the increase in the doping amount of Mg ions increases E$_a$. The coefficient of expansion of LSGM increases with the increase in doping degree and is positively correlated with the concentration of oxygen vacancies in LSGM, which results in a decrease in the mechanical strength of LSGM. Due to its active chemical nature, the stability of LSGM as a solid electrolyte for SOEC over a long period of time still needs to be further improved. The main issue that restricted the application of LSGM materials is the evaporation of elemental Ga in a reducing atmosphere.

SOEC electrolyte materials required excessive sintering temperatures (>1400°C) to achieve completed densification [31]. When the cathode layer is co-sintered with the electrolyte layer, physical strain may be generated on the porous electrode structure, leading to high manufacturing costs. Adding a suitable proportion of sintering additives (such as Bi$_2$O$_3$) can not only reduce the sintering temperature of the electrolyte, but also increase its ionic conductivity. In the study of Liu et al., the YSZ was doped with Bi_2O_3 allowing the sintering temperature of YSZ to be as low as 1000°C [32]. As shown in Figure 13.3d, Wang et al. used 5 mol% Bi_2O_3 as sintering additives, resulting in a relative densification of ScSZ of 95.7% at 1100°C [24]. In addition, Toor et al. used Cu doping causing a decrease in grain boundary conductivity and a decrease in sintering temperature [33].

13.3.2 CATHODE MATERIALS

During SOEC operation, the cathode supported the diffusion of CO_2 and CO gases and provides the active site for the CO_2RR reaction. The cathode material needed to have both high ionic and electronic conductivity, good stability and heat resistance with an expansion coefficient compatible with the electrolyte, suitable porosity, and high catalytic activity.

13.3.2.1 Metal-Oxide Cermet

Metal-oxide ceramics contain mainly precious metals (such as Pt) and non-precious metals (such as Ni) [36]. Since precious metal materials are more expensive, they are not suitable for SOEC cathode materials. Although the metallic Ni exhibits high reactivity, the cathodic reaction can only take place at the cathode/electrolyte triple-phase boundary (TPB) since Ni is a pure electronic conductor. Therefore, in order to extend the electrochemical reaction zone, Ni can be combined with ion-conductive materials (such as YSZ) to form Ni-oxide cement cathode [37]. As shown in Figure 13.4a, Ryu et al. prepared Ni-GDC cathode and achieved 11 times higher performance than Ni anodes [34]. At present, Ni-YSZ cermet has been widely used as cathode material for SOEC. However, metal Ni can be easily oxidized to NiO during the SOEC process, which will cause the degradation of the cell performance. Therefore, it is often necessary to maintain a reducing atmosphere at the cathode side when using Ni-based cement cathode materials. Wu et al. investigated the effect of different protective gases on the CO_2 electrolysis process and found that better electrochemical performance and stability can be achieved by using H_2 as a protective gas than CO [38]. When the CO_2/H_2 ratio was 3:1, no methane was generated during the reaction and the CO selectivity can reach 100%. Another important reason for the decrease in the stability of Ni-YSZ is the agglomeration of Ni particles. To suppress the coarsening and agglomeration issues of the Ni particles, the metal Ni particles should be finer and uniformly distributed on the matrix. The infiltration/wet impregnation method has been considered a promising approach to improve the performance and strength of conventional Ni-YSZ cathode. Chen et al. infiltrated CeO_2 in the form of nanoparticles on the surface of a Ni-YSZ electrode, which not only improved the electrochemical performance by three times, but also prevented the fracture of the Ni-YSZ electrode during the reduction process [37]. Yu et al. developed nickel–iron bimetallic-based cement cathode material by adding iron into nickel, which effectively inhibited the agglomeration of particles and enhanced the performance of Ni-based fuel electrodes [39]. Up to now, numerous researchers have been working on the modification of Ni-YSZ cermet materials to make them more suitable for electrolysis [40].

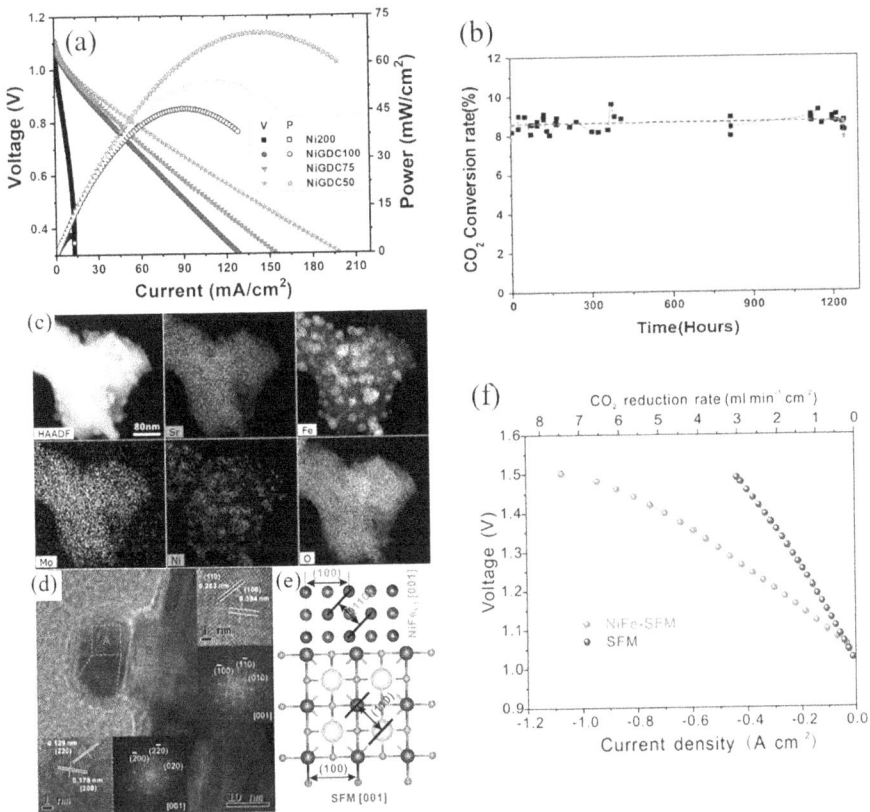

FIGURE 13.4 (a) Voltage–current–power curves of Ni200, NiGDC100, NiGDC75, and NiGDC50 with the ScSZ electrolyte support examined at 500°C [34], (b) long-term stability of LSCM at 700°C in the co-electrolysis environment [35], (c) HADDF image and corresponding EDS mapping of the NiFe-SFM powders [8], (d) high-resolution TEM micrograph of a typical NiFe particle grown on the SFM perovskite surface, (e) atom model of a well-defined crystal orientation relationship at the NiFe-SFM interface, and (f) electrochemical performance comparison between the NiFe-SFM cell and the SFM cell at 800°C under SOEC mode.

Cu has low catalytic activity, but has long-term stability due to its resistance to carbon build-up material. Therefore, it was often introduced into other fuel electrode materials, such as Cu-Ni-YSZ cermet [41]. The precious metal materials such as Pt and Ag had fewer practical applications due to their low stability [42].

13.3.2.2 Perovskite Oxide

Although Ni-YSZ is the most widely used cathode material, carbon deposition inevitably occurs during the electrolysis of CO_2, which leads to a blockage of gas diffusion inside the electrode [43]. Some perovskite materials that can be used as cathode for CO_2 electrolysis have also been found with high coking tolerance. As shown in Figure 13.5a, the chemical molecular formula of perovskite oxide is ABO_3, and in

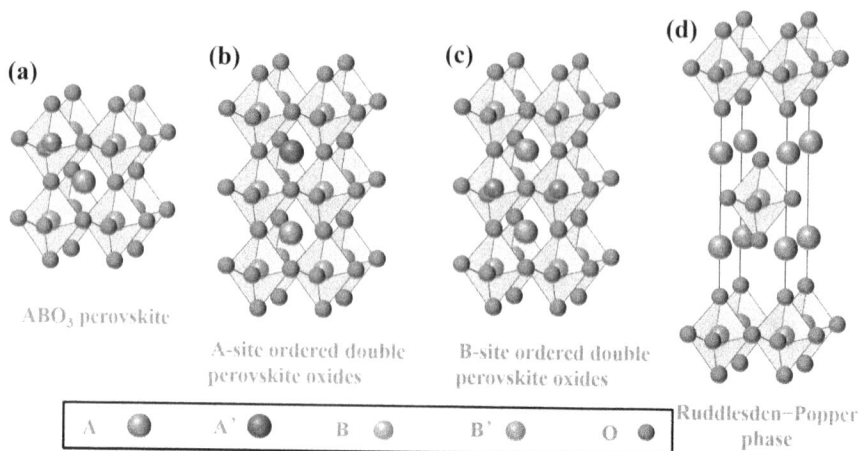

FIGURE 13.5 Crystal structure of (a) perovskite oxide (ABO$_3$), (b) double perovskite oxides (AA'B2O5$_{+\delta}$), (c) double perovskite oxides (A$_2$BB'O6$_{-\delta}$), and (d) Ruddlesden–Popper phase (A$_{n+1}$B$_n$O3$_{n+1}$, n = 1).

general, the A-site is occupied by rare earth element and alkaline earth metal element, while the B-site was occupied to transition metal element [44]. Generally, the perovskite materials are mixed ionic–electronic conducting (MIEC) ceramic after doping different metal elements at the A-site and B-site, respectively. Overall, the metal elements doped at the B-site are the key factor influencing the electrocatalytic activity of perovskites. When Fe, Co, Ni, Mn, or Cu elements were used as the B-site, the resulting perovskite material exhibited good oxygen reduction activity. The perovskite material demonstrated good stability and catalytic activity for carbon dioxide under the reducing environment when the B-site was Sc, Ti, V, or Cr.

As shown in Figure 13.4b, Kwon et al. demonstrated that La$_{0.75}$Sr$_{0.25}$Cr$_{0.5}$Mn$_{0.5}$O$_{3-\delta}$ (LSCM) was a perovskite oxide material that can be used for SOEC cathode with good long-term stability [35]. However, drawbacks such as poor catalytic activity and low electronic conductivity hindered the application of LSCM as a SOEC cathode. LSCM-GDC composite cathode also showed encouraging performance for CO$_2$ electrolysis at high current densities without the use of a reducing atmosphere and exhibited good stability at 5 ppm SO$_2$ in the feed gas [45]. LaFeO$_3$-based perovskite oxides have received much attention from researchers due to their excellent electrocatalytic activity and stability. Liu et al. prepared novel Ni-doped La(Sr)FeO$_{3-\delta}$ (LSF) cathode materials that exhibited excellent CO$_2$ electrolysis performance (1.21 A/cm^2, 1.55 V, 850°C) and outstanding long-term stability [46]. Yang et al. adopted the F-ion doping method to prepare novel La$_{0.6}$Sr$_{0.4}$Fe$_{0.8}$Ni$_{0.2}$O$_{2.9-\delta}$F$_{0.1}$ (LSFNF) cathode materials and demonstrated that F-ion doping can effectively improve the CO$_2$ adsorption capacity and the surface oxygen vacancy concentration at high temperatures [47]. Choi et al. developed La$_{0.6}$Sr$_{0.4}$Mn$_{0.2}$Fe$_{0.8}$O$_{3-\delta}$ (LSMF)-GDC electrodes for carbon dioxide electrolysis and achieved a current density of 1.4 A/cm^2 at 800°C and 1.5 V [48]. Lee et al. successfully enhanced the electrochemical performance of LSMF electrodes by infiltrating metal nanoparticles (Co, Fe, Ni, or Ru) and demonstrated that metal nanoparticles enhance the electrocatalytic activity for the CO_2RR process in the order of $Fe > Ru > Co > Ni$ [49].

Double perovskite oxide materials strontium ferrate has high ionic and electronic conductivities, as well as suitable coefficient of thermal expansion, but exhibits phase instability in a reducing atmosphere. One of the promising alternatives for SOEC cathode materials is a class of strontium ferrite materials doped with metal ions with good electrical properties and stability, such as $Sr_2Fe_{1.5}Mo_{0.5}O_{6-\delta}$(SFM) [50,51]. By doping SFM with the Co element, Lv et al. created a highly catalytically active double perovskite oxide $Sr_2Fe_{1.35}Mo_{0.45}Co_{0.2}O_{6-\delta}$ (SFMC) with excellent reversibility [52]. As shown in Figure 13.4c and d, Li et al. developed a novel heterostructured, NiFe-SFM cathode, and demonstrated that the strong coherent interface of NiFe-SFM can simultaneously enhance surface/interface oxygen exchange kinetics and CO/CO_2 activation processes.

Many scholars had been interested in in situ exsolution since it was such an interesting phenomenon. Researchers have demonstrated that in situ exsolved metal nanoparticles (mainly Fe, Co, Ni, Cu, and Ru) were very effective for promoting the CO_2RR reaction. As many as several dozen types of perovskite oxides that can in situ exsolve metal nanoparticles have been identified, such as $Sr_2Fe_{1.35}Mo_{0.45}Ni_{0.2}O_{6-\delta}$ (SFMN) [53], $SrMo_{0.8}Co_{0.1}Fe_{0.1}O_{3-\delta}$ (SMCFO) [54], $La_{0.7}Sr_{0.2}Ni_{0.2}Fe_{0.8}O_3$ [55] (LSNF), $La_{0.6}Sr_{0.4}Fe_{0.95}Ru_{0.05}O_{3-\delta}$ (LSFR) [56], and $Sr_2Ti_{0.8}Co_{0.2}FeO_{6-\delta}$ (STCF) [57]. Compared to other ways of attaching metal nanoparticles to the electrode surface (such as infiltration), the in situ exsolved metal nanoparticles are firmly anchored to the electrode surface so that they are not easy to be detached. Therefore, researchers have designed various heterostructured cathode materials to electrolysis CO_2, such as Fe/MnO_x-$(Pr, Ba)_2Mn_{2-y}Fe_yO_{5+\delta}$ [58], NiFe-SFM [8], and $Cu-La_{0.43}Sr_{0.37}TiO_{3-\delta}$ [59]. There are also a number of factors that can influence the formation of the heterogeneous structure, including the stoichiometric ratio of the A-site, as well as the time and temperature of the reduction treatment [60]. In order to obtain highly active and stable surfaces with fast reaction kinetics, more studies on the preparation and construction techniques of heterogeneous structures are essential. Table 13.1 summarizes the high-temperature CO_2 electrolysis performance measured on different SOECs exposing to different atmosphere.

13.3.3 ANODE MATERIALS

In a SOEC system, the anode was the place for OER to occur, and it serves to provide a place for the oxidation reaction of oxygen ions and a channel for the conduction of electrons. Therefore, it needs to have good catalytic activity, suitable porosity, high enough conductivity, good chemical stability, compatibility with other materials, and suitable coefficient of thermal expansion.

13.3.3.1 Precious Metal

Precious metal materials often did not match the thermal expansion coefficients of the commonly used YSZ electrolytes and their electrocatalytic activity decreases at high temperatures. Due to their high price, they were difficult to be used in large-scale production, so they have been gradually replaced by perovskite oxide materials.

TABLE 13.1

Comparison of Electrolysis Performance of SOECs with Different Electrode Materials

Cathode Materials	Cell Configuration	Test Condition	Current Density (A/cm^2)	ASR at OCV (Ω cm^2)	Refs
$Sr_2Ti_{0.8}Co_{0.2}FeO_{6-\delta}$(STCF)	STCF-SDC/ LSGM/ LSC–SDC–PrO$_x$	Pure CO_2	1.26 (1.6 V and 850°C)	0.33	[57]
$Sr_{1.97}Fe_{1.5}Mo_{0.5}Ni_{0.1}O_{6-\delta}$ (SFMN)	SFMN/SDC/ LSGM/ LSCF-SDC	30%CO– 70%CO$_2$	0.88 (1.46 V and 800°C)	-	[8]
$SrEu_2Fe_2O_7$(SEF)	SEF/LSGM/SEF	66.7%CO$_2$– 33.3%CO	1.27 (1.5V and 800°C)	0.6	[61]
$Pr_{0.4}Sr_{0.6}Fe_{0.8}Ni_{0.1}Nb_{0.1}O_{3-\delta}$ (PSFNNb)	PSFNNb-GDC/ LSGM/LSF-GDC	30% CO-70% CO$_2$	1.09 (1.5 V and 800°C)	0.70	[62]
$Pr_{0.5}Ba_{0.5}Fe_{0.8}Ni_{0.2}O_{3-\delta}$ (PBFN)	PBFN/GDC/YSZ/ GDC/PBFN	Pure CO_2	0.843 (2.0 V and 800°C)	2.56	[63]
$La_{0.66}Ti_{0.8}Fe_{0.2}O_{3-\delta}$ (LTF)	LTF/LDC/LSGM/ LDC/LSCF-GDC	Pure CO_2	0.4 (1.2 V and 800°C)	-	[64]
$(La_{0.2}Sr_{0.8})$ $Sr_{0.1}Ti_{1.0}O_{3+\delta}$ (LSST)	LSST/SDC/YSZ/ LSM-SDC	Pure CO_2	0.78 (1.7 V and 800°C)	-	[65]
$Sr_{1.9}La_{0.1}Fe_{1.5}Mo_{0.5}O_{6-\delta}$ (SLFM)	SLFM/LDC/ LSGM/LDC/ PBCC	Pure CO_2	1.992 (1.5 V and 800°C)	-	[66]
$La_{0.5}Ba_{0.5}Mn_{0.7}Co_{0.3}O_{3-\delta}$ (LBMCo)	LBMCo/LDC/ LSGM/BSCF	4%CO-96%CO$_2$	1.01 (1.2 V and 800°C)	-	[67]
$La_{0.6}Ca_{0.4}Fe_{0.8}Ni_{0.2}O_{3-\delta}$ (LCaFN)	LCaFN-GDC/ GDC/YSZ/GDC/ LCaFN	Pure CO_2	1.5 (2.0 V and 800°C)	0.71	[68]
LSF	LSF/SDC/LSGM/ LSCF-GDC	Pure CO_2	0.76 (1.5 V and 800°C)		[14]
$La_{0.4}Sr_{0.6}Co_{0.2}Fe_{0.7}Mo_{0.1}O_{3-\delta}$ (LSCFM)	LSCFM/LSGM/ LSCFM	Pure CO_2	1.45 (1.6 V and 800°C)	-	[69]
$(La_{0.75}Sr_{0.25})_{0.97}Cr_{0.5}Mn_{0.5}O_3$ (LSCM), $Ce_{0.6}Mn_{0.3}Fe_{0.1}O_2$(CMF)	LSCM-CMF/LDC/ LSGM/LDC/SSC	50%CO-50%CO$_2$	2.56 (1.5 V and 850°C)	0.08	[70]
$La_{0.9}Sr_{0.8}Co_{0.4}Mn_{0.6}O_{3.9-\delta}F_{0.1}$ (R.P.LSCoMnF)	R.P.LSCoMnF/ LSGM/ LSCF-GDC	30% CO-70% CO$_2$	0.499 (1.3 V and 850°C)	0.85	[71]
$La_{0.3}Sr_{0.7}Fe_{0.7}Cr_{0.3}O_{3-\delta}$ (LSFCr)	LSFCr/YSZ/ LSFCr	Pure CO_2	0.56 (1.5 V and 800°C)	0.47	[72]

(Continued)

TABLE 13.1 (*Continued*)
Comparison of Electrolysis Performance of SOECs with Different Electrode Materials

Cathode Materials	Cell Configuration	Test Condition	Current Density (A/cm²)	ASR at OCV (Ω cm²)	Refs
$La_{0.6}Sr_{0.4}Fe_{0.8}Ni_{0.2}O_{3-\delta}$ (LSFN)	LSFN/GDC/YSZ/ GDC/LSCF-GDC	30% CO-70%CO₂	0.85 (1.55 V and 800°C)	0.16	[46]
$La_{0.6}Sr_{0.4}Fe_{0.8}Ni_{0.2}O_{3-\delta}$ (LSFNF)	LSFNF-GDC/ YSZ/LSCF/GDC	Pure CO₂	1.93 (1.8 V and 850°C)	0.28	[47]
$La_{0.3}Sr_{0.7}Fe_{0.9}Ti_{0.1}O_{3-\delta}$ (LSFT)	LSFT/LSGM/ LSFT	Pure CO₂	1.9 (1.5 V and 800°C)	-	[73]
$La_{0.6}Sr_{0.4}Co_{0.5}Ni_{0.2}Mn_{0.3}O_3$ (LSCNM)	LSCNM/LSGM/ GDC/LSCF-GDC	30% CO-70%CO₂	0.423 (1.3 V and 800°C)	0.13	[74]

13.3.3.2 Perovskite Oxide

The first perovskite oxide material used as an anode in SOEC was $La_{1-x}Sr_xMnO_3$ (LSM) [75]. In the SOEC system with YSZ as the electrolyte operating at high temperature, LSM has been considered as a suitable electrode material SOEC due to its compatible coefficient of thermal expansion with YSZ electrolyte and chemical stability at high temperature [76]. However, since LSM is a pure electronic conductor, the OER could only happen at the TPB site of LSM/YSZ interface when pure LSM is utilized as the cathode material. The electrolyte material YSZ composite with LSM was frequently used by researchers to create LSM-YSZ composites as anodes [4,77]. As shown in Figure 13.6a, Song et al. employed Au nanoparticles loaded on the surface of the LSM-YSZ anode and achieved about 50% enhancement in electrochemical performance [78].

An increase in the number of reaction sites and better cell performance could be achieved by using MIEC materials that can conduct both oxygen ions and electrons as SOEC anode. Among the MIEC materials, Co-based perovskite oxides are often considered as more promising anode materials in SOEC due to their high electrocatalytic activity and oxygen ion conducting properties.

The electronic and ionic conductivities of $La_{1-x}Sr_xCoO_3$ (LSC) are well matched with the requirements of SOEC, but its phase structure is not stable, Co in LSC is easy to be evaporated or diffused at high temperature, leading to the poor long-term stability of the electrode [79]. Therefore, the Fe element is usually used to partially replace the Co element in the B-site to improve its stability, and at the same time, to make its coefficient of thermal expansion more compatible with the electrolyte materials.

When $La_{1-x}Sr_xCo_{1-y}Fe_yO_3$ (LSCF) was used as the anode material for SOEC, there was no obvious interfacial detachment, which confirmed its good interfacial stability [82,83]. Hjalmarsson et al. demonstrated that LSCF has superior electrochemical properties and stability than LSM-YSZ, making LSCF the most widely used SOEC

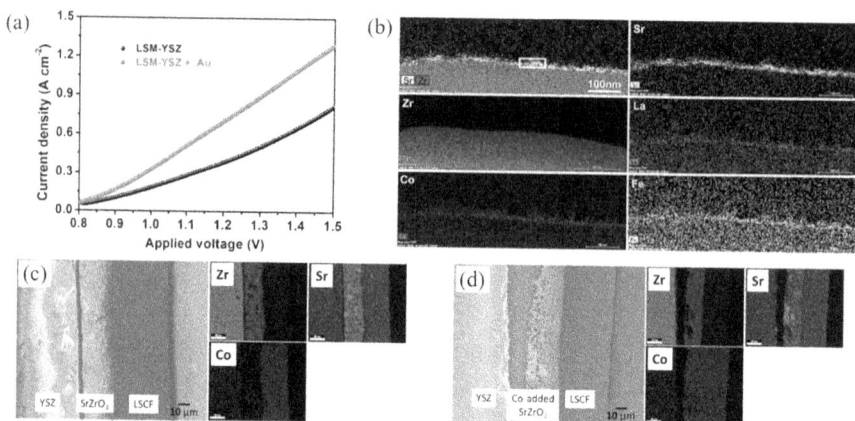

FIGURE 13.6 (a) Electrochemical performance of SOECs with LSM-YSZ and LSM-YSZ+Au anodes [78], (b) STEM-EDS maps of electrode/electrolyte interface after the directly assembled LSCF oxygen electrode was polarized at 0.5 A/cm² and 750°C for 100 hours under the electrolysis mode [80]; SEM images and EDX mapping of the cross-section of (c) the sample with pure SrZrO₃ interlayer and (d) the sample with Co added SrZrO₃ interlayer [81].

anode materials [84]. GDC barrier layers are usually used between the LSCF electrode YSZ electrolyte to avoid the formation of insulating phases such as $La_2Zr_2O_7$ and $SrZrO_3$ [80]. If the GDC barrier layers are not dense enough, the high mobility and high chemical driving force of strontium segregation on the surface of the LSCF electrode can lead to long-distance transport of SrO and the formation of $SrZrO_3$ at the GDC/YSZ interface. As shown in Figure 13.6b, on the surface of the LSCF electrode, strontium segregation was almost unavoidable, and the surface-segregated SrO layer occupied the active site of the OER, thus significantly reducing the electrocatalytic activity of the electrode [80]. In addition, surface-segregated Sr species may react with certain contaminants, such as gaseous chromium species from stainless steel interconnects and sulfur dioxide from the air, resulting in the reduced electrocatalytic activity of the anode [85,86].

In addition, the problem of delamination is a unique phenomenon that occurs at the anode/electrolyte interface of SOEC. As shown in Figure 13.6c and d, it is well known that the formation of high partial pressure of oxygen leads to electrode delamination during electrolysis operations, which is intensified by $La_2Zr_2O_7$ and $SrZrO_3$ at the interface. Shen et al. explored the effect of Co segregation on the long-term stability of SOEC and found that cobalt migration in perovskite oxide electrodes also substantially reduced the long-term stability of SOEC in addition to Sr segregation [81]. In recent years, a variety of high-performance anode materials such as $Ba_{0.5}Sr_{0.5}Co_{0.8}Fe_{0.2}O_{3-\delta}$ (BSCF) [87–89], $Ba_{1-x}Co_{0.7}Fe_{0.2}Nb_{0.1}O_{3-\delta}$ (BCFN) [90], and $Pr_{1-x}Ca_xFeO_{3-\delta}$ (PCF) [91] have been proposed to achieve excellent electrochemical performance.

Due to the high cost, high expansion coefficients, and poor phase stability of Co-based perovskite oxide materials, the development of novel Co-free perovskite oxide materials has risen to the forefront of research at this time [92]. Nowadays, a variety of Co-free anode materials such as $BaFe_{0.95}Pr_{0.05}O_{3-\delta}$ (BFP) [93], LSF [75], and $Sr_{0.9}Ti_{0.3}Fe_{0.7}O_{3-\delta}$ (STF) [94] with excellent performance and good stability have been developed.

13.3.3.3 Double Perovskite Oxides

In addition to the commonly used single perovskite oxide materials, there are also a lot of double perovskite oxide materials as alternative SOEC anodes, including A-site ordered double perovskite oxides ($AA'B2O5_{+\delta}$) and B-site ordered double perovskite oxides ($A_2BB'O5_{+\delta}$) (Figure 13.5b and c). In general, the A-site is occupied by rare earth or alkaline earth metal ions, the A'-site is occupied by alkali metal ions, while the B-site and B'-site are occupied by transition metal ions. Due to their high electronic conductivity and outstanding oxygen surface exchange and diffusion rates, double perovskite oxides were considered promising anode materials in SOEC. Wang et al. prepared $PrBa_{0.5}Sr_{0.5}Co_2O_{5+\delta}$-GDC composites to fulfill the needs of anodes for SOEC using a one-pot molten salt synthesis [95]. Tian et al. prepared the double perovskite oxide $PrBa_{0.5}Sr_{0.5}Co_{1.5}Fe_{0.5}O_{5+\delta}$ and evaluated its electrochemical performance as the anode in SOEC, achieving a current density of 3.694 A/cm² at 800°C and 2.0 V [96].

13.3.3.4 Ruddlesden–Popper Phase

In recent years, the development of Ruddlesden–Popper phase ($A_{n+1}B_nO_{3n+1}$) oxides as anodes for SOEC has become another direction of research, as shown in Figure 13.5d. Ruddlesden–Popper phase has a special crystal structure in which the perovskite structure of ABO_3 and the salt layer of AO alternate along the c-axis. Given the unique structure of Ruddlesden–Popper phase oxides, they exhibit excellent oxygen surface exchange and oxygen ion transfer properties compared to conventional SOEC air electrodes. Zheng et al. developed the novel Ruddlesden–Popper phase material $La_{1.5}Sr_{0.5}Ni_{0.5}Mn_{0.5}O_{4+\delta}$ (LSNM) as an SOEC anode and achieved a current density of 500 mA/cm² at 800°C and 1.4 V [97]. Vibhu et al. developed $Pr_2Ni_{0.8}Co_{0.2}O_{4+\delta}$ (PNC) anodes for SOEC and observed current densities of 3.0 and 1.9 A/cm² under an applied voltage of 1.5 V at 900°C and 800°C, respectively, which electrochemically outperforms conventional LSCF anodes [98].

There have been a lot of research findings in recent years regarding the modification of anode materials, and the process was basically the same as that of SOFC, mainly including infiltration [99–101], mechanical mixing [102], and in situ synthesis [103,104], all of which achieved very favorable results.

13.4 STACK OPERATION

The performance of single-cell SOEC mainly relies on the selection of novel electrode materials and the fabrication processes, while SOEC stacks have higher requirements for the design of sealant and interconnect components, which is a critical challenge in the industrialization of SOEC. The interconnects serve to separate the gases from the cathode and anode, and provide conductive pathways between all the stacks. High-chromium (Cr) containing stainless steel is the preferred choice as a connecting body, offering benefits such as cheapness, high mechanical strength, and superior thermal conductivity. However, the production of gaseous Cr species can cause performance degradation within the cell at high temperatures. As a result, researchers often use protective coatings to retard the evaporation of gaseous Cr species from stainless steel. As shown in Figure 13.6a–c, the most typical coating, $(Mn, Co)_3O_4$ spinel, was applied to Crofer 22 APU stainless steel, reducing the rate of Cr evaporation about four times after operation of 700 hours [74]. In addition to the interconnect corrosion, another potential risk during the operation of SOEC stack is the gas leakage. The sealant should be chemically, mechanically, and thermally stable under high-temperature operating conditions. In terms of the cost, glass and glass-ceramic composites are considered to be suitable sealant. However, the silica in the glass-based materials may permeate into the cell, clogging the pores of the cell and causing cell performance degradation [105].

In order to improve the conversion efficiency of SOEC and retard the cell performance degradation, Alenazey et al. investigated the influence of different operation conditions to the efficiency of six-cell SOEC stack [106]. During the H_2O–CO_2 co-electrolysis, the CO_2 content promotes higher current densities of the stack and the occurrence of reverse water gas shift (RWGS) also promotes the electrical conversion efficiency of the SOEC stack. Based on a 30-cell SOEC stack with LSM-YSZ/YSZ/Ni-YSZ configuration, Zhang et al. investigated the underlying mechanism of cell performance degradation during stack operation and suggested that delamination of the LSM-YSZ electrode and the slight agglomeration of Ni particles should be responsible for the cell performance degradation [107]. By comparing the operation process of large-size SOEC stacks with different anode materials, it was found that no significant delamination was observed for the stack with LSCF anode, which is in contrast to the stack with LSM or LSC anodes. Idaho National Laboratory has determined that optimization of electrode/electrolyte or interconnect coating/electrode interfacial microstructures can improve the durability of SOEC stacks [108]. As shown in Figure 13.6d, among the three electrolyte-supported SOEC stacks provided by Ceramatec, the degradation rates of the first two groups were 4.62%/kh and 6.87%/kh respectively, and the third group showed negative degradation in the 1900-hour test. The electrode-supported SOEC stacks supplied by MSRI showed good long-term durability with a total degradation rate of only 3.2%/kh. Fraunhofer Institute for Ceramic Technology and Systems (IKTS) tested 10 Sc1CeSZ electrolyte-supported stacks for SOFC, SOEC, and reversible operation (rSOC) condition [109]. An overall degradation rate of −0.5%/1000 hours was observed in a long-term test over 5000 hours. Mougin et al. designed a lightweight stack with high performance and durability, and the use of thin connectors reduces the cost of SOEC stack [110]. The degradation rate for the designed SOEC stacks was determined to be less than 3%/1000 hours (Figure 13.7).

Although flat cells are considered to be the most electrically efficient type, tubular cells have the advantage of being simple to seal. In addition to being sampling to seal, flat-tube cells can maintain the electrochemical performance of the SOEC stacks to the maximum extent. As shown in Figure 13.6e, Wu et al. found that the average degradation rate of the flat-tube SOEC stack was about 0.018%/cycle after 64 cycles (900 hours), demonstrating that $SrCrO_4$ generated by the reaction between anode and metal interconnect increased the resistance of electron transfer between the collector layer and the anode layer, which is the main reason for the cell performance degradation [112].

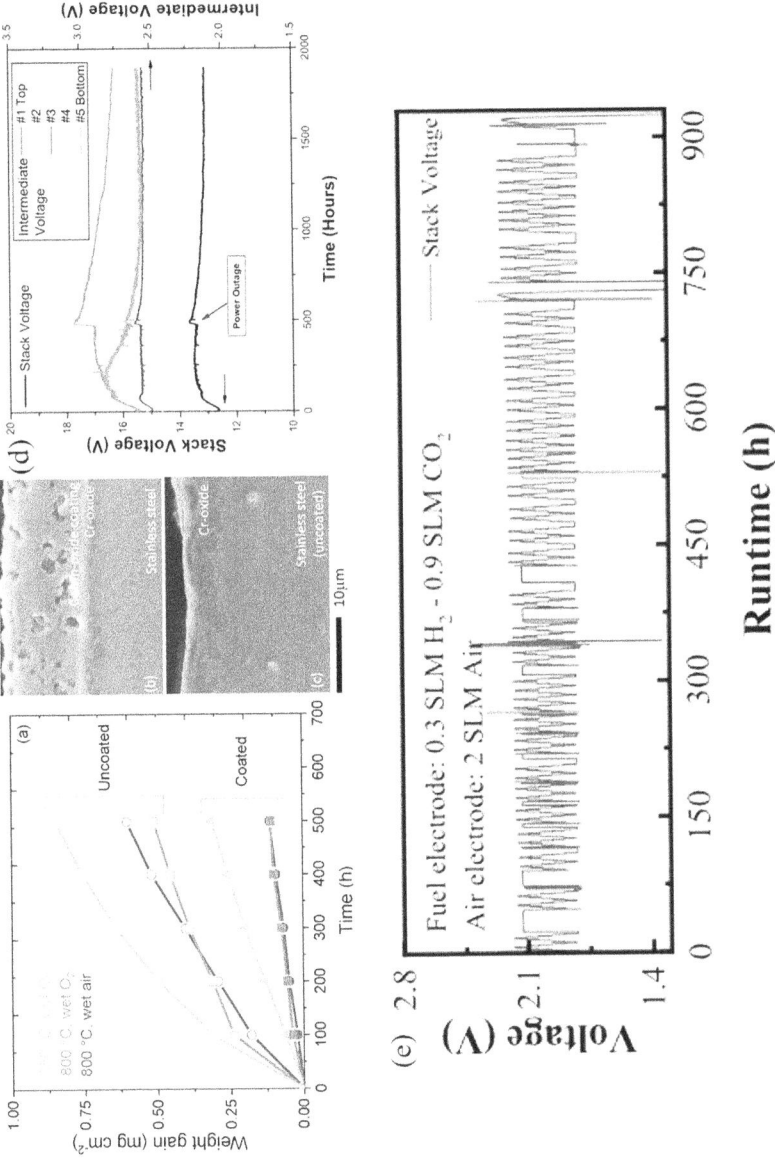

FIGURE 13.7 Oxidation behavior. (a) Intermittent weight gain data for coated (filled markers) and uncoated (open markers) coupons at 800°C (circles, squares) or 850°C (triangles) in humidified air (circles) or humidified O₂/N₂ (30:67) (squares). SEM cross-section images of (b) coated and (c) uncoated samples after 500 hours oxidation at 800°C in humidified O₂/N₂ (30:67) [111]; (d) Ceramatec stack #3 1900 hours test at 0.25 A/cm² in electrolysis mode [108]; and (e) long-term V–i curve of the stack in solid oxide electrolysis mode [112].

13.5   CONCLUSIONS AND FUTURE PROSPECTS

Research on SOEC has grown rapidly over the past decade. The focus of the related researches was mainly on the development of new materials, improving performance to reduce material costs, optimizing the production process, and preparing a new generation of highly efficient and durable SOEC. SOEC offered practical solutions for the production of easily storable chemicals from renewable resources. For SOEC operating at high temperatures, YSZ was the best choice of electrolyte due to its high ionic conductivity, high mechanical strength, and low cost. In addition, YSZ electrolytes were more compatible with most electrode materials in terms of physical and chemical properties. At moderate temperatures, LSGM seemed to be more suitable for use as electrolyte materials due to its high ionic conductivity for SOEC. The Ni-YSZ and LSM-YSZ cermet materials were the most widely used cathode and anode materials for SOEC. More work was needed to explore the compatibility of the electrodes with the electrolyte, the long-term stability of SOEC, and the performance degradation mechanism. Single cells could be in tubular or planar form. Although tubular SOEC has higher mechanical strength than planar SOEC, flat cells are mostly used in industrial production due to their better manufacturability and higher electrochemical performance.

At present, the development and application of SOEC were still mostly in its primary stage, and there were still many problems for many researchers to advance the path of its industrialization and commercialization, such as carbon deposition and sulfur toxicity. In addition, the process used in SOEC stacks needed to be further optimized, which includes the single-cell preparation and assembly method, the preparation of connector coatings, and the development of reliable sealing materials.

ACKNOWLEDGMENTS

The work is supported by the National Key R&D Program of China (Grant No. 2021YFE0107200) and the National Natural Science Foundation of China (52072405).

REFERENCES

1. Khan, M.S., et al., Air electrodes and related degradation mechanisms in solid oxide electrolysis and reversible solid oxide cells. *Renewable and Sustainable Energy Reviews*, 2021. 143: p.110918.
2. Vinchhi, P., et al., Recent advances on electrolyte materials for SOFC: A review. *Inorganic Chemistry Communications*, 2023. 152: p. 110724.
3. AlZahrani, A.A. and I. Dincer, Thermodynamic and electrochemical analyses of a solid oxide electrolyzer for hydrogen production. *International Journal of Hydrogen Energy*, 2017. **42**(33): pp. 21404–21413.
4. Li, Q., et al., Understanding the occurrence of the individual CO_2 electrolysis during H_2O-CO_2 co-electrolysis in classic planar Ni-YSZ/YSZ/LSM-YSZ solid oxide cells. *Electrochimica Acta*, 2019. **318**: pp. 440–448.
5. Mohebali Nejadian, M., P. Ahmadi, and E. Houshfar, Comparative optimization study of three novel integrated hydrogen production systems with SOEC, PEM, and alkaline electrolyzer. *Fuel*, 2023. 336: p. 126835.

6. Ni, M., M. Leung, and D. Leung, Technological development of hydrogen production by solid oxide electrolyzer cell (SOEC). *International Journal of Hydrogen Energy*, 2008. **33**(9): pp. 2337–2354.

7. Jiang, Y., F. Chen, and C. Xia, A review on cathode processes and materials for electro-reduction of carbon dioxide in solid oxide electrolysis cells. *Journal of Power Sources*, 2021. 493: p. 229713.

8. Li, Y., et al., Mutual conversion of CO-CO$_2$ on a perovskite fuel electrode with endogenous alloy nanoparticles for reversible solid oxide cells. *ACS Applied Materials & Interfaces*, 2022. **14**(7): pp. 9138–9150.

9. Song, Y., et al., High-temperature CO$_2$ electrolysis in solid oxide electrolysis cells: Developments, challenges, and prospects. *Advanced Materials*, 2019. **31**(50): p. e1902033.

10. Torrell, M., et al., Co-electrolysis of steam and CO$_2$ in full-ceramic symmetrical SOECs: A strategy for avoiding the use of hydrogen as a safe gas. *Faraday Discussion*, 2015. **182**: pp. 241–255.

11. Ren, R., et al., Boosting the electrochemical performance of Fe-based layered double perovskite cathodes by Zn(2+) doping for solid oxide fuel cells. *ACS Applied Materials & Interfaces*, 2020. **12**(21): pp. 23959–23967.

12. Gu, H., et al., Turning detrimental effect into benefits: Enhanced oxygen reduction reaction activity of cobalt-free perovskites at intermediate temperature via CO$_2$-induced surface activation. *ACS Applied Materials & Interfaces*, 2020. **12**(14): pp. 16417–16425.

13. Tsekouras, G., D. Neagu, and J.T.S. Irvine, Step-change in high temperature steam electrolysis performance of perovskite oxide cathodes with exsolution of B-site dopants. *Energy & Environmental Science*, 2013. **6**(1): pp. 256–266.

14. Yang, Y., et al., The electrochemical performance and CO$_2$ reduction mechanism on strontium doped lanthanum ferrite fuel electrode in solid oxide electrolysis cell. *Electrochimica Acta*, 2018. **284**: pp. 159–167.

15. Yu, Y., et al., CO$_2$ activation and carbonate intermediates: an operando AP-XPS study of CO$_2$ electrolysis reactions on solid oxide electrochemical cells. *Physical Chemistry Chemical Physics*, 2014. **16**(23): pp. 11633–11639.

16. Feng, Z.A., M.L. Machala, and W.C. Chueh, Surface electrochemistry of CO$_2$ reduction and CO oxidation on Sm-doped CeO$_{2-x}$: coupling between Ce^{3+} and carbonate adsorbates. *Physical Chemistry Chemical Physics*, 2015. **17**(18): pp. 12273–12281.

17. Zhang, L., et al., Electrochemical reduction of CO$_2$ in solid oxide electrolysis cells. *Journal of Energy Chemistry*, 2017. **26**(4): pp. 593–601.

18. Liu, T., et al., A review of zirconia-based solid electrolytes. *Ionics*, 2016. **22**(12): pp. 2249–2262.

19. Y. Arachi, H.S., O. Yamamoto, Y. Takeda, N. Imanishai, Electrical conductivity of the ZrO$_2$ -Ln$_2$O$_3$ (Ln=lanthanides) system. *Solid State Ionics*, 1999. **121**(1–4): pp. 133–139.

20. Fujimoto, T.G., et al., Mechanical and electrical characterization of 8YSZ-ScCeSZ ceramics. *Materials Research*, 2023. 26(suppl 1): p.e20220595.

21. Temluxame, P., et al., Phase transformation and electrical properties of bismuth oxide doped scandium cerium and gadolinium stabilized zirconia (0.5Gd0.5Ce10ScSZ) for solid oxide electrolysis cell. *International Journal of Hydrogen Energy*, 2020. **45**(55): pp. 29953–29965.

22. Arabaci, A., Effect of Sm and Gd dopants on structural characteristics and ionic conductivity of ceria. *Ceramics International*, 2015. **41**(4): pp. 5836–5842.

23. V.P. Gorelov, D.I.B., Ju. V. Sokolova,H. Na˙fe,F. Aldinger, The effect of doping and processing conditions on properties of La1–xSrxGa 1–yMgyO$_3$. *Journal of the European Ceramic Society* **21**(13): pp. 2311–2317.

24. Wang, H., et al., High-conductivity electrolyte with a low sintering temperature for solid oxide fuel cells. *International Journal of Hydrogen Energy*, 2022. **47**(21): p. 11279–11287.

25. Madhuri, C., et al., Effect of La3+, Pr3+, and Sm3+ triple-doping on structural, electrical, and thermal properties of ceria solid electrolytes for intermediate temperature solid oxide fuel cells. *Journal of Alloys and Compounds*, 2020. 849: p. 156636.

26. Liang, F., et al., Fabrication of Gd_2O_3-doped CeO_2 thin films through DC reactive sputtering and their application in solid oxide fuel cells. *International Journal of Minerals, Metallurgy and Materials*, 2023. **30**(6): pp. 1190–1197.

27. GÜÇTa Ş, D., V. Saribo ĞA, and M.A.F. ÖKsÜZÖMer, Microstructure and ionic conductivity investigation of samarium doped ceria(Sm0.2Ce0.8O1.9) electrolytes prepared by the templating methods. *Turkish Journal of Chemistry*, 2022. **46**(3): pp. 910–922.

28. Arabaci, A., Effect of the calcination temperature on the properties of Sm-doped CeO2. *Emerging Materials Research*, 2020. **9**(2): pp. 1–5.

29. Sepúlveda, E., et al., Preparation of LSGM electrolyte via fast combustion method and analysis of electrical properties for ReSOC. *Journal of Electroceramics*, 2022. **49**(2): pp. 85–93.

30. Zhang, Q., et al., Processing of perovskite $La_{0.9}Sr_{0.1}Ga_{0.8}Mg_{0.2}O_{3-\delta}$ electrolyte by glycine-nitrate combustion method. *International Journal of Hydrogen Energy*, 2021. **46**(61): pp. 31362–31369.

31. Hong, J.-E., S. Ida, and T. Ishihara, Effects of transition metal addition on sintering and electrical conductivity of La-doped CeO_2 as buffer layer for doped $LaGaO_3$ electrolyte film. *Solid State Ionics*, 2014. **262**: pp. 374–377.

32. Liu, S., et al., Preparation of Bi_2O_3-YSZ and YSB-YSZ composite powders by a microemulsion method and their performance as electrolytes in a solid oxide fuel cell. *Materials*, 2023. 16(13): p. 4673.

33. Toor, S.Y. and E. Croiset, Reducing sintering temperature while maintaining high conductivity for SOFC electrolyte: Copper as sintering aid for Samarium Doped Ceria. *Ceramics International*, 2020. **46**(1): pp. 1148–1157.

34. Ryu, S., et al., A self-crystallized nanofibrous Ni-GDC anode by magnetron sputtering for low-temperature solid oxide fuel cells. *ACS Applied Materials & Interfaces*, 2023. **15**(9): pp. 11845–11852.

35. Kwon, Y., et al., Long-term durability of $La_{0.75}Sr_{0.25}Cr_{0.5}Mn_{0.5}O^{-3}$ as a fuel electrode of solid oxide electrolysis cells for co-electrolysis. *Journal of CO_2 Utilization*, 2019. **31**: pp. 192–197.

36. Budiman, R.A., et al., Dependence of hydrogen oxidation reaction on water vapor in anode-supported solid oxide fuel cells. *Solid State Ionics*, 2021. 362: p. 115565.

37. Chen, D., et al., Ni-doped CeO_2 nanoparticles to promote and restore the performance of Ni/YSZ cathodes for CO_2 electroreduction. *Applied Surface Science*, 2023. 611: p. 155767.

38. Wu, A., et al., Performance of CO_2 electrolysis using solid oxide electrolysis cell with Ni-YSZ as fuel electrode under different fuel atmospheres. *International Journal of Green Energy*, 2021. **19**(11): pp. 1209–1220.

39. Yu, J., et al., Performance of Ni-Fe bimetal based cathode for intermediate temperature solid oxide electrolysis cell. *Solid State Ionics*, 2020. 346: p. 115203.

40. Tsekouras, G. and J.T.S. Irvine, The role of defect chemistry in strontium titanates utilised for high temperature steam electrolysis. *Journal of Materials Chemistry*, 2011. 21(25): pp. 9367–9376.

41. Lo Faro, M., et al., The role of CuSn alloy in the co-electrolysis of CO_2 and H_2O through an intermediate temperature solid oxide electrolyser. *Journal of Energy Storage*, 2020. 27: p. 100820.

42. Yang, Z., et al., Development of catalytic combustion and CO$_2$ capture and conversion technology. *International Journal of Coal Science & Technology*, 2021. **8**(3): pp. 377–382.

43. Sun, Y.-F., et al., Electrochemical performance and carbon deposition resistance of Ce-doped La$_{0.7}$Sr$_{0.3}$Fe$_{0.5}$Cr$_{0.5}$O$_{3-\delta}$ anode materials for solid oxide fuel cells fed with syngas. *Journal of Power Sources*, 2015. **274**: pp. 483–487.

44. Neagu, D., et al., In situ growth of nanoparticles through control of non-stoichiometry. *Nature Chemistry*, 2013. **5**(11): pp. 916–923.

45. Zheng, Y., et al., High-temperature electrolysis of simulated flue gas in solid oxide electrolysis cells. *Electrochimica Acta*, 2018. **280**: pp. 206–215.

46. Liu, S., Q. Liu, and J.-L. Luo, The excellence of La(Sr)Fe(Ni)O$_3$ as an active and efficient cathode for direct CO$_2$ electrochemical reduction at elevated temperatures. *Journal of Materials Chemistry A*, 2017. **5**(6): pp. 2673–2680.

47. Yang, C., et al., Anion fluorine-doped La$_{0.6}$Sr$_{0.4}$Fe$_{0.8}$Ni$_{0.2}$O$_{3-\delta}$ perovskite cathodes with enhanced electrocatalytic activity for solid oxide electrolysis cell direct CO$_2$ electrolysis. *ACS Sustainable Chemistry & Engineering*, 2022. **10**(2): pp. 1047–1058.

48. Choi, J., et al., Highly efficient CO$_2$ electrolysis to CO on Ruddlesden-Popper perovskite oxide with in situ exsolved Fe nanoparticles. *Journal of Materials Chemistry A*, 2021. **9**(13): pp. 8740–8748.

49. Lee, S.W., et al., Enhancing CO$_2$ electrolysis performance with various metal additives (Co, Fe, Ni, and Ru) - decorating the La(Sr)Fe(Mn)O$_3$ cathode in solid oxide electrolysis cells. *Inorganic Chemistry Frontiers*, 2023. **10**(12): pp. 3536–3543.

50. Liu, Q., et al., Perovskite Sr$_2$Fe$_{1.5}$Mo$_{0.5}$O$_{6-\delta}$ as electrode materials for symmetrical solid oxide electrolysis cells. *International Journal of Hydrogen Energy*, 2010. **35**(19): pp. 10039–10044.

51. Gao, X., L. Ye, and K. Xie, Voltage-driven reduction method to optimize in-situ exsolution of Fe nanoparticles at Sr$_2$Fe$_{1.5+x}$Mo$_{0.5}$O$_{6-\delta}$ interface. *Journal of Power Sources*, 2023. **561**: p. 232740.

52. Lv, H., et al., In situ investigation of reversible exsolution/dissolution of CoFe alloy nanoparticles in a Co-doped Sr$_2$Fe$_{1.5}$Mo$_{0.5}$O$_{6-\delta}$ cathode for CO$_2$ electrolysis. *Advanced Materials*, 2020. **32**(6): p. 1906193.

53. Lv, H., et al., In situ exsolved FeNi$_3$ nanoparticles on nickel doped Sr$_2$Fe$_{1.5}$Mo$_{0.5}$O$_{6-\delta}$ perovskite for efficient electrochemical CO$_2$ reduction reaction. *Journal of Materials Chemistry A*, 2019. **7**(19): pp. 11967–11975.

54. Liu, S., et al., Cogeneration of ethylene and energy in protonic fuel cell with an efficient and stable anode anchored with in-situ exsolved functional metal nanoparticles. *Applied Catalysis B: Environmental*, 2018. **220**: pp. 283–289.

55. Deka, D.J., et al., Investigation of hetero-phases grown via in-situ exsolution on a Ni-doped (La,Sr)FeO$_3$ cathode and the resultant activity enhancement in CO$_2$ reduction. *Applied Catalysis B: Environmental*, 2021. 286: p. 119917.

56. Marasi, M., et al., Ru-doped lanthanum ferrite as a stable and versatile electrode for reversible symmetric solid oxide cells (r-SSOCs). *Journal of Power Sources*, 2023. 555: p. 232399.

57. Sun, X., et al., Layered-perovskite oxides with in situ exsolved Co-Fe alloy nanoparticles as highly efficient electrodes for high-temperature carbon dioxide electrolysis. *Journal of Materials Chemistry A*, 2022. **10**(5): pp. 2327–2335.

58. Zhu, J., et al., Enhancing CO$_2$ catalytic activation and direct electroreduction on in-situ exsolved Fe/MnO$_x$ nanoparticles from (Pr,Ba)$_2$Mn$_{2-y}$Fe$_y$O$_{5+\delta}$ layered perovskites for SOEC cathodes. *Applied Catalysis B: Environmental*, 2020. 268: p. 118319.

59. Jo, S., et al., Stability and activity controls of Cu nanoparticles for high-performance solid oxide fuel cells. *Applied Catalysis B: Environmental*, 2021. 285: p. 119828.

60. Hu, T., et al., In situ/operando regulation of the reaction activities on hetero-structured electrodes for solid oxide cells. *Progress in Materials Science*, 2023. 133: p. 101050.

61. Huan, D., et al., Ruddlesden-Popper oxide $SrEu_2Fe_2O_7$ as a promising symmetrical electrode for pure CO_2 electrolysis. *Journal of Materials Chemistry A*, 2021. **9**(5): pp. 2706–2713.

62. Choi, J., et al., A highly efficient bifunctional electrode fashioned with in situ exsolved NiFe alloys for reversible solid oxide cells. *ACS Sustainable Chemistry & Engineering*, 2022. **10**(23): pp. 7595–7602.

63. Tian, Y., et al., Phase transition with in situ exsolution nanoparticles in the reduced $Pr_{0.5}Ba_{0.5}Fe_{0.8}Ni_{0.2}O_{3-\delta}$ electrode for symmetric solid oxide cells. *Journal of Materials Chemistry A*, 2022. **10**(31): pp. 16490–16496.

64. Hu, S., et al., Iron stabilized 1/3 A-site deficient La-Ti-O perovskite cathodes for efficient CO_2 electroreduction. *Journal of Materials Chemistry A*, 2020. **8**(40): pp. 21053–21061.

65. Ye, L., et al., Highly efficient CO_2 electrolysis on cathodes with exsolved alkaline earth oxide nanostructures. *ACS Applied Materials & Interfaces*, 2017. **9**(30): pp. 25350–25357.

66. Sun, C., et al., Boosting CO_2 directly electrolysis by electron doping in $Sr_2Fe_{1.5}Mo_{0.5}O_{6-\delta}$ double perovskite cathode. *Journal of Power Sources*, 2022. **521**: p. 230984.

67. Gan, J., et al., A high performing perovskite cathode with in situ exsolved Co nanoparticles for H_2O and CO_2 solid oxide electrolysis cell. *Catalysis Today*, 2021. **364**: pp. 89–96.

68. Tian, Y., et al., A novel electrode with multifunction and regeneration for highly efficient and stable symmetrical solid oxide cell. *Journal of Power Sources*, 2020. **475**: p. 228620.

69. Lv, H., et al., Atomic-scale insight into exsolution of CoFe alloy nanoparticles in $La_{0.4}Sr_{0.6}Co_{0.2}Fe_{0.7}Mo_{0.1}O_{3-\delta}$ with efficient CO_2 electrolysis. *Angewandte Chemie International Edition*, 2020. **59**(37): pp. 15968–15973.

70. Lin, W., et al., Enhancing electrochemical CO_2 reduction on perovskite oxide for solid oxide electrolysis cells through in situ A-site deficiencies and surface carbonate deposition induced by lithium cation doping and exsolution. *Small*, 2023. **19**(41): p. 2303305.

71. Park, S., et al., Improving a sulfur-tolerant Ruddlesden-Popper catalyst by fluorine doping for CO_2 electrolysis reaction. *ACS Sustainable Chemistry & Engineering*, 2020. **8**(16): pp. 6564–6571.

72. Ansari, H.M., et al., Deciphering the interaction of single-phase $La_{0.3}Sr_{0.7}Fe_{0.7}Cr_{0.3}O_{3-delta}$ with CO_2/CO environments for application in reversible solid oxide cells. *ACS Applied Materials & Interfaces*, 2022. **14**(11): pp. 13388–13399.

73. Hou, Y., et al., Excellent electrochemical performance of $La_{0.3}Sr_{0.7}Fe_{0.9}Ti_{0.1}O_3$-delta as a symmetric electrode for solid oxide cells. *ACS Appl Mater Interfaces*, 2021. **13**(19): pp. 22381–22390.

74. Park, S., et al., A sulfur-tolerant cathode catalyst fabricated with in situ exsolved CoNi alloy nanoparticles anchored on a Ruddlesden-Popper support for CO_2 electrolysis. *Journal of Materials Chemistry A*, 2020. **8**(1): pp. 138–148.

75. Kong, J., et al., Synthesis and electrochemical properties of LSM and LSF perovskites as anode materials for high temperature steam electrolysis. *Journal of Power Sources*, 2009. **186**(2): pp. 485–489.

76. Wang, R., et al., Chromium poisoning effects on performance of $(La,Sr)MnO_3$-based cathode in anode-supported solid oxide fuel cells. *Journal of the Electrochemical Society*, 2017. **164**(7): pp. F740–F747.

77. Liu, Z., et al., Electrochemical behaviors of infiltrated $(La, Sr)MnO_3$ and Y_2O_3-ZrO_2 nanocomposite layer. *International Journal of Hydrogen Energy*, 2017. **42**(8): pp. 5360–5365.

78. Song, Y., et al., Improving the performance of solid oxide electrolysis cell with gold nanoparticles-modified LSM-YSZ anode. *Journal of Energy Chemistry*, 2019. **35**: pp. 181–187.

79. Kushi, T., Effects of sulfur poisoning on degradation phenomena in oxygen electrodes of solid oxide electrolysis cells and solid oxide fuel cells. *International Journal of Hydrogen Energy*, 2017. **42**(15): pp. 9396–9405.

80. Ai, N., et al., Suppressed Sr segregation and performance of directly assembled La$_{0.6}$Sr$_{0.4}$Co$_{0.2}$Fe$_{0.8}$O$_{3-\delta}$ oxygen electrode on Y$_2$O$_3$-ZrO$_2$ electrolyte of solid oxide electrolysis cells. *Journal of Power Sources*, 2018. **384**: pp. 125–135.

81. Shen, J., et al., Activation of LSCF-YSZ interface by cobalt migration during electrolysis operation in solid oxide electrolysis cells. *International Journal of Hydrogen Energy*, 2022. **47**(90): pp. 38114–38123.

82. Laurencin, J., et al., Reactive mechanisms of LSCF single-phase and LSCF-CGO composite electrodes operated in anodic and cathodic polarisations. *Electrochimica Acta*, 2015. **174**: pp. 1299–1316.

83. Cacciuttolo, Q., et al., Influence of pressure on the electrical and electrochemical behaviour of high-temperature steam electrolyser La$_{0.6}$Sr$_{0.4}$Co$_{0.2}$Fe$_{0.8}$O$_3$ anode. *Journal of Solid State Electrochemistry*, 2018. **22**(12): pp. 3663–3671.

84. Hjalmarsson, P., et al., Influence of the oxygen electrode and inter-diffusion barrier on the degradation of solid oxide electrolysis cells. *Journal of Power Sources*, 2013. **223**: pp. 349–357.

85. Huang, Y.-L., et al., Chromium poisoning effects on surface exchange kinetics of La$_{0.6}$Sr$_{0.4}$Co$_{0.2}$Fe$_{0.8}$O$_{3-\delta}$. *ACS Applied Materials & Interfaces*, 2017. **9**(19): pp. 16660–16668.

86. Railsback, J.G., et al., Degradation of La$_{0.6}$Sr$_{0.4}$Fe$_{0.8}$Co$_{0.2}$O$_{3-\delta}$ oxygen electrodes on Ce$_{0.9}$Gd$_{0.1}$O$_{2-\delta}$ electrolytes during reversing current operation. *Journal of The Electrochemical Society*, 2017. **164**(10): pp. F3083–F3090.

87. Bo, Y., et al., Microstructural characterization and electrochemical properties of Ba$_{0.5}$Sr$_{0.5}$Co$_{0.8}$Fe$_{0.2}$O$_{3-\delta}$ and its application for anode of SOEC. *International Journal of Hydrogen Energy*, 2008. **33**(23): pp. 6873–6877.

88. Dey, S., et al., Synthesis and characterization of nanocrystalline Ba$_{0.6}$Sr$_{0.4}$Co$_{0.8}$Fe$_{0.2}$O$_3$ for application as an efficient anode in solid oxide electrolyser cell. *International Journal of Hydrogen Energy*, 2020. **45**(7): pp. 3995–4007.

89. Prasopchokkul, P., P. Seeharaj, and P. Kim-Lohsoontorn, Ba$_{0.5}$Sr$_{0.5}$(Co$_{0.8}$Fe$_{0.2}$)$_{1-x}$Ta$_x$O$_{3-\delta}$ perovskite anode in solid oxide electrolysis cell for hydrogen production from high-temperature steam electrolysis. *International Journal of Hydrogen Energy*, 2021. **46**(10): pp. 7023–7036.

90. Han, F., et al., Performance and reaction process of Ba$_{1-x}$Co$_{0.7}$Fe$_{0.2}$Nb$_{0.1}$O$_{3-\delta}$ cathode for intermediate-temperature solid oxide fuel cell. *Journal of Alloys and Compounds*, 2021. **886**: p. 161158.

91. Li, Y., et al., A promising strontium and cobalt-free air electrode Pr$_{1-x}$Ca$_x$FeO$_{3-\delta}$ for solid oxide electrolysis cell. *International Journal of Hydrogen Energy*, 2021. **46**(59): pp. 30230–30238.

92. Baharuddin, N.A., A. Muchtar, and M.R. Somalu, Short review on cobalt-free cathodes for solid oxide fuel cells. *International Journal of Hydrogen Energy*, 2017. **42**(14): pp. 9149–9155.

93. Gou, Y., et al., Pr-doping motivating the phase transformation of the BaFeO$_{3-\delta}$ perovskite as a high-performance solid oxide fuel cell cathode. *ACS Applied Materials & Interfaces*, 2021. **13**(17): pp. 20174–20184.

94. Zhang, J.-H., et al., A-site deficient Sr$_{0.9}$Ti$_{0.3}$Fe$_{0.7}$O$_{3-\delta}$ perovskite. A high stable cobalt-free oxygen electrode material for solid oxide electrochemical cells with excellent electrocatalytic activity and CO$_2$ tolerance. *Journal of the European Ceramic Society*, 2022. **42**(13): pp. 5801–5812.

95. Wang, H., et al., One-pot molten salt synthesis of $Ce_{0.9}Gd_{0.1}O_{2-\delta}$@$PrBa_{0.5}Sr_{0.5}Co_2O_{5+\delta}$ as the oxygen electrode for reversible solid oxide cells. *Materials Research Bulletin*, 2023. 160: p. 112115.

96. Tian, Y., et al., Preparation and properties of $PrBa_{0.5}Sr_{0.5}Co_{1.5}Fe_{0.5}O_{5+\delta}$ as novel oxygen electrode for reversible solid oxide electrochemical cell. *International Journal of Hydrogen Energy*, 2018. **43**(28): pp. 12603–12609.

97. Zheng, Y., et al., Mn-doped Ruddlesden-Popper oxide $La_{1.5}Sr_{0.5}NiO_{4+\delta}$ as a novel air electrode material for solid oxide electrolysis cells. *Ceramics International*, 2021. **47**(1): pp. 1208–1217.

98. Vibhu, V., et al., Cobalt substituted Pr2Ni1-Co O4+ (x = 0, 0.1, 0.2) oxygen electrodes: Impact on electrochemical performance and durability of solid oxide electrolysis cells. *Journal of Power Sources*, 2021. **482**: p. 228909.

99. Seyed-Vakili, S.V., et al., Enhanced performance of $La_{0.8}Sr_{0.2}MnO_3$ cathode for solid oxide fuel cells by co-infiltration of metal and ceramic precursors. *Journal of Alloys and Compounds*, 2018. **737**: pp. 433–441.

100. Qiu, P., et al., $LaCrO_3$-Coated $La_{0.6}Sr_{0.4}Co_{0.2}Fe_{0.8}O_3$-delta core-shell structured cathode with enhanced Cr tolerance for intermediate-temperature solid oxide fuel cells. *ACS Applied Materials & Interfaces*, 2020. **12**(26): pp. 29133–29142.

101. Fan, H., Y. Zhang, and M. Han, Infiltration of $La_{0.6}Sr_{0.4}FeO_{3-\delta}$ nanoparticles into YSZ scaffold for solid oxide fuel cell and solid oxide electrolysis cell. *Journal of Alloys and Compounds*, 2017. **723**: pp. 620–626.

102. Xi, X., et al., LSCF-GDC composite particles for solid oxide fuel cells cathodes prepared by facile mechanical method. *Advanced Powder Technology*, 2016. **27**(2): pp. 646–651.

103. Xu, N., et al., Co-synthesis of LSCFN-GDC electrode for symmetric solid oxide fuel cell running on propane. *Electrochimica Acta*, 2018. **265**: pp. 259–264.

104. Kang, D., et al., Synthesis of Cu/Ni-$La_{0.7}Sr_{0.3}Cr_{0.5}Mn_{0.5}O_3$. and its catalytic performance on dry methane reforming. *Journal of Rare Earths*, 2019. **37**(6): pp. 585–593.

105. Singh, K. and T. Walia, Review on silicate and borosilicate-based glass sealants and their interaction with components of solid oxide fuel cell. *International Journal of Energy Research*, 2021. **45**(15): pp. 20559–20582.

106. Alenazey, F., et al., Production of synthesis gas (H_2 and CO) by high-temperature Co-electrolysis of H_2O and CO_2. *International Journal of Hydrogen Energy*, 2015. **40**(32): pp. 10274–10280.

107. Zheng, Y., et al., Investigation of 30-cell solid oxide electrolyzer stack modules for hydrogen production. *Ceramics International*, 2014. **40**(4): pp. 5801–5809.

108. Zhang, X., et al., Improved durability of SOEC stacks for high temperature electrolysis. *International Journal of Hydrogen Energy*, 2013. **38**(1): pp. 20–28.

109. S. Megel, C.D., S. Rothe, C. Folgner, N. Trofimenko, A. Rost, M. Kusnezoff and M.J.E. Reichelt, A. Michaelis, C. Bienert, M. Brandner, Co-Electrolysis with CFY-stacks. *The Electrochemical Society*, 2017. **78**(1): pp. 3089–3102.

110. Mougin, J., et al., High temperature steam electrolysis stack with enhanced performance and durability. *Energy Procedia*, 2012. **29**: pp. 445–454.

111. Dogdibegovic, E., et al., Performance of stainless steel interconnects with $(Mn,Co)_3O_4$-based coating for solid oxide electrolysis. *International Journal of Hydrogen Energy*, 2022. **47**(58): pp. 24279–24286.

112. Wu, A., et al., Degradation of flat-tube solid oxide electrolytic stack for co-electrolysis of H_2O and CO_2 under pulsed current. *Journal of Power Sources*, 2023. 580: p. 233372.

Index

For Product Safety Concerns and Information please contact our EU
representative GPSR@taylorandfrancis.com
Taylor & Francis Verlag GmbH, Kaufingerstraße 24, 80331 München, Germany

www.ingramcontent.com/pod-product-compliance
Lightning Source LLC
Chambersburg PA
CBHW052011230326
41598CB00078B/2510